Solar Planetary Systems

Stardust to Terrestrial and Extraterrestrial Planetary Sciences

Solar Planetary Systems

Stardust to Terrestrial and Extraterrestrial Planetary Sciences

Asit B. Bhattacharya
University of Kalyani, India

Jeffrey M. Lichtman
Radio Astronomy Supplies, Whitesboro, Texas, USA

CRC Press
Taylor & Francis Group
Boca Raton London New York

CRC Press is an imprint of the
Taylor & Francis Group, an **informa** business

CRC Press
Taylor & Francis Group
6000 Broken Sound Parkway NW, Suite 300
Boca Raton, FL 33487-2742

Printed on acid-free paper
Version Date: 20161014

International Standard Book Number-13: 978-1-4987-6206-9 (Hardback)

Library of Congress Cataloging-in-Publication Data

Names: Bhattacharya, Asit B., author. | Lichtman, Jeffrey M., author.
Title: Solar planetary systems : stardust to terrestrial and extraterrestrial planetary sciences / Asit B. Bhattacharya and Jeffrey M. Lichtman.
Description: Boca Raton, FL : CRC Press, Taylor & Francis Group, [2017]
Identifiers: LCCN 2016040973| ISBN 9781498762069 | ISBN 1498762069
Subjects: LCSH: Solar system. | Planetary systems.
Classification: LCC QB501 .B53 2017 | DDC 523.2--dc23
LC record available at https://lccn.loc.gov/2016040973

Visit the Taylor & Francis Web site at
http://www.taylorandfrancis.com

and the CRC Press Web site at
http://www.crcpress.com

Printed and bound in the United States of America by Publishers Graphics, LLC on sustainably sourced paper.

Dedicated to these little astronomers and all others who have contributed to astronomy in the past, contributing at present, and will contribute in future.

Ansley and Megyn

Contents

Preface

The main objective of this book is to introduce our readers to the enormous amount of knowledge acquired by researchers and scientists in the field of astronomy and planetary science over the past centuries. In addition, we have written about our solar system with all the facts currently known and gathered since early civilizations (Greek, Roman, etc.). We have also tried to use information that would be of use in advancing the reader's knowledge and hopefully spark further curiosity of the reader to embark on further studies.

The second half of the twentieth century should be remembered more for the exploration of the planets in comparison to any other single achievement of mankind. In fact, no other generation has had the opportunity to go beyond our own world and observe, touch, and experience the forces that shape our universe. All our present knowledge about the solar system, as well as the universe, has been acquired using new space technology (space probes, Earth-based optical and radio telescopes, etc.) as well as books and records of early scientists. In addition, the technical information and data gathered by scientists have been used and written about for all to understand.

In writing this book, we have given priority to the student's understanding and learning. Students consider a book useful if certain criteria are met: ease in grasping concepts, lucid presentation, clarity of the content, and ability to simplify. This book qualifies on every count as an ideal textbook to meet the student's needs. Writing a text-cum-manual of this category poses several challenges: providing just enough content without sacrificing the essentials, highlighting the key concepts, presenting the concept in a novel format, and building informative assessment. All these have been accomplished throughout this book, and we are sure that more and more students will find it useful, benefiting from the authors' wide experience and knowledge. The original student-friendly format has been maintained but has also been updated, and thus it will be a companion throughout the career of an astronomer.

The content and curriculum in the text have been divided into different sections for the purpose of making it a primary as well as more advanced publication, keeping in mind that abstract concepts are almost always unacceptable and become too difficult for many young students with their varying abilities of acceptance. The universe, our galaxy, and the Earth–Moon double planet system well demonstrate some remarkable evidence of poetic and scientific design. We have attempted to focus on these intelligent designs, so that our readers can find the "astronomical forensics" of our solar systems including many fine points of our universe in a single book. No special expertise or knowledge is required for developing preliminary ideas; all technical and astronomical terms as well as proper names are mentioned and defined in the text at appropriate places and finally in the glossary. The bulk of this material should be familiar to experts in this field, particularly to planetary scientists and astronomers, but they may find a few interesting tidbits as well. All material used in this book was correct at the time of writing.

Dr. Asit B. Bhattacharya
University of Kalyani

Jeffrey M. Lichtman
Radio Astronomy Supplies

Acknowledgments

The authors thank the following people as well as organizations for allowing them to use already published or for other works and also used in portions of this book. In addition, they express their thanks to those who communicated via phone calls and those who offered assistance. The authors made no effort to use information from written/published sources without seeking the permission. In some cases, permission from a source was acknowledged. The authors did make sufficient attempts to locate or communicate with sources. In some cases, decisions were made some/part not to use certain information. Names or sources will be listed as an appreciation of their responses. Also, at the end of each chapter, a full list of sources (references) are given.

Dr. Imke De Patr and Dr. Samuel Gulkis, JPL (Jet Propulsion Laboratory)

NASA (National Association and Space Administration)

NRAO (National Radio Astronomy Observatory)

Dr. Rick Fisher (National Radio Astronomy Observatory) (ret.)

CSIRO (Commonwealth Scientific and Industrial Research Organization)

NSF (National Science Foundation)

JPL (Jet Propulsion Institute)

Peter Michaud, Gemini Observatory/AURA and Lynette Cook

Prof. Abel Méndez, Associate Professor of Physics and Astrobiology
Planetary Habitability Laboratory, University of Puerto Rico at Arecibo

Jeffrey Bennett, Astronomer, Writer

Lynette Cook, Gemini Observatory/AURA

Nancy Whilton, Courseware Portfolio Manager - Physics and Astronomy, Pearson Education

Peter Michaud, Gemini Observatory/AURA

Miteq

Bing (Photo Sources)

Wikipedia (Open Source)

Dr. David Morgan, UK

Jatila van der Veen, Ph.D., Project Scientist, UCSB Department of Physics, Lecturer, College of Creative Studies, UCSB Project Manager, Planck Mission Education and Public Outreach, JPL/NASA

Professor Schombert, University of Oregon

Robert Welsh, Assistant Professor: Physics and Astronomy

Glen I. Langston, PhD, Program Director, Electromagnetic Spectrum Management, National Science Foundation

R. Szabó and Astrobites, Nathan Sanders

Ryan Lane, University of Texas

Scholarpedia

Tom Skutsch, Miteq

Brenda Seise, Miteq

Stuart Sweet, Director of Technical Solutions, Signal Group, LLC // SolidSignal.com

Education-Portal.com

European Space Administration (ESA)

Jamie Stamm, Senior Editor at Remilon, LLC

Eva Kloostra, INSPIRE Advisor

Institutions having license with NASA for related material/text/artwork/photographic

Jonathan Wolf, Physics teacher at Scarsdale High School in Scarsdale, NY, Adjunct Assistant Professor of Physics, Hofstra University, NY, Seton Hall University, Fairleigh Dickinson University

Amelia Keller, Interim Coordinator, UConn Hartford Writing Center

Tom Deans, Associate Professor of English, Director, University Writing Center, University of Connecticut

Drs. M. Debnath, T. Das, J. Pandit, S. Mondal, A. Sarkar, and B. Raha, former and present scholars of KU, for helps in preparing a fraction of the manuscript

Finally, the authors express their sincere thanks to the editor Aastha Sharma, Marsha Pronin, and to the staff members of the CRC Press (Taylor & Francis Group), USA/UK, for their cooperation and support in course of publication of this book. They also thank Rina Bhattacharya and Cynthia Lichtman for their constant encouragement during the preparation of the manuscript.

About This Book

The main objective of this book is to introduce our readers to the enormous knowledge acquired by the researchers and scientists about our solar system since early civilizations. In addition, we have tried to use information that would be of use in advancing the reader's knowledge and hopefully spark further curiosity of the reader to embark on further studies. The content and curriculum in the text have been divided into different sections for the purpose of making it primary as well as more advanced material. No special expertise or knowledge is required for developing preliminary ideas; all technical and astronomical terms as well as proper names are mentioned and defined in the text at appropriate places and also in the glossary.

Authors

Asit B. Bhattacharya received his PhD from the Department of Physics, University of Calcutta, India, and postdoc from the Massachusetts Institute of Technology (MIT), Cambridge, Massachusetts. Presently, he is a professor in the Department of Physics, Kalyani University, and a fellow of the Institute of Electronics and Telecommunication Engineers. He worked in close collaboration with leading laboratories such as Lincoln Laboratory, Milestone Hill Observatory, and Earth, Atmospheric and Planetary Sciences of MIT. He has published more than 225 research papers in international journals and more than 130 papers in conferences, and guided more than 20 PhD students.

He is a member and reviewer of many scientific national and international journals, and his major field of work has been in antenna, astronomical radio spectrograph, solar–terrestrial physics, and remote sensing. He is the author of 17 textbooks on science and engineering that include *Astronomy and Astrophysics* (2008, Infinity Science Press, Hingham, MA), *Search for Extraterrestrial Intelligence* (2012, Lap Lambert Academic Publishing, Germany), *Particle Physics as a Building Block of the Universe* (2013, Lap Lambert Academic Publishing, Germany), *The Ionosphere and Its Transient Variations* (2014, Lap Lambert Academic Publishing, Germany), *Radar Systems and Radio Aids to Navigation* (2010, Khanna Publishers, New Delhi, India), and *Electronic Principles and Applications* (2006, New Central Book Agency, Kolkata, India). He obtained the best innovative paper award on communication in 2013. India International Friendship Society awarded him the prestigious "Shiksha Rattan Puraskar" and Certificate of Excellence for meritorious services, outstanding performance, and remarkable role in education. International Institute of Success Awareness honored him with their most coveted institutional and globally reputed "Glory of India Gold Medal" for remarkable contributions to India's national prestige.

Jeffrey M. Lichtman has been involved with radio astronomy since 1964. In the following years, he has added many subjects to his knowledge in this area. He is a U.S. military veteran (U.S. Army 1967–1970, Specialist 5th Class). His background includes schooling at Redstone Arsenal, Marshall Spaceflight Center, Huntsville, Alabama, and military experience in missile electronics (Nike Hercules Missile System, Radar and Computer maintenance) at the McGregor Missile Range, New Mexico, in addition to military and commercial avionics and RF systems. His education includes: Associate of Science (Broward College, Fort Lauderdale, FL), US ARMY School Certification, Electronics (Nike Hercules Missile System Radar and Computer), Redstone Arsenal (MSFC) (Huntsville, AL). Currently, he is the owner and founder of Radio Astronomy Supplies, which was founded in late 1994.

His functions concern marketing and engineering areas and is responsible for the development of new business and new product with the engineering staff. He has authored many papers as well as books and publications. He has worked with Grumman Aerospace on the LEM (Lunar Excursion Module), Hazeltine, Hamilton Standard (div. UTC), Bendix Avionics, and Lockheed Martin (C-130 Hercules H and J). His other areas of expertise include communications; technical, commercial, and military documentation; engineering; telecommunications; microwave and business marketing; FAA 747-200 commissioning for flight documentation; and cargo training.

1

Our Solar Planetary System at a Glance

1.1 Early Beliefs and Discoveries

For thousands of years, humanity did not recognize the existence of the solar system. They believed, in general, Earth to be stationary and at the center of the universe, whereas ethereal objects moved through the sky. Initially, Greek philosopher Aristarchus had speculated on a heliocentric reordering of the cosmos [1], but Nicolaus Copernicus was the first to develop a mathematically predictive heliocentric system [2]. His successors, Galileo Galilei, Johannes Kepler, and Isaac Newton, developed further speculations that led to the gradual acceptance of the idea that Earth moves around the Sun and that the planets are controlled by the same physical laws that governed Earth. Figure 1.1 shows Andreas Cellarius's illustration of the Copernican system, from the Harmonia Macrocosmica (1660).

Author's Note

With the invention of the telescope and better optics, eventually additional planets and moons were discovered. In addition, the use of unmanned spacecraft has enabled us to investigate geological phenomena, for example, mountains and craters, and seasonal meteorological phenomena, for example, clouds, dust storms, and ice caps on other planets.

1.1.1 Formation of the Solar System and Evolution

The solar system was formed 4.568 billion years ago from the gravitational collapse of a region within a large molecular cloud [3]. This initial cloud was several light years across and perhaps birthed several stars [4]. It is believed that hundreds of protoplanets may have existed in the early solar system. They either merged or were destroyed, leaving the planets as dwarf planets and leftover minor bodies. An artist's concept of the early solar system is presented in Figure 1.2.

Owing to their higher boiling points, only metals and silicates could exist in solid form in the warm inner solar system, and these would form the rocky planets of Mercury, Venus, Earth, and Mars. As metallic elements only comprised a small fraction of the solar nebula, the terrestrial planets could not grow very large. The giant planets, such as Jupiter, Saturn, Uranus, and Neptune, formed beyond the frost line. Leftover debris that never became planets congregated in asteroid belt, Kuiper belt, and Oort cloud.

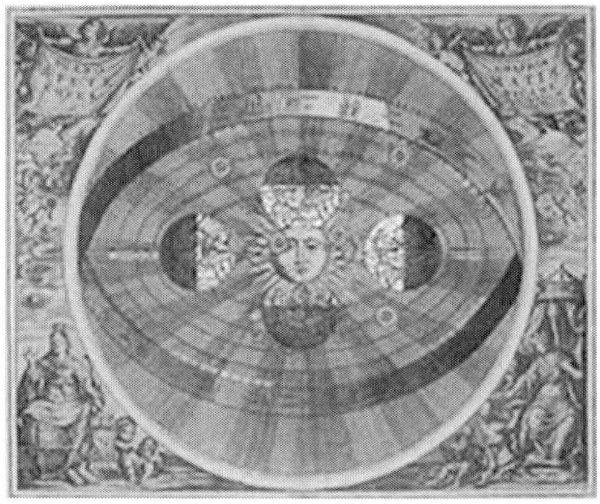

FIGURE 1.1
Andreas Cellarius's illustration of the Copernican system. (Courtesy of Wikipedia, http://en.wikipedia.org/wiki/File:Heliocentric.jpg [last modified on February 23, 2005].)

FIGURE 1.2
Artist's concept of the early solar system. (Courtesy of NASA, https://earthandsolarsystem.files.wordpress.com/2012/01/solarnebula.jpg [retrieved September 10, 2016].)

1.2 Our Solar System's Star

The Sun is a G2 spectral type, main-sequence star. In comparison to the majority of stars in the Milky Way, the Sun is rather large and bright [5]. Stars are well classified in the Hertzsprung–Russell diagram (Figure 1.3), which represents a graph plotting the brightness of stars with their surface temperatures. Hotter stars are brighter, in general. Stars following this pattern are called the main sequence and the Sun lies right in the middle of it. Stars brighter and hotter than the Sun are rare, whereas substantially dimmer and cooler stars, termed as red dwarfs, are common, producing 85% of the stars in the galaxy [5,6]. The Sun compared to the planets, in general, is illustrated in Figure 1.4.

The Sun is growing brighter with time; early in time, its brightness was 70% of what it is today [7]. The Sun was born in the later stages of the universe's evolution and contains more elements heavier than hydrogen and helium [8].

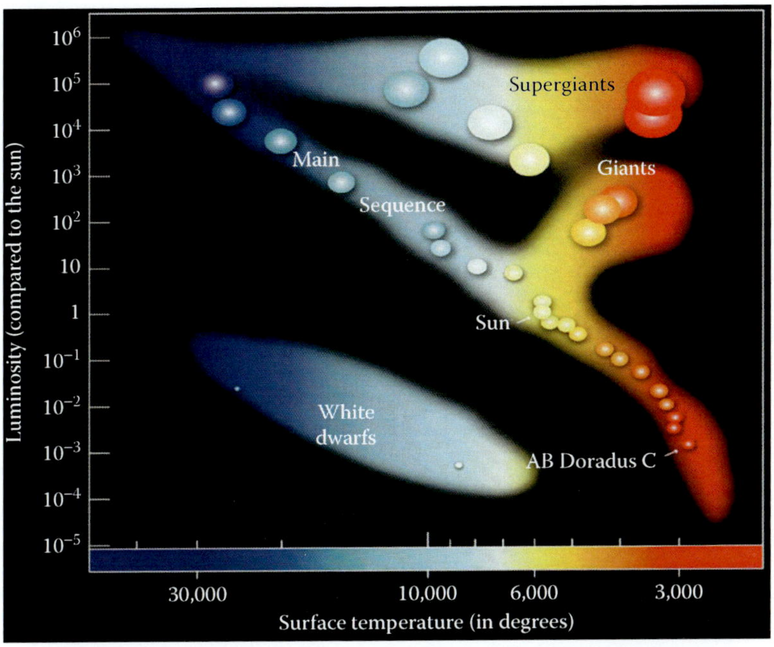

FIGURE 1.3
Hertzsprung–Russell diagram. (Courtesy of Yahoo Images, http://space-exploratorium.com/images/h-r-diagram.jpg [retrieved July 28, 2016].)

FIGURE 1.4
The Sun compared to the planets. (Courtesy of Wikipedia, http://en.wikipedia.org/wiki/File:Planets_and_sun_size_comparison.jpg [last modified on August 22, 2016].)

1.3 Interplanetary Medium

The majority of the solar system consists of a near-vacuum known as interplanetary medium. The Sun radiates a continuous stream of charged particles defined as the solar wind. This stream of charged particles spreads outward at about 1.5 million km/h [9], producing a tenuous atmosphere, the heliosphere, which permeates the interplanetary medium out to at least 100 AU (astronomical unit) [10]. Different features on the Sun's surface, such as solar flares and coronal mass ejections, disturb the heliosphere and thereby create space weather, causing geomagnetic storms [11]. The largest structure within the heliosphere is represented by the heliospheric current sheet, which is a spiral shape formed due to the actions of the Sun's rotating magnetic field on the interplanetary medium [12,13]. The heliospheric current sheet of spiral shape is shown in Figure 1.5. Figure 1.6 shows a NASA photograph of a magnetically formed flare.

The heliosphere and planetary magnetic fields shield the solar system partially from high-energy interstellar particles, the so-called cosmic rays. The interplanetary

FIGURE 1.5
The heliospheric current sheet. (Courtesy of Wikipedia, http://en.wikipedia.org/wiki/File:Heliospheric-current-sheet.gif [last modified on August 22, 2016].)

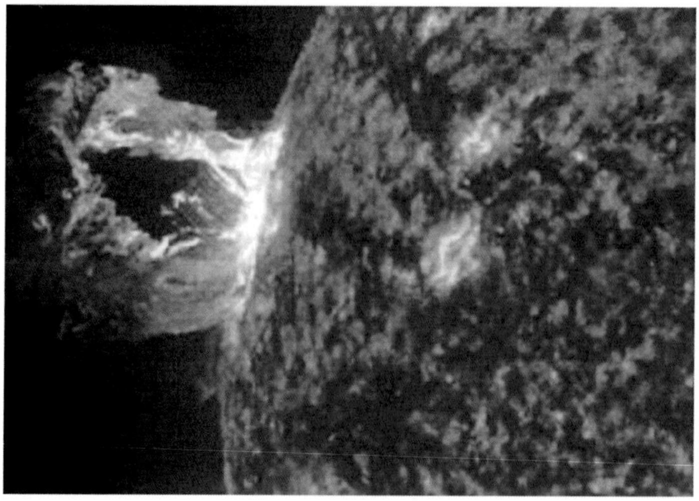

FIGURE 1.6
Solar flare. (Courtesy of NASA, Solar-Flare-NASA.jpg and http://howonearthradio.org/archives/4218 [retrieved September 10, 2016].)

medium has two disk-like regions of cosmic dust: (i) the zodiacal dust cloud lies in the inner solar system and causes the zodiacal light and (ii) second category of dust cloud extending from about 10 to about 40 AU is created by collisions within the Kuiper belt [14,15].

1.4 The Inner Solar System

The inner solar system is the region comprising the terrestrial planets and asteroids [16]. The objects of the inner solar system are relatively close to the Sun and composed mainly of silicates and metals. The radius of this whole region is shorter than the distance between the orbits of Jupiter and Saturn. Figure 1.7 illustrates the inner planets.

The four inner (terrestrial) planets have dense, rocky compositions with no ring systems. They are composed widely of refractory minerals like the silicates and metals like iron and nickel, which form their cores. Three of the four inner planets, such as Venus, Earth, and Mars, have sufficient atmospheres to generate weather. All of them have impact craters and tectonic surface features, such as rift valleys and volcanoes. The terms inner planet and inferior planet are not the same. By inferior planet we mean those planets that are closer to the Sun than Earth (i.e., Mercury and Venus). Figure 1.8 illustrates the size comparisons of Earth, Mars, Venus, and Mercury.

FIGURE 1.7
Inner planets and inferior planets. (Courtesy of Lichtman, Radio Astronomy Supplies, TX.)

FIGURE 1.8
The four planets Earth, Mars, Venus, and Mercury: left to right (sizes to scale). (Courtesy of Wikipedia, Solar system, http://en.wikipedia.org/wiki/Solar_System [last modified on August 22, 2016].)

1.5 Components and Structure of the Solar System

The overall structure of the solar system consists of the Sun, four relatively small inner planets surrounded by a belt of rocky asteroids, and four gas giants surrounded by the Kuiper belt consisting of icy objects. The inner solar system includes the four terrestrial planets and the asteroid belt, whereas the outer solar system is beyond the asteroids, including the four gas giants [17]. Since the discovery of the Kuiper belt, the outermost parts of the solar system are counted as a distinct region consisting of the objects beyond Neptune [18]. The orbits of the bodies in the solar system are presented in Figure 1.9 and can be seen in motion [85].

The Sun is the fundamental component of the solar system. The Sun is classified as a G2, main-sequence star comprised of hydrogen and helium and contains 99.86% of the solar system's known mass and dominates it gravitationally [19]. The gas giants, the four largest orbiting bodies of the Sun, account for 99% of the remaining mass, with Jupiter and Saturn together comprising more than 90%. The planets are situated very close to the ecliptic, but comets and Kuiper belt objects are at significantly greater angles to it [20,21]. Planets and most other objects orbit the Sun in the same direction as the rotation of the Sun [22]. However, there are exceptions, such as Halley's Comet.

Halley's Comet: Unusually for an object in the solar system, Halley's orbit is retrograde. It orbits the Sun in the opposite direction to the planets, or, clockwise from above the Sun's North Pole. The orbit is inclined by 18° to the ecliptic with much of it lying south of the ecliptic. Owing to the retrograde orbit, it has one of the highest velocities relative to the Earth of any object in the solar system. The 1910 passage was at a relative velocity of 70.56 km/s. As its orbit comes close to Earth's in two places, Halley is the parent body of two meteor showers: the Eta Aquarids in early May and the Orionids in late October.

Hilda: The Hilda (or Hildian) asteroids are a *dynamical group* of asteroids in a 3:2 orbital resonance with Jupiter. Hildas travel in their elliptical orbits in a manner that their

FIGURE 1.9
The orbits of the bodies in the solar system. (Courtesy of NASA/CXC/M.Weiss, Illustrations of objects from our solar system, http://chandra.harvard.edu/resources/illustrations/solarsystem.html [retrieved July 16, 2013].)

aphelia put them opposite Jupiter, or 60° ahead of or behind Jupiter. Over three successive orbits each Hilda asteroid traverses through all of these three points in sequence. As a matter of fact, a Hilda's orbit has a semi-major axis between 3.7 AU and 4.2 AU, an eccentricity less than 0.3, and an inclination less than 20°. Two collisional families exist within the Hilda group named as the Hilda family and the Shubart family.

As mentioned earlier, almost all the planets in the solar system possess secondary systems of their own, being orbited by planetary objects called natural satellites, or moons. The largest natural satellites are in synchronous rotation, with one face permanently turned toward their parent.

Kepler's laws of planetary motion explain the orbits of objects about the Sun, which suggests that each object travels along an ellipse with the Sun at one focus. Objects closer to the Sun move more rapidly, as they are more affected by the gravity of the Sun. On an elliptical orbit, the distance of the body from the Sun varies over the course of its year. The closest approach of a body to the Sun is called its *perihelion*, whereas its most distant point from the Sun is called *aphelion*. The orbits of the planets are almost circular, but many comets, asteroids, and Kuiper belt objects follow highly elliptical orbits. Solar system showing the plane of Earth's orbit around the Sun is shown in Figure 1.10. In both the panels, Mercury, Venus, Earth, and Mars are shown; the lower panel also reveals Jupiter making one full revolution with Saturn and Uranus making less than one full revolution.

Although the Sun dominates the system by mass, it accounts for only around 2% of the angular momentum [23] owing to the differential rotation within the gaseous Sun [24]. The planets, dominated by Jupiter, account for most of the rest of the angular momentum by

FIGURE 1.10
Solar system showing the plane of Earth's orbit around the Sun. (Courtesy of NASA, http://spaceplace.nasa.gov/ [retrieved September 10, 2016].)

the combination of their mass, orbit as well as distance from the Sun, with a considerable contribution from comets. The Sun is composed of nearly 98% hydrogen and helium [25]. Jupiter and Saturn comprise approximately all the remaining matter and possess atmospheres composed of about 99% of these elements [26,27]. Objects farther from the Sun are composed mainly of materials with lower melting points [28]. The boundary in the solar system beyond which the volatile substances condense is called the frost line. The frost line lies at about 5 AU from the Sun. The objects of the inner solar system are largely composed of rock [29], such as silicates, iron, or nickel, which remained solid under almost all conditions in the protoplanetary nebula [30].

1.6 Distances of the Planets in Astronomical Unit

The distance from Earth to the Sun is 1 AU (1.5×10^8 km). For comparison, the radius of the Sun may be noted, which is 0.0047 AU. The Sun therefore occupies 0.00001% (10^{-5}%) of the volume of a sphere with a radius the size of Earth's orbit. Jupiter is 5.2 AU from the Sun and has a radius of 0.00047 AU, whereas the most distant planet, Neptune, is 30 AU from the Sun. Figure 1.11 shows planets of the solar system to scale.

The planet Venus is about 0.33 AU farther out from the Sun than Mercury, whereas Saturn is 4.3 AU out from Jupiter. Neptune lies 10.5 AU out from Uranus.

If the Sun–Neptune distance is scaled to 100 m, then the Sun becomes about 3 cm in diameter; the gas giants will appear smaller than approximately 3 mm. The diameter of the Earth, along with the other terrestrial planets would be smaller than a flea (0.3 mm) at this scale [27].* Figure 1.12 reveals the distances of selected bodies of the solar system from the Sun. In the figure, the left and right edges of each bar correspond to the perihelion and

* The date is based on the oldest inclusions found to date in meteorites, and is thought to be the date of the formation of the first solid material in the collapsing nebula.

FIGURE 1.11
Planets of the solar system to scale. (Courtesy of Wikipedia, Solar system, https://en.wikipedia.org/wiki/Solar_System [last modified on August 22, 2016].)

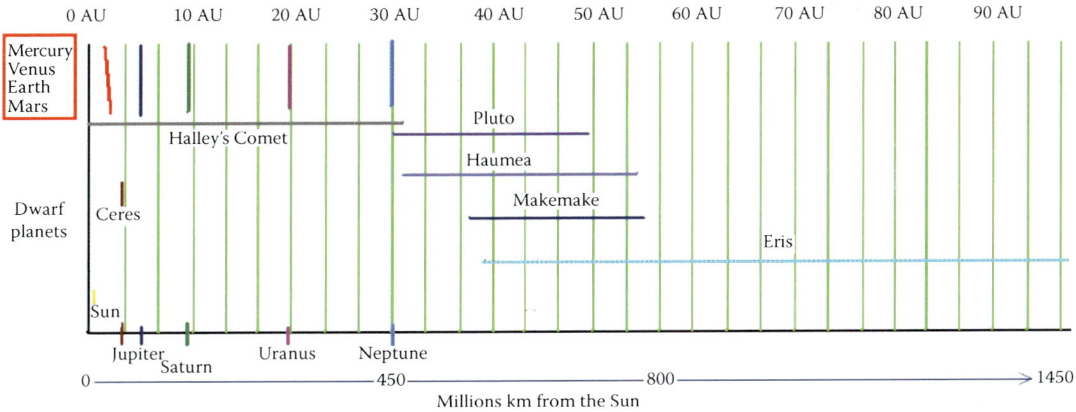

FIGURE 1.12
Distances of selected bodies of the solar system from the Sun. (Courtesy of Wikipedia, Solar system, https://en.wikipedia.org/wiki/Solar_System [last modified on August 22, 2016].)

aphelion of the body, respectively, whereas the long bars denote high orbital eccentricity. The radius of the Sun is 0.7 million km and the radius of Jupiter, the largest planet, is 0.07 million km, both too small to resolve on this image.

1.6.1 Mercury

Mercury (0.4 AU from the Sun) is the smallest planet in the solar system with no natural satellites [31]. It not only has a negligible atmosphere consisting of atoms blasted off its surface but also has a relatively large iron core. Its outer layers were stripped off by a giant impact [32,33]. Almost negligible atmosphere of Mercury consists of atoms blasted off its surface by the solar wind. Its relatively large iron core with thin mantle has not yet been fully explained. Hypotheses provide that its outer layers were stripped off by a giant impact; or, that it was prevented from fully accrediting by the energy of the Sun.

1.6.2 Venus

Venus (0.7 AU from the Sun) has a thick silicate mantle around an iron core, a substantial atmosphere, and internal geological activity, but has no natural satellites. It is the hottest planet, with surface temperatures more than 400 °C due to the amount of greenhouse gases in the atmosphere [34]. It has no magnetic field that would prevent depletion of its substantial atmosphere. This suggests that its atmosphere is frequently replenished by volcanic eruptions [35].

1.6.3 Earth

Earth (1 AU from the Sun) is the largest and densest of the inner planets with geological activity [36]. It is the only planet where plate tectonics has been observed. Earth's atmosphere contains 21% free oxygen [37] and has one natural satellite, the Moon.

1.6.4 Mars

Mars (1.5 AU from the Sun) is smaller than Earth and Venus and possesses an atmosphere of mostly carbon dioxide with a surface pressure of 6.1 millibars [38]. Its surface, peppered with vast volcanoes and rift valleys, reveals geological activity [39]. Its red color appears from iron oxide (rust) in its soil [40]. Mars has two tiny natural satellites, Deimos and Phobos, thought to be captured asteroids [41].

1.7 Asteroid

Asteroids are small solar system bodies composed largely by refractory rocky and metallic minerals with some ice [42]. The asteroid belt occupies the orbit in between Mars and Jupiter (between 2.3 and 3.3 AU from the Sun). It is considered as the remnants from the solar system's formation, which failed to coalesce due to the gravitational interference of Jupiter [43]. Asteroids range in size from hundreds of kilometers to microscopic. All asteroids (except the largest, Ceres) are classified as small solar system bodies [44]. An image of the asteroid belt (white), the Jupiter Trojans (green), Greeks (Dark Green), the Hildas (red), and near-Earth asteroids are shown in Figure 1.13.

The asteroid belt contains millions of objects over 1 km in diameter [45]. Asteroids with diameters between 10 and 10^{-4} m are known as meteoroids [46].

1.7.1 Ceres

Ceres (2.77 AU) is the largest asteroid with a diameter of slightly under 1000 km and a mass large enough for its own gravity to pull it into a spherical shape. Ceres was considered a planet in 1801 and was reclassified to asteroid in the 1850s [47]. It was classified as a dwarf in 2006 by the International Astronomical Union.

1.7.2 Asteroid Groups

Asteroids in the asteroid belt are classified into asteroid groups and families on the basis of their orbital characteristics. Asteroid belt also contains main-belt comets, which may have been the source of Earth's water [48].

The inner solar system is also dusted with rogue asteroids. Many of them cross the orbits of the inner planets [49].

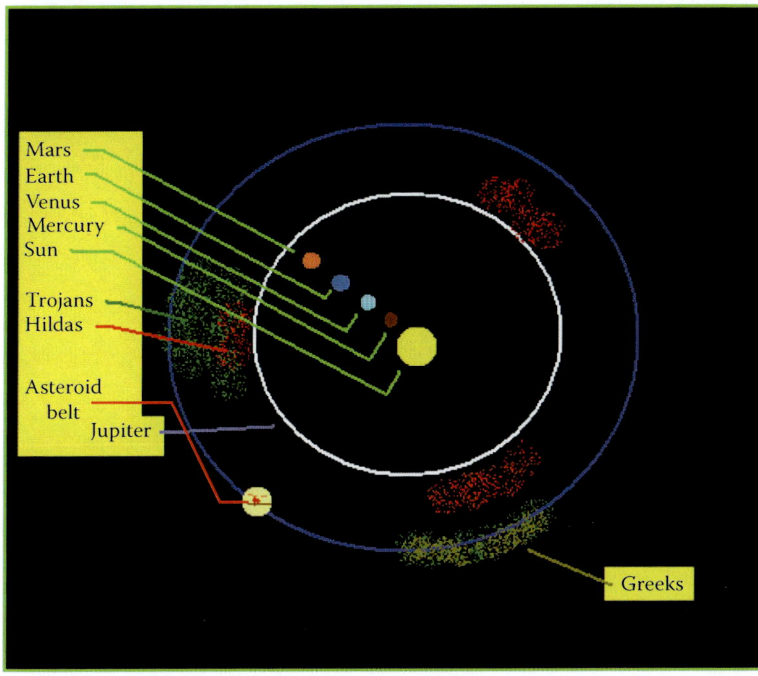

FIGURE 1.13
Image of the asteroid belt. (Courtesy of Wikipedia, Solar system, https://en.wikipedia.org/wiki/Solar_System [last modified on August 22, 2016].)

1.8 Outer Region of Solar System

The outer region of the solar system is home to the gas giants including their large moons. Many short-period comets as well as the centaurs also orbit in this region. Owing to their larger distances from the Sun, the solid objects in the outer solar system contain a higher proportion of volatiles, such as water, methane, and ammonia, than the rocky denizens of the inner solar system because the colder temperatures permit these compounds to remain solid.

1.9 Outer Planets as Gas Giants

The four outer planets or gas giants (also termed as Jovian planets), collectively make up 99% of the mass known to orbit the Sun. Both Jupiter and Saturn are many tens of times the mass of Earth and consist of hydrogen and helium; Uranus and Neptune are less massive (<20 Earth masses) and possess more ices in their makeup. For these reasons, many astronomers consider them to belong to their own category, *ice giants* [50]. In Figure 1.14, from top to bottom, we show Neptune, Uranus, Saturn, and Jupiter with approximate size.

FIGURE 1.14
Neptune, Uranus, Saturn, and Jupiter (from top to bottom; with approximate size). (Courtesy of Wikipedia, Solar system, http://en.wikipedia.org/wiki/Solar_System [last modified on August 22, 2016].)

It is interesting to note that all four gas giants have rings, although only Saturn's ring system can be observed easily from Earth. The term *superior planet* designates planets situated outside Earth's orbit and so we may include both the outer planets and Mars.

1.9.1 Jupiter

Jupiter (5.2 AU from the Sun), at 318 Earth masses, is composed mainly of hydrogen and helium. The strong internal heat of Jupiter creates semipermanent features in its atmosphere, for example, cloud bands and the Great Red Spot. Jupiter has 67 known satellites. The four largest Galilean satellites, Ganymede, Callisto, Io, and Europa, exhibit similarities to the terrestrial planets, such as volcanism and internal heating [51]. Ganymede is the largest satellite in the solar system; in fact, it is larger than Mercury.

In 1955, radio astronomers Burke and Franklin from the Carnegie Institution of Washington accidentally discovered naturally generated radio waves from Jupiter using a 96-acre antenna array. This discovery led to a greater understanding of planetary magnetic fields and plasmas and opened a new window in our exploration of the solar system. In addition, Jupiter is the only planet that emits strong radio signals. The synchrotron emission is monitored at high-frequency (HF) ranges of 18–24 MHz (megahertz) as well as few other HF ranges [52].

1.9.2 Saturn

Saturn (9.5 AU from the Sun), identified easily by its extensive ring system, has many similarities to Jupiter, for example, atmospheric composition and magnetosphere. Saturn has 60% of Jupiter's volume [53]. Its rings are made up of small ice and rock particles. Saturn has 62 confirmed satellites. Of those, Titan and Enceladus reveal signs of geological activity, although they are mainly made of ice [54]. Titan is the second-largest moon in the solar system, and it is also larger than Mercury. Titan is the only satellite in the solar system with a substantial atmosphere.

1.9.3 Uranus

Uranus (19.2 AU from the Sun), at 14 Earth masses, is the lightest of the outer planets. Uniquely, it orbits the Sun on its side; its axial tilt is more than 90° to the ecliptic. It has a colder core than other gas giants and radiates little heat into space [55]. Uranus has 27 known satellites. Of those, Titania, Oberon, Umbriel, Ariel, and Miranda are larger.

1.9.4 Neptune

Neptune (30 AU from the Sun) is slightly smaller than Uranus but more massive (equivalent to 17 Earths) and, therefore, more dense. It radiates more internal heat but less than Jupiter or Saturn [56]. Neptune has 14 known satellites. The largest satellite, Triton, is geologically active, with geysers of liquid nitrogen [57]. In fact, it is the only large satellite with a retrograde orbit. Neptune is accompanied in its orbit by several minor planets, called Neptune trojans.

1.10 Centaurs—The Minor Planet

The centaurs are icy comet-like bodies. Their orbits have semimajor axes larger than Jupiter's (5.5 AU from the Sun) and less than Neptune's (30 AU). The largest known centaur, Chariklohas, has a diameter of about 250 km [58]. The first centaur discovered, Chiron, has also been categorized as comet (95P), as it develops a coma similar to comets when they approach the Sun [59].

1.11 Comets

Comets are small bodies in the solar system. They are typically a few kilometers across, composed largely of volatile ices, and have highly eccentric orbits, usually a perihelion within the orbits of the inner planets and an aphelion far beyond Pluto. If a comet passes into the inner solar system, its proximity to the Sun causes its icy surface to sublimate and ionize, producing a coma that looks like a long tail of gas and dust visible to the naked eye (Figure 1.15).

Short-period comets have orbits lasting for a period of less than 200 years, whereas long-period comets have orbits lasting thousands of years. Short-period comets are believed to originate in the Kuiper belt, but long-period comets, such as Hale–Bopp, are assumed to originate in the Oort cloud. Some comets of hyperbolic orbits may originate outside the solar system, but it is difficult to find their precise orbits [60].

1.12 Trans-Neptunian Region

The area beyond Neptune, or the so-called trans-Neptunian region, is yet largely unexplored. It appears to consist of small worlds composed mainly of rock and ice. This region is also called as the *outer solar system*.

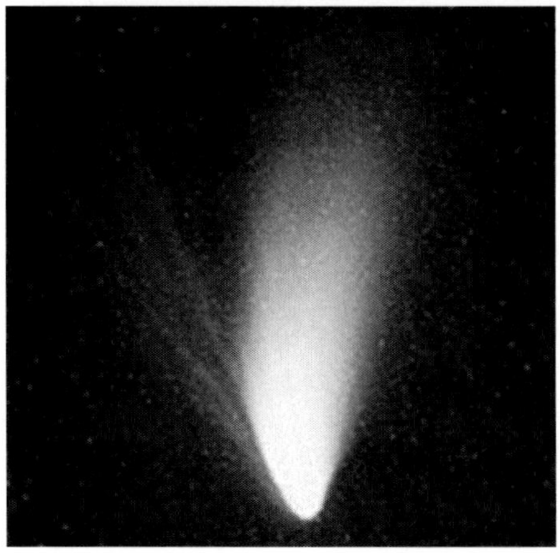

FIGURE 1.15
Comet Hale–Bopp, a long-period comet. (Courtesy of Wikipedia, Solar system, http://en.wikipedia.org/wiki/Solar_System [last modified on August 22, 2016].)

1.13 Kuiper Belt

The Kuiper belt is a large ring of debris similar to asteroid belt but consisting primarily of objects composed of ice [61]. The belt extends between 30 and 50 AU from the Sun. Although it contains anything from dozens to thousands of dwarf planets, it is composed mainly of small solar system bodies. Many larger Kuiper belt objects, such as Quaoar, Varuna, and Orcus, may prove to be dwarf planets with additional data. Estimated Kuiper belt objects are more than 100,000 in number with a diameter greater than 50 km, but their total mass is thought to be only a tenth or even a hundredth the mass of Earth [62]. Some Kuiper belt objects have multiple satellites [63], most of which have orbits that take them outside the plane of the ecliptic [64]. Figure 1.16 shows a plot of all Kuiper belt objects known in 2007, set against the four outer planets.

Generally, the Kuiper belt is divided into the *classical* belt and the resonances (see [61]). The first resonance starts within the orbit of Neptune itself, whereas the classical belt consists of objects having no resonance with Neptune [65].

1.14 Pluto and Charon

Pluto (39 AU [average] from the Sun), the dwarf planet, is the largest known object in the Kuiper belt. It was discovered in 1930 and considered as the ninth planet for a long time, which subsequently changed in 2006 with the adoption of a formal definition of planet.

NOTE: Not all agree with the 2006 decision.

Pluto has a relatively eccentric orbit and is inclined 17° to the ecliptic plane. It has a range from 29.7 AU from the Sun at perihelion within the orbit of Neptune to 49.5 AU at aphelion.

Charon is the largest moon of Pluto. It is a binary system with Pluto and in practice the two bodies orbit a barycenter (i.e., they appear to *orbit each other*) of gravity above their

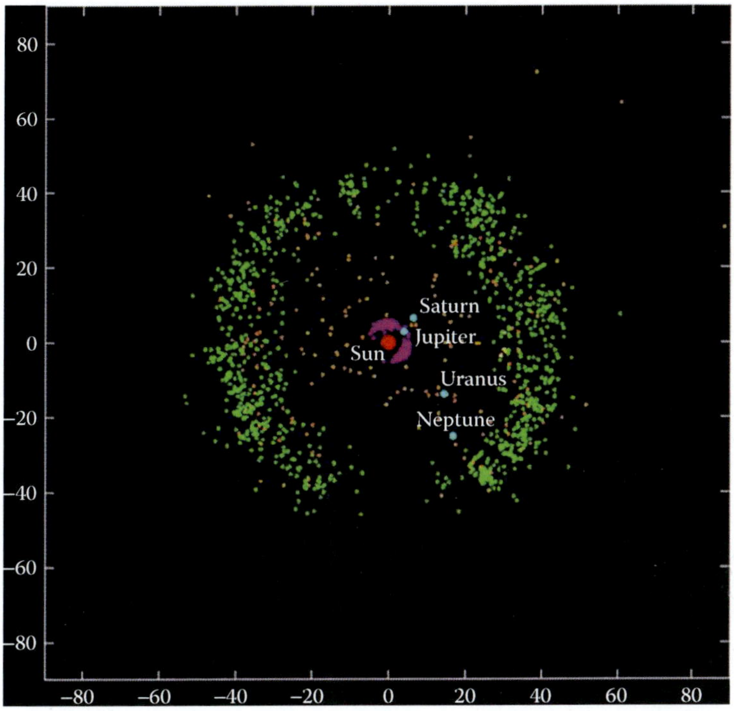

FIGURE 1.16
Plot of all Kuiper belt objects known in 2007, set against the four outer planets. (Courtesy of Wikipedia, Kuiper belt, https://en.wikipedia.org/wiki/Kuiper_belt [last modified on September 07, 2016].)

surfaces. Beyond Charon, four small moons, Styx, Nix, Kerberos, and Hydra are known to orbit within the system.

1.15 Makemake and Haumea

Makemake (45.79 AU [average] from the Sun), smaller than Pluto, is the largest known object in the classical Kuiper belt. It is the brightest object in the Kuiper belt after Pluto. It was designated a dwarf planet in 2008 [66].

Haumea (43.13 AU [average] from the Sun) is in an orbit similar to Makemake, but it is caught in a 7:12 orbital resonance with Neptune [66]. It is nearly the same size as Makemake and has two natural satellites. It was also designated a dwarf planet in 2008 [67].

1.16 Scattered Disk

The scattered disk overlaps the Kuiper belt but extends further outward. It is the source of short-period comets. Scattered disk have been ejected into erratic orbits by the gravitational influence of Neptune's early outward migration. Most scattered disk objects have perihelia within the Kuiper belt but aphelia far beyond it.

1.16.1 Eris

Eris (68 AU [average] from the Sun) is the largest scattered disk object. It is the most massive of the known dwarf planets and has one known moon, Dysnomia. Its orbit, like Pluto, is highly eccentric, with a perihelion of 38.2 AU and an aphelion of 97.6 AU.

1.17 Farthest Regions

The point where the solar system ends and interstellar space starts is not precisely defined, as its outer boundaries are shaped by two separate forces: the solar wind and the Sun's gravity. The outer limit of the solar wind's influence is about four times Pluto's distance from the Sun; this *heliopause* is considered the beginning of the interstellar medium.

1.17.1 Detached Objects

Sedna (provisionally designated 2003 VB12) was discovered by Mike Brown, Caltech Astronomer, and colleagues Chad Trujillo (Gemini Observatory) and David Rabinowitz (Yale University) on November 14, 2003. The object 90377 Sedna (520 AU [average] from the Sun) is a large, reddish object with elliptical orbit that takes it from about 76 AU at perihelion to 940 AU at aphelion. Taking 11,400 years to complete the orbit, Sedna is not part of the scattered disk or the Kuiper belt, as its perihelion is too distant to have been affected by Neptune's migration. Sometimes it is also termed *distant detached objects* (DDOs) [68]. Brown terms this population as the *inner Oort cloud* [69]. Sedna is similar to a dwarf planet, although its shape is yet to be determined.

1.18 The Oort Cloud

It is a hypothetical spherical cloud of up to a trillion icy objects (Figure 1.17). It is believed to be the source for all long-period comets and composed of comets that were ejected from the inner solar system by gravitational interactions with the outer planets. Oort cloud objects move slowly and can be perturbed by infrequent events, for example, collisions, the gravitational effects of a passing star, or the galactic tide, which is the tidal force exerted by the Milky Way [70,71].

1.19 Unknown Regions

There is much more to learn about the solar system and its workings. The gravitational field of the Sun is estimated to dominate the gravitational forces of surrounding stars out to about two light years (125,000 AU). By contrast, lower estimates for the radius of the Oort cloud do not place it farther than 50,000 AU [72]. The region between the Kuiper belt and the Oort cloud, an area tens of thousands of AU in radius, is yet unmapped. There are investigations of the region between Mercury and the Sun [73]. Objects may likely to be discovered in the solar system's uncharted regions.

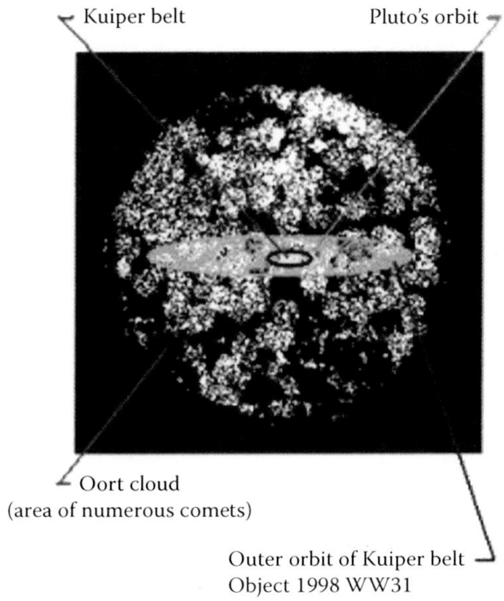

FIGURE 1.17
Artist's rendering of the Oort cloud and the Kuiper belt. (Courtesy of Wikipedia, Solar system, https://en.wikipedia.org/wiki/Solar_System [last modified on August 22, 2016].)

1.20 Solar System as Located in the Milky Way

The Sun is situated in one of the Milky Way's outer spiral arms, known as the Orion–Cygnus Arm or Local Spur [74]. It lies between 25,000 and 28,000 light years from the Galactic Centre [75], and its speed within the galaxy is about 220 km/s so that it completes one revolution every 225–250 million years. This revolution is termed as the solar system's galactic year [76]. Solar apex, the direction of the Sun's path through interstellar space, is near the constellation Hercules in the direction of the current location of the bright star Vega [77]. The plane of the ecliptic lies at an angle of nearly 60° to the galactic plane. Position of the solar system within the Milky Way is shown in Figure 1.18.

The location of the solar system in the galaxy is a factor in the evolution of life on Earth. Its orbit is close to circular and orbits near the Sun are at approximately the same speed as that of the spiral arms. Close to the center, gravity tugs the nearby stars, which could perturb bodies in the Oort cloud and send many comets into the inner solar system, creating collisions with potentially catastrophic implications for life on Earth. In addition, the intense radiation of the galactic center could also be responsible to interfere with the development of complex life [78].

1.20.1 Neighborhood

The solar system is presently situated in the local interstellar cloud. Figure 1.19 illustrates the interstellar medium. It is assumed to be near the neighboring G-Cloud [79]. However, it is unknown if the solar system is embedded in the local interstellar cloud or if it is in the region where the local interstellar cloud and G-Cloud are interacting [80,81]. The local interstellar cloud may be taken as an area of denser cloud in an otherwise sparse region

FIGURE 1.18
Position of the solar system within the Milky Way. (Courtesy of Lichtman, Radio Astronomy Supplies, TX.)

FIGURE 1.19
The interstellar medium beyond the heliosphere, consisting of various clouds of gases. (Courtesy of NASA, Goddard, Adler, U. Chicago, Wesleyan, http://www.nasa.gov/mission_pages/sunearth/news/gallery/solar-journey.html [retrieved September 10, 2016].)

known as the local bubble, which is suffused with high-temperature plasma, suggesting it as the product of several recent supernovae [82].

Within a distance of 10 light years, there are a relatively small amount of solar-type stars. The closest of those is the triple-star system Alpha Centauri, which is situated nearly 4.4 light years away. Alpha Centauri A and B are considered as a closely tied pair of Sun-like stars, whereas the small red dwarf Alpha Centauri C (also known as Proxima Centauri) orbits the pair at a distance of 0.2 light years. The next closest stars to the Sun are the red dwarfs Barnard's

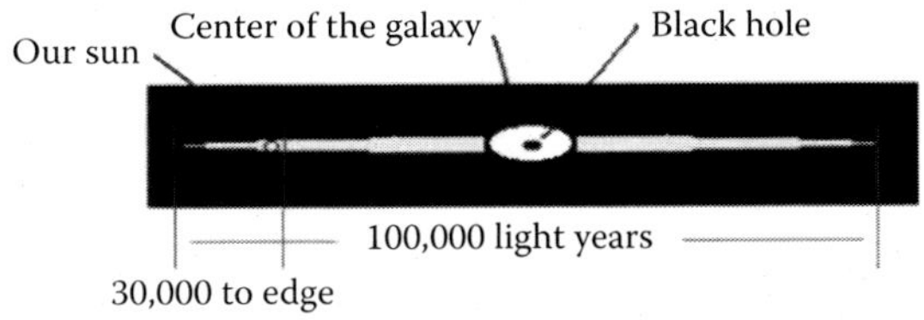

FIGURE 1.20
Earth's location in the observable universe. (Courtesy of Wikipedia, Solar system, https://en.wikipedia.org/wiki/Solar_System [last modified on August 22, 2016].)

Star (at 5.9 light years), Wolf 359 (7.8 light years), and Lalande 21185 (8.3 light years). The largest star within 10 light years is called Sirius, which is a bright main-sequence star having about twice the Sun's mass and orbited by a white dwarf called Sirius B, located 8.6 light years away. The nearest brown dwarfs are the binary Luhman 16 system (6.6 light years). The others within 10 light years are the binary red-dwarf system Luyten 726-8 (8.7 light years) and the solitary red dwarf Ross 154 (9.7 light years) [83]. The closest solitary Sun-like star in the Solar System is Tau Ceti (11.9 light years) [84]. Figure 1.20 illustrates the Earth's location in the observable.

References

1. F. Weinert (2009). *Copernicus, Darwin, & Freud: Revolutions in the History and Philosophy of Science.* Wiley-Blackwell, Hoboken, NJ. p. 21.
2. M. Woolfson (2000). The origin and evolution of the solar system. *Astronomy and Geophysics* 41 (1): 1–12.
3. A. Bouvier and M. Wadhwa (2010). The age of the solar system redefined by the oldest Pb-Pb age of a meteoritic inclusion. *Nature Geoscience* 3: 637–641. doi:10.1038/NGEO941.
4. A. Zabludoff (2003). Lecture 13: The nebular theory of the origin of the solar system. http://atropos.as.arizona.edu/aiz/teaching/nats102/mario/solar_system.html (retrieved December 27, 2006).
5. K. Than (January 30, 2006). Astronomers had it wrong: Most stars are single. SPACE.com. http://www.space.com/1995-astronomers-wrong-stars-single.html (retrieved August 1, 2007).
6. R. L. Smart et al. (2001). The second guide star catalogue and cool stars. In Hugh R. A. Jones, Iain A. Steele (eds). *Ultracool Dwarfs: New Spectral Types L and T.* Springer, Berlin & Heidelberg, Germany, p. 119.
7. N. J. Shaviv (2003). Towards a solution to the early faint sun paradox: A lower cosmic ray flux from a stronger solar wind. *Journal of Geophysical Research* 108 (A12): 1437. arXiv:astroph/0306477. Bibcode:2003JGRA.108.1437S. doi:10.1029/2003JA009997.
8. T. S. van Albada and N. Baker (1973). On the two Oosterhoff Groups of globular clusters. *The Astrophysical Journal* 185: 477–498. Bibcode:1973ApJ...185..477V. doi:10.1086/152434.
9. NASA Marshall Space Flight Center. Solar physics: The solar wind. http://solarscience.msfc.nasa.gov/SolarWind.shtml (retrieved October 3, 2006).
10. NASA. Voyager enters solar system's final frontier. http://www.nasa.gov/vision/universe/solarsystem/voyager_agu.html (retrieved April 2, 2007).
11. T. Phillips (2001). The Sun does a flip. Science@NASA. http://science1.nasa.gov/science-news/science-at-nasa/2001/ast15feb_1/ (retrieved February 4, 2007).
12. A star with two North Poles (2003). Science@NASA. https://science.nasa.gov/science-news/science-at-nasa/2003/22apr_currentsheet (retrieved September 28, 2016).
13. P. Riley (2002). Modeling the heliospheric current sheet: Solar cycle variations. *Journal of Geophysical Research* 107 (A7): SSH 8-1–SSH 8-6.
14. European Space Agency (ESA) (2002). ESA scientist discovers a way to shortlist stars that might have planets. *ESA Science and Technology.* http://sci.esa.int/home/29471-esa-scientist-discovers-a-way-to-shortlist-stars-that-might-have-planets/ (retrieved June 10, 2003).

15. M. Landgraf et al. (2002). Origins of solar system dust beyond Jupiter. *The Astronomical Journal* 123 (5): 2857–2861.
16. NASA Science (Planets). Inner solar system. http://solarsystem.nasa.gov/ (retrieved May 9, 2009).
17. A. Alexander (2006). New horizons set to launch on 9-year voyage to Pluto and the Kuiper Belt. The Planetary Society. http://www.planetary.org/news/2006/0116_New_Horizons_Set_to_Launch_on_9_Year.html (retrieved November 8, 2006).
18. L. Marochnik and L. Mukhin (1995). Is solar system evolution cometary dominated? In G. S. Shostak (ed.). *Progress in the Search for Extraterrestrial Life*. Astronomical Society of the Pacific Conference Series 74. p. 83.
19. H. F. Levison and A. Morbidelli (2003). The formation of the Kuiper belt by the outward transport of bodies during Neptune's migration. *Nature* 426 (6965): 419–421. doi:10.1038/nature02120. http://www.nature.com/nature/journal/v426/n6965/abs/nature02120.html#top (retrieved May 26, 2012).
20. H. F. Levison and M. J. Duncan (1997). From the Kuiper belt to Jupiter-family comets: The spatial distribution of ecliptic comets. *Icarus* 127 (1): 13–32.
21. L. Grossman (2009). Planet found orbiting its star backwards for first time. *New Scientist*. Daily News August 13, 2009. https://www.newscientist.com/article/dn17603-planet-found-orbiting-its-star-backwards-for-first-time/ (retrieved August 12, 2016).
22. Nineplanets.org (2007). An Overview of the Solar System. http//:www.nineplanets.org/overview.html (retrieved April 15, 2008).
23. S. L. Bi et al. (2011). Solar models with revised abundance. *The Astrophysical Journal* 731 (2): L42. arXiv:1104.1032.
24. Stanford Solar Center (2008). The Sun's vital statistics. http://solar-center.stanford.edu/vitalstats.html (retrieved 2008-07-29). Cited in Eddy, J. (1979). *A New Sun: The Solar Results From Skylab*, p. 37. NASA SP-402, NASA, Washington DC.
25. D. R. Williams (2006). Saturn fact sheet. NASA. http://nssdc.gsfc.nasa.gov/planetary/factsheet/saturnfact.html (retrieved July 31, 2007).
26. D. R. Williams (2004). Jupiter fact sheet. NASA. http://nssdc.gsfc.nasa.gov/planetary/factsheet/jupiterfact.html (retrieved August 8, 2007).
27. P. R. Weissman and T. V. Johnson (2007). *Encyclopedia of the Solar System*. Academic Press, Cambridge, MA. p. 615.
28. M. Podolak et al. (1995). Comparative models of Uranus and Neptune. *Planetary and Space Science* 43 (12): 1517–1522.
29. M. Podolak et al. (2000). Further investigations of random models of Uranus and Neptune. *Planetary and Space Science* 48 (2–3): 143–151.
30. NASA. Educator features, Solar System scale. http://www.nasa.gov/audience/foreducators/5-8/features/F_Solar_System_Scale.html (retrieved April 2, 2013).
31. P. Schenk and H. J. Melosh (1994). Lobate thrust scarps and the thickness of Mercury's lithosphere, Abstracts of the *25th Lunar and Planetary Science Conference*, March 14–18, Houston, TX.
32. B. Arnett (2006). Mercury. *The Nine Planets*. http://nineplanets.org/mercury.html (retrieved September 14, 2006).
33. A. G. W. Cameron (1985). The partial volatilization of Mercury. *Icarus* 64: 285–294.
34. M. A. Bullock (1997). *The Stability of Climate on Venus*. Southwest Research Institute, Boulder, CO. http://www.boulder.swri.edu/~bullock/Homedocs/PhDThesis.pdf (retrieved December 26, 2006).
35. P. Rincon (1999). *Climate Change as a Regulator of Tectonics on Venus*. Johnson Space Center Houston, TX, Institute of Meteoritics, University of New Mexico, Albuquerque, NM. http://www.boulder.swri.edu/~bullock/Homedocs/Science2_1999.pdf (retrieved November 19, 2006).
36. NASA Science. Big Questions - What are the characteristics of the Solar System that lead to the origins of life? http://science.nasa.gov/big-questions/ (retrieved August 30, 2011).
37. A. E. Egger. Earth's atmosphere: Composition and structure. VisionLearning.com. http://www.visionlearning.com/en/library/Earth-Science/6/History-of-Earth%E2%80%99s-Atmosphere-II/203 (retrieved December 26, 2006).
38. D. C. Gatling and C. Leovy (2007). Mars atmosphere: History and surface interactions. In L.-A. McFadden et al. (eds). *Encyclopedia of the Solar System*. Academic Press, New York. pp. 301–314.

39. D. Noever (2004). Modern Martian marvels: Volcanoes? *NASA Astrobiology Magazine*. http://www.astrobio.net/news-exclusive/modern-martian-marvels-volcanoes/ (retrieved July 23, 2006).
40. NASA. Mars: A kid's eye view. https://education.jsc.nasa.gov/explorers/p10_resources.html (retrieved May 14, 2009).
41. S. S. Sheppard et al. (2004). A survey for outer satellites of mars: Limits to completeness. *The Astronomical Journal* 128: 2542–2546. http://www2.ess.ucla.edu/~jewitt/papers/2004/SJ2004C.pdf (retrieved December 26, 2006).
42. Ask an Astronomer at Cornell University. Are Kuiper belt objects asteroids? Are large Kuiper belt objects planets? http://curious.astro.cornell.edu/physics/76-our-solar-system/comets-meteors-and-asteroids/the-kuiper-belt-and-oort-cloud/319-are-kuiper-belt-objects-asteroids-are-large-kuiper-belt-objects-planets-intermediate (retrieved March 1, 2009).
43. J.-M. Petit et al. (2001). The primordial excitation and clearing of the Asteroid Belt. *Icarus* 153 (2): 338–347.
44. International Astronomical Union (IAU) Planet Definition Committee (2006). http://www.iau.org/news/pressreleases/release/iau0601/newspaper/ (retrieved March 1, 2009).
45. European Space Agency (ESA). New study reveals twice as many asteroids as previously believed (2002). http://www.esa.int/Our_Activities/Space_Science/New_study_reveals_twice_as_many_asteroids_as_previously_believed (retrieved June 23, 2006).
46. M. Beech and D. I. Steel (1995). On the definition of the term meteoroid. *Quarterly Journal of the Royal Astronomical Society* 36 (3): 281–284.
47. NASA. History and discovery of asteroids. http://dawn.jpl.nasa.gov/DawnClassrooms/1_hist_dawn/ (retrieved August 29, 2006).
48. P. Berardelli (2006). Main-belt comets may have been source of Earths water. *SpaceDaily*. http://www.spacedaily.com/reports/Main_Belt_Comets_May_Have_Been_Source_Of_Earths_Water.html (retrieved June 23, 2006).
49. W. F. Bottke Jr et al. (ed.) (2002). Origin and evolution of near-Earth objects. *Asteroids III*. University of Arizona Press, Tucson, AZ, pp. 409–422.
50. J. J. Lissauer and D. J. Stevenson (2006). *Formation of Giant Planets*. NASA Ames Research Center, California Institute of Technology. University of Arizona Press, Tucson, AZ. pp. 591–605. https://en.wikipedia.org/wiki/Giant_planet (retrieved January 16, 2006).
51. R. T. Pappalardo (1999). *Geology of the Icy Galilean Satellites: A Framework for Compositional Studies*. Brown University. https://so.wikipedia.org/wiki/Bahda_Midaysay_Qoraxdu (retrieved January 16, 2006).
52. R. A. Weintraub (ed) (2005). How one night in a field changed astronomy. NASA. http://www.nasa.gov/vision/universe/solarsystem/radio_jupiter.html (retrieved November 30, 2007).
53. F. Cain. Density of Saturn. Universe Today. http://www.universetoday.com/15322/density-of-saturn/ (retrieved August 9, 2013).
54. J. S. Kargel (1994). Cryovolcanism on the icy satellites. *Earth, Moon, and Planets* 67: 101–113.
55. D. Hawksett et al. (2005). 10 Mysteries of the solar system. *Astronomy Now* 19 (8): 65–75.
56. M. Podolak et al. (1990). Post Voyager comparisons of the interiors of Uranus and Neptune. *Geophysical Research Letters* 17 (10): 1737.
57. N. S. Duxbury and R. H. Brown (1995). The plausibility of boiling geysers on triton. *Beacon eSpace*. http://adsabs.harvard.edu/cgi-bin/nph-bib_query?bibcode=1995LPI....26..347D&db_key=AST (retrieved January 16, 2006).
58. J. Stansberry et al. (2007). Physical properties of Kuiper Belt and Centaur objects: Constraints from Spitzer Space Telescope. *The Solar System Beyond Neptune*: 161. arXiv:astro-ph/0702538.
59. P. Vanouplines (1995). *Chiron Biography*. Vrije Universitiet Brussel. http://www.vub.ac.be/STER/www.astro/chibio.htm (retrieved June 23, 2006).
60. M. Królikowska (2001). A study of the original orbits of hyperbolic comets. *Astronomy and Astrophysics* 376 (1): 316–324.
61. S. C. Tegler (2007). Kuiper belt objects: Physical studies. In L.-A. McFadden et al. (eds). *Encyclopedia of the Solar System*. Elsevier Inc., Northern Arizona University. pp. 605–620.

62. A. Delsanti and D. Jewitt (2006). *The Solar System Beyond the Planets*. Institute for Astronomy, University of Hawaii. Archived from the original on January 29, 2007. https://web.archive.org/web/20070129151907/http://www.ifa.hawaii.edu/faculty/jewitt/papers/2006/DJ06.pdf (retrieved January 3, 2007).
63. M. E. Brown et al. (2006). Satellites of the largest Kuiper belt objects. *The Astrophysical Journal* 639 (1): L43–L46.
64. E. I. Chiang et al. (2003). Resonance occupation in the Kuiper Belt: Case examples of the 5:2 and Trojan resonances. *The Astronomical Journal* 126 (1): 430–443.
65. M. W. Buie et al. (2005). Procedures, resources and selected results of the deep ecliptic survey. *Earth, Moon, and Planets* 92 (1): 113.
66. M. E. Brown (2008). Individualities and peculiarities (Part VI): The largest Kuiper Belt objects. In M. A. Barucci, H. Boehnhardt, D. P. Cruikshank and A. Morbidelli (eds). *The Solar System Beyond Neptune*. The University of Arizona Press, Tucson, AZ.
67. International Astronomical Union (IAU, Paris). IAU names fifth dwarf planet Haumea (2008). http://www.iau.org/news/pressreleases/detail/iau0807/ (retrieved September 16, 2011).
68. D. Jewitt (2004). *Sedna—2003 VB12*. University of Hawaii, HI. http://web.archive.org/web/20040624224040/www.ifa.hawaii.edu/~jewitt/kb/sedna.html (retrieved June 23, 2006).
69. M. Brown (2004). Sedna. CalTech. http://web.gps.caltech.edu/~mbrown/sedna/ (retrieved May 2, 2007).
70. S. A. Stern and P. R. Weissman (2001). *Rapid Collisional Evolution of Comets during the Formation of the Oort Cloud*. Space Studies Department, Southwest Research Institute, Boulder, CO. http://www.nature.com/nature/journal/v409/n6820/full/409589a0.html (retrieved November 19, 2006).
71. B. Arnett (2006). The Kuiper belt and the Oort cloud. Nineplanets.org. http://www.astrobio.net/news-briefs/?op=modload&file=article&sid=1360&mode=thread&order=0&thold=0 (retrieved June 23, 2006).
72. T. Encrenaz et al. (2004). *The Solar System*. 3rd ed. Springer-Verlag, Berlin, Germany, p. 1.
73. D. D. Durda et al. (2004). A new observational search for Vulcanoids in SOHO/LASCO coronagraph images. *Icarus* 148: 312–315.
74. R. Drimmel and D. N. Spergel (2001). Three dimensional structure of the Milky Way disk. *The Astrophysical Journal* 556: 181–202.
75. F. Eisenhauer et al. (2003). A geometric determination of the distance to the galactic center. *The Astrophysical Journal* 597 (2): L121–L124.
76. S. Leong (2002). Period of the Sun's orbit around the Galaxy (cosmic year). *The Physics Factbook*. http://hypertextbook.com/facts/2002/StacyLeong.shtml (retrieved April 2, 2007).
77. C. Barbieri (2003). Elementi di Astronomia e Astrofisica per il Corso di Ingegneria Aerospaziale V settimana. IdealStars.com. http://dipastro.pd.astro.it/planets/barbieri/Lezioni-AstroAstrofIng04_05-Prima-Settimana.ppt (retrieved February 12, 2007).
78. L. Mullen (2001). Galactic habitable zones. *Astrobiology Magazine*. http://www.astrobio.net/news-exclusive/galactic-habitable-zones/ (retrieved June 23, 2006).
79. Wikipedia. Local interstellar cloud. https://en.wikipedia.org/wiki/Local_Interstellar_Cloud (last modified September 29, 2016).
80. NASA. Our local galactic neighborhood. https://en.wikipedia.org/wiki/Galaxy (last modified September 30, 2016).
81. Into the interstellar void. Centauri Dreams. https://en.wikipedia.org/wiki/Proxima_Centauri (last modified October 04, 2016).
82. NASA. Near-Earth Supernovas. https://en.wikipedia.org/wiki/History_of_supernova_observation (last modified September 09, 2016).
83. Stars within 10 light years. *SolStation*. https://en.wikipedia.org/wiki/List_of_nearest_stars_and_brown_dwarfs (last modified September 29, 2016).
84. Tau Ceti. *SolStation*. https://en.wikipedia.org/wiki/Tau_Ceti (last modified October 04, 2016).
85. Wikipedia. Solar System. http://en.wikipedia.org/wiki/Solar_System (last modified on August 23, 2016).

2

Is the Solar System Stable?

2.1 Introduction

The stability of the solar system is a subject of much interest in the field of astronomy. Although the planets have been stable historically and will be in the short term, their weak gravitational effects on one another can be added up to some extent in unpredictable ways. On account of this, besides others, the solar system is considered to be chaotic [1], and also for that even the most precise long-term models for the orbital motion of the solar system are not assumed valid over more than a few tens of millions of years [2]. However, the solar system is taken as stable in human terms, in that none of the planets will collide with each other or be ejected from the system in the next few billion years [3], and also the Earth's orbit will be relatively stable [4].

After Newton discovered his famous laws (1687) of motion and gravity, mathematicians and astronomers have given a tremendous effort to search for evidence for the stability of the planetary motions. This quest led to many mathematical developments and several successive *proofs* of stability for the solar system [5].

Newton used these laws to determine the motion of a single planet around the Sun and showed that the planet followed an ellipse with the Sun at one focus. However, the actual solar system contains eight planets, six of which were known to Newton, and each planet exerts small, periodically varying gravitational forces on all the others [6].

Attempts were first made to find regularity in the motion of these wanderers and hence their movements among the fixed stars could be predicted. The ideal model, for Hipparchus and Ptolemy, was a combination of uniform circular motions, the epicycles. Kepler, between 1609 and 1618, fixed the planets' trajectories: having assimilated the thoughts of Copernicus. He placed the Sun at the center of the universe and, from the consideration of the observations of Brahe, was able to show that the planets followed elliptical path around the Sun. At the end of revolution, each planet found itself back to its starting point and thus retraced the same ellipse. This vision of a perfectly stable solar system where all orbits were periodic was not unchallenged for long. In 1687, Newton announced the law of universal gravitation. Restricting the law to the interactions of planets with the Sun alone, one may get Kepler's phenomenology. But Newton's law is applicable to all interactions. Jupiter is attracted by the Sun, and also attracting Saturn. However, both the planets, Jupiter and Saturn, attract each other. According to Newton, the perturbations among the planets were sufficiently strong enough to destroy the stability of the solar system.

Yet, Newton's law did not enjoy its present status and astronomers wondered if it was really enough to account for the observed movements of bodies in the solar system. The problem of solar system stability was a basic problem because Kepler and Halley were able to analyze the Chaldean observations transmitted by Ptolemy. This observation stated that Saturn was moving away from the Sun at the time when Jupiter was moving closer. By extrapolating these observations, one may obtain that 6 million years ago Jupiter and Saturn were at identical distance from the Sun.

2.2 Stability as the Oldest Problem

The stability of the solar system is an oldest problem in theoretical physics, dating back to Isaac Newton. After the laws of motion and gravity, Newton used these to find the motion of a single planet around the Sun and showed that the planet followed an ellipse with the Sun at one focus. However, the actual solar system contains eight planets, six of which were known to Newton, and each planet exerts small, periodically varying gravitational forces on all the others. Other mathematicians have also made philosophical speculation by the problem of the stability of the solar system. Many illustrious mathematicians and physicists worked on this problem in the three centuries since Newton, including Vladimir Arnold, Boris Delaunay, Carl Friedrich Gauss, Andrei Kolmogorov, Joseph Lagrange, Laplace, Jürgen Moser, Henri Poincaré, Siméon Poisson, and others. Somel's *proofs* of stability have been noted, which are based on approximations that are not completely accurate for our own solar system and thus do not prove its stability.

The straightforward way to solve the problem of the stability of the solar system is to follow the planetary orbits for a few billion years on a computer. At the present time, following planetary systems over billion-year intervals is difficult mostly owing to serial problem, as we have to follow the orbits from 2011 to 2020 before we can follow them from 2021 to 2030, whereas most of the computational speed gains of the last few years have been achieved by parallelization, which is the distribution of a computing problem among hundreds or thousands of processors that work simultaneously.

2.3 Laplace–Lagrange Stability of the Solar System

A serious difference between computations and observations remained open till the end of the eighteenth century, when Lagrange and Laplace established the equations of motion. Lagrange considered that the motion of a planet remains close, over a short duration, to a Keplerian ellipse. This ellipse was used as the basis for a coordinate system as shown in Figure 2.1. Lagrange then expressed the differential equations that govern the variations in this elliptic motion under the effect of perturbations from other planets.

Laplace and Lagrange successfully calculated the long-term variations in the planets' semimajor axes under the effects of perturbations by other planets. Their calculations revealed that, up to first order in the masses of the planets, these variations disappear. Poisson and Poincaré subsequently established that this result remains true through second order in the masses of the planets but not through third order. Figure 2.1 clearly shows the elliptical elements. At any given time, a planet (J) can be moved on an elliptical orbit,

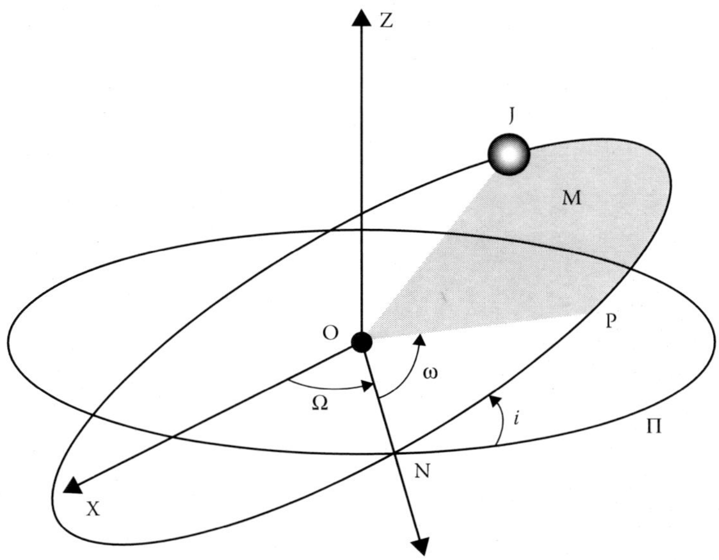

FIGURE 2.1
Elliptical elements. (Courtesy of Scholarpedia, Stability of the solar system, http://www.scholarpedia.org/article/Stability_of_the_solar_system [last modified June 26, 2016].)

with semimajor axis a and eccentricity e, with the Sun at one focus (O). The orientation of this ellipse with respect to a fixed plane Π, and a direction of reference OX, is considered by three angles: the inclination i, the longitude of the node Ω, and the longitude of perihelion $\varpi = \Omega + \omega$, where ω is the argument of perihelion (P). The position of the planet on this ellipse can be represented by the mean longitude $\lambda = M + \varpi$, where the mean anomaly M represents an angle proportional to the area OPJ, the third law of Kepler.

This result seemed to contradict Ptolemy's findings from antiquity. However, by examining the periodic perturbations between Jupiter and Saturn, Laplace found a quasiresonant term (2λJupiter–5λSaturn) in their longitudes, which has amplitude of 46′50″ in Saturn's longitude and a period of about 900 years. This indicates why observations taken in 228 BC and then in 1590 and 1650 could give the impression of a secular term. Laplace next calculated some other periodic terms and obtained a theory of motion for Jupiter and Saturn, which was in good agreement with eighteenth-century observations. Using the same theory, he was able to account for Ptolemy's observations to within one minute of arc, without considering further terms in his calculations. He revealed that Newton's law was in itself sufficient to explain the movement of the planets throughout known history and this exploit no doubt partly accounted for Laplace's determinism. This result, where Laplace and Lagrange demonstrated that the planets' semimajor axes undergo small oscillations only and do not have secular terms, was the first major result of stability for the solar system. Laplace also, at the same time, firmly established Newton's law as the universal explanation for the motion of the celestial bodies.

Although the solar system as a whole could be affected by these events, the Sun and planets are not expected to be disturbed [7]. However, the cumulative probability of a chance encounter with a star increases and disruption of the planets becomes all but inevitable.

2.4 Marginal Stability of the Solar System

Summarizing the reported results on a plane graph, representing the zone swept by the planetary orbits for the maximum values of their eccentricity, we find that the solar system interns is *full*, and there is no space for any additional body. There is a necessity of at least 3.5 billion years to permit a collision between Mercury and Venus, but an additional body placed in this system will collide more rapidly with one of the existing planets. Estimates of the zones possibly occupied by the inner planets of the solar system over 5 Ga (Gegaannus) is presented in Figure 2.2. The circular orbits correspond to the bold lines, whereas the zones visited by each planet resulting from the possible increase in eccentricity are the shaded zones. For Mercury and Venus, these shaded zones overlap. Mars can go as far as 1.9 AU. This approximately corresponds to the inner limit of the asteroid belt [8].

This finding provides the concept of marginal stability for the solar system. It indicates that the solar system is unstable, but catastrophic phenomena leading to the destruction of the system in its current form can take place only within a time that is comparable with its age of approximately 5 billion years. The observation of the present state then makes it possible to assume that it always was thus for the solar system, since the end of its formation. We thus achieve a self-organization of the system toward an increasingly stable state in which there exists a state of marginal stability.

This observation leads then to the concept of marginal stability for the solar system: the solar system is unstable, but catastrophic phenomena leading to the destruction of the system in its current form can take place only in a time comparable with its age, that is to say approximately 5 billion years. The observation of this present state then makes it possible to suppose that it always was thus for the solar system, since the end of its formation. At that time, it could have remained some other bodies than the current planets, but in this case, the system would have been much more unstable, and a collision or an ejection could have taken place (an example could be the impactor of the Earth, which was at the origin of the formation of the Moon). After this event, the remaining system becomes much more stable. We thus obtain a self-organization of the system toward increasingly stable states, which are always states of marginal stability. This is in close agreement with the models of formation of planets by accretion of planetesimals [9], as it reveals how the residual bodies could disappear, in particular in the internal solar system.

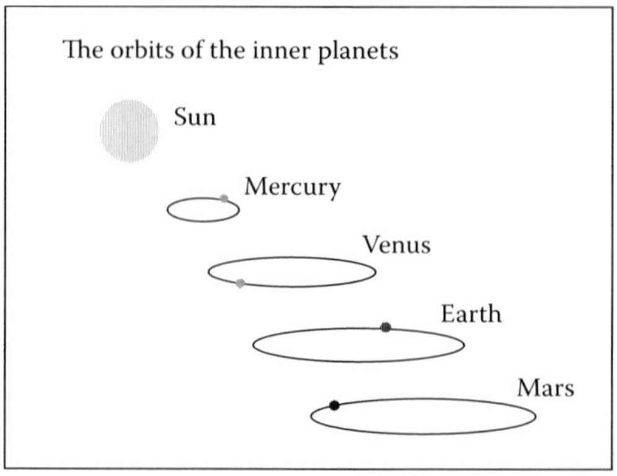

FIGURE 2.2
Estimates of the zones/orbits (not to scale) possibly occupied by the inner planets of the solar system. (Courtesy of Lichtman, Radio Astronomy Supplies, TX.)

It is interesting to note that the zone swept by the orbit of Mars to its maximum eccentricity attains the limits of the asteroid belt. Concerning the system of the outer planets, the things are significantly different, as the direct gravitational short-period perturbations are more dominant. These observations also suggest an idea of the general aspect of a planetary system around a star. If the process of formation from planetary to planetesimals is correct, it becomes possible that the planetary systems will be in a state of marginal stability, like our own solar system.

2.5 Galactic Collision and Planetary Disruption

The vast majority of galaxies in the universe are moving away from the Milky Way. The Andromeda Galaxy, the largest member of our local group of galaxies, is heading toward it at about 120 km/s [10]. In 4 billion years, Andromeda and the Milky Way will collide, causing deformation as tidal forces. If this disruption occurs, astronomers determine a 12% chance that the solar system will be pulled outward into the Milky Way's tidal tail and a 3% chance that it will become gravitationally bound to Andromeda [10].

After a series of glancing blows, during which the likelihood of the solar system's ejection rises to 30% [11], the galaxies' super massive black holes will merge and eventually, in roughly 6 billion years, the Milky Way and Andromeda will complete their merger into a giant elliptical galaxy, which may lead to a short period of intensive star formation called a starburst [11]. The force of these interactions will push the solar system into the new galaxy's outer halo, leaving it relatively unscathed by the radiation from these collisions [10,11].

2.6 Subsequent Studies

In the last few decades, the problem of solar system stability has advanced considerably, mainly due to technological improvements, which permit rigorous analytic calculations and numerical integrations over model timescales approaching the age of the solar system. These efforts consist of direct numerical integration of the equations of motion sometimes, with additional relativistic corrections or perturbations due to the Moon. Preliminary studies were limited to the motion of the outer planets, from Jupiter to Pluto. The more rapid the orbital movement of a planet, the more difficult it becomes to numerically integrate its motion. To integrate the orbit of Jupiter with a conventional integrator, a step-size of 40 days will suffice, whereas a step-size of 0.5 days is essential to integrate the motion of the whole solar system [12–14]. These investigations, reaching 100 million years, confirmed the stability of the system and the validity of the old approach of Laplace and Lagrange. At the same time, calculations of the same system were carried out at MIT (Massachusetts Institute of Technology) by a vectorized computer designed for the task [15,16]. The integration revealed that the motion of Pluto is chaotic, showing exponential divergence with respect to the initial conditions, with a characteristic time of 20 Ma. But as the mass of Pluto is very small to that of the Sun, this does not induce macroscopic instabilities in the rest of the solar system which appeared relatively stable in these numerical studies.

The system of equations so obtained comprises some 150,000 terms and does not model the motion of the planets but rather the averaged motion of their orbits. It can be integrated numerically on a computer using a very large step-size, on the order of 500 years. An integration over 200 million years revealed that the solar system, and more particularly the system of inner planets (Mercury, Venus, Earth, and Mars), is chaotic, with a Lyapunov time of 5 million years [17]. The improvement in computer speed and the development of new methods for numerical integration confirmed most of the results by the direct integration of Newton's equations [18–20], which shows the eccentricity of the Earth (a) and Mars (b) during a 6 Matime span centered at the present.

Author's Note

The Julian year, as used in astronomy and other sciences, is a time unit defined as exactly 365.25 days. This is the normal meaning of the unit *year* (symbol "a" from the Latin annus) used in various scientific contexts. The Julian century of 36,525 days and the Julian millennium of 365,250 days are used in astronomical calculations. In the Unified Code for Units of Measure, the symbol a (without subscript) always refers to the Julian year aj of exactly 31,557,600 s.

365.25 days of 86,400 s = 1 a = 1 aj = 31.5576 Ms

The SI multiplier prefixes may be applied to it to form ka (kiloannus), Ma (megaannus), and so on.

References

1. J. Laskar (1994). Large-scale chaos in the Solar System. *Astronomy and Astrophysics* 287: L9–L12.
2. J. Laskar et al. (2004). A long-term numerical solution for the insolation quantities of the Earth. *Astronomy and Astrophysics* 428 (1): 261.
3. W. B. Hayes (2007). Is the outer Solar System chaotic? *Nature Physics* 3 (10): 689–691.
4. J. Gribbin (2004). *Deep Simplicity: Bringing Order to Chaos and Complexity*. Random House, New York.
5. J. Laskar (2001). Solar system: Stability. In P. Murdin (ed.). *Encyclopedia of Astronomy and Astrophysics*. Institute of Physics Publishing, Bristol, UK. doi: 10.1888/0333750888/2198.
6. S. Tremaine (2011). Is the Solar System stable? Institute for Advanced Study, Princeton, NJ. https://www.ias.edu/ideas/2011/tremaine-solar-system (accessed September 28, 2016).
7. NASA (2012). NASA's Hubble shows Milky Way is destined for head-on collision. http://www.nasa.gov/mission_pages/hubble/science/milky-way-collide.html (accessed August 12, 2016).
8. J. Laskar (1995). Large scale chaos and marginal stability of the Solar System. In the *XIth ICMP Meeting*, Paris, July 1994. International Press. Inc., Somerville, MA. pp. 75–120.
   J. Laskar (1996). Large scale chaos and marginal stability in the solar system. *Celestial Mechanics and Dynamical Astronomy* 64: 115–162.
9. V. S. Safronov (1969). *Evolution of the Protoplanetary Cloud and Formation of the Earth and Planets*. Translated from Russian in 1972. Keter Publishing House, Israel, 212 pp.
10. F. Cain (2007). When our galaxy smashes into Andromeda, what happens to the Sun? *Universe Today*. http://www.universetoday.com/1604/when-our-galaxy-smashes-into-andromeda-what-happens-to-the-sun/ (accessed October 03, 2016).
11. J. T. Cox and A. Loeb (2007). The collision between the milky wayand Andromeda. *Monthly Notices of the Royal Astronomical Society* 386 (1): 461.

12. S. N. Cohen et al. (1973). Construction of biologically functional bacterial plasmids in vitro. *Proceedings of the National Academy of Sciences of the United States of America* 70: 3240–3244.
13. H. Kinoshita and H. Nakai. (1984). Motions of the perihelions of Neptune and Pluto. *Celestial Mechanics* 34: 203–217.
14. A. Carpino et al. (2004). Aromatase immunolocalization in human ductuliefferentes and proximal ductus epididymis. *Journal of Anatomy* 204 (3): 217–220.
15. J. H. Applegate et al. (1986). The outer Solar System for 200 million years. *The Astronomical Journal* 92: 176.
16. G. J. Sussman and J. Wisdom (1988). Numerical evidence that the motion of Pluto is chaotic. *Science* 241: 433–437.
17. J. Laskar (1989). A numerical experiment on the chaotic behavior of the Solar System. *Nature* 338: 237–238.
18. T. R. Quinn et al. (1991). A three million year integration of the Earth's orbit. *The Astronomical Journal* 101: 2287–2305.
19. J. Laskar et al. (1992). Confirmation of resonant structure in the Solar System. *Icarus* 88: 266.
20. G. J. Sussman and J. Wisdom (1992). Chaotic evolution of the Solar System. *Science* 257: 56–62.

3

Origin of the Solar System

3.1 Introduction

The twenty-first century has become a golden era for planetary systems. In addition, new planetary systems are being seen by those involved in this research, such as Gulkis [1], de Pater [2], Geoffrey Marcy [3], as well as other research institutions. A report supports that a total of 1800 such planets have been identified as of June 29, 2014 [4]. Besides, hundreds more systems are unconfirmed. In terms of total confirmed planets, the solar system with eight planets remains the largest known planetary system. However, HD 10180 has nine planets, of which two are yet unconfirmed [5], and is the largest known exoplanetary system including Kepler-90, which also has seven confirmed planets [6]. In fact, today, a wide range of planetary systems have been discovered, with different orbital arrangements around stars with their variety in types. Of all, the closest confirmed system is Gliese 832c [7] at 14.8 light years (ly) with one confirmed planet, whereas the closest unconfirmed system is Alpha Centauri [8] at 4.37 ly with a planet of Earth mass. On the other hand, the closest multiplanet system is Gliese 876 [9] at 15.3 ly with four confirmed planets. The present-day knowledge of planetary systems including that of the solar system has significantly increased beyond doubt, but much is still unknown about their origin and evolution. Scientists believe that the current theories of their formation and evolution are relatively new to the field of planetary science. A subject of particular interest is the field of astrobiology that is largely concerned with the habitable zone of planetary systems. This zone is believed to have the most potential to develop and sustain extraterrestrial life.

3.2 Orbits

The orbits of the planets are elliptical in shape with the Sun at one focus, though all except Mercury are very nearly circular. The orbits of the planets are nearly in the same plane, which are known as the ecliptic and defined by the plane of the Earth's orbit. The ecliptic is inclined only 7° from the plane of the Sun's equator. The orbits of all the planets are in the same direction. They are moving in a counterclockwise motion if one is looking down from above the Sun's North Pole. The solar system consists of the Sun and the eight official planets including at least three *dwarf planets* besides more than 130 satellites belonging to the planets as well as a large number of small bodies such as comets, asteroids, and the vast interplanetary medium. It may be assumed that there might be many more planetary satellites that have not yet been discovered. The inner solar system contains the Sun and

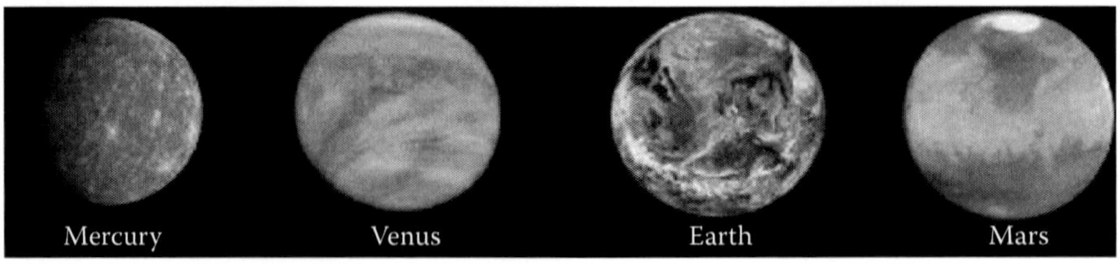

FIGURE 3.1
The four planets of the inner solar system. (Courtesy of Wikipedia, Solar system, https://en.wikipedia.org/wiki/Solar_System [last modified on August 19, 2016].)

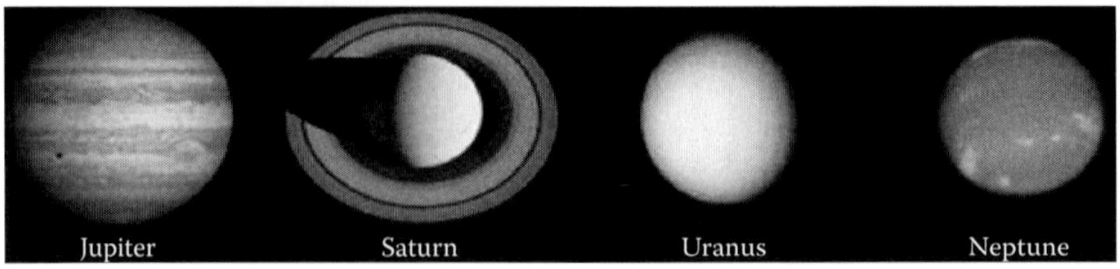

FIGURE 3.2
The planets of the outer solar system. (Courtesy of Wikipedia, Solar system, https://en.wikipedia.org/wiki/Solar_System [last modified on August 19, 2016].)

the four planets Mercury, Venus, Earth, and Mars, as shown in Figure 3.1. The main asteroid belt, not shown in the figure, lies between the orbits of Mars and Jupiter.

The planets of the outer solar system are shown in Figure 3.2. They are Jupiter, Saturn, Uranus, and Neptune. Pluto is now categorized as a dwarf planet.

It is important to note in this connection that the solar system is mostly empty space. The planets are very small in comparison to the space between them. Even the dots on the diagrams above are to be taken as very big to be in proper scale with respect to the sizes of the orbits.

3.3 Gravitationally Bound Nonstellar Objects

A planetary system is considered as a set of gravitationally bound nonstellar objects in orbit around a star or system. Planetary systems describe systems with one or more planets, although such systems may also consider dwarf planets, asteroids, natural satellites, meteoroids, comets, planetesimals [10,11], and other features. As already pointed out, the Sun and its planetary system, which includes our Earth, is known as the solar system [12,13]. The interchangeable terms extrasolar system and exoplanetary system are used for other planetary systems. Individually, they are referred to as system prefixed following the name of the star or star system. For a long time after the sixteenth century and Copernican heliocentric concept, human understanding of planetary systems was limited. Despite the discovery and exploration of the solar system and centuries of conjecture, it remained usual with no extraordinary attraction. The groundbreaking discovery of the unusual PSR B1257 + 12 system [14] and its extrasolar planets, as well as a number of significant records and the first pulsar in a multistar planetary system made this very promising to the scientists working in different fields. Figure 3.3 presents an artist's concept of a planetary system.

FIGURE 3.3
An artist's concept of a planetary system. (Courtesy of Wikipedia, Planetary system, https://en.wikipedia.org/wiki/Planetary_system [last modified on August 19, 2016].)

3.4 Heliocentric Solar System

By the word *heliocentrism*, we mean that the Sun was at the center of the universe. This was opposed to *geocentrism*, which placed the Earth at the center of the universe. Historically, the notion of a heliocentric solar system, with the Sun at the center, was possibly first proposed in the Vedic literature [15] of ancient India, which often refers to the Sun as the *center of spheres*. Some of the authors interpret Aryabhatta's [16] writings in Āryabhaṭīya as implicitly heliocentric. The heliocentric solar system idea was proposed in western philosophy and Greek astronomy as early as the third century BC by Aristarchus of Samos [17] but had received practically no support from most other ancient astronomers.

3.5 Classification of a Planetary System

With the new discoveries of planetary systems, astronomers have started to classify them by type. In particular, certain types of stars are now known for producing specific type of planetary systems; they are classified by the spectral type of the host star. Main-sequence stars such as the Sun (spectral type G or K) have represented most of the planetary system discovered so far. They are also classified by the size and types of planets as well as on the basis of their orbital configuration. Planetary systems organized by spectral type of host star are shown in Figure 3.4.

It may be mentioned that hot Jupiter systems with a gas giant very close to the star are one of the most common found so far and hot Neptune type systems have also been observed. Some theories, such as scattering, have been suggested for the formation of large

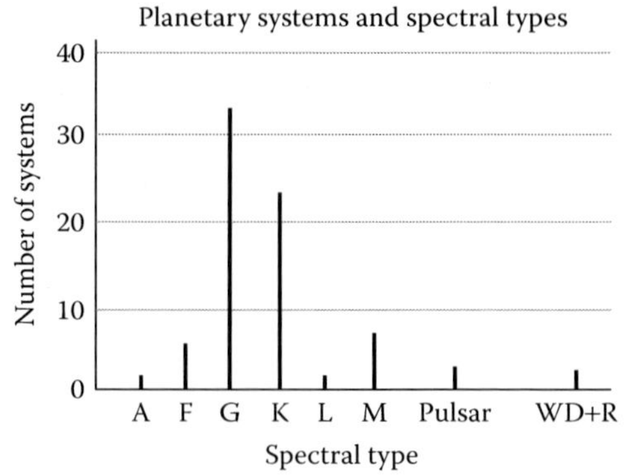

FIGURE 3.4
Planetary systems organized by spectral type of host star. (Courtesy of Wikipedia, Planetary system, https://en.wikipedia.org/wiki/Planetary_system [last modified on August 19, 2016].)

planets close to their parent stars [18,19]. Dusty disks with a wide ring of dust and comets are considered another common type of system [20]. Presently, few systems have been observed to be analogous to the solar system with terrestrial planets close to the parent star. Systems consisting of multiple super-Earths have also been identified [21].

3.6 Orbital Particulars

Unlike the solar system, which has convincingly shown coplanar and circular orbits, many of the known planetary systems exhibit much higher orbital eccentricity [22]. Some of them consist of planets in dramatic elliptical orbits, whereas some other consists of planets on inconsistent orbital inclination [23]. A very good example of such a system is 16 Cygni. It is true that no planets in the solar system are completely tidally locked. Planets with regular orbits, close to their primary stars, are assumed to typically experience tidal locking. Some other systems appear to display orbital resonances. For example, Gliese 876 has a three-body Laplace resonance [24,25]. Such a resonance is also found in Kepler-9, which has two planets in mean motion resonance.

3.7 Formation: Presolar Nebula

The nebular hypothesis clearly indicates that the solar system formed from the gravitational collapse of a fragment of a giant molecular cloud [26] that had a size of about 20 parsec (pc), while the fragments were roughly 1 pc across [27]. The further collapse of the fragments led to the formation of dense cores 0.01–0.1 pc in size [28]. Presolar nebula is one of these collapsing fragments that in turn would form in the solar system [29]. The composition of this was about the same as that of the Sun today, with hydrogen along with helium and trace amounts of lithium produced by Big Bang nucleosynthesis forming

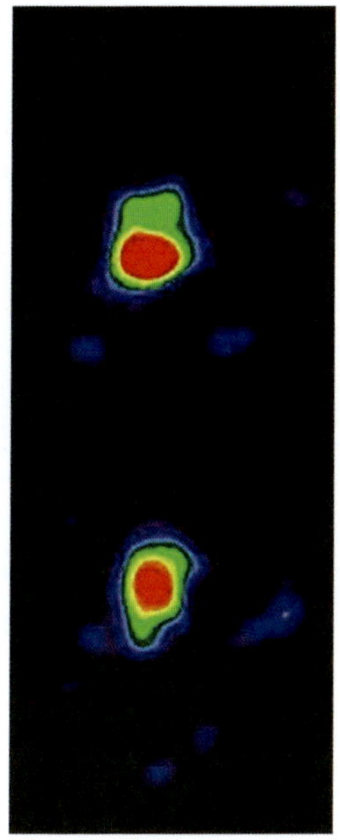

FIGURE 3.5
False-color radio picture of twin disks in a double star system. (Courtesy of L. F. Rodriguez (UNAM) et al., NRAO, AUI.)

about 98% of its mass. The remaining 2% of the mass consisted of heavier elements that were created by nucleosynthesis in earlier generations of stars [30]. Late in the life, they ejected heavier elements into the interstellar medium [31]. Figure 3.5 shows a radio image of a pair of planetary nebulae that very well harbor planetary material as well. Figure 3.6 shows Hubble and ESA images of protoplanetary disks in the Orion Nebula, which is a light-year wide. This is probably very similar to the primordial nebula from which the Sun formed.

Explanation of the above [32]: Sun-like stars are forming—and probably planets too—hidden inside Lynds 1551, an interstellar cloud of molecular gas and dust in the constellation Taurus. Using new receivers, coordinated radio telescopes at the Very Large Array near Socorro, New Mexico, can now sharply image the dusty protoplanetary disks surrounding these young stars at radio wavelengths. Just announced, this exciting example shows a false-color radio picture of twin disks in a double star system! A yellow bar indicates the scale in astronomical units (AUs) where 1 AU is the average distance between the Earth and Sun. The stars (unseen near the center of each disk) are about 45 AUs apart, comparable to the radius of the orbit of Pluto. Similar protoplanetary disks have been seen around single stars, but these twin disks are much smaller, each limited in size by the gravity of the nearby companion star. In fact, if large planets form orbiting near the edges of these disks, they may be ejected from the binary system. Figure 3.6 is a Hubble image of protoplanetary disks in the Orion Nebula.

Protoplanetary disks in the Orion Nebula (ESA), the oldest inclusions (Referencing the Allende Meter) observed in meteorites, assumed to trace the first solid material (Allende contains chondrules and CAIs) for forming in the presolar nebula, are 4.567 billion years old [33,34]. Investigations of ancient meteorites show traces of stable daughter nuclei of

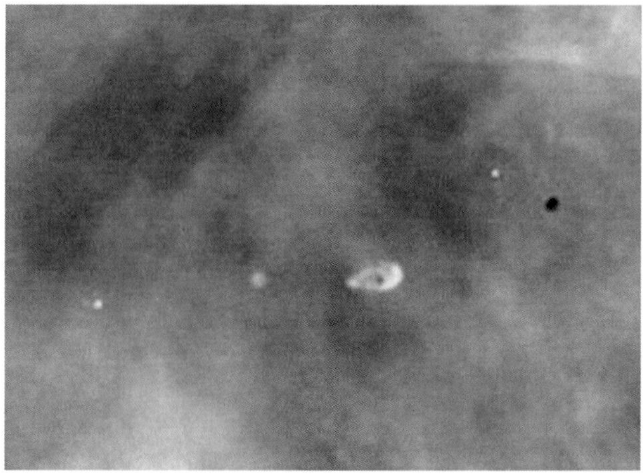

FIGURE 3.6
Hubble image of protoplanetary disks in the Orion Nebula. (Courtesy of European Space Agency [ESA], Protoplanetary disks in the Orion Nebula, https://www.spacetelescope.org/images/opo9229c/ [retrieved August 22, 2016].)

short-lived isotopes (e.g., iron-60) that only form in exploding, short-lived stars. This study suggests that one or more supernovae occurred near the Sun at the time of its formation. A shock wave from a supernova may have triggered the formation of the Sun producing regions of overdensity within the cloud that caused those regions to collapse [35]. As only massive, short-lived stars create supernovae, the Sun must have formed in a large star-forming region that produced massive stars, perhaps similar to the Orion Nebula [36–38]. Careful analyses of the structure of the Kuiper belt and of anomalous materials within it conclude that the Sun formed within a cluster of between 1,000 and 10,000 stars with a diameter of between 6.5 and 19.5 ly and a collective mass equivalent to 3,000 Suns. The cluster so formed started to break apart between 135 and 535 million years after formation [39,40]. Simulations of our Sun interacting with close-passing stars over the first 100 million years of its life create similar orbits found in outer solar system, such as detached objects [40]. Owing to the conservation of angular momentum, the nebula spun faster as it collapsed. When the material within the nebula condensed, the atoms within it started to collide with increasing frequency to convert their kinetic energy into heat. The center where majority mass collected became gradually hotter than the surrounding disk [41]. About 100,000 years [42], the competing forces of gravity, magnetic fields, gas pressure, and rotation caused the contracting nebula to flatten into a spinning protoplanetary disk [41] with a diameter of ~200 AU and form a hot, dense protostar (in which hydrogen fusion has not yet started) at the center [43]. During this time of its evolution, the Sun is assumed to have been a T Tauri star [44]. Investigations of T Tauri stars reveal that they are usually accompanied by disks of preplanetary matter with masses of 0.001–0.1 solar masses [45]. Hubble Space Telescope has found protoplanetary disks of up to 1000 AU in diameter in star-forming regions, in the case of the Orion Nebula [46]. After 50 million years, the temperature and pressure at the core of the Sun became so large that its hydrogen began to fuse, producing an internal source of energy that countered gravitational contraction until hydrostatic equilibrium was attained [29, p. 301]. This allowed the Sun's entry into the prime phase of its life, called the main-sequence star that derives energy from the fusion of hydrogen into helium in their cores [47].

3.8 Formation of Various Planets

Planets are assumed to have formed from the solar nebula, the disk-shaped cloud of gas, and dust left over from the Sun's formation [48]. The presently accepted technique for the formation of the planets is called accretion, in which the planets started as dust grains in orbit around the central protostar. By direct contact, these grains created clumps up to 200 m in diameter, which in turn collided to produce larger bodies, known as planetesimals, of ~10 km in size [49]. These gradually increased through further collisions, growing at the rate of centimeters per year over the next few million years [49]. An artist's concept of the solar nebula is reproduced in Figure 3.7.

The inner solar system was too hot for volatile molecules such as water and methane to condense so that the planetesimals that were created there could only form from compounds with high melting points, such as metals (e.g., iron, nickel, and aluminum) and rocky silicates. The rocky bodies would become the terrestrial planets (Mercury, Venus, Earth, and Mars). These compounds are very rare in the universe, comprising only 0.6% of the mass of the nebula and that is why the terrestrial planets could not grow very large [41]. The terrestrial embryos grew to nearly 0.05 Earth masses and ceased accumulating matter around 100,000 years after the formation of the Sun [50]. The gas giants (Jupiter, Saturn, Uranus, and Neptune) formed further out beyond the frost line, the point between the orbits of Mars and Jupiter where the material is sufficiently cool for volatile icy compounds to remain solid. At present time, the four gas giants comprise below 99% of all the mass orbiting the Sun. As the frost line accumulated large amounts of water through evaporation from icy material, it formed a region of lower pressure that increased the speed of orbiting dust particles and halted their motion toward the Sun. Uranus and Neptune are believed to have formed after Jupiter and Saturn did, when the strong solar wind had blown away much of the disk material. Consequently, the planets accumulated little hydrogen and helium [51]. The problem with the formation theories for these planets is the timescale of their formation. It would have taken a hundred million years for their cores to accrete, which means that Uranus and Neptune probably formed closer to the Sun.

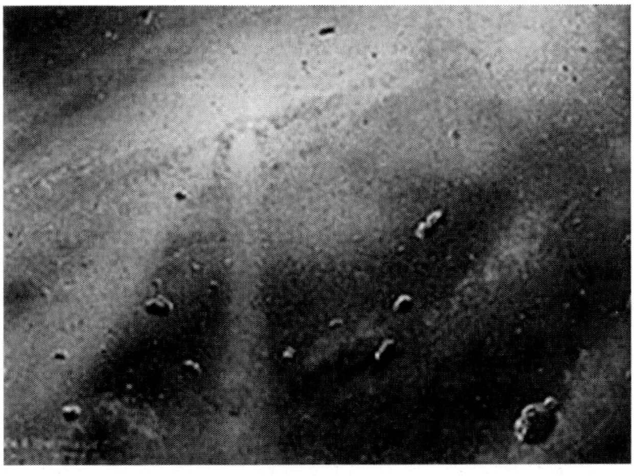

FIGURE 3.7
An artist's concept of the solar nebula. (Courtesy of Wikipedia, Formation and evolution of the solar system, https://en.wikipedia.org/wiki/Formation_and_evolution_of_the_Solar_System#Formation_of_the_planets [last modified on July 12, 2016].)

3.9 Subsequent Evolution

Planets were originally assumed to have created in or near their current orbits. However, this concept underwent radical change during the late twentieth and early twenty-first centuries. Presently, it is believed that the solar system looked largely different after its initial formation. Many objects at least as massive as Mercury were present in the inner solar system, while the outer solar system was much more compact than it is presently observed and also the Kuiper belt was much closer to the Sun [52]. Figure 3.8 shows an artist's concept of the giant impact thought to have formed the Moon.

3.10 Terrestrial Planets

After the planetary formation epoch, the inner solar system was populated by 50–100 Moon- to Mars-sized planetary embryos [53,54]. The next growth cycle was possible as these bodies collided and merged, which took about 100 million years. The giant collision occurred among the 50–100 Moon- to Mars-sized planetary embryos. A gaseous envelope developed surrounding the Mercury was responsible to form the Moon, while another removed the outer envelope of the Mercury [55]. The model, however, was not able to explain how the initial orbits of the prototerrestrial planets was produced [53]. One hypothesis for the *eccentricity dumping* is that the terrestrials planets formed in a disk of gas still had not yet been expelled by the Sun. The *gravitational drag* [56] of this residual gas would have lowered the energy of the planets, smoothing out their orbits [54]. If such gas existed, it would have prevented the terrestrial orbits of the planets from becoming so eccentric in the first place [52]. Another hypothesis suggests that gravitational drag occurred not between the planets and residual gas but between the planets and the remaining small bodies. As the large bodies passed

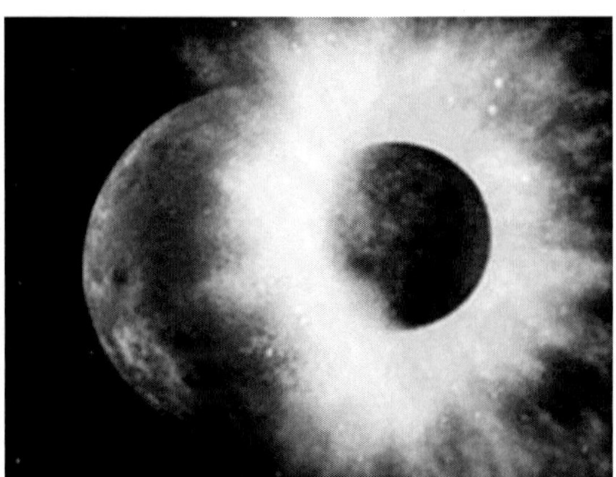

FIGURE 3.8
An artist's concept of the giant impact thought to have formed the Moon. (Courtesy of NASA, Planetary smash-up, http://www.nasa.gov/multimedia/imagegallery/image_feature_1454.html [retrieved October 21, 2015].)

through the crowd of smaller objects, they are attracted by the larger planets' gravity, forming a region of greater density, a *gravitational wake*, in the larger path of the objects. As a result, the increased gravity of the wake slowed the larger objects down into more regular orbits [57].

3.11 Asteroid Belt

The outer edge of the terrestrial region, between 2 and 4 AUs from the Sun, is known as the asteroid belt, which initially contained more than enough matter to produce two to three Earth-like planets and a large number of planetesimals. Planetesimals in this region later coalesced and formed 20–30 Moon- to Mars-sized planetary embryos [58]. Orbital resonances with Jupiter and Saturn are strong in the asteroid belt. Gravitational interactions with more massive embryos scattered many planetesimals into those resonances. The gravity of Jupiter increased the velocity of objects within these resonances, creating them to shatter on collision with other bodies rather than accrete [59]. As Jupiter migrated inward following its formation, resonances would have swept across the asteroid belt [60]. The cumulative action of the resonances and the embryos has two roles. It either scattered the planetesimals away from the asteroid belt or excited their orbital inclinations and eccentricities [58,61]. Some of those massive embryos were ejected by Jupiter, while others may have migrated to the inner solar system playing a role in the final accretion of the terrestrial planets [58,62]. During this period, the effects of the giant planets and planetary embryos left the asteroid belt that is still 10–20 times more than the current mass in the main belt [63]. The water was delivered by planetary embryos and small planetesimals thrown out of the asteroid belt by Jupiter [62]. A population of main-belt comets, discovered in 2006, also suggested a possible source for Earth's water [64,65].

3.12 Planetary Migration

Nebular hypothesis suggests that the outer two planets are in the *wrong place*. Uranus and Neptune, the so-called ice giants, exist in a region where the reduced density of the solar nebula and longer orbital times render their formation highly implausible [66]. The two are assumed to have formed in orbits near Jupiter and Saturn, where more material was available, while migrating outward to their present positions over hundreds of millions of years [52]. Figure 3.9 reveals simulations of the outer planets and Kuiper belt: (a) before Jupiter/Saturn 2:1 resonance, (b) scattering of Kuiper belt objects into the solar system after the orbital shift of Neptune, and (c) after ejection of Kuiper belt bodies by Jupiter [67].

After the formation of the solar system, the orbits of the giant planets continued to alter slowly due to their interaction with the large number of remaining planetesimals. Eventually, friction within the planetesimal disk made the orbits of Uranus and Neptune circular again [68,69]. In contrast to the outer planets, the inner planets are not assumed to have migrated largely as their orbits have remained stable following the period of giant

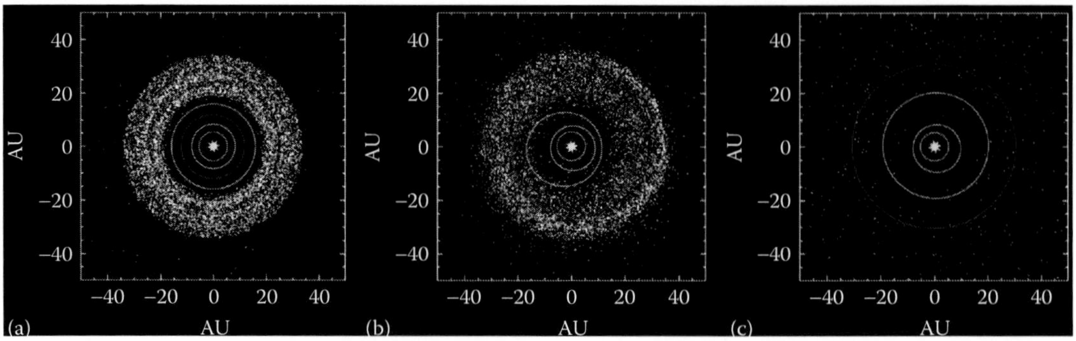

FIGURE 3.9
Simulations showing outer planets and Kuiper belt. (a) Before Jupiter/Saturn 2:1 resonance, (b) scattering of Kuiper belt objects into the solar system after the orbital shift of Neptune, and (c) after ejection of Kuiper belt bodies by Jupiter. (Courtesy of Wikipedia, https://en.wikipedia.org/wiki/Kuiper_belt#/media/File:Lhborbits.png [last modified on August 22, 2016].)

impacts [51]. Another interesting question is why Mars came out so small in comparison to Earth. A study published in 2011 suggested that Jupiter had migrated inward to 1.5 AUs and when Saturn formed, Jupiter migrated back to its present position. In this way, Jupiter consumed much of the material that would have created a bigger Mars. Simulations also reproduce the characteristics of the modern asteroid belt, with dry asteroids and water-rich objects more or less analogous to comets [70,71].

3.13 Late Heavy Bombardment

Gravitational disruption from the outer planets' migration would have sent large numbers of asteroids into the inner solar system [61]. This may have triggered the Late Heavy Bombardment that occurred nearly 4 billion years ago, 500–600 million years after the formation of the solar system [67,72]. This period of heavy bombardment lasted many hundred million years. This is responsible for the cratering still visible on geologically dead bodies of the inner solar system such as the Moon and Mercury [67,73]. The oldest known evidence for life on Earth dates to 3.8 billion years ago; nearly after the end of the Late Heavy Bombardment [74]. Figure 3.10 shows meteor crater in Arizona, which was created 50,000 years ago by an impact or only 50 m across; it is a stark reminder indicating that the accretion of the solar system is not over.

Impacts are assumed to be a regular part of the evolution of the solar system. That they continue to happen can be assured by the collision of Comet Shoemaker–Levy 9 with Jupiter in 1994, the 2009 Jupiter impact event, the Chelyabinsk meteor, the Tunguska event, as well as the impact feature Meteor Crater in Arizona. Thus, the process of accretion is not complete, and may still pose a threat to life on Earth [75,76]. After Late Heavy Bombardment, the evolution of the asteroid belt was mainly controlled by collisions [77]. As a result, many larger objects have been broken and sometimes newer objects have been forged from the remnants in less violent collisions [77].

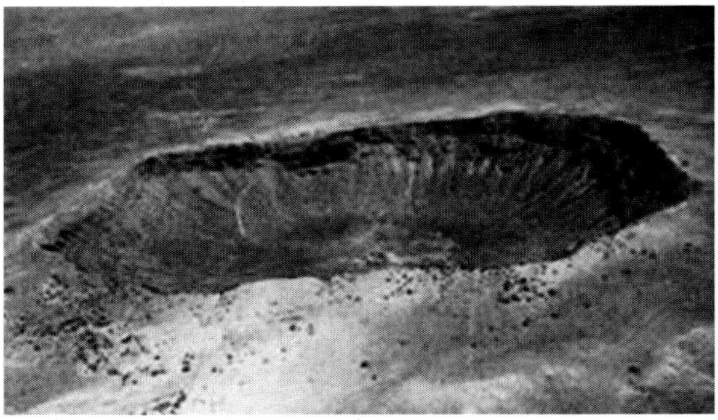

FIGURE 3.10
Meteor Crater in Arizona. (Courtesy of NASA, Images of Meteor Crater, http://neo.jpl.nasa.gov/images/meteorcrater.html [retrieved August 23, 2005].)

3.14 Giant Impact Hypothesis

Moons have come to exist around most of the planets and many other solar system bodies. These natural satellites originated by following any one of three possible mechanisms. (i) Conformation from a circumplanetary disk in the cases of the gas giants. (ii) Formation from impact debris given a sufficiently large impact at a shallow angle. (iii) Capture of a passing object. Jupiter and Saturn have some large moons, such as Io, Europa, Ganymede, and Titan, which may have originated from disks around each giant planet in similar way that the planets formed from the disk around the Sun [78]. The outer moons of the gas giants appear to be small and have eccentric orbits with arbitrary inclinations. These are the desired characteristics captured bodies [29, pp. 118–120, 79]. The largest irregular moon is Neptune's moon Triton, which is assumed to be a captured Kuiper belt object [76]. Moons of solid solar system bodies are created by both collisions and capture. The two small moons of Mars, Deimos and Phobos, are assumed to be captured asteroids [80]. The Moon of the Earth formed as a result of a single, large oblique collision [81,82]. The moons of trans-Neptunian objects Pluto (Charon) and Orcus (Vanth) may also have formed owing to a large collision. Pluto–Charon, Orcus–Vanth, and Earth–Moon systems are unusual in the solar system in that the satellite's mass is at least 1% that of the larger body [83,84].

3.15 Long-Term Stability

The solar system has remained in a state of chaos for over a billion years although, the chaos of the early solar system has been less, we still witness events that remind us of the unrest [85]. One example of this chaos is the Neptune–Pluto system, which lies in a 3:2 orbital resonance. Another example is Earth's axial tilt, which will be incomputable at some point between 1.5 and 4.5 billion years from now [86]. The outer orbits of the planets are chaotic over longer timescales in the range of 2–230 million years [87]. Such chaos manifests strongly as changes in eccentricity, with some orbits of the planets becoming significantly elliptical [88]. Ultimately, the solar system is stable and it is unlikely that none of the planets will collide with each other or be ejected from the system in the next few billion years [87]. Beyond this, within 5 billion years or so the eccentricity of the Mars may grow to around 0.2, so that it lies on an Earth-crossing orbit, leading to a potential collision. In the same timescale, the eccentricity of the Mercury may grow even more and a close encounter with Venus could eject it from the solar system altogether [85] or send it on a collision with Venus or Earth [89]. According to numerical simulations, this could happen within a billion years, in which Mercury's orbit is perturbed [90].

3.16 Moon–Ring Systems

The evolution of Moon systems is driven by the tidal forces. If a Moon revolves in the same direction as the rotation of the planet and the planet is rotating faster than the orbital period of the Moon, the bulge will be pulled constantly ahead of the Moon. In this situation, the angular momentum is transferred from the rotation of the primary to the revolution of the satellite. The Moon gains energy and thereby spirals outward, while the Earth rotates more slowly over time. The Moon is now tidally locked to the Earth; one of its revolutions around the Earth is same as one of its rotations about its axis. Thus, it always shows one face to the Earth. The Moon will continue to recede from Earth and the spin of the Earth will continue to slow gradually. If they survive in about 50 billion years, the expansion of the Sun, the Earth, and Moon will become tidally locked to each other. In fact, each will be caught up in what is called a *spin–orbit resonance* when the Moon will circle the Earth in about 47 days and both Moon and Earth will rotate around their axes in the same time, each only visible from one hemisphere of the other [91,92]. Other examples are the Galilean moons of Jupiter and also many of Jupiter's smaller moons [93] and the larger moons of Saturn [94]. Figure 3.11 reveals Neptune and its moon Triton. The photograph was taken by *Voyager 2*. Triton's orbit will eventually take it within Neptune's Roche [95] limit, tearing it apart and forming a new ring system.

A different scenario occurs when the tidal bulge lags behind the Moon in its orbit. In this case, the angular momentum of the rotation and revolution has opposite signs, so transfer leads to decreases in the magnitude of each. Such a fate awaits the Mars moon Phobos [96], Triton of Neptune (in 3.6 billion years) [97], Metis and Adrastea of Jupiter [98], and at least 16 small satellites of Uranus and Neptune [99]. A third possibility is where the primary and Moon are tidally locked to each other when the tidal bulge stays directly under the Moon and there is no transfer of angular momentum. Also the orbital period will not change. Pluto and Charon are examples of the configuration [100].

3.17 Stellar Evolution and Future of the Earth

The large changes in the solar system will come from changes in the Sun itself as it ages. As the Sun burns through its supply of hydrogen fuel, it gets hotter and burns the remaining fuel faster. Consequently, the Sun is growing brighter at a rate of 10% every 1.1 billion

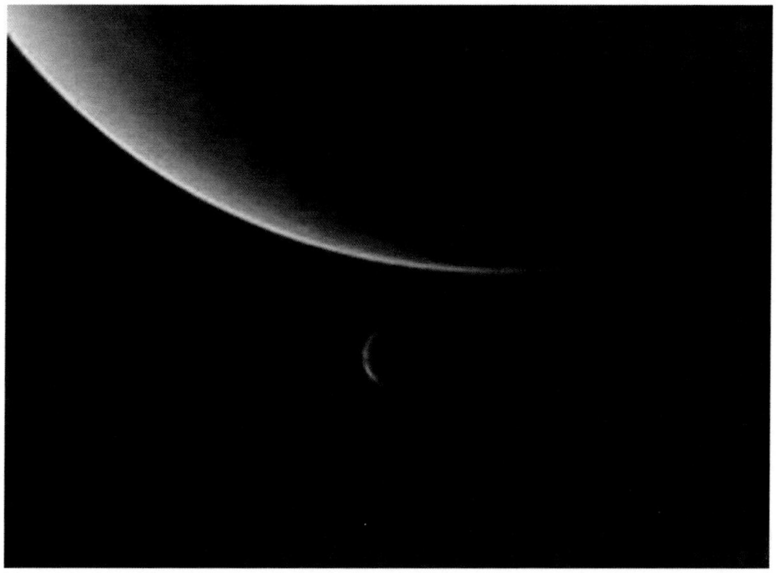

FIGURE 3.11
Neptune and its moon Triton, taken by Voyager 2. (Courtesy of NASA, Montage: Neptune and Triton, http://www.nasa.gov/multimedia/imagegallery/image_feature_2344.html [retrieved January 6, 2016].)

years [101]. Evaporation of water from the surface of ocean could accelerate temperature increase, ending all life on Earth [102]. It is possible that as Mars's surface temperature increases, carbon dioxide and water frozen under the surface regolith will release into the atmosphere providing a potential future abode for life [103]. By 3.5 billion years from now, it is predicted that the Earth's surface conditions will be similar to those of Venus today [104]. The relative size of the Sun as it is now (inset) compared to its estimated future size as a red giant is shown in Figure 3.12.

Around 5.4 billion years from now, the core of the Sun will become hot enough to trigger hydrogen fusion in its surrounding shell [105]. Over the course, the Sun's remaining outer layers will fall away, ejecting a vast stream of matter into space. This will form a halo known as a planetary nebula. The ejected material will contain the helium and carbon produced by the nuclear reactions, continuing the enrichment of the interstellar

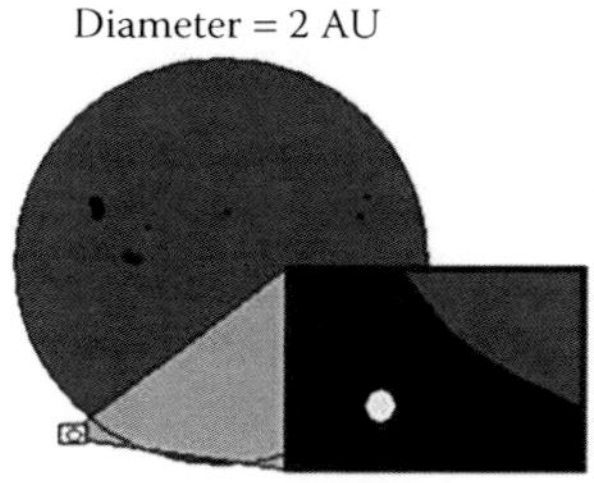

FIGURE 3.12
Relative size of the Sun as it is now (inset) compared to its estimated future size as a red giant. (Courtesy of Wikipedia, Future of Earth, https://en.wikipedia.org/wiki/Future_of_the_Earth [last modified on June 27, 2016].)

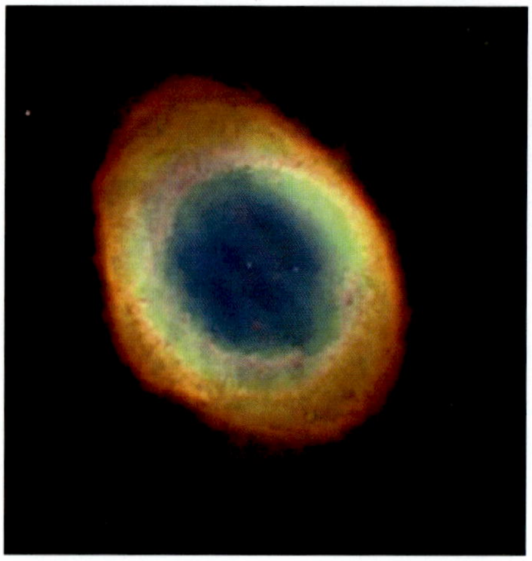

FIGURE 3.13
Ring nebula (NGC 7293, The Helix Nebula), a planetary nebula similar to what the Sun will become. (Courtesy of Wikipedia, Planetary nebula, https://en.wikipedia.org/wiki/Planetary_nebula [last modified on June 27, 2016].)

medium with heavy elements [106]. Figure 3.13 reveals a Ring nebula, which is similar to what the Sun will become.

As the Sun dies, its gravitational pull on the orbiting bodies will weaken due to its mass loss. All remaining orbits of the planets will expand; if Venus, Earth, and Mars still exist, their orbits will lie roughly at 1.4, 1.9, and 2.8 AU [107]. After billions more years, the Sun will ultimately cease to shine altogether, becoming a black dwarf [108].

3.18 Galactic Interaction

The solar system moves alone through the Milky Way galaxy in a circular orbit approximately 30,000 ly from the galactic center with speed about 220 km/s. The period required for the solar system to complete one revolution around the galactic center is in the range of 220–250 million years. From its formation, the solar system has completed at least 20 such revolutions [108]. Location of the solar system within our galaxy is shown in Figure 3.14.

According to mathematical models, by increasing the flux of Oort cloud comets into the solar system by a factor of 4 leads to a massive increase in the likelihood of a devastating impact [109].

3.19 Properties of Our Solar System

To identify the true origin of the solar system, we need to identify the conditions put forth by the existing theory. Any convincing theory dealing with the origins of our solar system should be able to explain the various properties therein. The primary conditions that must

FIGURE 3.14
Location of the solar system within our galaxy. (Courtesy of NASA, Milky Way and our location, http://www.nasa.gov/mission_pages/sunearth/news/gallery/galaxy-location.html [retrieved August 22, 2016].)

be explained include the following. (i) Placement of the Sun at the center of the solar system. (ii) The procession of the planets around the Sun in a counterclockwise direction as viewed from above the North Pole of Earth. (iii) The placement of the small rocky worlds nearest to the Sun, with the large gas giants (the Jovian planets) further out. (iv) All the planets appear to have formed around the same time as the Sun. (v) The chemical composition of the Sun and planets. (vi) The existence of comets and asteroids.

3.20 Identifying a Suitable Theory

The only theory that meets all the requirements is known as the solar nebula theory, which suggests that the solar system arrived at its current form after collapsing from a molecular gas cloud some 4.568 billion years ago. The region of space near the Sun still contain enough temperatures that the lighter elements cannot yet precipitate, but where the temperatures are cooler, the lighter elements will dominate. As the small, hot globs of material collide together, they form larger blocks of material. These objects will become large enough to become spherical under the force of their own gravity.

3.21 Problem with the Solar Nebula Theory

Here are some facts we should know. Where did our own solar system come from? Is our Sun just an accident? Are the precisely balanced orbits of its moons and planets just the result of random explosions and outer space traffic jams? Is it all the result of haphazard chance? There are many questions still unanswered, hopefully awaits for development of suitable theories.

References

1. NASA. Comet feature named after late NASA scientist Claudia Alexander. Jet Propulsion Laboratory, Pasadena, CA. http://science.jpl.nasa.gov/people/Gulkis/JPL (retrieved August 23, 2016).
2. I. de Pater. Department of Astronomy, University of California, Berkeley, CA. http://astro.berkeley.edu/people/faculty/depater.html (retrieved August 23, 2016).
3. G. Marcy. Exoplanets. http://astro.berkeley.edu/~gmarcy/ (retrieved August 23, 2016).
4. J. Schneider (2011). Interactive extra-solar planets catalog. *The Extrasolar Planets Encyclopedia.* http://exoplanet.eu/catalog/ (retrieved July 13, 2012).
5. *USA Today* (2012). http://content.usatoday.com/communities/sciencefair/post/2012/04/hd-10180-recordsetting-solar-system-spotted/1#.T4DuANnHGC4 |url= missing title (help) (retrieved August 23, 2016).
6. S. Alpha (2013). Kepler-90 (KOI-351) - Seven planet, compact solar system analogue. http://solar-flux.forumandco.com/t1419-kepler-90-koi-351-seven-planet-compactsolar-system-analogue (retrieved September 28, 2016).
7. SciTechDaily (2014). Astronomers discover a potentially habitable nearby super-Earth, Gliese 832 c. http://scitechdaily.com/astronomers-discover-potentially-habitable-nearby-super-earth-gliese-832-c/ (retrieved September 28, 2016).
8. M. Wall (2012). Discovery! Earth-size alien planet at Alpha Centauri is closest ever seen. http://www.space.com/18089-earth-size-alien-planet-alpha-centauri.html (retrieved August 12, 2016).
9. Wikia Inc., A new dawn, Gliese 876. http://thexuniverse.wikia.com/wiki/Gliese_876 (retrieved August 12, 2016).
10. D. J. Dsrling (2004). *The Universal Book of Astronomy, from the Andromeda Galaxy to the Zone of Avoidance.* Hoboken, NJ: Wiley. p. 394.
11. V. Illingworth (2000). *Collins Dictionary of Astronomy.* London: Collins. p. 314.
12. V. Illingworth (2000). *Collins Dictionary of Astronomy.* London: Collins. p. 382.
13. I. Ridpath (2003). *A Dictionary of Astronomy.* Oxford, New York: Oxford University Press. p. 420.
14. B. Dunbar. Exoplanet, House of Horror, NASA. http://www.nasa.gov/topics/universe/features/exoplanetHouseOfHorrors.html (retrieved October 29, 2009).
15. About Education. http://archaeology.about.com/od/vterms/g/vedic.htm (retrieved January 18, 2016).
16. Encyclopaedia Britannica. https://www.britannica.com/biography/Aryabhata-I (last modified February 03, 2015).
17. C. M. Linton (2004). *From Eudoxus to Einstein: A History of Mathematical Astronomy.* Cambridge, UK: Cambridge University Press. pp. 38–39.

    J. L. E. Dreyer (1906). *History of the Planetary Systems from Thales to Kepler.* Cambridge, UK: Cambridge University Press. pp. 135–148.
18. S. J. Weidenschilling and F. Marzari (1996). Gravitational scattering as a possible origin for giant planets at small stellar distances. *Nature* 384 (6610): 619–621.
19. UK Astronomy Technology Centre. http://www.roe.ac.uk/roe/support/pr/rsexhibition/typesofsolarsystem.pdf (retrieved September 28, 2016).
20. R. Barnes (1996). Rory Barnes' Homepage. http://faculty.washington.edu/rkb9/ (retrieved September 28, 2016).
21. R. Dvorak et al. (2010). *Dynamical Habitability of Planetary Systems.* Vienna, Austria: Institute for Astronomy, University of Vienna.
22. NASA—Out of whack planetary system offers clues to a disturbed past (2010). Nasa.gov. http://www.nasa.gov/mission_pages/hubble/science/outofwack.html (retrieved August 17, 2012).
23. G. W. Marcy et al. (2001). A pair of resonant planets orbiting GJ 876. *The Astrophysical Journal* 556 (1): 296–301.

24. E. J. Rivera et al. (2010). The Lick-Carnegie exoplanet survey: A Uranus-mass fourth planet for GJ 876 in an Extrasolar Laplace configuration. *The Astrophysical Journal* 719: 890–899. doi:10.1088/0004-637X/719/1/890.
25. T. Montmerle et al. (2006). Solar system formation and early evolution: The first 100 million years. *Earth, Moon, and Planets* 98 (1–4): 39–95.
26. A. Zabludoff (2003). Lecture 13: The Nebular Theory of the origin of the Solar System. http://atropos.as.arizona.edu/aiz/teaching/nats102/mario/solar_system.html (retrieved August 12, 2016).
27. J. J. Rawal (1986). Further considerations on contracting Solar Nebula. *Earth, Moon, and Planets* 34 (1): 93–100.
28. W. M. Irvine (1983). The chemical composition of the pre-solar nebula. In T. I. Gombosi (ed.). *Cometary Exploration: Proceedings of the International Conference.* Akademiai Kiado, Budapest, Hungary. vol. 1, pp. 3–12.
29. M. Zeilik and S. A. Gregory (1998). *Introductory Astronomy & Astrophysics.* Philadelphia, PA: Saunders College Publishing.
30. C. H. Lineweaver (2001). An estimate of the age distribution of terrestrial planets in the universe: Quantifying metallicity as a selection effect. *Icarus* 151 (2): 307.
31. NASA Astronomy Picture of the Day (September 25, 1998). http://apod.nasa.gov/apod/ap980925.html (retrieved September 28, 2016).
32. A. Bouvier and M. Wadhwa (2010). The age of the solar system redefined by the oldest Pb-Pb age of a meteoritic inclusion. *Nature Geoscience* 3: 637–641.
33. Y. Amelin and K. Alexander (2007). Pb isotopic age of the Allende chondrules. *Meteoritics & Planetary Science* (University of Arizona) 42 (7/8): 1321–1335. Bibcode:2007M&PS...42.1321A. doi:10.1111/j.1945 - 5100.2007.tb00577.x.
34. J. Williams (2010). The astrophysical environment of the solar birthplace. *Contemporary Physics* 51 (5): 381–396.
35. J. J. Hester et al. (2004). The cradle of the Solar System. *Science* 304 (5674): 1116–1117.
36. M. Bizzarro et al. (2007). Evidence for a late supernova injection of 60Fe into the protoplanetary disk. *Science* 316 (5828): 1178–1181.
37. M. Kelly. Slow-moving rocks better odds that life crashed to Earth from space. News at Princeton. https://www.princeton.edu/main/news/archive/S34/82/42M30/ (retrieved September 24, 2012).
38. S. F. P. Zwart (2009). The lost siblings of the Sun. *The Astrophysical Journal* 696 (L13–L16): L13.
39. N. A. Kaib and T. Quinn (2008). The formation of the Oort cloud in open cluster environments. *Icarus* 197 (1): 221–238.
40. A. Zabludoff (Spring 2003). Lecture 13: The Nebular Theory of the origin of the Solar System. http://atropos.as.arizona.edu/aiz/teaching/nats102/mario/solar_system.html (retrieved December 27, 2006).
41. T. Montmerle et al. (2006). Solar System formation and early evolution: The first 100 million years. *Earth, Moon, and Planets* 98 (1–4): 39–95.
42. J. S. Greaves (2005). Disks around stars and the growth of planetary systems. *Science* 307 (5706): 68–71.
43. M. W. Caffe et al. (1987). Evidence in meteorites for an active early sun. *The Astrophysical Journal, Part 2—Letters to the Editor* 313: L31–L35.
44. M. Momose et al. (2003). Investigation of the physical properties of protoplanetary disks around T Tauri stars by a high-resolution imaging survey at lambda = 2 mm. In S. Ikeuchi, J. Hearnshaw, T. Hanawa (eds.). *The Proceedings of the IAU 8th Asian-Pacific Regional Meeting,* vol. I, 289. Astronomical Society of the Pacific Conference Series. p. 85.
45. D. L. Padgett et al. (March 1999). Hubble space telescope/NICMOS imaging of disks and envelopes around very young stars. *The Astronomical Journal* 117 (3): 1490–1504.
46. S. Yi et al. (2001). Toward better age estimates for stellar populations: The isochrones for solar mixture. *The Astrophysical Journal Supplement* 136: 417.

47. A. P. Boss and R. H. Durisen (2005). Chondrule-forming shock fronts in the solar nebula: A possible unified scenario for planet and chondrite formation. *The Astrophysical Journal* 621 (2): L137–L140.
48. P. Goldreich and W. R. Ward (1973). The formation of planetesimals. *The Astrophysical Journal* 183: 1051.
49. D. N. C. Lin (May 2008). The genesis of planets (fee required). *Scientific American* 298 (5): 50–59.
50. E. W. Thommes et al. (2002). The formation of Uranus and Neptune among Jupiter and Saturn. *The Astronomical Journal* 123 (5): 2862.
51. M. Brown. Dysnomia, the moon of Eris. Personal web site. http://web.gps.caltech.edu/~mbrown/planetlila/moon/ (retrieved February 1, 2008).
52. J.-M. Petit and A. Morbidelli (2001). The primordial excitation and clearing of the asteroid belt. *Icarus* 153 (2): 338–347.
53. J. Kominami and S. Ida (2001). The effect of tidal interaction with a gas disk on formation of terrestrial planets. *Icarus* 157 (1): 43–56.
54. S. C. Solomon (2003). Mercury: The enigmatic innermost planet. *Earth and Planetary Science Letters* 216 (4): 441–455.
55. Wikipedia. Gravity drag. http://en.wikipedia.org/wiki/Gravity_drag (last modified May 10, 2016).
56. P. Goldreich and Y. Re'em Sari (2004). Final stages of planet formation. *The Astrophysical Journal* 614 (1): 497.
57. W. F. Bottke et al. (2005). Linking the collisional history of the main asteroid belt to its dynamical excitation and depletion. *Icarus* 179 (1): 63–94.
58. R. Edgar and P. Artymowicz (2004). Pumping of a planetesimal disc by a rapidly migrating planet. *Monthly Notices of the Royal Astronomical Society* 354 (3): 769–772.
59. E. R. D. Scott (2006). Constraints on Jupiter's age and formation mechanism and the Nebula lifetime from chondrites and asteroids. In *Proceedings of the 37th Annual Lunar and Planetary Science Conference*. League City, TX: Lunar and Planetary Society.
60. D. O'Brien et al. (2007). The primordial excitation and clearing of the asteroid belt—Revisited. *Icarus* 191 (2): 434–452.
61. S. N. Raymond et al. (2007). High-resolution simulations of the final assembly of Earth-like planets 2: Water delivery and planetary habitability. *Astrobiology* 7 (1): 66–84.
62. G. A. Krasinsky et al. (July 2002). Hidden mass in the asteroid belt. *Icarus* 158 (1): 98–105.
63. H. H. Hsieh and D. Jewitt (2006). A population of comets in the main asteroid belt. *Science* 312 (5773): 561–563.
64. F. Reddy (2006). New comet class in Earth's backyard. Astronomy.com. http://www.astronomy.com/not-found.aspx?item=%2fasy%2fdefault&user=extranet\Anonymous&site=website (retrieved April 29, 2008).
65. G. J. Taylor (2001). Uranus, Neptune, and the mountains of the moon. In: *Planetary Science Research Discoveries*. Honolulu, HI: Hawaii Institute of Geophysics & Planetology.
66. R. Gomes et al. (2005). Origin of the cataclysmic Late Heavy Bombardment period of the terrestrial planets. *Nature* 435 (7041): 466–469.
67. H. F. Levison et al. (2007). Origin of the structure of the Kuiper Belt during a dynamical instability in the orbits of Uranus and Neptune. *Icarus* 196 (1): 258–273.
68. M. J. Fogg and R. P. Nelson (2007). On the formation of terrestrial planets in hot-Jupiter systems. *Astronomy and Astrophysics* 461 (3): 1195.
69. Jupiter may have robbed Mars of mass, new report indicates (2011). Southwest Research Institute, San Antonio, TX (Press release). June 6.
70. K. J. Walsh et al. (2011). A low mass for Mars from Jupiter's early gas-driven migration. *Nature* 475 (7355): 206–209.
71. K. Hansen (2005). Orbital shuffle for early solar system. *Geotimes*. http://www.geotimes.org/june05/WebExtra060705.html.
72. NASA History Division. Chronology of planetary surfaces. http://history.nasa.gov/SP-467/ch3.htm (retrieved March 13, 2008).

73. University of California-Los Angeles (UCLA). University of Colorado scientists strengthen case for life more than 3.8 billion years ago (Press release), July 21, 2006. http://newsroom.ucla.edu/releases/UCLA-University-of-Colorado-Scientists-7203 (retrieved August 12, 2016).
74. C. R. Chapman (1996). The risk to civilization from extraterrestrial objects and implications of the Shoemaker-Levy 9 comet crash. *Abhandlungen der GeologischenBundeanstalt, Wien* 53: 51–54.
75. C. B. Agnor and H. P. Douglas (2006). Neptune's capture of its moon Triton in a binary-planet gravitational encounter. *Nature* 441 (7090): 192–194.
76. W. F. Bottke et al. (2005). The origin and evolution of stony meteorites. *Proceedings of the International Astronomical Union*. Dynamics of Populations of Planetary Systems, vol. 197, pp. 357–374.
77. N. Takato et al. (2004). Detection of a deep 3-μm absorption feature in the spectrum of Amalthea (JV). *Science* 306 (5705): 2224–2227.
78. D. C. Jewitt et al. (2004). Jupiter's outer satellites and Trojans. In F. Bagenal, T. E. Dowling, W. B. McKinnon (eds.). *Jupiter. The Planet, Satellites and Magnetosphere*. Cambridge, UK: Cambridge University Press. pp. 263–280.
79. S. S. Sheppard. The giant planet satellite and moon page. Departamenteso e pa pajaro of Terrestrial Magnetism at Carniege Institution for science. https://home.dtm.ciw.edu/users/sheppard/satellites/ (retrieved December 19, 2014).
80. R. M. Canup and E. Asphaug (2001). Origin of the Moon in a giant impact near the end of the Earth's formation. *Nature* 412 (6848): 708–712.
81. D. J. Stevenson (1987). Origin of the moon—The collision hypothesis. *Annual Review of Earth and Planetary Sciences* 15 (1): 271.
82. R. M. Canup (2005). A giant impact origin of Pluto-Charon. *Science* 307 (5709): 546–550.
83. M. E. Brown et al. (2010). The size, density, and formation of the Orcus-Vanth system in the Kuiper Belt. *The Astronomical Journal* 139 (6): 2700.
84. J. Laskar (1994). Large-scale chaos in the solar system. *Astronomy and Astrophysics* 287: L9–L12.
85. O. N. de Surgy and J. Laskar (1997). On the long term evolution of the spin of the Earth. *Astronomy and Astrophysics* 318: 975–989.
86. W. B. Hayes (2007). Is the outer Solar System chaotic? *Nature Physics* 3 (10): 689–691.
87. I. Stewart (1997). *Does God Play Dice?* 2nd ed. London, UK: Penguin Books. pp. 246–249.
88. D. Shiga (2008). The solar system could go haywire before the sun dies. *New Scientist*, Daily News, 23 April 2008. https://www.newscientist.com/article/dn13757-solar-system-could-go-haywire-before-the-sun-dies/ (retrieved August 12, 2016).
89. K. Batygin and G. Laughlin (2008). On the dynamical stability of the Solar System. *The Astrophysical Journal* 683 (2): 1207.
90. C. D. Murray and S. F. Dermott (1999). *Solar System Dynamics*. Cambridge, UK: Cambridge University Press. p. 184.
91. T. Dickinson (1993). *From the Big Bang to Planet X*. Camden East, Ontario: Camden House. pp. 79–81.
92. A. Gailitis (1980). Tidal heating of Io and orbital evolution of the Jovian satellites. *Monthly Notices of the Royal Astronomical Society* 201: 415.
93. R. Bevilacqua et al. (1980). Resonances and close approaches. I. The Titan-Hyperion case. *Earth, Moon, and Planets* 22 (2): 141–152.
94. Encyclopaedia Britannica. https://www.britannica.com/topic/Roche-limit (last modified February 05, 2016).
95. B. G. Bills et al. (2006). Improved estimate of tidal dissipation within Mars from MOLA observations of the shadow of Phobos. *Journal of Geophysical Research* 110 (E7): E07004.
96. C. F. Chyba et al. (1989). Tidal evolution in the Neptune-Triton system. *Astronomy and Astrophysics* 219: 23.
97. J. A. Burns et al. (2004). Jupiter's Ring-Moon system. In F. Bagenal, T. E. Dowling, W. B. McKinnon (eds.). *Jupiter: The Planet, Satellites and Magnetosphere*. Cambridge, UK: Cambridge University Press. p. 241.

98. M. J. Duncan and J. J. Lissauer (1997). Orbital stability of the Uranian satellite system. *Icarus* 125 (1): 1–12.
99. M. Buie et al. (2006). Orbits and photometry of Pluto's satellites: Charon, S/2005 P1, and S/2005. *The Astronomical Journal* 132 (1): 290.
100. J. Hecht (1994). Science: Fiery future for planet Earth. *New Scientist* (1919). p. 14.
101. R. Knut Jørgen (2004). Our changing solar system. Center for International Climate and Environmental Research.
102. J. S. Kargel (2004). *Mars: A Warmer, Wetter Planet.* Southwest Research Institute, San Antonio, TX.
103. S. Leong (2002). Period of the Sun's orbit around the galaxy (cosmic year). In G. Elert (ed.). *The Physics Factbook (self-published).* http://hypertextbook.com/facts/2002/StacyLeong.shtml (retrieved September 28, 2016).
104. K. P. Schroder and R. C. Smith (2008). Distant future of the Sun and Earth revisited. *Monthly Notices of the Royal Astronomical Society* 386 (1): 155–163.
105. B. Balick (2006). Planetary nebulae and the future of the Solar System. Personal Web Site. https://faculty.washington.edu/balick/ (retrieved June 23, 2006).
106. I. J. Sackmann et al. (1993). Our Sun. III. Present and future. *The Astrophysical Journal* 418: 457.
107. G. Fontaine et al. (2001). The potential of white dwarf cosmochronology. *Publications of the Astronomical Society of the Pacific* 113 (782): 409–435.
108. S., Michael. Perturbing the Oort cloud. *American Scientist.* The Scientific Research Society. http://www.americanscientist.org/issues/pub/perturbing-the-oort-cloud (retrieved March 25, 2008).
109. M. Williams (July 19, 2016). How was the solar system formed? – The Nebular hypothesis. *Universe Today.* http://www.universetoday.com/38118/how-was-the-solar-system-formed/ (retrieved September 28, 2016).

4

Hypotheses Concerning the Formation of Solar System

4.1 Introduction

To answer scientific questions, many times we need the formulation of a hypothesis that is usually tested against the facts, so we look for contradictions that rule out or demand modification to the hypothesis. The process for hypothesis formulation is to build a proper theory that is lengthy and career dependent. That is why the sociology of science always prefers testing a hypothesis and getting it confirmed through works of many scientists, as the creator of the hypothesis has a strong psychological attachment to his work. There was no attempt to link theories to the existence of a *solar system*, as it was not generally thought that the solar system, in the sense as we now understand it, existed. The first attempt toward a theory of solar system formation and evolution was the general acceptance of heliocentrism. This concept placed the Sun at the center of the system and the Earth in an orbit around it. The first recorded use of the term *solar system* dates back to 1704 [1]. The present standard theory for solar system formation, the *nebular hypothesis* [2], has fallen in and out of favor since its formulation in the eighteenth century. The criticism of the hypothesis was its apparent inability for explaining the Sun's relative lack of angular momentum when compared to the planets [3]. In the early 1980s, increased studies of young stars have shown them to be surrounded by cool disks of dust and gas, exactly as the nebular hypothesis predicts, which led to the reacceptance of the hypothesis [3]. To understand how the Sun will continue to evolve, we must have knowledge of the power source. Eddington's confirmation of Einstein's theory of relativity led to his realization that the energy of the Sun comes from nuclear fusion reactions in its core [4]. In 1935, Eddington suggested that other elements might be involved in the power production within stars [5]. Hoyle elaborated on this premise by arguing that evolved stars, known as red giants, formed many elements heavier than hydrogen and helium in their cores. When a red giant finally casts off its outer layers, the elements would be recycled for producing other star systems [5].

4.2 Encounter Hypothesis: An Early Theory

There are many problems with the existing hypotheses and models that are used to support the naturalistic story of how the solar system formed. This naturalistic story of the solar system goes back 4.5 billion years. In this view, the solar system started its journey as a cloud of dust and gas that collapsed on it. It is considered that a star began to form as the cloud was compressed by some type of unknown forces. As the star spun and gathered more mass, a disk of dust began for creating the planets. The young solar system was formed by taking over millions of years and from the process much debris was left over, which are present today as asteroids and large belts and clouds of material. There was no intervention in the process and it is still continuing. What we really know about the origin of the solar system is highly interesting. According to the secular scenario, energy was converted slowly into matter as the universe continued to expand after the big bang. That matter accumulated to form the first atoms. The first stages of the naturalistic universe contained energy only, but as the expanding universe slowed and cooled, helium, hydrogen, and small amounts of lithium were formed from the energy. The gases thus gathered into nebulae to give birth to stars and planets. As the earliest stars produced their energy through nuclear fusion, heavier elements were assembled and star systems with planets began to form. It is a fact that the star formation and stellar evolution have never been observed, as they are supposed to occur over many millions of years. Evolutionists propose that the stars form as gases, hydrogen and helium in the early universe, collapse due to gravity. According to them, as the nebula collapses, the gases heat up and the nebula spins itself into a flattened disk. But one major problem with this scenario is that as the gases are heated, the pressure also increases, which would tend to cause the nebula to expand and counteract the gravitational collapse. With a view to counter this problem, it was further suggested that some type of *shock* overcomes the gas pressure exactly at the right time. The so-called shock may come owing to the explosion of a nearby supernova or may come from some other sources. In fact, the problem has presently become a circular argument. There would have been other stars attaining the supernova stage to cause the first stars to form. Consequently, the argument may work for later generations of stars but unable to explain the first generation formed. Interestingly, if the first generation could not form, then they could not have created the materials and force for later generations to form. Evolutionists then suggest that a protostar would form, which continued to gain mass as it attracted nearby gases and particles. If the pressure in the core of the star causes to increase the temperature to 10 million Kelvin, nuclear fusion can start making the star stable, which appears when the inward gravitational force balances the outward gas pressure. It is clear that depending on the size and composition of the star, it demands a multibillion-year process until it consumes all of its nuclear fuel. Finally, at the end of its life, it will either contract and become a white dwarf or explode in a spectacular supernova, leaving behind a black hole or a neutron star. How this process exactly occurred in our solar system may be assumed as a matter of inferring events in the past from what we find in the present. It is true that there are certain assumptions that must be taken as the basis when creating historical theories. The widely acceptable ideas are mainly based on uniformitarian assumptions and have many problems.

4.3 Nebular Hypothesis

The formation of our solar system is found to be very typical in comparison to other star systems in the universe. An interstellar cloud of dust and gases started to collapse on itself about 5 billion years ago, as the force of gravity pulled the particles together.

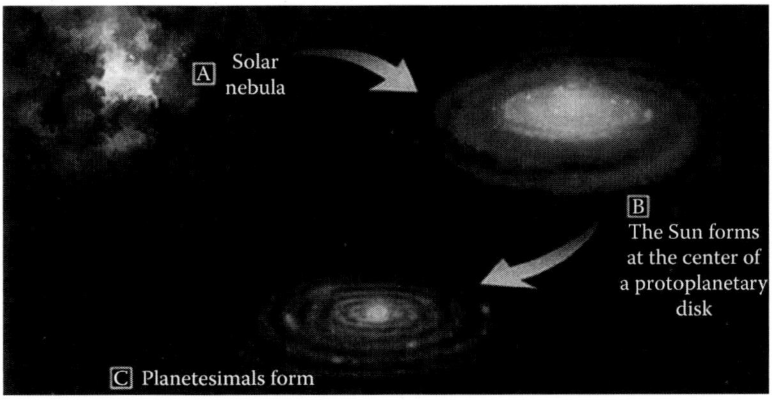

FIGURE 4.1
Formation of the universe: According to nebular hypothesis, the solar system was formed from a rotating cloud of dust and gas (marked by A); the Sun was formed at the center of rotating disk (marked by B); and planetesimals collided, gaining eventually sufficient mass to be planets (marked by C). (Data from Patterson R., Origin of the solar system, *Evolution Exposed: Earth Science*, Answers in Genesis, Petersburg, KY, 2010.)

This scenario created a remarkable problem. There is no doubt that the force of gravity is pulling the particles together, but other forces, such as the pressure exerted by the gases in a balloon, are pushing the particles apart. As gravity is a relatively weak force, this hypothesis faced a great difficulty to explain how the stars and planets are formed in an actual situation. The model used by evolutionists for explaining the formation of star systems is known as the nebular hypothesis. The hypothesis suggests that the solar system formed from a spinning, collapsing nebula; the Sun formed at the center with the planets forming at different distances. The idea was initially put forth in the 1700s but subsequently has undergone minor revisions. Our solar system started as a nebula, which began to spin and finally collapse due to gravity. As the nebula spun faster and faster, the protostar started to produce, as pointed out, and a disk of dust and gases surrounded the protostar. The spinning nebular disk became warmer at the center than at the edges and planetesimals started to form slowly as the dust particles stuck to one another, as illustrated in Figure 4.1. According to the nebular hypothesis, the solar system formed from a rotating cloud of dust and gas as marked by "A" in the figure; the Sun formed at the center of the rotating disk, which is marked by "B" in the same figure; planetesimals collided and as a result gained sufficient mass to become planets, as marked by "C."

4.4 Protoplanet Hypothesis

The present day working model for the formation of the solar system is known as the protoplanet hypothesis [6], which incorporates many of the components of the nebular hypothesis but adds some new aspects from current knowledge of fluids and states of matter [7,8]. The protoplanet hypothesis is illustrated in Figures 4.2 and 4.3.

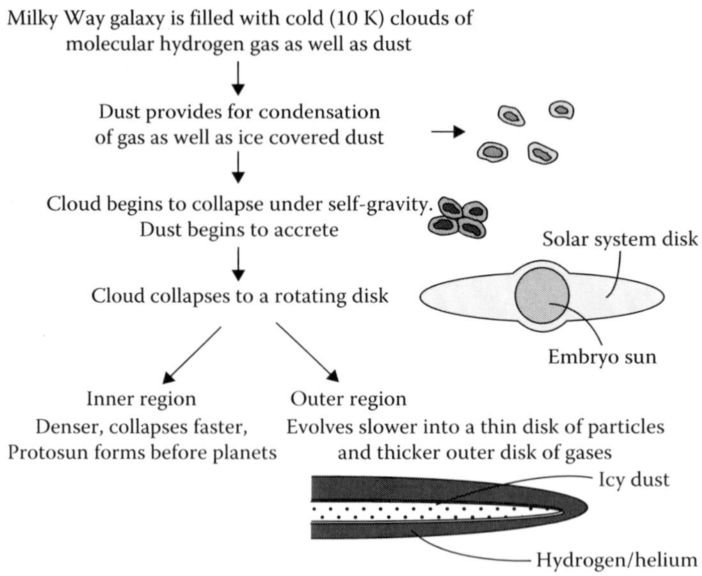

FIGURE 4.2
The protoplanet hypothesis. Support for the protoplanet hypothesis has been observed by the detection of disk material around of stars, such as Beta Pictoris, and by the Hubble images of the Orion Nebula (Figure 4.3). (Courtesy of Professor Schombert, http://abyss.uoregon.edu/~js/copyright.html and http://abyss.uoregon.edu/~js/ast121/lectures/lec24.html [retrieved August 18, 2016].)

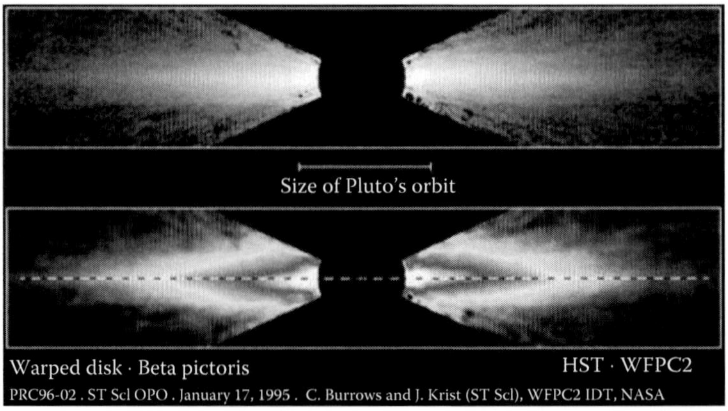

FIGURE 4.3
Beta Pictoris as a support for the protoplanet hypothesis. (Courtesy of Professor Schombert and NASA, http://abyss.uoregon.edu/~js/copyright.html and http://abyss.uoregon.edu/~js/ast121/lectures/lec24.html [retrieved August 22, 2016].)

4.5 Early Solar System Disk

In the early stage of the planet formation, the mass starts increasing by accreting planetesimals. As the force of gravity is directly proportional to mass, the largest planetesimals are accreted first. The early formation period of protoplanets was able to sweep the early

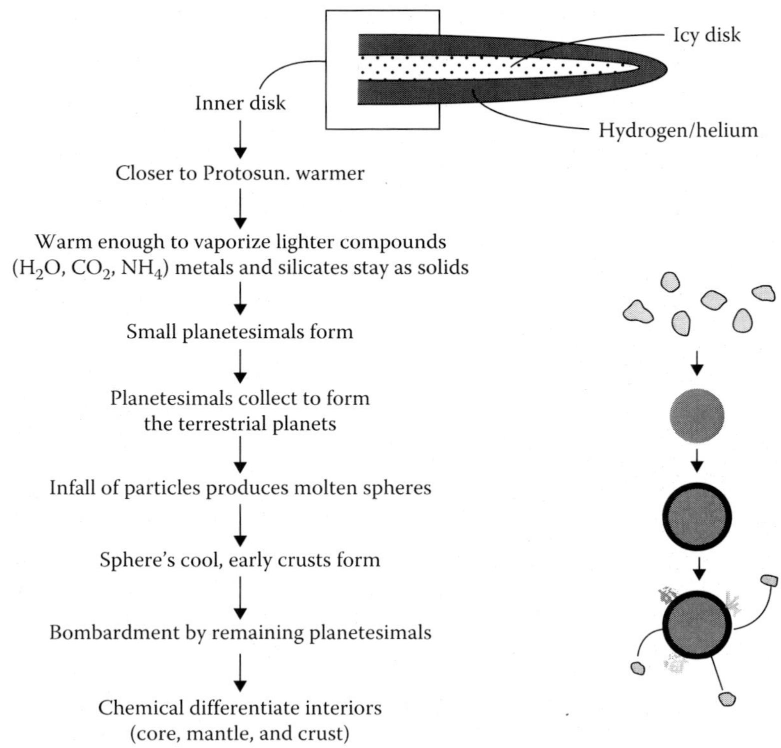

FIGURE 4.4
The early solar system disk. (Courtesy of Professor Schombert, http://abyss.uoregon.edu/~js/copyright.html and http://abyss.uoregon.edu/~js/ast121/lectures/lec24.html [retrieved August 18, 2016].)

solar system clean of large bodies [9]. It is interesting to note that the lighter compounds are vaporized in the inner solar system. So the question that arises is that where did all the out gassing material come from? The answer to this question is definitely comets that fall from the outer solar system, after the planets form [7]. Figure 4.4 illustrates the early solar system disk.

4.6 Outer Solar System

The Jovian worlds were considered as the most efficient at capturing planetesimals, which had an early edge on gathering mass in the colder outer solar disk. They only served to increase their already large masses. When the planetesimals shrunk into an average size, collisions with protoplanets led to fragmentation [10]. As a matter of fact, the solar system was divided very fast into large protoplanets and smaller planetesimals, which in due course became the numerous meteors we find presently. Any leftover large bodies were assumed to be captured as moons or ejected by gravity to assist into the Oort cloud [11]. The onset of thermonuclear fusion in the core of the Sun produced sufficient luminosity so that the remaining hydrogen and helium gases in the solar disk

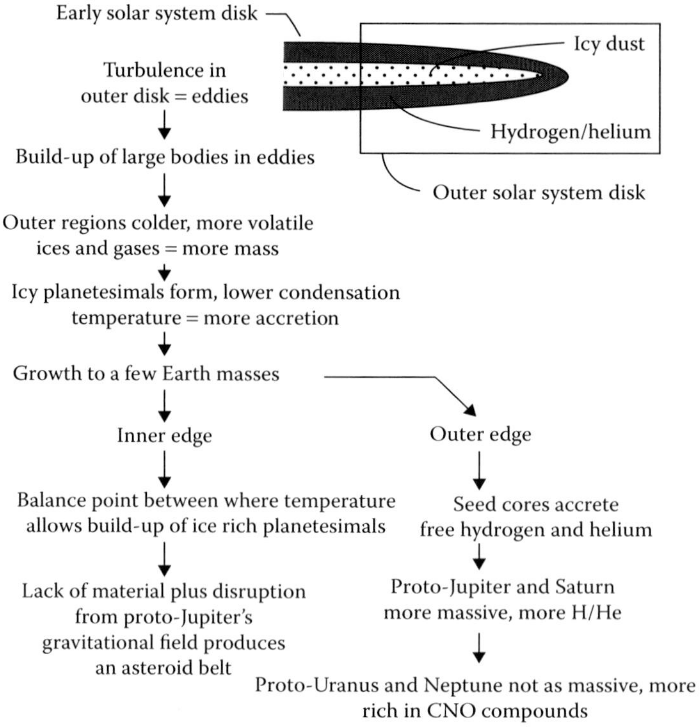

FIGURE 4.5
Solar system divided quickly into large protoplanets and smaller icy planetesimals, which were tossed into the Oort cloud by Jovian worlds. (Courtesy of Professor Schombert, http://abyss.uoregon.edu/~js/copyright.html and http://abyss.uoregon.edu/~js/ast121/lectures/lec24.html [retrieved August 18, 2016].)

were replaced by radiation pressure [11,12]. The only problem that remains is the distribution of angular momentum. The present day explanation for the fact that most of the angular momentum is in the outer planets is that, by means of some mechanism, the Sun has lost angular momentum. Figure 4.5 reveals the solar system, which is divided quickly into large protoplanets and smaller icy planetesimals, tossed into the Oort cloud by Jovian worlds.

4.7 Magnetic Braking

The early Sun had a greater flow of particles via the solar winds. The charged particles are affected by both the laws of motion and electromagnetic forces. When the solar wind leaves the solar surface, particles are *dragged* due to the magnetic field. This, in turn, slows down the rotation of the Sun [7,13]. Figure 4.6a shows how the solar wind (stream of charged particles) moves outward from the Sun in a radial direction. It may be noted that the magnetic field of the Sun also travels outward but is twisted due to rapid rotation of the Sun. Figure 4.6b illustrates that the charged particles in the solar wind

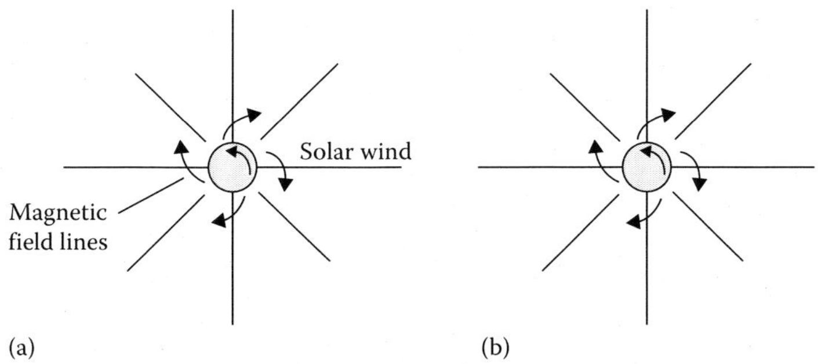

FIGURE 4.6
(a) The solar wind moves outward from the Sun in the radial direction. (b) Charged particles in the solar wind are trapped in the Sun's magnetic field. (Courtesy of Professor Schombert, http://abyss.uoregon.edu/~js/ast121/lectures/lec24.html and http://abyss.uoregon.edu/~js/copyright.html [retrieved August 18, 2016].)

become trapped in the Sun's magnetic field and are dragged along within the field lines. This results in angular momentum, which is transferred from the Sun to the charged particles by the magnetic field [7].

4.8 Migrating Planets

The protoplanet hypothesis explains a number of features dealing with the solar system. The outer solar system still exhibits some unusual properties such as that of Pluto/Charon. Scientists believe that the solar system was not born in the configuration that we find today. The planets in the outer solar system are assumed to have migrated to their present positions. For migration, there is a need of some interactions between the planet and another fairly large body. In the formation of the early solar system, there were many Moon- and Mars-sized bodies. They were more recognized in the outer solar system [14]. A large planetesimal that crosses near at the time of crossing near Neptune will lose some energy and fall down near Jupiter. It then gains energy to be ejected into the Oort cloud. There is one very frequently asked question: Did the solar system always look the way it does now? Figure 4.7 illustrates that the outer planets may have migrated to their present orbits. In the solar planetary system, there is a reduction in the size of Jupiter's orbit and expansion in the orbits of Saturn, Uranus, and Neptune [7]. As Neptune moves outward, it begins to perturb the orbits of the trans-Neptunian objects (large ice-covered asteroids of which Pluto/Charon are a member). This causes the orbit of Pluto/Charon to be highly eccentric to assume a resonant inclination of 3:2 in which it resides at this time.

Figure 4.8 illustrates the planetary migration. The upper part of the figure is related to the solar system at the time when the planets were formed, whereas the lower part shows the present situation [14]. The orbit of Jupiter is believed to have shrunk slightly, whereas the orbits of Saturn, Uranus, and Neptune have expanded. It may, however, be noted that the inner planetary regions were not significantly affected by this process. According to the existing theory, Pluto was originally in a circular orbit. However, as

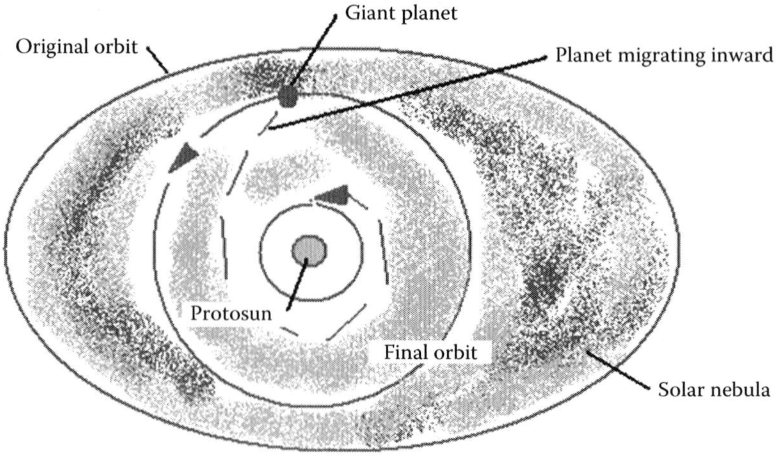

FIGURE 4.7
Evidence indicating that the outer planets may have migrated to their present orbits. (Courtesy of Lichtman, Radio Astronomy Supplies, TX.)

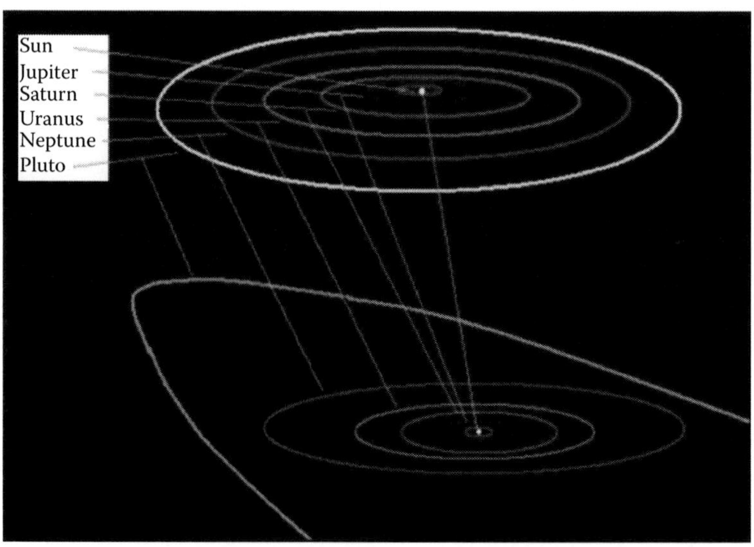

FIGURE 4.8
Illustrations of planetary migration. (Courtesy of Professor Schombert, http://abyss.uoregon.edu/~js/ast121/lectures/lec24.html and http://abyss.uoregon.edu/~js/copyright.html [retrieved August 18, 2016].)

Neptune migrated outward, it swept Pluto into a 3:2 resonant orbit whose period became proportional to Neptune's. In fact, the gravity of Neptune forced the orbit of Pluto to become more eccentric and inclined to the plane of the other planets' orbits.

Figure 4.9 illustrates time since the formation of the solar system (millions of years) versus mean distance from the Sun (AU) plotted for Neptune and Pluto for 3:2 orbit and 2:1 orbit.

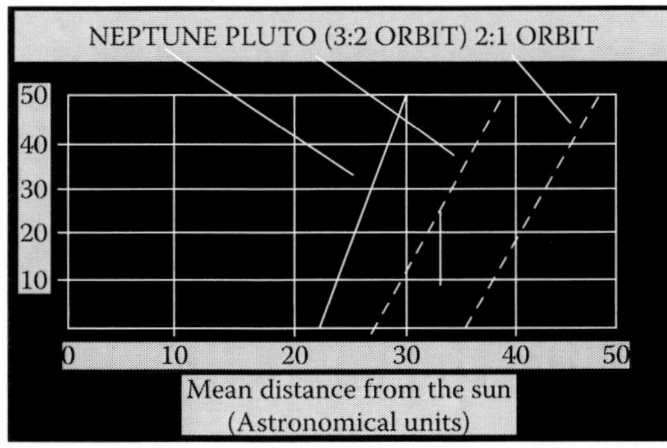

FIGURE 4.9
Time since formation of the solar system versus the mean distance from the Sun for Neptune and Pluto. (Redrawn for clarity by Lichtman; Courtesy of Professor Schombert, http://abyss.uoregon.edu/~js/ast121/lectures/lec24.html and http://abyss.uoregon.edu/~js/copyright.html [retrieved August 18, 2016].)

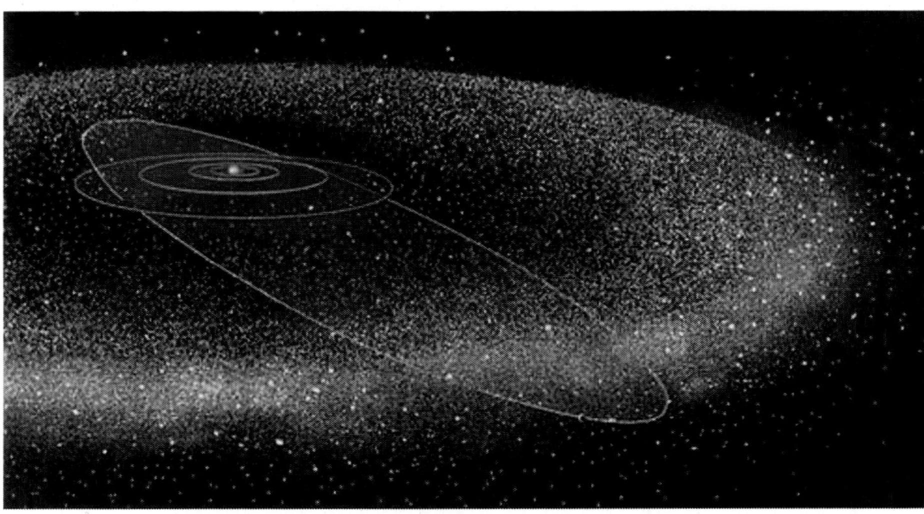

FIGURE 4.10
Planetesimals close to the orbit of Neptune pushed as the Kuiper belt. (Courtesy of Professor Schombert, http://abyss.uoregon.edu/~js/ast121/lectures/lec24.html and http://abyss.uoregon.edu/~js/copyright.html [retrieved August 18, 2016].)

As shown in Figure 4.10, all the leftover planetesimals close to the orbit of Neptune are pushed into a torus-shaped region termed as the Kuiper belt [15]. Smaller planetesimals that are comparatively small are thrown out into the Oort cloud.

4.9 Stellar Spectrum

Doppler spectroscopy is utilized for detecting the periodic velocity shift of the stellar spectrum due to the effects of an orbiting giant planet [16]. This method is also known as the radial velocity method. Ground-based studies can measure Doppler shifts greater

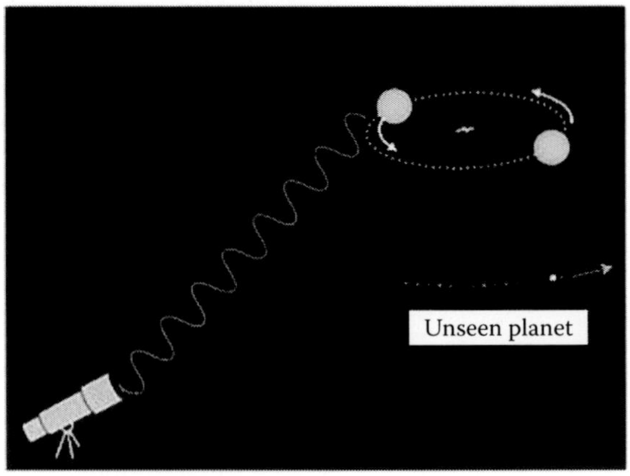

FIGURE 4.11
Doppler shift caused by the stellar wobble. (Modified for clarity by Lichtman; Courtesy of Professor Schombert, http://abyss.uoregon.edu/~js/ast121/lectures/lec24.html and http://abyss.uoregon.edu/~js/copyright.html [retrieved August 18, 2016].)

than 3 m/s due to the reflex motion of the star. This measurement corresponds to a minimum detectable mass of 33 Earth masses for a planet at 1 AU from a one solar-mass star. Figure 4.11 presents the Doppler shift caused by the stellar wobble.

All the stars reveal Doppler shift at least to some extent because all stars are travelling in the gravitational field of our galaxy. However, the stars that travel toward us and then away from us and so on are gravitationally bound to another object. The magnitude of this varying velocity, together with the period of the motion, shows the mass of an invisible companion to a visible star [7]. Our Sun moves with a speed of about 12 m/s owing to the most massive planet in our solar system, Jupiter. The periodicity of the motion of the Sun becomes the same as the orbital period of Jupiter. It has been reported that one cycle of Jupiter takes nearly 12 years. In Figure 4.12,

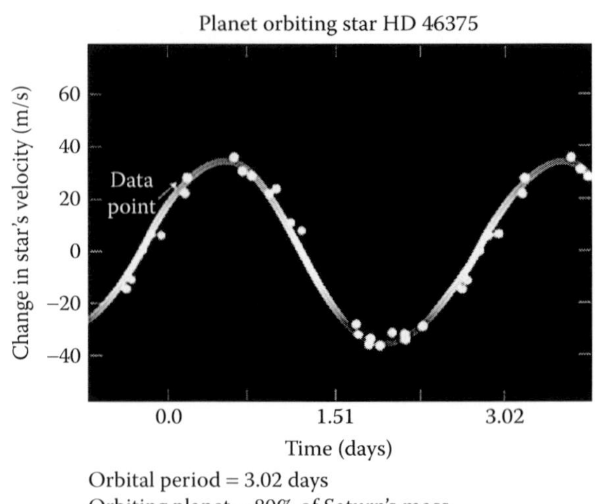

FIGURE 4.12
Plot showing the change in the velocity of a star against time. (Courtesy of Professor Schombert, http://abyss.uoregon.edu/~js/ast121/lectures/lec24.html and http://abyss.uoregon.edu/~js/copyright.html [retrieved August 18, 2016].)

we have plotted the change in the velocity of a star against time. The orbital period is shown as 3.02 days.

Astronomers suggested that the planets would cross the face of the star if the planets' orbital planes move between Earth and the star. On November 7, 1999, an automatic telescope found a 1.7% dip in the star's brightness [7]. With the orbital plane of the planet known, the astronomers for the first time were able to successfully find the mass of the planet and also determine its density from the size of the planet. It was found that although the mass of the planet is only 63% of Jupiter's mass, its radius is 60% larger than that of Jupiter. Planets are unable to absorb all incidents of light; rather, a major part gets reflected. The albedo means the fraction of incident light reflected and not absorbed. The albedo of the Earth is 0.37, whereas that of Venus is 0.65 and that of the Moon is about 0.12. Another type of complication arises because planets are not ideal black bodies. Carbon dioxide, water vapor, and other atmospheric gases are opaque in the near-IR where the maximum of the black body emission occurs. A less-than-ideal radiator must be hotter than a black body for radiating the same amount of luminosity that extends the habitable zone outward [17].

4.10 Collapsing Clouds of Gas and Dust: Nebular Hypothesis

Earlier, we have introduced the nebular hypothesis on which some elaborate discussions are made here. In the eighteenth century, Kant and Laplace proposed the nebular hypothesis in its original form. The primary steps are indicated in Figure 4.14. A great cloud of gas and dust, termed as nebula, starts to collapse because the gravitational forces that would like to collapse it overcome the forces associated with gas pressure that would like to expand it [18]. The initial collapse might be triggered by different perturbations; a supernova blast wave, density waves in spiral galaxies, and so on. Figure 4.13 details the primary steps of the nebular hypothesis.

In this hypothesis, a cloud of gas and dust collapsed due to gravity. The cloud starts to spin faster due to conservation of angular momentum. It is unlikely that such a nebula would be produced with no angular momentum, so initially it was thought to be probably spinning slowly [19]. Due to the conservation of angular momentum, the cloud spins faster as it contracts.

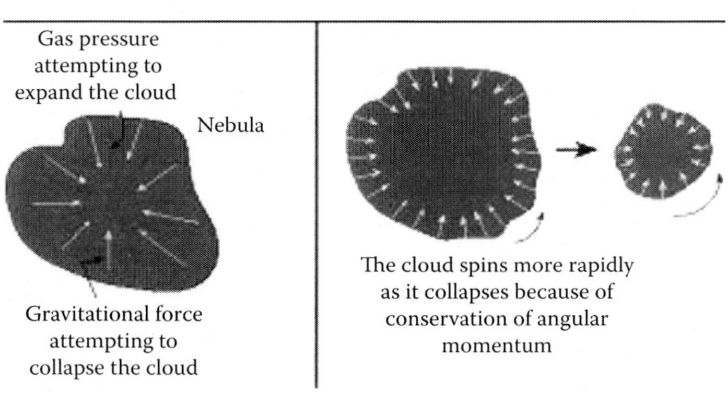

FIGURE 4.13
The primary steps of the nebular hypothesis. (Anonymous, University of Tennessee Knoxville, http://csep10.phys.utk.edu/astr161/lect/solarsys/nebular.html [retrieved August 22, 2016].)

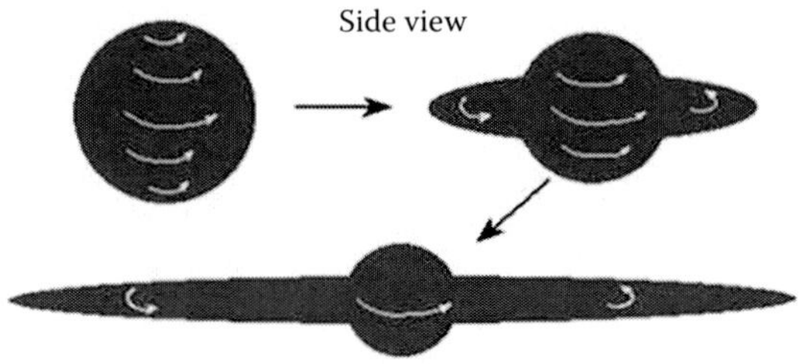

FIGURE 4.14
The collapsing, spinning nebula begins to flatten into a rotating pancake. (Anonymous, University of Tennessee Knoxville, http://csep10.phys.utk.edu/astr161/lect/solarsys/nebular.html [retrieved August 22, 2016].)

4.10.1 Further Effects of Gravity

Due to the competing forces related to gravity, gas pressure, and rotation, the contracting nebula starts to flatten into a spinning pancake shape with a bulge at the center [18,20]. This is illustrated in Figure 4.14.

4.11 Condensation of Protosun and Protoplanets

As the nebula collapses, further rotating clouds cause local regions to contract gravitationally. These local regions of condensation will initially become the Sun and the planets, along with their moons and other debris [21]. The upper part of Figure 4.15 shows the side view of the protosun and the protoplanets, whereas lower part shows the top view of the collapsing protoplanet.

As they are still condensing, the incipient Sun and planets are known as the protosun and protoplanets, respectively.

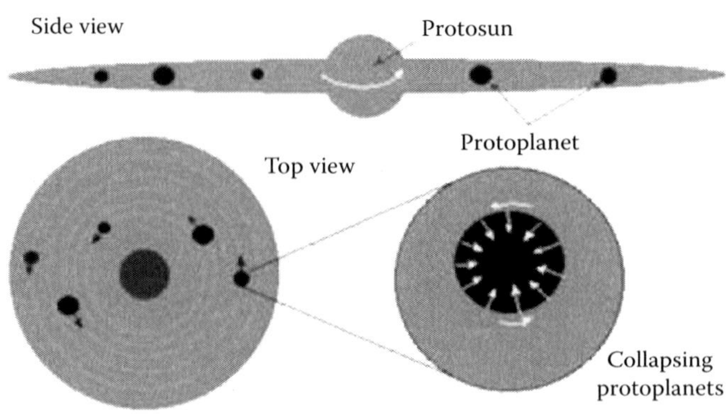

FIGURE 4.15
The side view of the protosun and the protoplanets and the top view of the collapsing protoplanet. (Anonymous, University of Tennessee Knoxville, http://csep10.phys.utk.edu/astr161/lect/solarsys/nebular.html [retrieved August 22, 2016].)

4.11.1 Evidence for the Nebular Hypothesis

Due to original angular momentum and subsequent evolution of the collapsing nebula, this hypothesis provides a natural explanation of some basic facts about the solar system [22]. First of all, the orbits of the planets lie approximately in a plane with the sun at the center and all the planets revolve in the same direction with rotation axes almost perpendicular to the orbital plane. The nebular hypothesis justifies many of the basic features of the solar system, but we still do not understand clearly how all the details are accounted for by this hypothesis [23].

References

1. Solar system (2008). *Merriam Webster Online Dictionary*. http://www.merriam-webster.com/dictionary/solar%20system (retrieved August 23, 2016).
2. M. M. Woolfson (1984). Rotation in the solar system. *Philosophical Transactions of the Royal Society* 313 (1524): 5.
3. N. Henbest (1991). Birth of the planets: The Earth and its fellow planets may be survivors from a time when planets ricocheted around the Sun like ball bearings on a pinball table. *New Scientist*. https://www.newscientist.com/article/mg13117837-100-birth-of-the-planets-the-earth-and-its-fellow-planets-may-be-survivors-from-a-time-when-planets-ricocheted-around-the-sun-like-ball-bearings-on-a-pinball-table/ (retrieved August 23, 2016).
4. D. Whitehouse (2005). *The Sun: A Biography*. John Wiley & Sons, Hampshire, UK.
5. S. Mitton (2005). Origin of the chemical elements. In *Fred Hoyle: A Life in Science*. Aurum Press, London, UK. pp. 197–222.
6. E. Howell. How are planets formed?. *Universe Today*. http://www.universetoday.com/85761/protoplanet-hypothesis/ (retrieved December 23, 2015).
7. University of Oregon. Origin of the Solar System lecture 24. http:/abyss.uoregon.edu/~js/ast121/lectures/lec24.html (retrieved August 23, 2016).
8. B. G. Elmegreen (1979). On the disruption of a protoplanetary disc nebula by a T Tauri like solar wind. *Astronomy and Astrophysics* 80: 77.
9. K. Hansen (2005). Orbital shuffle for early solar system. *Geotimes* 435 (7041): 466–9.
10. W. B. Hayes (2007). Is the outer Solar System chaotic? *Nature Physics* 3 (10): 689–691.
11. A. Morbidelli (2008). Origin and dynamical evolution of comets and their reservoirs. arXiv:astro-ph/0512256. https://arxiv.org/pdf/astro-ph/0512256v1.pdf (retrieved August 12, 2016).
12. B. E. Clark and R. E. Johnson (1996). Interplanetary weathering: Surface erosion in outer space. *Eos, Transactions, American Geophysical Union* 77 (15): 141.
13. M. Küker et al. (2003). Magnetic star-disk coupling in classical T Tauri systems. *The Astrophysical Journal* 589 (1): 397.
14. R. Edgar and P. Artymowicz (2004). Pumping of a planetesimal disc by a rapidly migrating planet. *Monthly Notices of the Royal Astronomical Society* 354 (3): 769–772.
15. M. E. Brown et al. (2010). The size, density, and formation of the Orcus-Vanth system in the Kuiper Belt. *The Astronomical Journal* 139 (6): 2700.
16. E. W. Thommes et al. (2002). The formation of Uranus and Neptune among Jupiter and Saturn. *The Astronomical Journal* 123 (5): 2862.
17. G. Fontaine et al. (2001). The potential of white dwarf cosmochronology. *Publications of the Astronomical Society of the Pacific* 113 (782): 409–435.
18. E. G. Blackman (Spring 2006). The origin of the Solar System. http://www.pas.rochester.edu/~blackman/ast104/nebular.html (retrieved September 28, 2016).

19. M. M. Woolfson (1993). Solar system—Its origin and evolution. *The Quarterly Journal of the Royal Astronomical Society* 34: 1–20.
20. T. Devitt (2001). What puts the brakes on madly spinning stars? University of Wisconsin-Madison, Madison, WI. http://news.wisc.edu/what-puts-the-brakes-on-madly-spinning-stars/ (retrieved August 12, 2016).
21. T. Nakamoto and Y. Nakagawa (1994). Formation, early evolution, and gravitational stability of protoplanetary disks. *The Astrophysical Journal* 421: 640–650.
22. D. Brewster (1876). *More Worlds Than One: The Creed of the Philosopher and the Hope of the Christian.* Chatto & Windus, Piccadilly, London. p. 153.
23. E. Kokubo and S. Ida (2002). Formation of protoplanet systems and diversity of planetary systems. *The Astrophysical Journal* 581 (1): 666–680.
24. R. Patterson (2010). Origin of the Solar System. In *Evolution Exposed: Earth Science.* Answers in Genesis, Petersburg, KY.

5

Solar System Exploration across the Universe

5.1 Introduction

Exploration of the solar system means searching information or resources, visitation, or observation of Sun and planets of the solar system as well as smaller bodies like comets, asteroids, and dust [1]. In fact, the solar system, from that point of view, may be taken as humankind's cosmic backyard.

5.2 Relative Sizes of the Objects

Figure 5.1 illustrates the Sun and planets of the solar system. The relative sizes of the objects have been drawn to scale but not the distances between them. Pluto and the other dwarf planets are not shown in this figure.

Figure 5.2 is chosen for presenting a photograph of Earth as indicated by the circle. The photograph was taken by Voyager 1 from a distance of 6.4 billion kilometers. The streaks of light in the figure are diffraction spikes radiating from the Sun (off frame to the left). This photograph is popularly named as *Pale Blue Dot* [2].

5.3 Early Observations

Many civilizations throughout the ages, except a few remarkable exceptions, did not recognize the existence of the solar system. It was thought that the Earth was stationary and at the center of the universe and different from the ethereal objects that moved through the sky. Indian astronomer and mathematician Aryabhata and the Greek philosopher Aristarchus of Samos made a speculation on a heliocentric reordering of the cosmos [3]. It was Nicolaus Copernicus who first developed a mathematically predictive heliocentric method that was further contributed to by seventeenth-century successors, Galileo Galilei, Johannes Kepler, and Isaac Newton. They developed an understanding of physics and an idea that the Earth moves around the Sun and the planets are governed by the same physical laws that govern the Earth was gradually accepted. Later on, this led to the investigation of geological phenomena such as mountains, craters, and seasonal meteorological phenomena, for example, clouds, dust storms, and ice caps on the other planets.

FIGURE 5.1
The Sun and planets of the solar system. (Courtesy of Wikipedia, Solar system, https://en.wikipedia.org/wiki/Solar_System [last modified on August 19, 2016].)

FIGURE 5.2
Photograph by Voyager 1 called "Pale Blue Dot," indicating the Earth by a circle. (Courtesy of NASA, Solar system portrait - Earth as 'pale blue dot', http://visibleearth.nasa.gov/view_rec.php?id=601 [retrieved August 19, 2016].)

5.4 First Exploration of the Solar System

When astronomers first started to investigate the solar system, the problem was the size and faintness of the objects when observing with the naked eye.

5.4.1 The Benefits of the Telescope

The first exploration of the solar system was made by using a simple telescope (the first telescope appeared in 1608) (Figure 5.3) [4]. Galileo was the first to identify physical details about the individual bodies of the solar system. By his observations, he was able to discover that the Moon was cratered and that the Sun was marked with sunspots. He further identified the four satellites orbiting Jupiter [5]. Huygens followed Galileo's observations by discovering Saturn's moon Titan and also marking the shape of the rings of Saturn [6]. Subsequently, Cassini discovered four more moons of Saturn, as well as the Cassini division in Saturn's rings [7]. A replica of Isaac Newton's telescope is presented in Figure 5.4 [8].

FIGURE 5.3
The Sun photographed by French physicists Louis Fizeau and Leon Foucault. The first photograph of the Sun (shown) was taken in 1845. (Courtesy of National Science Foundation, High Altitude Observatory, Boulder, CO.)

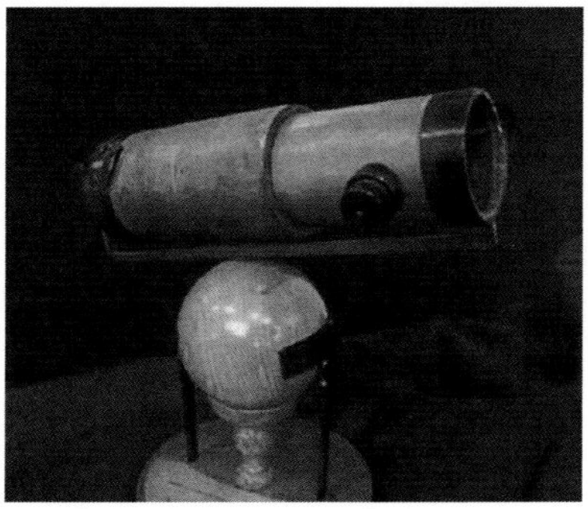

FIGURE 5.4
A replica of Isaac Newton's telescope. (Courtesy of Wikipedia, Timeline of telescope technology, https://en.wikipedia.org/wiki/Timeline_ of_telescope_technology [last modified on August 18, 2016].)

With the early telescope, in 1705, Halley realized that repeated sightings of a comet were recording of the same object that returned regularly once in every 75–76 years. Halley's observation was the first report that anything other than the planets orbited the Sun [9]. In 1781, Herschel while looking for binary stars in the constellation of Taurus observed an object that he thought was a new comet. Its orbit showed that it was a new planet, Uranus [10]. In 1801, Piazzi discovered Ceres, a small world between Mars and Jupiter, which initially was considered a new planet. In due course, discoveries of thousands of small worlds in the same region led to their eventual reclassification as asteroids [11].

5.4.2 The Benefits of Spectroscopy

In 1859, Joseph von Fraunhofer invented the spectroscope [12]. Using the newly invented instrument, Bunsen and Kirchhoff examined the spectral signature of the Sun and discovered

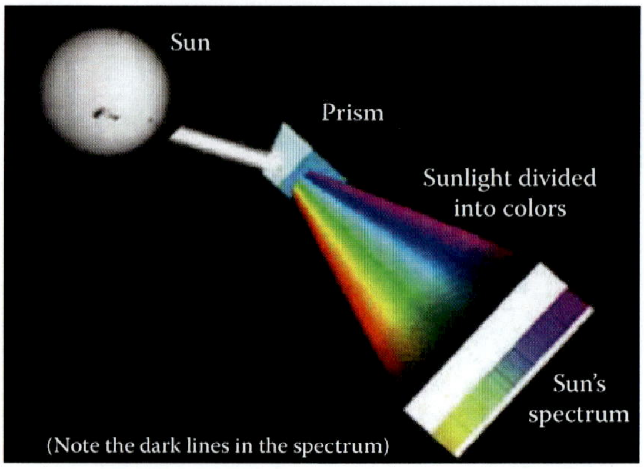

FIGURE 5.5
Solar spectrum.

that it was composed of the same elements as existed on Earth. This discovery established a physical link between the Earth and the heavens [13]. Sun Earth Connection Coronal and Heliospheric Investigation (SECCHI) [14] compared the spectral signature of the Sun with other stars and observed them to be virtually identical. Figure 5.5 shows the solar spectrum.

5.5 Observations by Spacecraft

From the very beginning of the Space Age, a great deal of exploration has been done by manned and robotic spacecraft missions. The operation launching of the spacecrafts missions were first done by NASA (National Association and Space Administration) and later on by ESA (European Space Agency) and ultimately by different countries with their own programs as well as private companies such as, SpaceX. The first artificial object sent into space in 1957 was the Soviet satellite *Sputnik 1*, which successfully orbited the Earth [15]. The American probe, *Explorer 6* was launched in 1959. This was the first satellite to image the Earth from space. Another American probe, *Pioneer 10* passed the orbit of Pluto in 1983, and its last transmission was received in January 2003, sent from nearly 82 AU away. An artist's conception of *Pioneer 10* is presented in Figure 5.6. The space probe is receding from the Sun at over 43,400 km/h [16].

Planets in the solar system have been visited to varying degrees by spacecrafts launched from Earth. Through these unmanned missions, we have been able to get close-up photographs of all the planets and also perform tests of the soils and atmospheres through robotic rovers.

5.5.1 Flybys

Another probe, *Luna 1*, sped past the Moon in 1959. This was the first successful probe to fly by another solar system body. Initially, it was meant to impact with the Moon, but instead, it missed its target and became the first artificial object to orbit the Sun. *Mariner 2* was the first probe to fly by another planet, Venus, in 1962. In 1965, the first successful flyby of Mars was made by *Mariner 4*, whereas *Mariner 10* first passed Mercury in 1974. *Pioneer 10*, which flew by Jupiter in 1973, was the first probe to explore the outer planets.

FIGURE 5.6
An artist's conception of Pioneer 10. (Courtesy of NASA, NASA Astronomy picture of the day [June 20, 1998], http://apod.nasa.gov/apod/ap980620.html [retrieved August 19, 2016].)

In 1979, *Pioneer 11* first visited Saturn. Following their launch in 1977, the *Voyager* probes performed a grand tour of the outer planets. Both the probes passed Jupiter in 1979 and Saturn in 1980–1981. *Voyager 2* made close approaches to Uranus in 1986 and Neptune in 1989. The *Voyager* probes are now far beyond Neptune's orbit and are on the course to study the termination shock, heliosheath, and heliopause. Both *Voyager* probes, according to NASA, have encountered the termination shock at a distance of nearly 93 AU from the Sun [17]. When the International Cometary Explorer (ICE) passed by the comet Giacobini-Zinner [18] in 1985, the first flyby of a comet occurred, whereas the first flybys of asteroids were conducted by the *Galileo* space probe, which imaged both 951Gaspra in 1991 and 243Ida in 1993 on its way to Jupiter. Lineae (lines) on Europa by the *Galileo* spacecraft is presented in Figure 5.7.

No Kuiper belt object has yet been found by a spacecraft. In 2006, the *New Horizons* probe was launched, which became the first man-made spacecraft to explore this area. This unmanned mission is presently scheduled to fly by Pluto in July 2015. With its success, the mission will be extended to find a number of other Kuiper belt objects [19]. Scientists have been concerned about exploration beyond the Asteroid Belt since 2011.

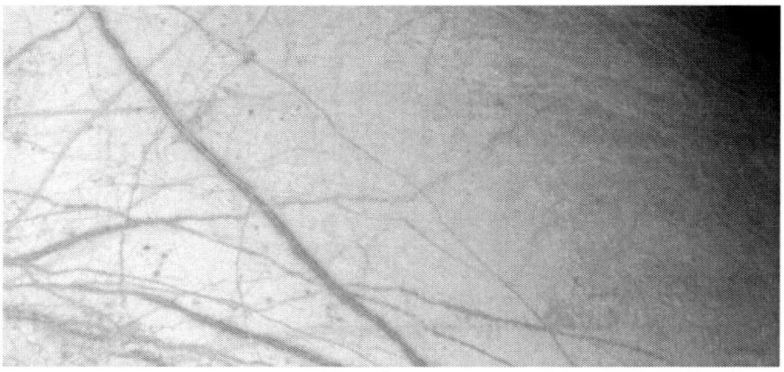

FIGURE 5.7
Lineae on Europa by the *Galileo* spacecraft. (Courtesy of NASA, http://solarsystem.nasa.gov/europa/home.cfm [retrieved August 18, 2016].)

5.6 Orbiters, Rovers, and Landers

Our Moon became the first solar system body in 1966 to be orbited by an artificial satellite (*Luna 10*), followed by Mars in 1971 (*Mariner 9*), Venus in 1975 (*Venera 9*), Jupiter in 1995 (*Galileo*), the asteroid 433 Eros in 2000 (*NEAR Shoemaker*), Saturn in 2004 (*Cassini–Huygens*), and Mercury and Vesta in 2011 (*MESSENGER* and *Dawn*, respectively). Dawn is currently set to orbit the asteroid–dwarf planet Ceres in 2015. Soviet *Luna 2* probe was the first to land on a different solar system body. This impacted the Moon in 1959. Since then, there has been an increase in the number of increasingly distant planets being visited by man-made probes landing on or impacting the surfaces of Venus in 1966 (*Venera 3*) and Mars in 1971 (*Mars 3*). A successful landing, however, did not occur until *Viking 1* in 1976. Further success came with the study of the asteroid 433Eros in 2001 (*NEAR Shoemaker*) and Saturn's moon Titan (*Huygens*), as well as the comet Tempel 1 (*Deep Impact*) in 2005. Moreover, the *Galileo* orbiter dropped a probe into the atmosphere of Jupiter in 1995. As Jupiter has no physical surface, it was destroyed due to increasing pressure and temperature during its descent [20]. It is important to note that up to the present time, only two worlds in the solar system have been visited by mobile rovers, the Moon and Mars. The first rover to visit a separate celestial body was the Soviet *Lunokhod 1*, which landed on the Moon in 1970. The first to visit a new planet in 1997 was Sojourner, which traveled 500 m across the surface of Mars. NASA's Lunar rover, which traveled with Apollos 15, 16, and 17 between 1971 and 1972, was the only manned rover to visit another world [20]. Figure 5.8

FIGURE 5.8
Curiosity rover self-portrait at *Rocknest* (October 31, 2012) with the rim of Gale Crater and the slopes of Aeolis Mons in the distance. (Courtesy of NASA, http://www.nasa.gov/image-feature/jpl/pia20316/curiosity-self-portrait-at-martian-sand-dune [retrieved August 19, 2016].)

exhibits the Curiosity rover self-portrait at *Rocknest* (October 31, 2012) with the rim of Gale Crater and the slopes of Aeolis Mons in the distance.

5.7 Overview of Spacecraft Exploration

In Table 5.1 we show an overview of some missions to the solar system. The names of the spacecrafts, launch year, and end year as well as the particulars flyby of the planets are shown in the table.

5.8 Human Spaceflight

Presently, manned exploration of the solar system is mainly confined to Earth's immediate environs. In 1961, the first human in space and to orbit the Earth was Yuri Gagarin. He was a Soviet cosmonaut launched into space on *Vostok 1*. The year 1969 saw the first person to

TABLE 5.1

Overview of Some Missions to the Solar System

Spacecraft	Launch	Mercury	Venus	Mars	Jupiter	Saturn	Uranus	Neptune	End
Pioneer 10	1972				Flyby				2003
Mariner 10	1973	Flyby	Flyby						1975
Voyager 1	1977				Flyby	Flyby			—
Voyager 2	1977				Flyby	Flyby	Flyby	Flyby	—
Galileo	1989		Flyby		Orbiter				2003
Ulysses	1990				Flyby				2009
Cassini	1997		Flyby		Flyby	Orbiter			—
Mars Odyssey	2001			Orbiter					—
MER-A/B	2003			Rovers					2010
Mars Express	2003			Orbiter					—
MESSENGER	2004	Orbiter	Flyby						—
MRO	2005			Orbiter					—
Venus Express	2005		Orbiter						—
New Horizons	2006				Flyby				—
Juno	2011				Orbiter				—
Curiosity (MSL)	2011			Rover					—

Source: Courtesy of Wikipedia, Discovery and exploration of the solar system, http://en.wikipedia.org/wiki/Discovery_and_exploration_of_the_Solar_System# cite_note-exp-1 (last modified August 10, 2016).

FIGURE 5.9
Owen Garriott on an Earth orbit EVA in 1973. (Courtesy of NASA, Spacewalk outside skylab - August 1973, http://www.nasa.gov/audience/foreducators/spacesuits/historygallery/sky-aug73.html [retrieved August 18, 2016].)

walk on the surface of another solar system body (our moon), Neil Armstrong followed by Buzz Aldrin in the Apollo 11 mission. Within 3 years' time, five more Moon landings occurred through 1972 [21]. Figure 5.9 presents a photograph showing Owen Garriott on an Earth orbit EVA in 1973.

The US Space Shuttle debuted in 1981 as the only reusable spacecraft to make multiple orbital flights successfully. The five shuttles have flown a total of 121 missions. Of those, two were destroyed in accidents. NASA's Skylab was the first orbital space station to host more than one crew. This was able to hold three crews successfully from 1973 to 1974. The Soviet space station Mir was the first true human settlement in space, which was continuously occupied for a period of approximately 10 years, from 1989 to 1999. It was subsequently decommissioned in 2001. Its successor, called the International Space Station, has followed a continuous human presence in space from that time [22]. SpaceShipOne, the first privately funded vehicle, started its journey in 2004 to reach space on a suborbital flight. In the same year, US President George Bush declared the Vision for Space Exploration. He announced a replacement for the aging Shuttle, a return to the Moon, and a manned mission to Mars [22,23].

5.9 Mapping the Baby Universe

NASA's WMAP (Wilkinson Microwave Anisotropy Probe) mission reveals conditions as they existed in the early universe by measuring the properties of the Cosmic Microwave Background radiation over the full sky in greater detail. At the very beginning of the universe, and for the first 300,000 years or so, there was huge amount of dust and debris after the Big Bang. During this long period of time, there was a dense, white fog of free electrons and protons. Light could not pass even a trillionth of a millimeter without bumping into an electron and so visibility was practically nil. Interestingly, after such a long time something

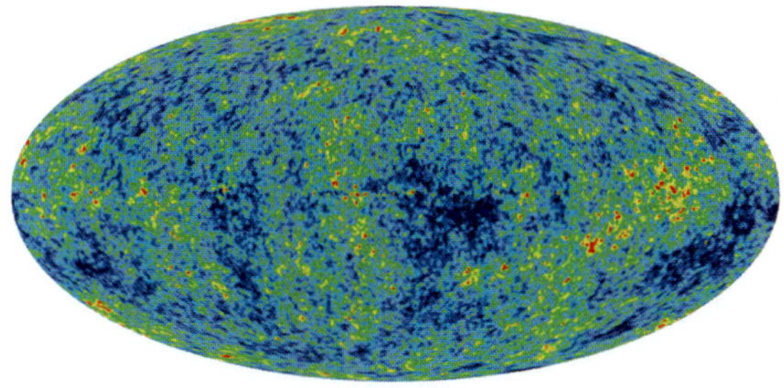

FIGURE 5.10
COBE mapped the CMB microwave background in all directions. (Courtesy of NASA, CMB images, http://map.gsfc.nasa.gov/media/990166/index.html [retrieved August 18, 2016].)

happened. The universe cooled to a mere 2700°C, cold enough to permit electrons to cling to protons, and particles combined to form hydrogen atoms. Also, light broke free and shone across the universe to show the inside structure. This first light is still with us and the expansion of the universe has *stretched* it, alternately called redshifted [24], into the microwave band [25]. COBE (Cosmic Background Explorer) [26] mapped the interesting and highly informative microwave background in all directions, as shown in Figure 5.10.

The Cosmic Microwave Background is essentially the afterglow of the Big Bang, which came from an era 1 billion years or so before anything Hubble or Chandra can see. NASA planned to launch the WMAP with a view to survey the ancient radiation in unprecedented detail. The Microwave Anisotropy Probe was responsible for *WMAP*, slight temperature fluctuations (0.00001°C) within the microwave background, which now averages 2.73°C above absolute zero [27]. The noted temperature differences point back to density differences in the fiery baby universe in which there was a slightly more matter here and a slightly less matter there. Areas of enhanced density by a small amount had stronger gravity in comparison to low-density areas. The areas of high density *pulled back* on the background radiation, forming it slightly cooler in those directions. In fact, the small difference in density led to our current structure of galaxies, galaxy clusters, and voids of seemingly empty space [28]. Figure 5.11 exhibits much higher image resolution by WMAP.

The Cosmic Microwave Background was discovered in 1964 by Penzias and Wilson of Bell Laboratories by accident [29]. They were attempting to eliminate static interfering with radio experiments and for that they tuned out local radio stations. Despite their best efforts, they were not able to eliminate a constant low-level noise. From this, they came

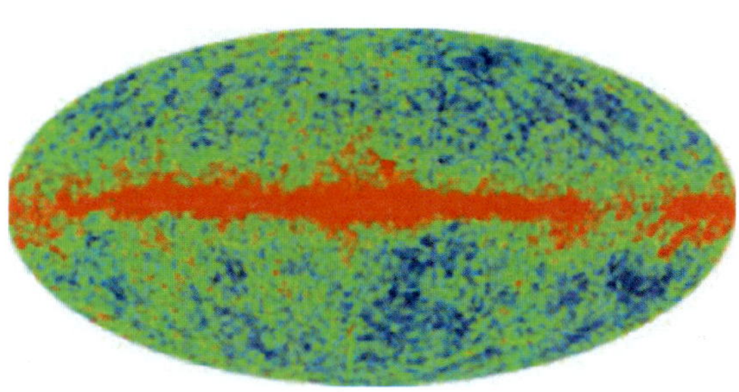

FIGURE 5.11
High-resolution image by WMAP. (Courtesy of NASA, CMB images, http://map.gsfc.nasa.gov/media/060913/index.html [retrieved August 18, 2016].)

to realize from some discussion with cosmologists that this noise has been lingering for 10–15 billion years, as predicted by the Big Bang theory [30]. COBE was the first satellite used to investigate the cosmic microwave fossil. In 1992, COBE noted that there were fluctuations in the temperature. By making a comparison with the earliest structure of the universe to the present universe, WMAP determines how and when the first galaxies formed. WMAP is also involved in examining the geometry of the universe as flat or curved in shape. Three balloon-borne experiments, BOOMERANG, MAXIMA, and TopHat investigated the microwave background at high angular resolution, though they could only survey a tiny patch of sky during their time of brief flights [30]. Detailed studies by WMAP smoothed out the slight discrepancies among data from some recent non-satellite experiments and remove the chaff of cosmological theories. WMAP is one of the most important space-science missions and credited with excellent findings, yielding a single 2-year image of the baby universe. But while a Hubble picture says a thousand words, the WMAP survey can speak of 15 billion years of glorious static.

5.10 Unlocking the Secrets of the Giant Planet Jupiter

The present day exploration of Jupiter started with the Voyager Mission. Two flybys successfully provided the first encounters with this giant planet, preparing the way for further studies. After the extraordinary encounters of Voyager, curiosity about the planet and its intriguing atmosphere deepened [31]. The Galileo Mission, with its orbiter and probe, started elaborate investigations of not only Jupiter but also its mysterious moons. Scientists believe that the Juno mission [32] will give us the first close-up view of Jupiter's poles. Figure 5.12 shows the planet's poles and aurora.

The Galileo Mission on its way to Jupiter made the first discovery of an asteroid with a moon. It became the first spacecraft to orbit the powerful gas giant Jupiter and was the first to launch a probe into the atmosphere of Jupiter. The harvest of new knowledge gathered by Galileo still left many mysteries. This archetypical giant planet holds a number of extraordinary secrets concealed under its dense layer of clouds. In fact, a clear understanding of the fundamental processes involved in the formation of our solar system is a subject of great interest.

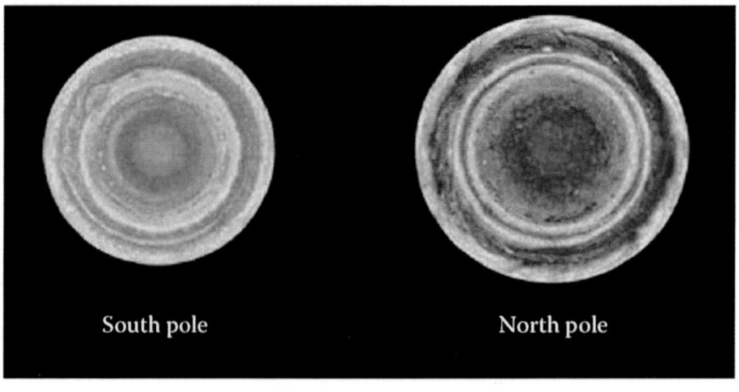

FIGURE 5.12
View of Jupiter's poles. (Courtesy of NASA/JPL/Space Science Institute, Cassini's best maps of Jupiter [PIA07783 and PIA07784], http://photojournal.jpl.nasa.gov/catalog/PIA07783 [retrieved August 18, 2016].)

5.11 Continued Exploration

The new and enriched information obtained on the planet's structure, atmosphere, interior, and magnetosphere after the completion of the Juno mission might solve some of the many unresolved questions. It will answer complicated science questions and test present theories regarding the formation of our solar system. In addition, it could also provide possible new insights into understanding planetary formation in other stellar systems across the universe. The Juno spacecraft will arrive at Jupiter in 2016, entering the first polar orbit around the planet. With its suite of nine science instruments, Juno will begin the first measurement of the amount of water on Jupiter besides the interior structure of the planet, including the first search for the existence of a high-density planetary core [31]. It will also make the first detection of ammonia clouds in the deep atmosphere and will simultaneously map the intense magnetic field, observing the Jupiter's auroras from positions above the planet's poles for the first time. Figure 5.13 shows an artist's rendition of the Juno spacecraft.

5.11.1 Interesting Findings of Jupiter

Many scientists think that Jupiter originated early in the formation of the solar system. This is mainly because it is constituted primarily by hydrogen and helium, like the Sun.

NOTE: There are some thoughts that if Jupiter had more mass that it might have been a star, similar to our Sun [33]. According to most scientists, Jupiter captured most of the material left over from the giant collapsing interstellar cloud of dust and gas from the formation of the infant Sun [30,31]. Jupiter was able to maintain its original composition, the composition of the solar nebula, due to the gravity generated by its enormous mass. Jupiter harbors many hidden secrets from the earliest whispers of the planetary system. Jupiter's Red Spot and colorful clouds as observed by the Hubble Space Telescope's Wide-Field Planetary are interesting. A black and white presentation of the same can be seen in Figure 5.14.

Telescope's looking at a couple of thoughts, if Jupiter is very wet, it must have captured more of the icy planetesimals than a simple formation theory would predict. On the other hand,

FIGURE 5.13
An artist's rendition of the Juno probe and Jupiter. (Courtesy of NASA/JPL-Caltech, Juno spacecraft: NASA's new mission to Jupiter, http://www.space.com/32742-juno-spacecraft.html [retrieved August 18, 2016].)

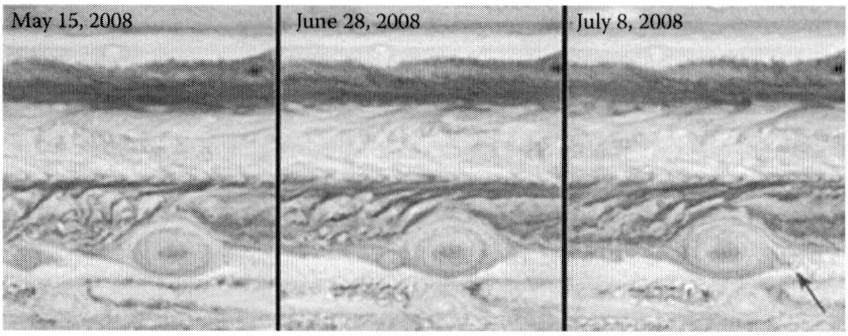

FIGURE 5.14
Black and white presentation of Jupiter's Red Spot and clouds as observed by the Hubble Wide-Field Planetary. (Courtesy of NASA, Hubble's planetary portrait captures new changes in Jupiter's great Red Spot, http://www.nasa.gov/press-release/goddard/hubble-s-planetary-portrait-captures-new-changes-in-jupiter-s-great-red-spot [retrieved October 13, 2015].)

a very dry Jupiter would also challenge the existing theories. Another scenario indicates that with its huge gravitational influence and sheer size, Jupiter may have deflected damaging intruders like comets and asteroids away from the inner solar system where Earth and other terrestrial planets orbit. Planetesimals that escaped these deflections brought water and other necessary compounds for the origin of life to Jupiter, Earth, and other habitable zones. The Juno spacecraft has a very important mission and one of its tasks is to find the existence of water on Jupiter for the first time. This will be done by detecting microwave radiation from the atmosphere at great depths. It will further show the structure of the interior of the planet in more detail than ever before. Also, it will determine if the planet has a dense core or not. All these measurements will help to examine which of the competing theories about the origin of the planet are correct. Again, there will be better mapping of gravitational and magnetic fields of Jupiter than in previous observations. Other fascinating features that will be studied are the colorful zones of Jupiter's belts and other cloud phenomena such as Great Red Spot. Other observations that Juno will observe is the global structure and motion of the planet's atmosphere below the cloud tops as well as mapping variations in the composition of the atmosphere, temperature, clouds, and patterns of movement down to unprecedented depths [31]. A spectacular NASA Hubble Space Telescope close-up view of an electric-blue aurora that is glowing one-half billion miles away on the giant planet Jupiter was an attractive phenomenon to the scientists. We have presented a black and white photograph of the same in Figure 5.15.

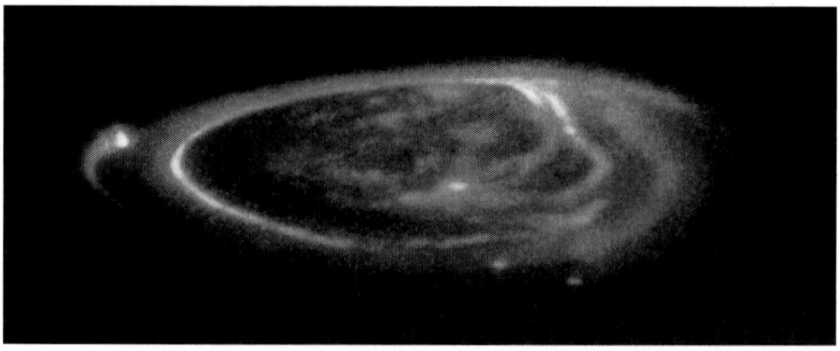

FIGURE 5.15
A black and white photograph of NASA Hubble Space Telescope close-up view of an electric-blue aurora. (Courtesy of NASA, Satellite footprints seen in Jupiter's Aurora, http://www.nasa.gov/mission_pages/juno/multimedia/pia03155.html [retrieved June 30, 2016].)

5.11.2 Magnetic Properties

At huge depths and under the enormous pressure in Jupiter's atmosphere, hydrogen gas is squeezed into fluid called liquid metallic hydrogen. At these pressures, the hydrogen behaves like an electrically conducting metal and this region is assumed to be the source of the planet's intense magnetic field, the magnetosphere [34]. The powerful magnetic environment produces the brightest auroras in our solar system. Juno will directly sample the charged particles and magnetic fields close to the Jupiter's poles while simultaneously determining the auroras in ultraviolet and infrared light. Results derived from Juno's observation will improve our understanding of these remarkable phenomena significantly. Juno will also permit us to take a giant step forward in our knowledge on how giant planets form and the role that process plays in producing our solar system. Figure 5.16 illustrates Jupiter's magnetosphere.

Another phenomenon of Jupiter is radio emission. The radio emission was discovered by Burke and Franklin in 1955. The decametric radio waves have frequencies between 10 and 40 MHz. These types of radio waves from Jupiter are never heard above 40 MHz. This seems to be the maximum frequency. From our knowledge of the cause of the radio waves and knowing that the frequency depends on the strength of the magnetic field, we can estimate the maximum strength of Jupiter's magnetic field. It turns out that Io, one of the larger moons of Jupiter, has a very big effect on whether we hear any Jupiter radio emission or not. Scientists have found that Io enhances Jupiter's emission of decameter radio waves. As Io orbits Jupiter, there are only certain positions in its orbit where our chances of hearing radio emissions become much greater. Jupiter's magnetic field moves rapidly past Io as it orbits. Io is very unique because it is the most volcanically active body in the solar system. Io is continually flexed by the gravitational pulls of Jupiter and the other satellites. This flexing causes Io to be molten, and volcanoes on its surface are almost continually erupting. Tons of material, mostly sulfur compounds, are ejected each second. Some fraction of this material escapes Io and travels into space. Once in space, the molecules soon lose their electrons, becoming ionized, and are trapped within Jupiter's magnetic field. These ions form a vast donut-shaped ring around Jupiter called the Io Torus [35]. Figure 5.17 illustrates the Torus region.

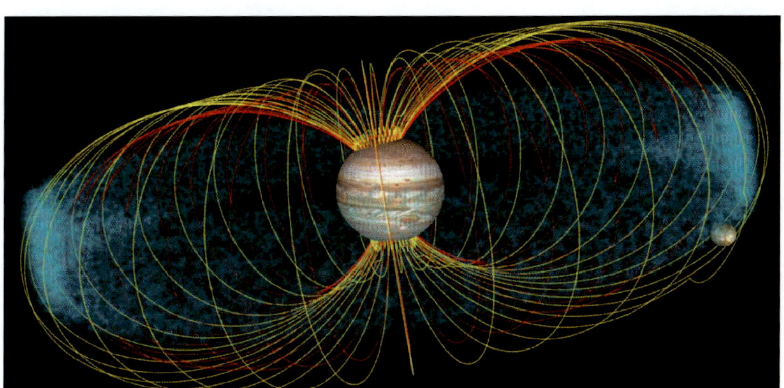

FIGURE 5.16
Schematic of Jupiter's magnetosphere and the components influenced by Io (near the center of the image): the plasma torus (in red), the neutral cloud (in yellow), the flux tube (in green), and magnetic field lines (in blue). (Courtesy of Wikipedia, Io (moon), https://en.wikipedia.org/wiki/Io_(moon)#/media/File:Jupiter_magnetosphere_schematic.jpg [last modified September 09, 2016].)

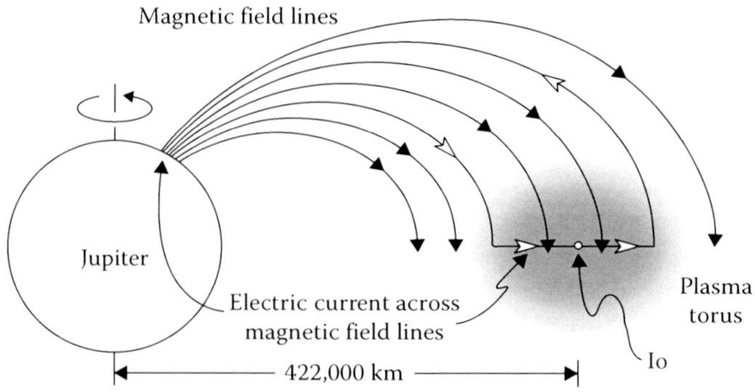

FIGURE 5.17
Torus region of Jupiter. (Courtesy of Duane Pontius, A Nebula in our own back yard, http://faculty.bsc.edu/dpontius/Files/torus_article/torus.html [retrieved August 30, 2016].)

5.12 Exploration through a Huge City of Stars

Solar system exploration across the universe has provided us with many interesting and valuable bits of information about our star and its planets, which are a tiny part of the Milky Way galaxy [36]. Our journey to interstellar space started with Voyager 1 [37] and has been followed by many others since. Figure 5.18 shows an artist's rendition of Voyager 1 and a photograph in assembly phase.

The Milky Way is a huge city of stars, so large that even at the speed of light, it would take 105 years to cross it. All the stars in the night sky including our Sun are simply some of the residents of this galaxy [38]. The Europa Jupiter System Mission [39] is likely to launch

FIGURE 5.18
Artist's rendition of Voyager 1 and a photograph of the spacecraft construction. (Courtesy of NASA, http://www.nasa.gov/centers/goddard/multimedia/largest/jupiter_magnetosphere.jpg_prt.htm [retrieved August 18, 2016].)

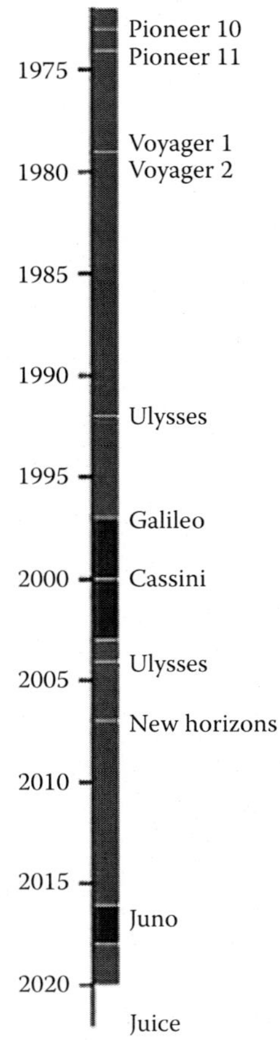

FIGURE 5.19
Spacecraft launch milestones (years).

sometime around 2020. The purpose is to continue with an extended study of the moons of Jupiter, particularly Europa and Ganymede. The observation may be able to settle the long-running scientific debate over whether an ocean of liquid water exists under the icy surface of Europa. Milestones indicating the years for launching of different spacecrafts is shown in Figure 5.19.

From the reports, it appears that on February 8, 1992, the Ulysses solar probe flew past the North Pole of Jupiter at a distance of 451,000 km. This swing-by maneuver was needed for Ulysses to attain an extremely high-inclination orbit around the Sun, increasing its inclination to the ecliptic to 80.2°. The gravity of the giant planet bent the spacecraft's flight path downward and away from the ecliptic plane, keeping it into a final orbit around the north and south poles of the Sun. The size and shape of the probe's orbit were adjusted to a smaller degree, so that its aphelion remained at nearly 5 AU, which is essentially the Jupiter's distance from the Sun, while its perihelion lay somewhat beyond 1 AU, equivalent to Earth's distance from the Sun. During its Jupiter encounter, the probe did some measurements of the planet's magnetosphere. As the probe had no cameras, there was no possibility of taking any image. In February 2004, the probe arrived further at the vicinity of Jupiter when the distance from the planet was sufficiently greater (about 240 million km), but it started observations of Jupiter again. A considerable and major problem facing space probes to Jupiter is that the planet has no solid surface to land. If the probe can endure the crushing force of the atmosphere and its fluid interior [40], there will be a wealth of information obtained for planetary science. Below is a summarized list of knowledge gained by instrumentation and techniques developed by theoretical consideration as well as through simulation [40]. First, our universe is expanding. Scientists believe that about 14 billion years ago, the universe was compressed into a single point in space. Second, there are at least 1011 galaxies in the universe. A galaxy is full of stars. Our Sun is just one of at least 1011 stars in our own Milky Way galaxy. Each of those stars could have their own planetary system. Third, more or less 68% of the universe is made of dark energy, whereas dark matter makes up about 27%. The rest are everything on Earth, everything ever observed with our instruments, and all normal matter adds up to less than 5% of the universe. Fourth, our universe has a *foamy* structure. The galaxies are concentrated in vast sheets and filaments surrounding cosmic voids. The Milky Way galaxy is in the neighborhood of about 30 galaxies. Our nearest major galaxy is named as Andromeda. Fifth, more than 1700 extrasolar planets or exoplanets have been confirmed. There are thousands of potential exoplanets that are yet to be confirmed. Other planetary systems could have the potential for life, but they are yet to be ascertained. Sixth, two-thirds of the galaxies within the universe are spiral shaped similar to Milky Way galaxy. They have generally elliptical shapes, but a few have unusual shapes like toothpicks or rings. Seventh, a black hole is a huge amount of matter packed into a very small area and never an empty space in the universe. A black hole results in a gravitational field so strong that nothing (not even light) can escape. Eighth, the Hubble Space Telescope found a tiny patch of sky for 11.6 days and observed nearly 104 galaxies of all sizes, shapes, and colors.

References

1. Solar System Exploration. https://www.lpi.usra.edu/vexag/road_final.pdf (retrieved August 23, 2016).
2. C. Sagan (September 9, 1990). The Earth from the frontiers of the Solar system—The Pale, Blue Dot. *PARADE Magazine*.
3. R. Garg (2012). *Realm of Space Science*. S. K. Kataria & Sons, New Delhi, India. 225 pp. http://www.scribd.com/doc/51978691/realm-of-space-science1 (retrieved August 23, 2016).
4. First photo of sun (1845). National Geographic. http://photography.nationalgeographic.com/photography/enlarge/first-solar-photophotography.html (retrieved April 7, 2016).
5. E. W. Weisstein (2006). Galileo Galilei (1564–1642). Wolfram Research. http://scienceworld.wolfram.com/biography/Galileo.html (retrieved November 8, 2006).
6. European Space Agency (ESA). Discoverer of Titan: Christiaan Huygens (2005). ESA Space Science. http://www.esa.int/About_Us/Welcome_to_ESA/ESA_history/Christiaan_Huygens_Discoverer_of_ Titan (retrieved October 27, 2010).
7. G. D. Cassini (June 8, 1625–September 14, 1712). SEDS.org. http://messier.seds.org/xtra/Bios/cassini.html (retrieved October 27, 2010).
8. Wikipedia. History of the telescope. http://en.wikipedia.org/wiki/History_of_the_telescope (retrieved May 10, 2016).

   L. Cox (2013). Who Invented the Telescope? SPACE.com.http://www.space.com/21950-who- invented-the-telescope.html (retrieved August 10, 2016).

9. Comet Halley. University of Tennessee. http://csep10.phys.utk.edu/astr161/lect/comets/halley.html (retrieved December 27, 2006).
10. W. Herschel (1738–1822). enotes.com. http://science.enotes.com/earth-science/herschel-sir-william (retrieved November 8, 2006).
11. INAF - Astronomical Observatory of Palermo (October 11, 2000). Discovery of Ceres: 2nd Centenary, January 1, 1801–January 1, 2001, About Giuseppe Piazzi. http://www.astropa.unipa.it/Asteroids2001/ (retrieved September 28, 2016).
12. Wikipedia. Spectrometer. http://en.wikipedia.org/wiki/Spectrometer (last modified on April 7, 2016).
13. *Spectroscopy and the Birth of Astrophysics*. Center for History of Physics, a Division of the American Institute of Physics. https://www.aip.org/history/exhibits/cosmology/tools /tools-spectroscopy.htm (retrieved April 30, 2008).
14. Sun Earth Connection Coronal and Heliospheric Investigation (SECCHI). http://secchi.nrl.navy.mil/ (retrieved July 30, 2016).
15. NASA. Sputnik 1. https://www.nasa.gov/multimedia/imagegallery/image_feature_924.html (retrieved July 30, 2009).
16. D. Savage and M. Mewhinney (February 25, 2003). Farewell Pioneer 10. NASA. http://www.nasa.gov/centers/ames/news/releases/2003/03_13AR.html (retrieved July 11, 2007).
17. R. Culp (2002). Time Line of Space Exploration. http://www.rocketmime.com/space/timeline.html (retrieved July 1, 2006).
18. Comet Space Missions. http://spider.seds.org/spider/Comets/c_missions.html (retrieved November 23, 2007).
19. NASA. New Horizons NASA's Pluto-Kuiper Belt Mission (2006). https://www.nasa.gov/mission_pages/newhorizons/main/ (retrieved July 1, 2016).
20. Wikipedia. Solar System. http://en.wikipedia.org/wiki/Discovery_and_exploration_of_the_Solar_System#cite_note -exp-1 (last modified on July 6, 2016).
21. FY 2011 Budget. NASA. http://www.nasa.gov/news/budget/2011.html (retrieved February 14, 2011).
22. NASA. NASA Hails Success of Commercial Space Program. http://www.nasa.gov/content/nasa-hails-success-of-commercial-space-program (retrieved July 24, 2014).
23. A. Siddiqi. *Challenge to Apollo the Soviet Union and the Space Race, 1945–1974*. NASA. p. 832.
24. Wikipedia. Red shift. http://en.wikipedia.org/wiki/Redshift (last modified on July 13, 2016).
25. H. C. Chiang et al. (2009). Measurement of cosmic microwave background polarization power spectra from two years of BICEP data. *The Astrophysical Journal* 711: 1123–1140.
26. NASA. Cosmic Background Explorer (1989). http://lambda.gsfc.nasa.gov/product/cobe/ (retrieved August 18, 2016).
27. D. J. Fixsen (2009). The temperature of the cosmic microwave background. *The Astrophysical Journal* 707: 916–920.
28. Goddard Space Flight Center. *Special Exhibits*. http://imagine.gsfc.nasa.gov/docs/features/exhibit/WMAP_exhibit.html (retrieved November 6, 2014).
29. A. A. Penzias. The origin of elements. *Nobel Lecture*. http://www.nobelprize.org/nobel_prizes/physics/laureates/1978/penzias-lecture.html (retrieved October 4, 2006).
30. NASA. http://solarsystem.nasa.gov/news/ (retrieved September 28, 2016).
31. NASA. http://www.nasa.gov/press-release/nasas-juno-spacecraft-in-orbit-around-mighty-jupiter (retrieved July 5, 2016).
32. NASA. Juno. https://www.nasa.gov/mission_pages/juno/main/index.html (retrieved July 7, 2016).
33. Jupiter: The Star that Might Have Been. http://starryskies.com/articles/dln/1-02/jupiter.star.html (retrieved July 7, 2016).
34. Wikipedia. Magnetosphere of Jupiter. http://en.wikipedia.org/wiki/Magnetosphere_of_Jupiter (last modified on June 30, 2016).

35. L. N. Garcia. The Jovian Decametric Radio Emission. How scientists learn about Jupiter by observing its radio emissions. http://radiojove.gsfc.nasa.gov/library/sci_briefs/decametric.htm (retrieved August 18, 2016).
36. K. Than (January 30, 2006). Astronomers had it wrong: Most stars are single. SPACE.com. http://www.space.com/1995-astronomers-wrong-stars-single.html (retrieved September 28, 2016).
37. NASA. NASA Space Science Data Coordinated Archive. Voyager 1. http://nssdc.gsfc.nasa.gov/nmc/spacecraftDisplay.do?id=1977-084A (retrieved April 27, 2016).
Images and Video Voyager, the Intersteller Mission. NASA. http://voyager.jpl.nasa.gov/imagesvideo/imagesofvoyager.html (retrieved April 27, 2016).
38. NASA. Junocam. https://www.missionjuno.swri.edu/ (retrieved August 18, 2016).
39. NASA. Europa Mission. http://www.jpl.nasa.gov/missions/europa-clipper/ (retrieved August 18, 2016).
40. T. Guillot (1999). A comparison of the interiors of Jupiter and Saturn, *Planetary and Space Science* 47: 1183–1200.

6

Solar-Like Planetary Systems

6.1 Introduction

Throughout recorded history and perhaps even before, we have wondered about the probable existence of other worlds, similar or dissimilar to our own. The earliest understanding of the solar system indicates that there were indeed other worlds in orbit about our Sun. Our growing understanding of their nature reveals that all are dramatically different from our Earth and also very different from one another. As we gathered knowledge that the stars in the sky are other Suns that are part of distant galaxies, consisting of billions of stars; it appears as a near certainty that other planets must orbit those other stars [1]. Until the early 1990s, this could not be proven, but, through the efforts through research, radio and optical astronomers detected small changes in stellar emission to show the presence of planetary systems around other stars. These planets are named as *exoplanets* [2] for the purpose of distinguishing them from our own solar system neighbors. Extrasolar planets, or popularly called exoplanets, are planets orbiting stars outside our solar system. The first discovery of the exoplanet was confirmed in 2003 [2]. It is very difficult to detect these planets directly, as they are very dim and hard to see due to the brightness of the stars they orbit. As an example, we may mention that the Sun is a billion (10^9) times brighter than Jupiter and 10 billion (10^{10}) times brighter than our Earth.

6.2 Planets around Other Stars

Most of the detected exoplanets have shown their presence by small effects they have on their star. As the planet follows its orbital path, the star follows a complementary motion of its own. This becomes a tiny effect proportional to the planet/star mass ratio. For the solar system, the Sun moves in synch with the Earth at a speed that seems to be slow dance to readily detect in a distant system. However, the motion of the Sun in synch with Jupiter is closer to a fast run and can be detected by applying different methods. The motion of the host star can be determined as a Doppler shift in its spectrum or as a change in its position in the sky [1]. In both the cases, these are highly challenging measurements and need sensitive instruments. Exoplanet orbits have random orientations and sometimes the orbit carries the planet between us and its star. The exoplanet might be detected by the reduction of light from the star. Such transits have been found and thereby a number of planets have been discovered by applying this technique. A second effect that can show the presence of a planet around another star is the bending of light from background stars owing to the gravitational field of an intervening star. If the intervening star has an orbiting planet, it may change the gravitational lensing effect significantly, known as microlensing [3]. A good number of exoplanets have been discovered by the Doppler method and other techniques, and the number continues to increase steadily. It is assumed that stars like the Sun form due to spinning protostellar disks of gas and dust. The Earth and other planets of the solar system are developed from the remains of that disk and the same process would be effective throughout the galaxy. A preliminary assumption might be that other planetary systems would be like the solar system. However, the first detections of exoplanets showed bodies that are unlike any solar system planet and that many exoplanet systems are very dissimilar from ours. In some exosystems, planets are so close to their star that they are heated to high temperature and their upper atmospheres are swept into space, whereas in some other systems, planets follow elongated orbits in contrast to the approximately circular orbits of the solar system.

6.3 Nearby Solar-Like Planetary System

Astronomers have produced evidence that Vega, one of the brightest stars in the sky, has a planetary system around it that is similar to our own solar system, more than any other so far discovered. The planets discovered around other stars have been reported as very large gaseous (Jupiter-like) planets orbiting close to their star. This is quite unlike our own solar system. New computer modeling techniques have suggested that observations of the structure of a faint dust disk around Vega may be best explained by a Neptune-like planet orbiting at a similar distance to Neptune in our own solar system and having similar mass. The large orbit of the Neptune-like planet means that there is enough room inside it for small rocky planets similar to the Earth [4]. Figure 6.1 shows the world's most sensitive

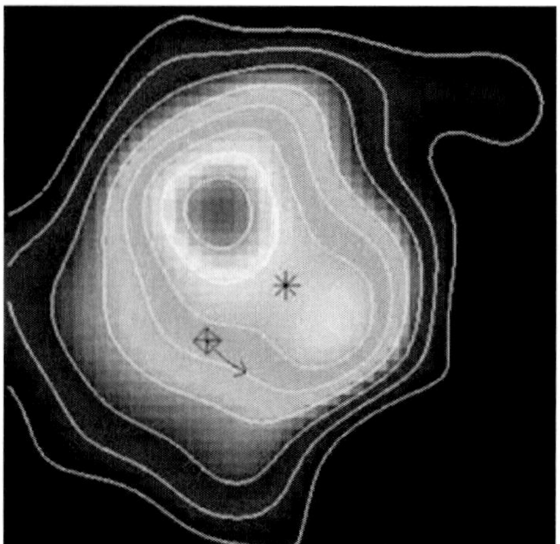

FIGURE 6.1
SCUBA image with the position of the star (*) and the predicted position and direction of the planet (x) marked. (Courtesy of PPARC, Wikipedia, http://www.pparc.ac.uk/ [last modified January 27, 2016].)

submillimeter camera SCUBA [5] image with the position of the star (*) and the predicted position and direction of the planet (x). The distance between the star and the planet is equivalent to twice that between the Sun and Neptune.

The SCUBA camera, built at the ATC (Astronomy Technology Center [UK]), was operated on the James Clerk Maxwell Telescope in Hawaii. The SCUBA image exhibits a disk of very cold dust (–180°C) in orbit around the star. The irregular shape of the disk was the clue that it is likely to contain planets. Although one cannot directly find the planets, which were created from clumps in the disk of dust around the star. Suitable modeling suggested that the Neptune-like planet actually formed much closer to the star than its current position, but as it moved out to its current wide orbit over about 56 million years ago, many comets were swept out with it, resulting in the dust disk to appear jumbled. It is believed that a similar process happened in our solar system. Neptune was *pushed* away from the Sun due to the presence of Jupiter orbiting inside it. Thus, it appears that similar to a Neptune-like planet, Vega may also have a more massive Jupiter-like planet in a smaller orbit.

6.3.1 Facts about Vega

- Vega is the fifth brightest star and the third brightest visible in the Northern hemisphere.
- It has a diameter three times bigger than that of the Sun.
- It is 58 times brighter than the Sun.
- Together with Deneb and Altair, Vega forms the summer triangle.
- It is 25 light years away from the Sun.
- It is the brightest star in the constellation Lyra, or the Harp, which is supposed to have been created by the Greek God Hermes, who gave it to his half-brother Apollo, who then gave it to his son Orpheus, the musician of the Argonaughts [4]. Vega was the first star ever to be photographed. During the night of July 16–17, 1850, the historic picture was taken at the Harvard Observatory using a 15-in. refractor telescope during a 100-s exposure. A sky map showing the position of Vega is illustrated in Figure 6.2.

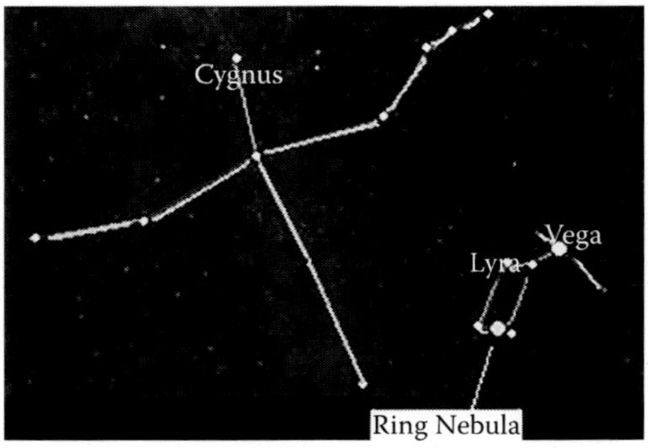

FIGURE 6.2
Sky map showing the position of Vega. (Courtesy of Lichtman, Radio Astronomy Supplies, TX.)

6.4 Planets Orbiting Distant Stars

Whether our solar system is the only planetary system in the universe or not has intrigued scientists for hundreds of years. Within the last few decades, many planets orbiting distant stars have been detected [5,6]. An artist's impression of a distant solar system is presented in Figure 6.3, whereas Figure 6.4 reveals a giant planet orbiting close to a solar-type star.

6.5 Spacecraft Kepler

The NASA/JPL spacecraft mission, known as Kepler [7], looked at about 156,000 stars simultaneously in a section of the Cygnus constellation with the task to search for Earth-sized exoplanets during a 4-year period [8]. The spacecraft primarily focused on exoplanets

FIGURE 6.3
An artist's impression of a distant solar system. (Courtesy of NASA, A distant solar system (artist's concept animation), http://www.jpl.nasa.gov/spaceimages/details.php?id=PIA07097 [retrieved August 18, 2016].)

FIGURE 6.4
Artist's rendering of a possible exoplanetary system with a gas-giant planet orbiting close to his parent star which is more massive than our sun. (Courtesy of Gemini Observatory/AURA [artwork by Lynette Cook], Gas-Giant exoplanets cling close to their parent stars. http://www.gemini.edu/node/12025 [retrieved August 18, 2016].)

that could be in the stars' habitable zones. It was noted by Kepler that only 0.5% of the stars are expected to have their planets' orbits in the habitable zones. A terrestrial exoplanet with mass in the range 0.5–10 Earth masses will cause its star to dim by a fractional amount of between 0.00005 and 0.0004, respectively, and the transits will last for only few hours. To solidly verify detection, each exoplanet transit must be repeatable at least two more times after the first detected transit and with the same time interval between transits and depth of transit. This repeatability of the transits indicates that something was orbiting the star. A solar-type star with an exoplanet in the habitable zone would transit the star once a year. But this assumes that the stars are calm and steady like our Sun. The Kepler team observed that a number of stars are a bit more active and more variable than our Sun. Hence, they started more observations to tease out the dimmings due to transiting exoplanets due to the intrinsic variability of the stars themselves. The hardware failure of some of its stabilizer reaction wheels brought the observing of the section region of Cygnus to an end in early 2013. However, the Kepler team still has many thousands of transits to shift through and hence there are plenty of discoveries still to be made. In June 2014, the Kepler mission started using a technique that uses the pressure of sunlight to help point the telescope with the remaining two stabilizer reaction wheels. This mission called K2 will be utilized to find objects near the ecliptic [8]. In the K2 phase, Kepler will find other objects in addition to hunting for exoplanets, such as variable stars, star clusters, and galaxies. The Kepler team has also pointed out how the exoplanet detection works can be done successfully and how the different planet parameters can be derived. Figure 6.5a shows the Kepler spacecraft (NASA), whereas Figure 6.5b presents Kepler's original field of view in Cygnus (NASA).

Up to June 2014, Kepler had observed over 4200 planetary candidates, with more than 20% of the planetary systems having multiple planets. Candidate exoplanets are those that have not been verified yet through follow-up observations. Around 980 exoplanets have been confirmed as of June 2014, making the transit technique the most prolific method so far. Using the number of confirmed exoplanets from the candidate list, we have acquired a very good idea of transit process and its working in finding possible planets. It is interesting to note that at least 38 of the confirmed exoplanets found by Kepler are definitely rocky exoplanets with densities greater than three times that of water. The first rocky exoplanet discovered, named as Kepler-10b, has a density 5.8 times that of water. The oceans there would be liquid iron. Another confirmed exoplanet is Kepler-22b, which is the first one from the Kepler mission known to be in its star's habitable zone and is 2.4 times the diameter of Earth [8]. Subsequently, from the information gathered, Kepler-22b is considered to

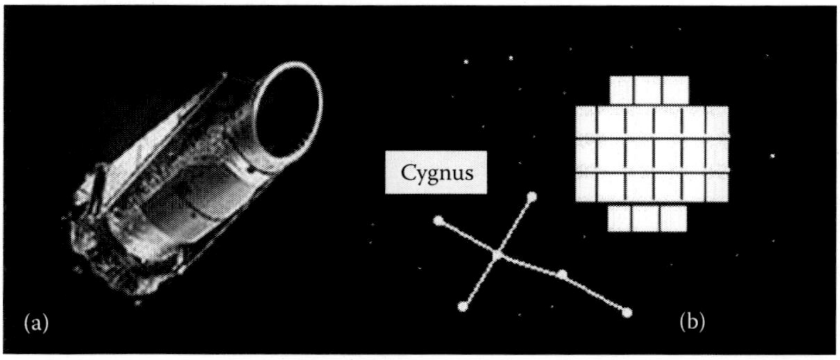

FIGURE 6.5
(a) Kepler spacecraft. (b) Kepler's original field of view in Cygnus. (Redrawn for clarity by Lichtman; Courtesy of NASA, https://www.nasa.gov/mission_pages/kepler/multimedia/images/kepler-field-of-view-photo.html [retrieved July 2, 2014] and http://kepler.nasa.gov/ [retrieved August 12, 2016].)

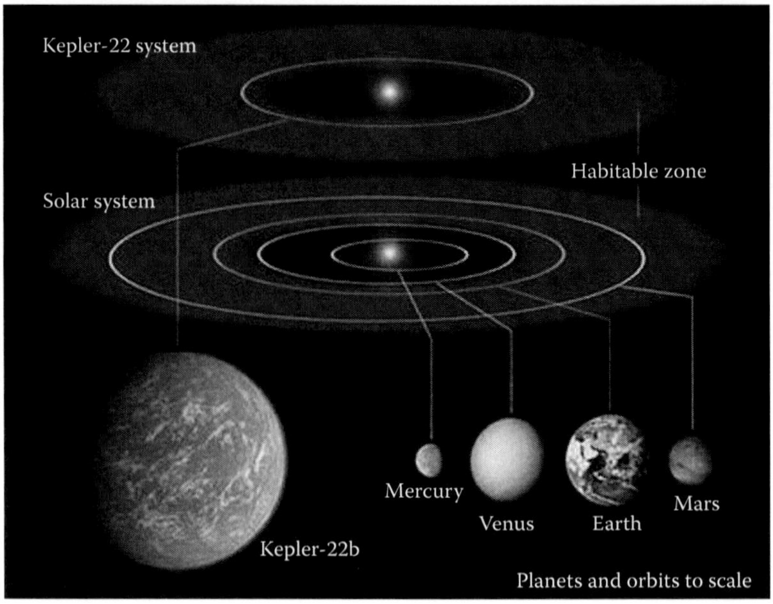

FIGURE 6.6
A comparison of Kepler-22b and its star's habitable zone with our solar system. (Courtesy of NASA/Ames/JPL-Caltech, http://www.astronomynotes.com/solfluf/s12.htm [retrieved July 2, 2014] and http://kepler.nasa.gov/ [retrieved August 12, 2016].)

be a *mini-Neptune* instead of a *super-Earth*. An important comparison of Kepler-22b and its star's habitable zone with our solar system is shown in Figure 6.6. The picture of Kepler-22b is nothing but an artist's imagination of what Kepler-22b might look like. As the star Kepler-22 is slightly cooler than our Sun, its habitable zone is also slightly smaller than Kepler-22 Sun's habitable zone.

The smallest confirmed exoplanet in a star's habitable zone found so far by Kepler is Kepler-186f, which is just 1.1 times the size of the Earth. If we assume that the Kepler-186f has the same density as that of the Earth, it would be a little over 1.3 times the mass of the Earth. The statistics of the known densities of exoplanets as well as sophisticated modeling concludes that about 75% of the exoplanets smaller than 1.5 Earth-radii are rocky worlds. It is likely that Kepler-186f is a rocky world composed mostly of silicates, iron, nickel, and magnesium, like the Earth. A comparison of the five planets of Kepler-186 system with the inner three planets of our solar system is shown in Figure 6.7 [9]. It is important to note that the star Kepler-186 is cooler and only 4% the luminosity of the Sun and hence its habitable zone is significantly smaller than that of the Sun.

Another confirmed exoplanet system is Kepler-20 [10], which has five exoplanets including two that are the size of Earth. It has been found that Kepler-20e is smaller than Venus and Kepler-20f is just 3% larger than the Earth. But both of them orbit well inside Kepler-20's habitable zone. This discovery confirmed that the Kepler spacecraft can detect Earth-sized exoplanets and that such exoplanets definitely exist. An example of another confirmed exoplanet system is Kepler-37 [10], which has one exoplanet slightly greater than the Moon, a second slightly smaller than Venus, and a third twice the size of the Earth. Out of the confirmed exoplanets (as of June 2014), over 140 have diameters less than 1.25 that of the Earth. Of the candidate exoplanets, at least 100 candidates are small worlds with diameters from 0.8 to 2 times the diameter of the Earth and situated in the star's habitable zone.

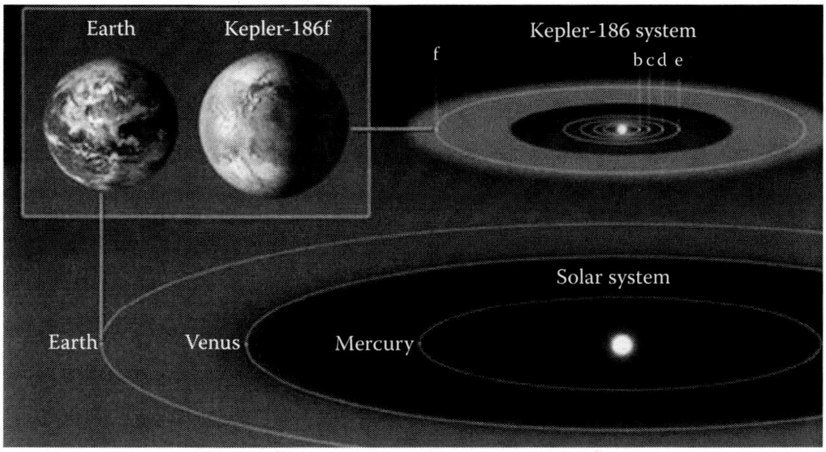

FIGURE 6.7
Comparison of the five planets of the Kepler 186 system with the inner three planets of our solar system. (Courtesy of NASA/Ames/SETI Institute/JPL-Caltech, http://kepler.nasa.gov/news/index.cfm?FuseAction=ShowNews&NewsID=190 [retrieved August 18, 2016].)

6.6 Theory of *Hot Jupiters*

According to the theory of solar system formation, massive Jupiter-like planets can form outside of the frost region of the star system and have nearly circular orbits. Hot Jupiters, on the other hand, are massive Jovian planets that have highly elliptical orbits [11]. Astronomers believe that planetary migration is one of the explanations for the formation of Hot Jupiters. This theory explains that Hot Jupiters are formed in the outer regions of their solar system and then migrate inward. This migration is caused by the propagation of waves through the gaseous disk around the young planet. These young Jovian planets get massive enough so that their gravity is strong to clear the orbital path of the gas disk. As a result, the planet creates waves that propagate through the disk, which causes the material to collect at one place. Then this collected material gains enough gravitational force that reduces the orbital energy of the planet causing the planet to migrate inward [11]. Figure 6.8 shows an image from a numerical simulation of the interaction of a massive planet with a gaseous protoplanetary disk [12,13].

Another theory that explains the highly eccentric orbits of Hot Jupiters is based on gravitational perturbation. Gravitational perturbation happens when two young Jovian planets have a close gravitational encounter. This counter can send one planet out of the star system and the other inward toward the star into a highly elliptical orbit [11]. Another feature of the Hot Jupiter is that it is not formed at the place where it appears ultimately but it locates at a much larger distance from the central star. The planet then moved to its current position. Figure 6.9 reveals an artist's conception of one of the *Hot Jupiter*.

6.6.1 Hot Jupiter

Hot Jupiters are a class of extrasolar planets that are really close to the parent star, which is only a few stellar radii away and has an orbital period of 3 days or, maybe even 1.5 days for some cases. Jupiter with fairly cold weather has a very long period of about 12 years

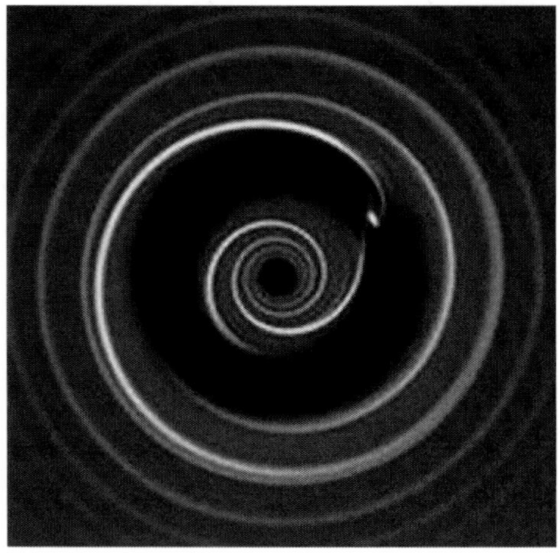

FIGURE 6.8
Numerical simulation of the interaction of a massive planet with a gaseous protoplanetary disk. (Courtesy of P. Armitage, Planetary formation and migration, Scholarpedia, 3 (3), 4479, 2008.)

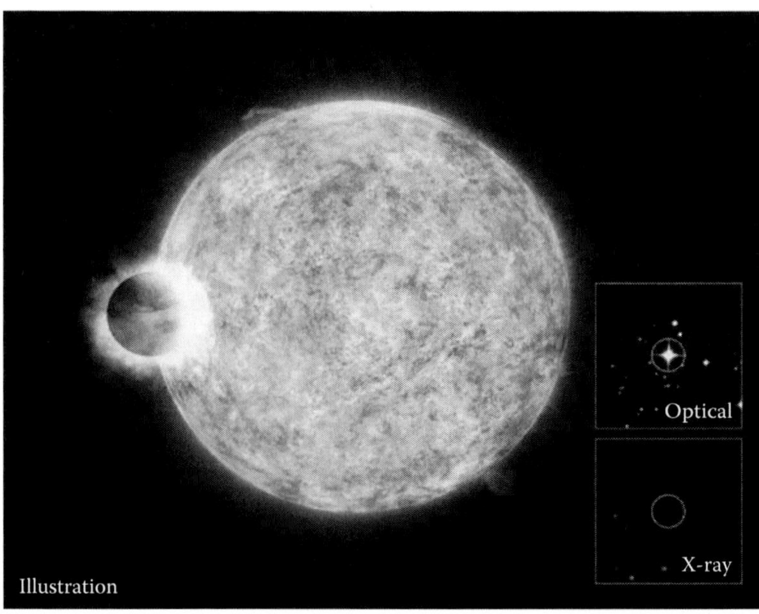

FIGURE 6.9
WASP-18b: A *Hot Jupiter*. (Courtesy of NASA, http://www.nasa.gov/chandra/multimedia/wasp18-2014.html [retrieved July 31, 2015].)

orbits at ~5 AU from the Sun. Being close to parent star, the temperature of Hot Jupiter lies between 1000 and 2000 K. They have the largest gravitational pull on their stars, so their name in the Doppler method of planet detection is the strongest. In 1995, Michel Mayor and Didier Queloz [14] discovered the first planet and Hot Jupiter around a Sun-like star using the radial velocity technique. They used the spectrum of the star 51 Pegasi for detecting periodic Doppler shifts caused by the planet's gravitational pull on the star. This technique contributed toward finding Hot Jupiters around less massive stars. Hot Jupiters

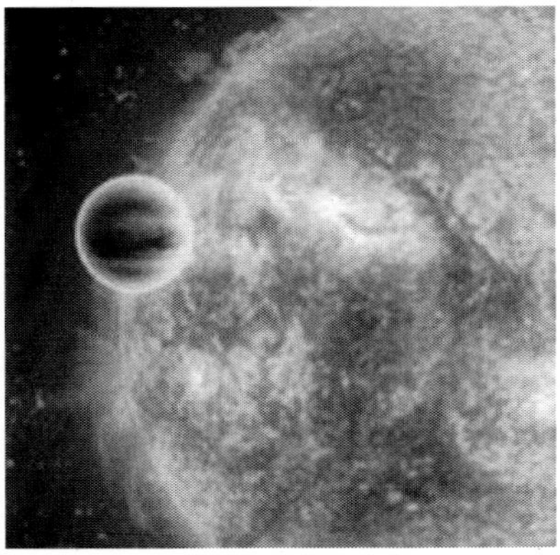

FIGURE 6.10
Artist's impression of a gas-giant exoplanet transiting across the face of its star. (Courtesy of NASA, http://www.jpl.nasa.gov/spaceimages/wallpaper.php?id=PIA10364 [retrieved August 19, 2016].)

have an orbital period of a few days and are much easier to detect than Earth-size planets very far from their stars [15]. Until a few decades ago, exoplanets and their solar systems were the matter of theory and assumptions. As our knowledge of the universe moved from the area of guess work to hard data, we came to see our Sun as one of countless stars. In 1995, the newly observed first Hot Jupiter 51 Pegasi b was discovered from the spectrum of the star 51 Pegasi due to the detection of periodic Doppler shifts caused by the planet's gravitational pull on the star [14]. Till now, more than 415 Hot Jupiters have been discovered [15]. Figure 6.10 [16] shows an artist's impression of a gas-giant exoplanet transiting across the face of its star.

Most of the discovered exoplanets have been detected by radial velocity studies of the host star. The second-largest group has been detected through planetary transits. Twenty-six planets have been detected by imaging and 12 have been discovered through planet-lens signatures detected during gravitational lensing events, in which the host star serves as the primary lens. In 2006, the European Space Agency launched the COROT [17] spacecraft, which was the first satellite used for searching extrasolar planets. The COROT spacecraft has successfully discovered many extrasolar planets. Subsequently, in 2009, NASA launched Kepler spacecraft, which also discovered more than 25 confirmed planets and around 1250 eligible candidates [18]. Both of these missions used the transit method when the planet passes in front of its star, blocking a very small proportion of the starlight [19]. The lensing method provides a potentially important complement to the radial velocity and transit studies that have already been discovering Hot Jupiters. It allows planet discovery even if the central star is too dim, for detailed spectral studies and for all orbital inclinations, in contrast to transit studies. Furthermore, lensing provides a direct measure of the lens mass, at least in cases in which the mass of the central star can be determined. Lensing searches for Hot Jupiters can be effective for nearby stars [20], allowing detailed follow-up studies. Companions of different Hot Jupiters are shown in Table 6.1 with reference to their distances and associated semi-major axis, orbital period, and the corresponding temperature.

Kepler-13Ab (= KOI-13.01) is a unique transiting Hot Jupiter. It is one of very few known short-period planets orbiting a hot A-type star, making it one of the hottest planets currently known. The availability of Kepler data allows us to measure the planet's occultation (secondary eclipse) and phase curve in the optical, which we combine with occultation's observed by warm Spitzer at 4.5mic and 3.6mic and a ground-based occultation observation

TABLE 6.1

Hot Jupiters with Companions

Planet	Distance (pc)	Semi-Major Axis (AU)	Orbital Period P_{orb} (days)	M_p (MJ)	Temperature (K)
Sources with eccentricities less than or equal to 0.1%					
HD 187123b	48.26	0.042	3.10	0.51	1320
HD 209458b	49.63	0.047	3.52	0.69	1316
55Cnc b	12.34	0.116	14.65	0.84	661
55Cnc b	12.34	0.240	44.3	0.84	661
HAT-P-13b	214	0.0426	2.91	0.85	1504
ρ CrB b	17.24	0.226	39.84	1.06	—
u And e	13.47	5.25	3848	1.06	376
HD 189733b	19.45	0.031	2.22	1.13	1100
υ And b	13.49	0.059	4.62	1.4	1440
Qatar-2b	1.14	0.0215	1.33	2.49	1180
HD 195019b	38.52	0.137	18.20	3.58	—
τ Boo b	15.62	0.048	3.31	6.5	6375
Sources with eccentricities greater than 0.1%					
HAT-P-17b	90.0	0.088	10.3	0.53	707
HIP 14810d	52.9	1.9	962	0.57	175.6
HD 38529b	39.28	0.131	14.31	0.86	—
HIP 14810c	52.9	0.55	147.73	1.3	327
HAT-P-17c	90.0	2.75	1798	1.400	127.5
HD 217107b	19.72	0.073	7.12	1.85	960
HD 187123c	48.26	4.89	3810	1.990	122.7
HAT-P-31b	354	0.055	5.00	2.2	1325
HD 217107c	19.72	5.27	4210	2.5	111.1
HD 37605b	43.98	3.82	54.23	2.86	414
HD 37605c	43.98	0.261	2720	3.38	116.7
HIP 14810b	52.9	0.069	6.67	3.9	918
HD 178911Bb	42.59	0.345	71.48	7.29	470
70Vir b	22.0	0.484	116.69	7.46	479
HD 114762b	39.5	0.363	83.89	11.68	487
HAT-P-13c	214	1.19	448.2	14.5	276

in the Ks band (2.1mic). We derive a day-side hemisphere temperature of 2750 ± 160 K as the effective temperature of a black body showing the same occultation depths [21]. Comparing the occultation depths with one-dimensional planetary atmosphere models suggests the presence of an atmospheric temperature inversion. Our analysis shows evidence for a relatively high geometric albedo, Ag = 0.33 + 0.04 − 0.06. Although measured with a simplistic method, a high Ag is supported also by the fact that the one-dimensional atmosphere models underestimate the occultation depth in the optical. We use stellar spectra to determine the dilution, in the four wide bands where occultation was measured, due to the visual stellar binary companion 1.15 ± 0.05-in. away. The revised stellar parameters measured using these spectra are combined with other measurements leading to revised planetary mass and radius estimates of Mp = 4.94 − 8.09Mjup and Rp = 1.406 ± 0.038Rjup. Finally, we measure a Kepler mid-occultation time that is 34.0 ± 6.9 s earlier than expected based on the mid-transit time and the delay due to light travel time, and discuss possible scenarios [22]. The revised stellar parameters when combined with other measurements, leading to revised planetary mass and radius, which can be estimated as,

$$\mathrm{Mp} = 4.94 - 8.09\mathrm{MJ}$$

and

$$\mathrm{Rp} = 1.406 \pm 0.038\mathrm{RJ}$$

Kepler's mid-occultation time was measured as (34.0 ± 6.9) s earlier than expected based on the mid-transit time and the delay due to light travel time [21].

6.6.2 The Relationship of Planet Jupiter to a Hot Jupiter and Its Parent Star

Scientists have found that most known exoplanets have many similarities with the Jovian planets in our solar system, such as size, density, and composition. Exoplanets are probably made of hydrogen and helium gas. These planets are very close to the star, so they experience a higher surface temperature than on the Jovian planets [23]. Table 6.2 shows a comparison between our planet Jupiter and *Hot Jupiters* exoplanets.

It appears from the table that the distance of our planet Jupiter is much greater than the distance of the Hot Jupiter and its corresponding parent star. The cloud-top temperature of

TABLE 6.2

Comparative Studies between Jupiter and *Hot Jupiters*

Dominant Features	**Jupiter**	***Hot Jupiters***
Appearance	 Jupiter	 *Hot Jupiters* orbiting other stars
Composition	Composed primary of H and He	Composed primary of H and He
Distance and orbit	5 AU from the Sun; orbital period: 10475.8 Jupiter solar days	As closes as 0.03 AU to their stars; orbit as short as 1.2 Earth days
Cloud-top temperatures	~130 K	Up to 1300 K
Cloud composition	Clouds of various H compounds	Mainly consists of *rock dust*
Radius	1 Jupiter radius	Up to 1.3 Jupiter radii
Mass	1 Jupiter mass	0.2–2 Jupiter masses
Average density	1.33 g/cm³	As low as 0.3 g/cm³
Moons, rings, magnetosphere	Present	Unknown

Jupiter is only about one-tenth of that of the exoplanet Hot Jupiter. It is of further interest to note that the average density of Jupiter is much greater in comparison to that of the Hot Jupiter. HD 189733b, which was first mapped in 2007 by the Spitzer Space Telescope and HAT-P-7b, was recently observed by the Kepler mission. Hot Jupiter probably has cloud layers but due to high levels of insulation different elements could condense therein. It is far too hot for ammonia, methane, and water to condense, like they do in Jupiter's atmosphere. These high temperatures might even allow for clouds made from materials we would normally think of as solids on Earth (e.g., some metals). Most of these exhibit high-speed winds distributing the heat from the day side to the night side, thus the temperature difference between the two sides is relatively low. They are more common around F- and G-type stars and somewhat less common around K-type stars. Hot Jupiters around red dwarfs are very rare [24,25]. Main criteria for detecting Hot Jupiters from earth-based observatories are as follows [26]:

- Proximity to solar system is the first criterion, as radio flux intensity decreases quadratically to distance.
- Earth has an ionospheric cutoff frequency of 5–10 MHz. To radiate at a higher frequency, Hot Jupiter mass should be large, as magnetic moment is roughly proportional to the mass of the planet.
- Proximity of Hot Jupiters to their host stars increases the probability of receiving radio bursts, as emitted radio power is proportional to energy delivered to planetary magnetosphere.
- Young stars can transport high energetic radio flux to a planetary system, which, in turn, increases radio emission power.

Note: The flux unit or jansky (symbol Jy) is a non-SI unit of spectral flux density, or spectral irradiance, equivalent to 10–26 $W/m^2/Hz$. The jansky is sometimes called the flux unit. It was noticed that due to the strong decameter-wave galactic background, Jupiter's intense decametric radio emission (~106 Jy, where 1 Jy = 10–26 $W/m^2/Hz$) is undetectable beyond 0.5pc [27]. Even emissions from Hot Jupiters can be approximately 102 times more intense than Jupiter's emission. Received flux density received on Earth will not exceed a few nJy, whereas currently the minimum detectable flux reaching earth-based observatories is of the order of a few μJy. Jovian radio emissions mostly depend on internal processes like Io internal plasma torus, ionosphere–magnetosphere coupling, and so on. By extrapolating this physics of Jupiter's rotation-dominated magnetosphere to the case of Hot Jupiters, Nichols [28] proved that rapidly rotating giant planets irradiated by a strong stellar X-UV flux and orbiting beyond ~1 AU from their parent star should generate intense ionosphere–magnetosphere current systems and associated radio emissions up to 105 times greater than Jupiter's. According to Rosner et al. [29], stellar X-ray luminosity is proportional to stellar rotation, so host stars like τ Boo and 51 Pegasi with higher rotational velocity emit stronger and frequent radio bursts, making them the prime target for ground-based observation.

6.6.3 Characteristics

Hot Jupiters are very close to their stars, so they are receiving very intense levels of starlight causing their cloud-top temperature to be much warmer than Jupiter's [30]. The flux of sunlight that a planet is receiving is inversely proportional to the square of distance of separation. The closer the planet, the greater the flux, thus producing a higher intensity level of sunlight. Intensity being proportional to the fourth power of temperature, the greater intensity of sunlight creates higher temperature. Jupiter has a cloud-top temperature of

130 K, whereas Hot Jupiters' cloud-top temperatures can be up to 1300 K [30]. The high temperatures of Hot Jupiters can affect the composition of the clouds. Jupiter's clouds are made up of ammonia ice flakes and water because the cloud-top temperature of Jupiter is low enough for these compounds to condense. Hot Jupiters, however, are too hot for these gases to condense. Rocks can only condense at these high temperatures; therefore, the clouds of Hot Jupiters are made up of rock dust. The high intensity of solar heat causes the planet's atmosphere to inflate, resulting in a larger radius and lower density. Hot Jupiters' upper atmospheres can extend beyond three times the radius of the planet. The expansion of the atmosphere is due to the hydrodynamic state, where intense heating of the upper atmosphere gives gas particles very high kinetic energy and pushes the gas in an upward direction [31,32]. The average velocity of the gas particles is increased by

$$V_{avg} = \sqrt{\frac{8kT}{\pi M}}.$$

An increase in temperature with increment in altitude makes the upper atmosphere hydrostatically unstable and the velocity of the particles then becomes larger than the escape velocity [33]. Therefore, the hydrogen gas in the atmosphere flows upward and escapes causing a blow-off situation, where the light gases drag the heavier gases to escape with them. This blow-off effect is maintained by constant energy input from the source star. The closeness to the star also alters the shape of the gravitational field. We see the gravitational field re-forms from spherical to an elongated shape as altitude increases. This elongated shape allows more gas to escape, making the escaping components resemble a comet tail as it is moving away from the planet and the star [31].

6.6.4 Transit Timing Variations

It was believed that Hot Jupiters tend to be the only planets orbiting their host stars. Scientists searched for evidence of other planets and exomoons in the Hot Jupiter systems in the Kepler data. This was done by inspecting the times at which the Hot Jupiter transits its host star [34]. Transit times should be strictly periodic for a system consisting only of the host star and the Hot Jupiter. However, a massive exomoon orbiting the Hot Jupiter or an additional planet orbiting the host star would gravitationally tug on the Hot Jupiter and cause the transits to occur periodically. There are several other possibilities why the transit times might appear to vary, it was found that the observed transit timing variations of most of the Hot Jupiters can be explained by the timing of the Kepler observations and the variability of their host stars. The most recent list of Kepler planet candidates contains more than 160 Hot Jupiters with at least 12 transits and reliable orbital periods. Szabó et al. [34] obtained the transit times for each of the Hot Jupiters and calculated transit time variations by taking the peaks in the frequency distribution data through Fourier transform. Fourier transforms for six Hot Jupiters are shown in Figure 6.11. Peaks found in the Fourier transforms show that the transit timing variations are being periodically perturbed though explanations vary. In figure 6.11, the transit timing variation method considers whether transits occur with strict periodicity, or if there is a variation. When multiple transiting planets are detected, they can often be confirmed with the transit timing variation method. This is useful in planetary systems far from the Sun, where radial velocity methods cannot detect them due to the low signal-to-noise ratio. If a planet has been detected by the transit method, then variations in the timing of the transit provide an extremely sensitive method of detecting additional non-transiting planets in the system with masses comparable to Earth's. It is easier to detect transit-timing

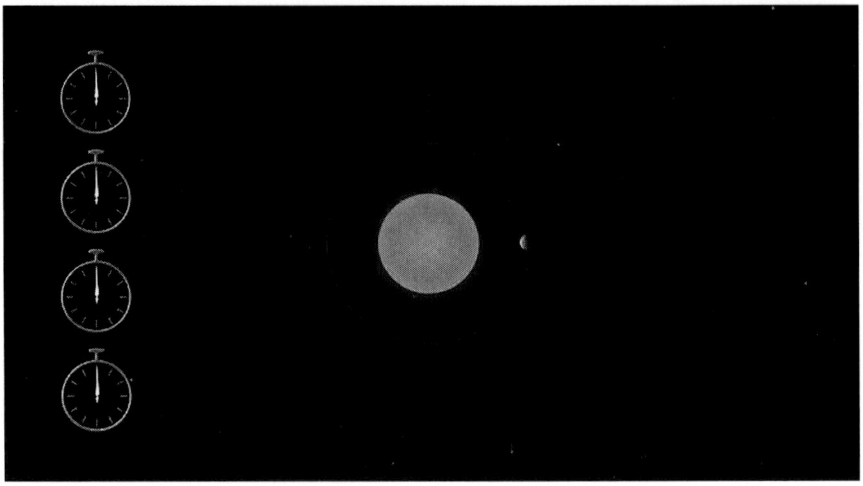

FIGURE 6.11
Animation showing difference between planet transit timing of 1-planet and 2-planet systems. Note: For full visual understanding, refer to this URL. (Courtesy of NASA, http://kepler.nasa.gov/multimedia/animations/scienceconcepts/?ImageID=98 [retrieved September 21, 2016].)

variations if planets have relatively close orbits, and when at least one of the planets is more massive, causing the orbital period of a less massive planet to be more perturbed. The main drawback of the transit timing method is that usually not much can be learned about the planet itself. Transit timing variation can help to determine the maximum mass of a planet. In most cases, it can confirm if an object has a planetary mass, but it does not put narrow constraints on its mass. There are exceptions though, as planets in the Kepler-36 and Kepler-88 systems orbit close enough to accurately determine their masses.

These transit timing variations can also be induced due to the particular sampling rate. Some transit timing variations can also be attributed to the magnetic activity of the host star. Solar planetary Jovian planets can only form in the cold outer regions of the star system and have nearly circular orbits. Hot Jupiters, on the other hand, are massive Jovian planets that are close-in and have highly elliptical orbits. They are thought to form at a great distance from the star beyond the frost line, where the planet can form from rock, ice, and gases. The planets then migrate inward to the star, where they eventually form a stable orbit [35]. One possible explanation is planetary migration: Hot Jupiters are formed in the outer regions of their solar system and then migrate inward. In a typical system, a gas giant orbiting 0.02 AU around its parent star loses 5%–7% of its mass during its lifetime, but orbiting closer than 0.015 AU can mean evaporation of the whole planet except for its core [36].

6.6.5 Structure and Evolution

A major question arose along with the amazing discovery of 51 Pegasi: Is it possible for a gas-giant planet to survive that close to its parent star without losing its atmosphere? Retention of a given atomic or molecular species in a planet's atmosphere is that the RMS (root mean square) velocity of the species at the planet's effective temperature should not be more than about one-sixth of the planet's surface escape velocity. Balancing the flux received from the star against the black-body flux re-radiated from the planet, allowing for the Bond albedo A, representing the fraction of incoming flux reflected directly back into space, we obtain the planet's equilibrium temperature by using

$$T_{eq} = T_s \left(\frac{R_s}{2a} \right)^{\frac{1}{2}} (1-A)^{\frac{1}{4}}.$$

In the equation, the star's effective temperature is T_s, radius is R_s, and the planet's orbital radius is a. If the planet consists mainly of molecular hydrogen, the most likely escape route is as atomic hydrogen, following photo dissociation in the upper atmosphere, so we require that

$$\sqrt{\frac{3kT_{eq}}{m_H}} < \frac{1}{6}\sqrt{\frac{2GM_P}{R_p}}.$$

Assuming the planet's Bond albedo is zero, we can see the escape velocity is roughly seven times the RMS velocity of a hydrogen atom for 51 Pegasib with a Jupiter-like radius. It shows those massive Hot Jupiter s can retain their atmosphere despite the intense irradiation from their star. The radii of massive planets depend strongly on their masses and compositions. The equation of state in the deep interior of a cold Jupiter-mass body is governed by balancing the Coulomb and gravitational forces against electron degeneracy pressure. For planets less than our Jupiter-mass, the Coulomb forces that bind outer electrons to their nuclei balance the degeneracy pressure, so the core densities of such planets are independent of mass, giving $R_p \propto M_p^{(1/3)}$. At Jupiter's mass and above, gravitational forces are sufficient to overcome the degeneracy pressure, leading to the standard nonrelativistic degenerate equation of state, which yields $R_p \propto M_p^{-(1/3)}$. A simple analytic mass-radius relation for cold nonrelativistic bodies can be derived as

$$\left(\frac{4}{3}\pi\rho_0 \right)^{\frac{1}{3}} R_p = \frac{M_P^{\frac{1}{3}}}{1+\left(\frac{M_P}{M_1} \right)^{\frac{2}{3}}},$$

where $M_1 \simeq 3.3$ MJ is the mass of the planet with maximum radius [37]. This model predicts for planet composed of 75% hydrogen and 25% helium with core density $\rho_0 = 419$ kg/m^3, a good fit to the radii of Jupiter. For rocky bodies made primarily of iron and silicates, the Earth's mean density $\rho_0 = 5500$ kg/m^3 gives a good match to the Earth's radius. Applying this to a body of 51 Pegasi b's minimum mass, we find that a cold hydrogen-helium body should $R_p = 0.87$ RJ, whereas an iron-silicate composition yields $R_p = 0.377$ RJ. This approximate treatment is relevant only to the end-state reached by a planet cooling in isolation. More sophisticated structural and evolutionary models have been constructed by several authors. Guillot et al. [38] found that a gas-giant protoplanet remains fully convective through an initial phase of rapid contraction, followed by a longer intermediate period of cooling. Once the internal luminosity due to contraction drops below the power absorbed from the parent star, an outer radiative zone develops, which allows the planet's luminosity and radius to decrease below the values predicted by fully convective models. The discovery of transits in a Hot Jupiter orbiting the 4.5 Gyr old F8V star HD 209458 provided the first direct test of these structural models and revealed the planet to have a mass $M_p = 0.69 \pm 0.05$ MJ and radius $R_p = 1.35 \pm 0.06$ RJ [39,40]. This large radius confirmed the gas-giant nature of HD 209458b [41], leading Burrows et al. [42] to propose that irradiation from the primary was responsible for inhibiting the planet's contraction. This almost certainly rules out any substantial rocky core in HD 209458b and suggests that the additional internal

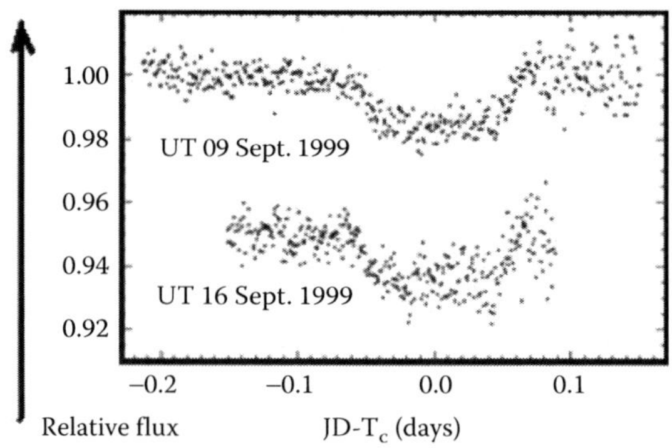

FIGURE 6.12
Image of a planet transiting a star, for illustration of transit method of discovery of extrasolar planets. Sizes of the star, the planet and its orbit are roughly like HD 209458b. (Courtesy of Wikipedia, HD 209458 b, https://en.wikipedia.org/wiki/HD_209458_b [last modified on September 29, 2016].)

heat source could be provided by the dissipation of tidal energy within a young planet during spin synchronization and orbit circularization [43]. This contribution becomes insignificant for older planets if the spin is locked and the orbit is circularized, unless other planets further out can continue to perturb the orbit. Recent studies show that 1% of the intense incident radiation is converted to kinetic energy in the planet's atmosphere and subsequently transported to the deep interior [44] to slow down the planet's contraction. Figure 6.12 [45] shows the light curve of the transit of HD. This light curve led to precise determinations of the planet's mass and radius and of the parent star's radius and limb-darkening profile. The steep ingress and egress rule out the presence of rings around the HD 209458b and the repeatability of the eclipse timing rules out the presence of exomoons with masses greater than three Earth masses [42].

6.6.6 Atmospheres and Albedos

Here, we will focus on the formation of Hot Jupiters with their strange attracting behaviors as well as examining critically their atmospheres and other mysterious characteristics. The core accretion theory [46] for giant planet formation predicts enrichment super-solar abundances like C (carbon), N (nitrogen), P (phosphorus), S (sulphur), Xe (xenon), and Ar (argon). Although, the atmospheres of gas-giant planets are predominantly composed of molecular hydrogen and atomic helium, the abundance of oxygen, which is expected to be the most dominant constituent of planetesimals, is unknown. The absorption spectra are determined by the atomic and molecular trace species that strongly absorb (*the absorption of incident radiation is governed by Beer's, law which can be readily derived by considering an amount of absorbing material diluted in a nonabsorbent solvent*) in the gas phase or else form cloud decks of condensates that scatter and absorb stellar and planetary radiation. The emission spectrum was measured with broadband filters by the Spitzer Space Telescope using photometry and at higher spectral resolution in various ranges of wavelength [47–50]. The variation in brightness temperature at different wavelengths is due to either altitude effect or day-night energy redistribution. Figure 6.13 shows dayside emission spectrum of HD 189733b.

Common molecules and metals that are expected to be present in the gas phase include NH_3, H_2O, N_2, CO, CH_4, silicates such as $MgSiO_3$, and alkali metals. The abundances

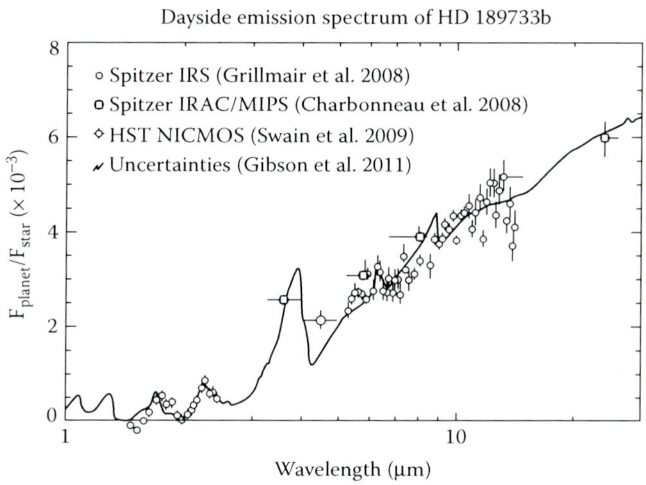

FIGURE 6.13
Dayside emission spectrum of HD 189733b. (Data from N. Gibson et al., *Monthly Notices of the Royal Astronomical Society* 411, 2199–2213, 2011; A. B. Bhattacharya et al., *International Journal of Application or Innovation in Engineering and Management* 2, 162–171, 2013; D. Charbonneau et al., *The Astrophysical Journal* 686, 1341, 2008 A. Bhattacharya et al., *International Journal of Innovative Research in Science, Engineering and Technology* 3, 54–58, 2014.)

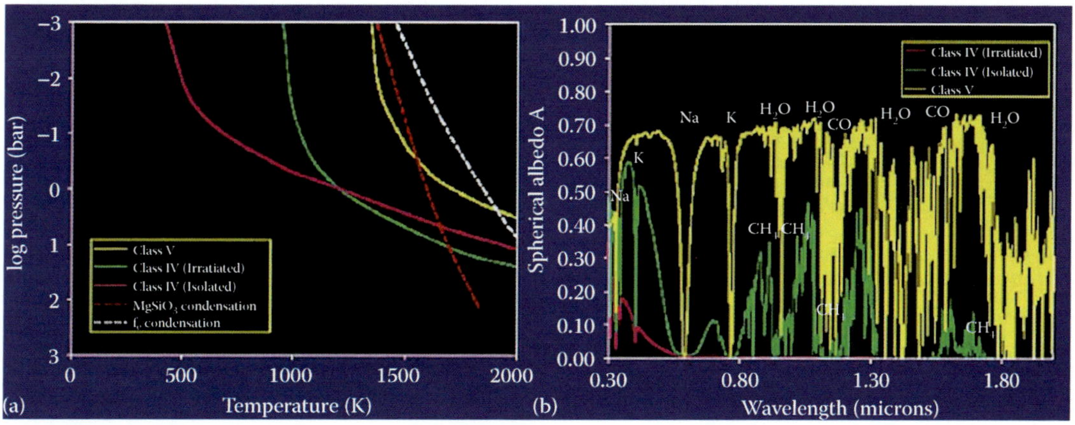

FIGURE 6.14
(a) Temperature–pressure profiles and (b) albedo spectra for model atmospheres for close-orbiting giant exoplanets. (Courtesy of NASA, https://exep.jpl.nasa.gov/presentations/giantsSession/AAS224Burrows [retrieved August 12, 2016].)

of these species at different atmospheric altitude depend on the local chemical equilibrium and the relative abundances of their atomic constituents. At different levels in the atmosphere, individual elements may be consumed by condensation and refilled by convective upwelling of enriched material from lower atmospheric layers. Figure 6.14a shows the condensation curves for the most common cloud-forming species that cross the temperature–pressure profiles of different planetary models at different heights in the atmosphere and Figure 6.14b shows the albedo spectra for model atmospheres. Hot Jupiters possess an upper cloud deck of silicate dust at pressures around 0.3 bar and a deeper cloud deck of iron at around 2 bar. As the distance from parent star increases, the deeper the clouds of a given condensate are expected to form [51].

The effect on the albedo spectrum of varying cloud formation depth is dramatic, with the sodium and potassium resonance lines absorbing most of the optical spectrum in the less-irritated models, as shown in Figure 6.14b. Most current models assume that above each cloud deck, the atmosphere is depleted of species that are *mopped up* in the cloud formation process. For example, it is expected that all the silicon is removed from the atmosphere above an $MgSiO_3$ cloud deck. On a Hot Jupiter, in the clear atmosphere above the silicate cloud deck, the dominant surviving gas-phase species are expected to include CO, CH_4, NH_3, N_2, and H_2O. At the high temperatures, CO is expected to dominate over CH_4 and N_2 over NH_3, making CO and H_2O the most likely absorbers in the infrared. The alkali metals, particularly sodium and potassium, are not subject to rainout at these high temperatures, and so are expected to remain in the gas phase at more or less solar abundance. A cloudless atmosphere is expected only if temperature is optimal for condensation of metallic species like Mg, Ca, and Al. Above 1250 K, these elements are expected to condense high in the atmosphere, keeping the Hot Jupiter atmosphere mostly cloudless. If there is a dust layer, the composition and size distribution of the particles determine the planet's optical albedo and phase function [52,53]. The particle size distribution is determined by the relative timescales for nucleation and grain growth, the rate at which particles of a given size precipitate out to lower atmospheric levels, and the rate at which convective upwelling replenishes the supply of the elements concerned [54] and given that the clouds themselves may alter the thermal structure of the atmosphere. The composition and plausibility of hazes over the large temperature range of Hot Jupiters needs to be investigated theoretically. In the beginning of the space age and exploration, astronomers were learning and discovering new information about the planets outside our solar system. It was in 1995 when the first extrasolar planet was discovered orbiting a Sun-like star. The planets were given the name Hot Jupiters [55]. The word Jupiter was used because they are giant planets, even more massive than Jupiter. The term *hot* was used because they orbit close to their stars, even closer than Mercury. These planets orbit their Sun every 2 or 3 days and have high surface temperatures [56] between approximately 0.015 and 0.5 AU to their parent stars [57,58], whereas our Jupiter orbits its parent star at 5.2 AU, thus causing low surface temperatures.

6.6.7 The Best-Known Hot Jupiter

One of the most favorable targets for measuring the structure and composition of an exoplanets atmosphere is the transiting Hot Jupiter, HD 189733b. Some of the reasons are as follows: when it passes behind the star during a secondary eclipse, the brightness of the system drops by about 0.5% in the near-infrared region [59]. Furthermore, with a K-band magnitude of 5.5, the brightness of the host star provides the large number of photons that are required to perform these high-precision measurements. HD 189733b is one of the more favorable targets for planetary atmosphere studies, the other being HD 209458b, which was the first transiting planet discovered [59]. The planets, HD 189733 and HD 209458 are easier to measure due to their atmospheric features that facilitate spectroscopy transmission. The dotted lines show curves of constant signal-to-noise ratio, S/N. The planets to the right of the top curve have an atmosphere detectable with 90-min integration (~1 transit) with the Hubble Space Telescope [60]. The planets to the right of the remaining two lines would have detectable atmospheres if the amount of incident photons were increased by a factor 5 and 10, respectively. Figure 6.15 shows the photometric signal (Stellar V magnitude) versus atmospheric signal [60].

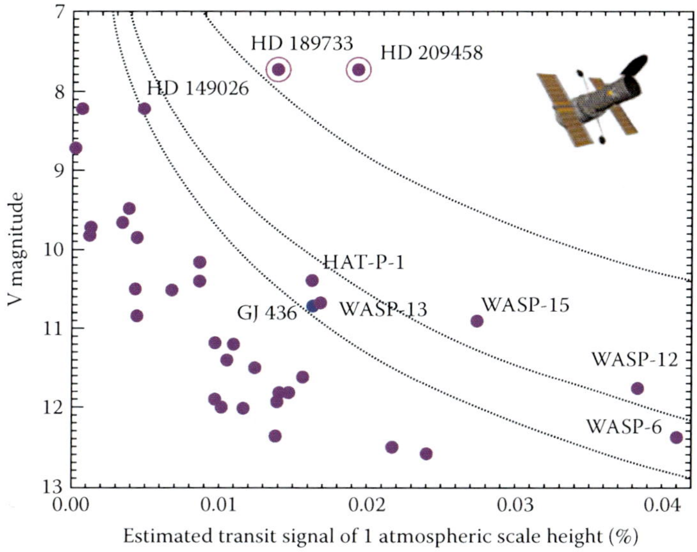

FIGURE 6.15
Photometric signal (Stellar V magnitude) versus atmospheric signal. (Data from F. Pont, The Hot Jupiter HD 189733b, ExoClimes.com, 2011. http://www.exoclimes.com/topics/the-hot-jupiter-hd-189733-b/ [retrieved August 12, 2016].)

6.6.7.1 Hot Jupiter Tidal Forces

The atmospheres of Hot Jupiters are very unusual. They have very high temperatures that measure in the thousand to several thousand Kelvin. At these temperatures, the planets could have clouds of molten rock. They are actually more similar to the compositions of relatively cool stars [61]. It is believed that they are tidally locked, which means that one side of the planet is getting all of the heat and the other side is cold. Tidal forces occur when the gravitational force on one side of a large object is stronger than that on the other side, causing the object to stretch. As the Hot Jupiters orbit around their stars, the strength and direction of tidal forces change due to the elliptical orbit. This variation causes the planet to be flexed in different directions, generating much friction inside it. This friction tends to warm up the planet. Tidal forces have also locked Hot Jupiters and their stars into synchronous rotation. A synchronous rotation is the one in which the rotation period and orbital period of planets are equal; therefore, the same side of the planet is always facing the sun. This means that Hot Jupiters are rotating rapidly because their rotational period is only a few days long [11]. Figure 6.16 shows the tidal effect of Earth and Moon.

Astronomers can detect which gases are present in the atmospheres of Hot Jupiters by analyzing the spectrum of starlight filtered through the atmosphere of the planet when it passes in front of the star. These Hot Jupiters are expected to have a very different composition from planets in our own solar system like Jupiter, where temperatures at the cloud tops are around 150°C (302°F). The first hot planet observation shows an unexpected absence of titanium oxide. With this said, current 3D models of atmospheres propose that small part of this heavy molecule may be circulated by fast winds allowing the gaseous titanium oxide to reach the upper atmosphere [62]. If the gas is not detected, then it may suggest that either the winds are not strong enough or the molecule is forming much larger parts that are too heavy be lifted. Thermal gradients in Hot Jupiters drive winds at the rate

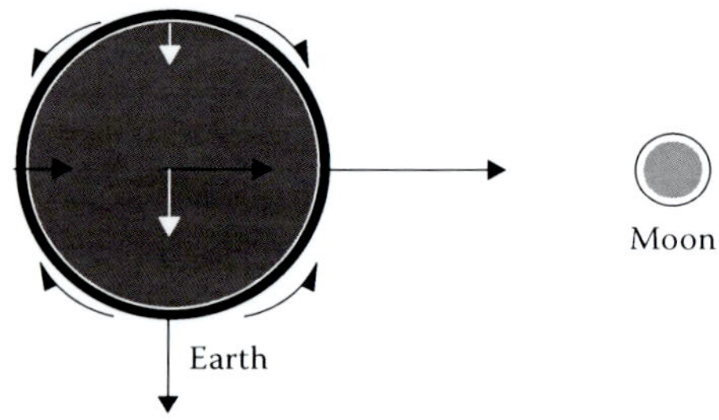

FIGURE 6.16
The action of the Earth–moon tidal force.

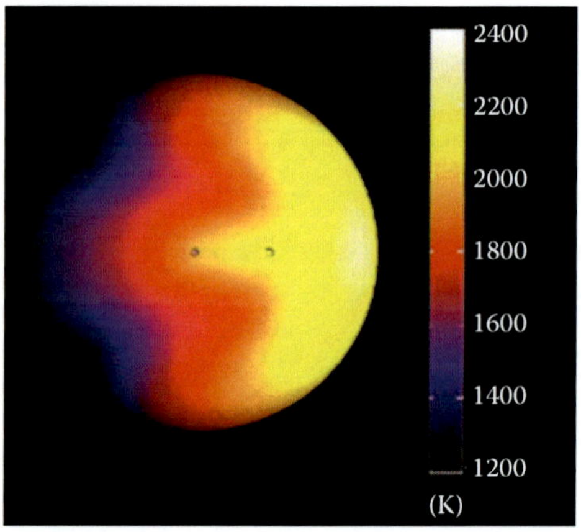

FIGURE 6.17
The temperature variation of Hot Jupiter. (Courtesy of Nikole Lewis/MIT.)

of 6000 mph, which, in turn, carries suffocating heat around the globe. The blue color of the planet may be caused by silicate particles in the atmosphere of the planet, which scatter blue wavelengths of light from the parent star. Figure 6.17 shows the temperature variation of Hot Jupiter. It appears from the figure that the star-facing side features a daytime temperature of 2127°C (35,829°F), whereas the night-facing side experiences a temperature of 927°C. So, even at night, this planet is 10 times hotter than Jupiter.

6.6.7.2 Detection of Water Vapor in Hot Jupiter

Researchers have used a new technique to analyze and detect the gaseous atmospheres of Jupiter-type exoplanets. Using this new technique, the first detection of atmospheric water of a planet orbiting the nearby star of Tau Boötis was done [63]. Figure 6.18 shows simulated data for the method used for detecting water vapor detected around the Hot Jupiter around Tau Boötis b.

Figure 6.18 clearly shows that planetary signal has been increased in strength by several orders of magnitude relative to the actual signal. The dotted lines show the blue and red shifts of the planetary and stellar lines in the data, respectively. The method utilized the radial velocity (RV) technique. This technique is commonly used in the visible region of the spectrum for discovering nontransiting exoplanets. Using the Doppler effect, RV detection traditionally determines the motion of a star due to the gravitational pull of a companion planet. The star moves opposite that of the orbital motion of the planet and the stellar features

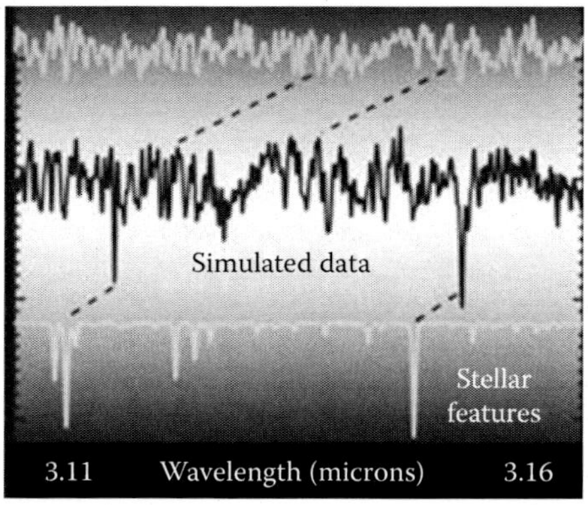

FIGURE 6.18
Simulated data for the method used for detecting water vapor features detected around the Hot Jupiter Tau Boötis b. (Courtesy of Alexandra Lockwood/Caltech.)

shift in wavelength. A large planet or a planet closer to its host star provides a larger shift. This RV technique was expanded into the infrared to find the orbit of Tau Boötis b around its star and added further analysis of the light shifts via spectroscopy. Since every compound emits a different wavelength of light, this unique light signature allows the researchers to analyze molecules that make up the planet's atmosphere. Using data of Tau Boötis b from the Near Infrared Echelle Spectrograph (NIRSPEC) for comparison of the molecular signature of water to the light spectrum emitted by the planet, we confirmed that the atmosphere contains water vapor [64]. However, this current state of the technique cannot detect Earth-like planets around stars like the Sun. Future telescopes such as the James Webb Space Telescope and the Thirty Meter Telescope will enable us to examine much cooler planets that are more distant from their host stars and where liquid water is more likely to exist [64].

6.6.8 The Strange Attraction

Sizzling gas giants circling close to their host stars keep them young and active. Hot young stars are wildly active and emit giant, violent flares as well as huge eruptions of charged particles from their sizzling surfaces. This bright emission is assumed to arise from intense magnetic fields driven by their rapid rotation. As stars age, they naturally become less active, their X-ray emission weakens, and their rotation becomes slow [33]. Astronomers have theorized that a Hot Jupiter circling extremely close to the host star's surface might be able to sustain this activity. If the planet is orbiting faster than the star rotates, it will transfer angular momentum to the star. This will inhibit a star's spin down, causing it to continue rotating rapidly and hence appear young even while it ages. Binary systems have been used for examining only one star known as an exoplanet. Both stars in a binary system are expected to be of the same age. So, if the star with the Hot Jupiter emits more X-rays than its twin, there is no doubt that its added youth is triggered by the planet's involvement. Figure 6.19 shows an artist's conception of a Hot Jupiter circling an active star.

6.6.9 Hot Jupiter Exoplanets Are Not Eaten by Their Stars

It has been found that huge, burning-hot alien planets when come close to their parent stars are not consumed by them. The inward migration of the Hot Jupiters tends to halt before they spiral down to their end. As a matter of fact, all Hot Jupiters get closer and

FIGURE 6.19
An artist's conception of a Hot Jupiter circling an active star. (Courtesy of NASA, Knocking on heaven's door, http://www.nasa.gov/missions/science/planetsurvey.html [retrieved September 10, 2016].)

closer to their stars, but this process stops before the stars get too close [65]. The planets mostly stabilize once their orbits become circular, whipping around their stars every few days. Figure 6.20 exhibits the inward migrations of Hot Jupiter exoplanets obstruction when their orbits are circularized by their parent stars' gravity.

6.6.9.1 The Blow-Out Effect

All Hot Jupiter atmospheres are blown out on account of the intense solar winds. NASA's Chandra and the ESA's XMM Newton chronicled HD 189733b as it transited its star. The two probes detected a drop in X-rays three times deeper than the corresponding decrease in optical light. This would imply that the subject had an incredible sized atmosphere

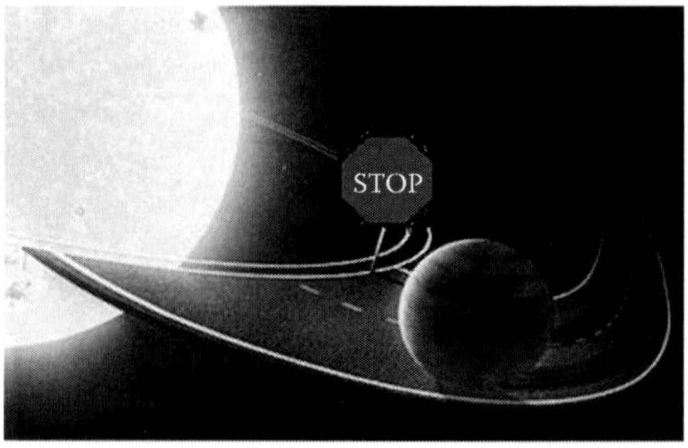

FIGURE 6.20
The inward migrations of *Hot Jupiter* exoplanets stop when their orbits are circularized by their parent stars' gravity.

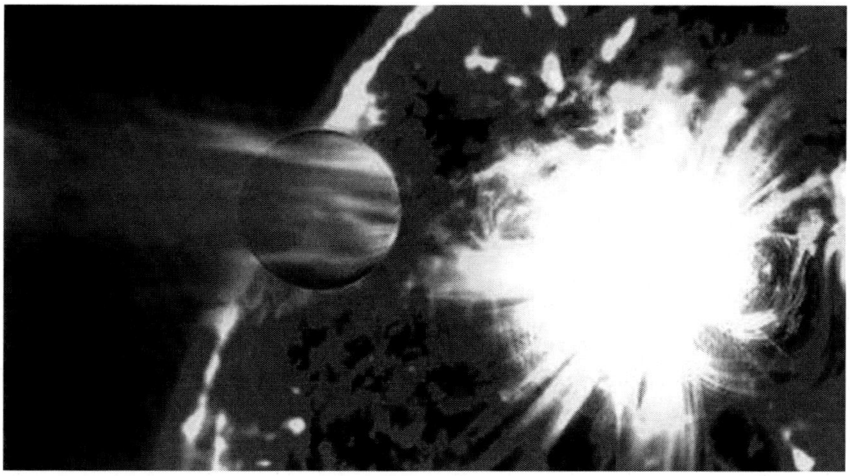

FIGURE 6.21
Artistic impression of blow-out effect of Hot Jupiter. (Courtesy of NASA, Hot Jupiters are the most astounding planets in the galaxy, http://io9.gizmodo.com/hot-jupiters-are-the-most-astounding-planets-in-the-gal-1177931435 [retrieved August 20, 2016].)

Figure 6.21 reveals an artistic impression of a blow-out effect on a Hot Jupiter. It is thought that a Hot Jupiter will get smashed by an intense stellar flare [66].

Overall, the prospects for direct observational study of the physical properties of a Hot Jupiter appear extremely promising. The process of formation and evolution of these planets will be invaluable in the search for systems with properties more similar to those of our own solar system [67]. Furthermore, many of the techniques pioneered in studying the Hot Jupiters will surely come in useful when space-based interferometers, coronagraphs, and large ground-based telescopes (30- to 100-m aperture) bring the characterization of terrestrial exoplanets within our grasp. At the same time, new optical and infrared spectral separation techniques will uncover the role played by the chemistry and physics of irradiated gas-giant atmospheres in regulating their cooling and contraction.

6.7 Giant Planets Near Snow Lines

Architecture is largely involved with giant planets having greater orbital periods, typically about 300 days to about 4 years. These systems contain some planets significantly larger than the Earth but smaller than Neptune. They are tightly packed, all orbiting closer to their star than the Earth is to our Sun. Their orbits are relatively circular. Figure 6.22 shows the presence of a giant planet near the snow lines.

Some of these planets have also formed farther out in the disk and migrated to their current location. It may not be a coincidence that many of these planets locate near the water-snow line, which is essentially the location in the protoplanetary disk and where the solid surface density increases due to condensation of water ice. This is due to the fact that the snow line affects the formation of planetesimals, migration of giant planets.

FIGURE 6.22
Giant planet near the snow lines. (Courtesy of ESO/M. Kornmesser/Nick Risinger.)

6.7.1 Long-Period Giant Planets

Astronomers have also discovered giant planets with orbital distances ranging from a couple of to several astronomical units. These systems contain nearly two to four giant planets orbiting their star with a distance comparable to that of Venus to our Sun, usually with significantly elongated orbits. These wide, eccentric orbits show that some of these planets have ejected others from the host planetary system. Investigations have shown that this *planet-scattering* effect can produce a broad range of eccentric orbits. Another possibility is that the planets in wide orbits formed around a star from a different solar system eventually make their way to a new host star.

6.8 Proportion of Stars with Planets

An extrasolar planet is a planet that does not orbit Earth's Sun but a different star, stellar remnant, or brown dwarf. Besides, free-floating planets, which are not orbiting any star, also are to be considered separately, especially if they are free-floating gas giants. Over 1800 extrasolar planets have been discovered, including 467 multiple planetary systems as of September 29, 2014 [15]. The overall proportion of stars with planets is uncertain, as not all planets can be detected. The radial velocity method and the transit method are most sensitive to large planets in small orbits. Many known exoplanets are Hot Jupiter planets of Jovian mass or larger that move in very small orbits with periods of approximately a few days. An elaborate survey of planets detected by radial velocity in 2005 found that about 1.2% of Sun-like stars have a Hot Jupiter, where *Sun-like star* refers to any main-sequence star of spectral classes late-F, G, or early-K without a close stellar companion [68]. This value of 1.2% is more than double the frequency of Hot Jupiters as detected by the Kepler spacecraft. With that said, the Kepler field of view covers a different region of the Milky Way where the metallicity of stars is different [69]. It is also estimated that 3%–4.5% of

Sun-like stars possess a giant planet with an orbital period of 100 days or less where *giant planet* means a planet of at least 30 Earth masses [70]. It is seen that small planets of Earth-like mass or slightly larger are more common than giant planets [71]. There are more planets in large orbits than in small ones. On the basis of this finding, it has been estimated that about 20% of Sun-like stars have at least one giant planet, whereas at least 40% may have planets of lower mass [71–73]. Gravitational microlensing data observed between 2002 and 2007 suggested the proportion of stars with planets is much higher and estimates an average of 1.6 planets orbiting between 0.5 and 10 AU per star in the Milky Way. It is suggested by many astronomers that "stars are orbited by planets as a rule, rather than the exception" [74]. It was announced in November 2013 that 22 ± 8% of Sun-like stars have an Earth-size planet in the habitable zone [75,76]. Whatever the proportion of stars with planets, the total number of exoplanets must be sufficiently large. Since the Milky Way has at least 200 billion stars, it must also contain tens or hundreds of billions of planets.

6.9 Spectral Classification

Most of the known exoplanets orbit the corresponding star, which is very similar to the Sun: main-sequence stars of spectral categories F, G, or K. Planet search programs have concentrated on such stars. In addition, statistical analysis shows that lower-mass stars, such as red dwarfs of spectral category M, are less likely to have planets massive enough to be detected [72,77]. Stars of spectral category A rotate very quickly, in general, which makes it very difficult to measure the small Doppler shifts induced by orbiting planets because the spectral lines are very broad. This type of massive star evolves into a cooler red giant, which rotates more slowly and hence can be measured using the radial velocity method. Till early 2011, nearly 30 Jupiter-class planets had been found around spectral K giants, including Pollux, Gamma Cephei, and Iota Draconis. Doppler surveys conducted on a wide variety of stars indicate about one in six stars having twice the mass of the Sun are orbited by one or more Jupiter-sized planets. Alternatively, one in 16 for Sun-like stars and only one in 50 for red dwarfs are noted. On the contrary, microlensing surveys suggest that long-period Neptune-mass planets are found around one in three red dwarfs [78]. Figure 6.23 illustrates spectral classifications.

Experimental findings using the Spitzer Space Telescope reveal that extremely massive stars of spectral category O, which are largely hotter than the Sun, create a photo evaporation effect that inhibits planetary formation [79]. If the O-type star goes supernova, any

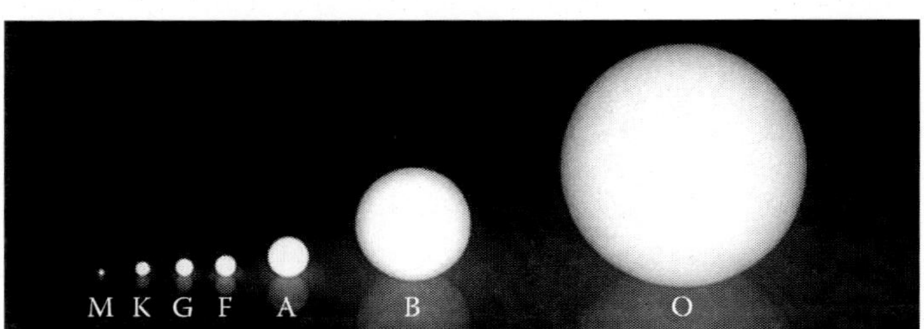

FIGURE 6.23
Spectral classifications using the Spitzer Space Telescope.

planet that had formed would become free floating owing to the loss of stellar mass till the natal kick of the resulting remnant pushes it in the same direction as that of an escaping planet [80].

6.10 Metallicity

Mainly ordinary stars are composed of light elements hydrogen and helium, as well as a small proportion of heavier elements, and this fraction is termed as star's metallicity, even if the elements are not metals in the traditional sense [68]. It is usually denoted by m/H and expressed on a logarithmic scale, where zero is the Sun's metallicity. A 2012 study of the Kepler spacecraft data observed that the smaller planets with radii less than that of Neptune were found around stars with metallicities in the range $-0.6 < [m/H] < +0.5$, whereas the larger planets were noted mainly around stars with metallicity at the higher end of this range. Kepler can only observe planets very close to their star and the detected gas giants migrated from further out [82].

6.11 Multiple Stars

Most of the known planets orbit single stars. However, some are found to orbit one member of a binary star system [83], and few circum-binary planets have been discovered that orbit around both members of binary star. Several planets in triple-star systems are also known [84], and one in the quadruple system Kepler-64. The Kepler results show circum-binary planetary systems are relatively common. But none of the close binary stars surveyed appear to have eclipsing planets. Half of the stars have orbital periods of 2.7 days or less, but none of the binaries with planets have a period less than 7.4 days. Kepler findings also reported that circum-binary planets tend to orbit their stars close to the critical instability radius. Theoretical calculations indicate the minimum stable separation is about two to three times the size of the stars' separation [85].

6.11.1 Open Clusters

An open cluster, also called galactic cluster, is a group of up to a few thousand stars formed from the same giant molecular cloud having nearly same age. Figure 6.23 reveals star cluster NGC 3572 and its surroundings [86]. More than 1100 open clusters have been discovered within the Milky Way Galaxy and many more are assumed to exist [87]. These open clusters are loosely bound to each other with mutual gravitational attraction and become disrupted by close encounters with other clusters and clouds of gas [88]. Figure 6.23 shows the star cluster NGC 3572 and its surroundings [87]. This led to the hypothesis that the open-cluster environment hinders planet formation. However, a 2011 study found that there have been an insufficient number of surveys of clusters for developing such a hypothesis [89]. Recent studies of both giant planets [90] and low-mass planets [91] in open clusters have been found to be consistent. The open cluster NGC 6811 contains two known planetary systems Kepler-66 and Kepler-67.

6.12 Galactic Distribution of Planets

The Milky Way is 105 light years across, but 90% of planets with known distances lie approximately 2000 light years from Earth. The Kepler spacecraft has identified many planet candidates further away than this, but many of them have not been confirmed. One of the methods being used to detect planets further away is microlensing [92]. The WFIRST (Wide-Field Infrared Survey Telescope) [93] could use microlensing for measuring the relative frequency of planets in the galactic bulge versus galactic disk [94]. It has been found that planets are more common in the disk than in the bulge [95]. Determining distances using microlensing events is highly difficult. The first planet considered with high probability of being in the bulge is MOA-2011-BLG-293Lb at a distance of nearly 25,000 light years [96]. A critical survey shows that 90% of planets with known distances lie within about 2000 light years of Earth, as of July 2014. Figure 6.24 illustrates the area of planet investigation.

Population I Stars

Metal-rich stars are those young stars whose metallicity is the highest. The high metallicity makes them more likely to possess planetary systems than older populations because planets form by the accretion of metals. The Sun is an example of a metal-rich star and is considered an intermediate population I star. Population I stars have regular elliptical orbits around the Galactic Center with a low relative velocity [97].

Population II Stars

Metal-poor stars are those with relatively low metallicity, hundreds or thousands times less than that of the Sun. These objects formed at the earlier time of the universe. Intermediate population II stars are common in the bulge near the center of the Milky Way, whereas population II stars found in the

FIGURE 6.24
Our planet hunting neighborhood.

galactic halo are older and more metal-poor. Globular clusters also contain high numbers of population II stars [98]. In 2014, the first planets around a halo star were found around Kapteyn's star, the nearest halo star to Earth, around 13 light years away.

6.12.1 Orbital Parameters

Most extrasolar planet candidates have been discovered using indirect methods and so only a few physical and orbital parameters can be found out. For example, out of the six independent parameters that define an orbit, the radial velocity method can determine four: semi-major axis, eccentricity, longitude of periastron, and time of periastron. The other two parameters remain unknown: inclination and longitude of the ascending node.

6.12.2 Distance from Star, Semi-Major Axis, and Orbital Period

Figure 6.25 reveals how a planet and a star orbit their common center of mass (gray cross).

Figure 6.26 presents a scatter plot showing masses and orbital periods of all extrasolar planets discovered through 2013, with colors indicating the different methods of detection. In the figure, the solar system planets are marked as gray circles. The horizontal axis plots the log of the semi-major axis, whereas the vertical axis plots the log of the mass. The orbit of a planet is not centered on the star but on their common center of mass as shown on the right of the figure. For circular orbits, the semi-major axis is the distance between the planet and the center of mass of the system. For elliptical orbits, the planet–star distance changes over the course of the orbit, in which case, the semi-major axis is the average of the largest and smallest distances between the planet and the center of mass of the system. When the sizes of the star and planet are relatively small compared with the size of the orbit and the orbit is approximately circular and the center of mass is not too far from the star's center, such as in the Earth–Sun system, then the distance from any point on the star to any point on the planet is nearly the same as the semi-major axis. But when the radius of a star expands and it turns into a red giant,

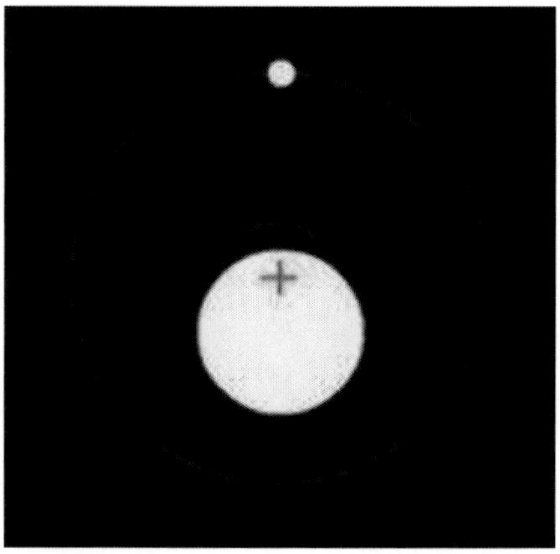

FIGURE 6.25
A planet and a star orbit their common center of mass.

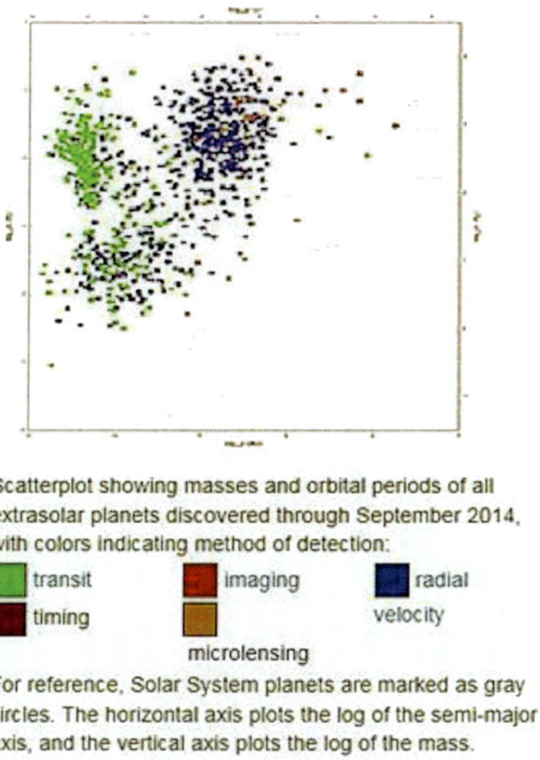

FIGURE 6.26
Scatter plot showing masses and orbital periods of all extrasolar planets discovered up to 2013 (colors indicating method of detection). (Courtesy of Wikipedia, Exoplanet, http://en.wikipedia.org/wiki/Exoplanet [last modified September 07, 2016].)

then the distance between the planet and the star's surface can become close to zero or even less than zero if the planet has been engulfed by the expanding red giant, whereas the center of mass from which the semi-major axis is calculated will still be near the center of the red giant.

There are exoplanets that are much closer to their parent star than any planet in our solar system. In addition, there are also exoplanets that are much further from their star. Mercury, the closest planet to the Sun at 0.4 AU, takes 88 days to orbit, but the shortest known orbits for exoplanets take only a few hours, for example, Kepler-70b. The Kepler-11 system has five of its planets in shorter orbits than that of Mercury. Neptune is 30 AU from the Sun and takes 165 years to orbit, but there are exoplanets that are hundreds of astronomical units from their star and take more than a thousand years to orbit, for example, 1RXS1609b.

6.12.3 Eccentricity

The eccentricity of an orbit is a measurement of its elliptical (elongated) shape. Most of the exoplanets with orbital periods of 20 days or less have near-circular orbits, or very low eccentricity, which is believed to be due to tidal circularization [99].

6.13 Inclination versus Spin–Orbit Angle

Orbital inclination is the angle between an orbital plane of a planet and another plane of reference. For exoplanets, the inclination is usually stated with respect to an observer on Earth and the angle used is that between the planet's normal orbital plane and the line of sight from Earth to the star. For this reason, most of the planets observed by the transit method are close to 90° [100]. As the word *inclination* is utilized in exoplanet studies for this line-of-sight inclination, the angle between the planet's orbit and the star's rotation use a different word and is termed the spin–orbit angle or spin–orbit alignment. In majority of the cases, the orientation of the star's rotational axis is unknown. The Kepler spacecraft has observed a few hundred multiplanet systems. In most of these systems, the planets all orbit in almost the same plane and are similar to the solar system [101]. However, a combination of astrometric and radial velocity measurements reveals that some planetary systems contain planets whose orbital planes are significantly tilted relative to each other [102]. More than 50% of the Hot Jupiters have orbital planes substantially misaligned with their parent star's rotation. A substantial fraction of Hot Jupiters even have retrograde orbits, which means that they orbit in the opposite direction from the star's rotation [103].

6.13.1 Rotation and Axial Tilt

The first measurement of a planet's rotation period was announced in April 2014. On the basis of the assumption that the axial tilt of the planet is small, the length of day for the super-Jupiter gas-giant Beta Pictoris b was estimated to be 8 h [104,105]. With an equatorial rotational velocity of 25 km/s, this spin is faster than that of the gas giants of the solar system. Beta Pictoris b's distance from its star is 9 AU. At such distances, the rotation of Jovian planets cannot be slowed by tidal effects [105]. Beta Pictoris b is still warm and young. It is assumed that over the next hundreds of millions of years, it will cool down and shrink to about the size of Jupiter. Figure 6.27 shows the plot of equatorial spin velocity versus mass for planets comparing Beta Pictoris b to the solar system planets.

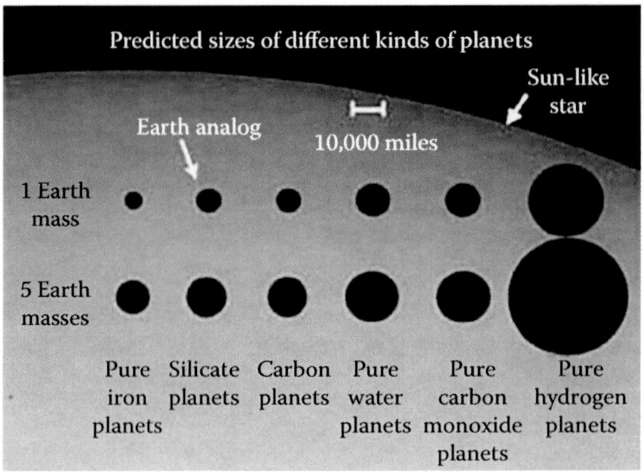

FIGURE 6.27
Comparison of planet compositions. (Courtesy of Marc Kuchner/NASA GSFC, Scientists model a cornucopia of Earth-sized planets, http://imagine.gsfc.nasa.gov/news/27sep07.html [retrieved August 27, 2016].)

6.13.2 Origin of Spin and Tilt of Terrestrial Planets

Giant impacts have a significant effect on the spin of terrestrial planets. On an average, the spin angular velocity will be about 70% of the velocity, which would cause the planets to break up and fly apart. The natural outcome of planetary embryo impacts at speeds slightly more than escape velocity. In later stages, terrestrial planet spin is affected owing to impacts with planetesimals. During the stage of giant impact, the thickness of a protoplanetary disk becomes far larger than the size of planetary embryos, so collisions are equally likely to come from any direction in three dimensions. This causes in the axial tilt of accreted planets ranging from 0 to 180° for any direction with both prograde and retrograde spins equally probable. Hence, prograde spin with small axial tilt is common for the solar system's terrestrial planets except for Venus; however, it is not generally common for terrestrial planets built by giant impacts. The initial axial tilt of a planet sustaining giant impacts can be substantially changed by stellar tides if the planet is near to its star or by satellite tides if the planet has a large satellite [106].

6.14 Tidal Effects

For most planets, the rotation period and axial tilt are unknown; however, a large number of planets have been detected with very short orbits and will have reached an equilibrium rotation that can be predicted. Tidal effects are due to the forces acting on a body differing from one part to another [105]. For example, the gravitational effect of a star changes with distance from one side of a planet to another. Heat from a star produces a temperature gradient between the day and night sides, which is another source of tides. For example, on Earth the air pressure variations on the ground are affected more by temperature differences than by gravitational ones. Tides modify the rotation and orbit of planets until an equilibrium is attained. When the rotation rate is slowed, there is an increase of the orbit semi-major axis owing to the conservation of angular momentum. Most of the large moons in the solar system, including the Moon, are tidally locked to their host planet. The same side of the moon is always facing the planet means the moons' rotation periods are synchronous with their orbital period. However, when the orbit is eccentric, as in many exoplanets' orbits of their host stars, another condition, referred to as equilibrium, results in spin–orbit resonances that are far more likely than synchronous rotation. A spin–orbit resonance occurs when the rotation period and the orbital period are in an integer ratio. This is known as commensurability. Nonresonant equilibriums, for example, the retrograde rotation of Venus, can also occur when both gravitational and thermal atmospheric tides are both significant.

6.14.1 Mass

When a planet is found by the radial velocity method, its orbital inclination becomes unknown and can range from 0 to 90°. This method is unable to find the true mass (M) of the planet but provides a lower limit for its mass. In some cases, an apparent exoplanet may be a more massive object such as a brown dwarf or red dwarf. If a planet's orbit is nearly perpendicular to the line of vision, then the planet can be detected through the transit method. Further, astrometric observations and dynamical considerations in multiple-planet systems can provide an upper limit to the planet's true mass. The mass of a transiting exoplanet can also be obtained from the transmission spectrum of its atmosphere, as

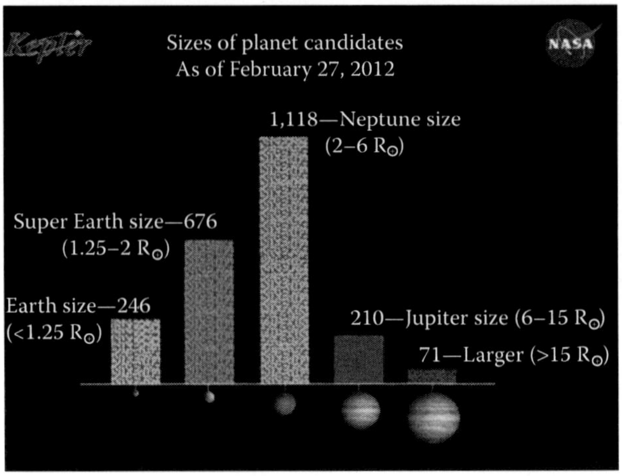

FIGURE 6.28
Comparison of sizes of planets with different compositions. (Courtesy of NASA, NASA Kepler results usher in a new era of astronomy, http://www.nasa.gov/content/nasa-kepler-results-usher-in-a-new-era-of-astronomy [retrieved September 10, 2016].)

it can be utilized to constrain independently the atmospheric composition, temperature, pressure, and scale height [107].

6.14.2 Radius, Density, and Bulk Composition

Prior to recent results from the Kepler spacecraft, most of the confirmed planets were gas giants comparable in size to Jupiter or greater, as those may be most easily detected. The planets detected by Kepler are mainly between the size of Neptune and Earth [100]. If a planet is detectable by both radial velocity and transit methods, then both its true mass and radius can be determined. One can then obtain the planet's density. Planets with low density are composed mainly of hydrogen and helium, whereas planets of intermediate density have water as a major constituent. A planet of high density is inferred to be rocky, like Earth and the other terrestrial planets of the solar system. Gaseous planets are usually hot, as they are close to their star or their formation is expanded by the heat. For colder gas planets, there is a maximum radius slightly larger than that of Jupiter that occurs when the mass attains a few Jupiter masses. Adding mass beyond this point causes the radius to shrink [108–110]. Comparison of sizes of planets with different compositions is made in Figure 6.28.

6.15 Catalog Names of Exoplanets

Most exoplanets have catalog names. In 2014, the IAU (International Astronomical Union) launched a process for assigning proper names to exoplanets [111,112]. This task involved public nomination and voting for the new names. The IAU plans to declare the new names in August 2015 [99]. The decision to give the planets new names followed the private company Uwingu's exoplanet naming contest, which the IAU criticized seriously [113]. Previously, a few planets had received unofficial names: Notably, Osiris (HD 209458b), Bellerophon (51 Pegasi b), and Methuselah (PSR B1620-26b). The convention for naming exoplanet is an extension of that used by the Washington Multiplicity Catalog (WMC) for multiple-star systems and then later adopted by the IAU [84]. The brightest member of the star system receives the letter "A." Distinct components not contained within "A" are labeled "B," "C," and so on. Subcomponents are designated by one or more suffixes with the primary label, starting with lowercase letters for the second hierarchical level and then numbers for the third [114]. If there is a triple-star system in which two stars orbit each other closely with a third star in a more distant orbit, the two closely orbiting stars would be named Aa and Ab, whereas the distant star would be named B. For historical reasons, this standard is not always maintained, for example, Alpha Centauri A, B, and C are *not* labeled Alpha Centauri Aa, Ab, and B.

6.16 Extrasolar Planet Standard

Following an extension of the above standard, an exoplanet's name is given by taking the name of its parent star and adding a lowercase letter. The first planet discovered in a system is given the designation "b" and later planets are given subsequent letters. If some planets in the same system are discovered at the same time, the closest one to the star gets the next letter, followed by the other planets in order of orbital size. For example, in the 55Cancri system, the first planet, 55Cancri b, was discovered in 1996; two additional planets were simultaneously discovered in 2002 with the nearest to the star being named 55Cancri c and the other 55Cancri d; a fourth planet was claimed in 2004 and named 55Cancri e despite lying closer to the star than 55Cancri b, and the most recently discovered planet, in 2007, was named 55Cancri f despite lying between 55Cancri c and 55Cancri d [115]. In April 2012, the highest letter in use is "j," for the unconfirmed planet HD 10180j, and with "h" being the highest letter for a confirmed planet, belonging to the same host star) [15].

6.17 Confirmed Discoveries

The first published discovery to receive subsequent confirmation was made in 1988 by Canadian astronomers [116]. Although they were cautious about claiming a planetary detection, their radial velocity observations showed that a planet orbits the star Gamma Cephei. Partly due to the fact that observations were at the very limits of instrumental capabilities at the time, astronomers remained skeptical for several years about this and other similar observations. It was assumed some of the apparent planets might instead have been brown dwarfs, objects intermediate in mass between planets and stars. In 1990, additional observations were published that supported the existence of the planet orbiting Gamma Cephei [117], but subsequent work in 1992 again raised serious doubts [118]. Finally, in 2003, improved techniques permitted the planet's existence to be confirmed [119].

6.18 Unconfirmed Extrasolar Planets

October 17, 2012, brought news of an unverified planet, Alpha Centauri Bb, which is orbiting Alpha Centauri B. This is one of the three stars in a triple-star system nearest to Earth's Sun [120]. Alpha Centauri Bb is an Earth-size planet but not in the habitable zone within which liquid water can exist [121]. As of March 2014, NASA's Kepler mission had identified more than 2900 planetary candidates, several of them being nearly Earth-sized and located in the habitable zone, some around Sun-like stars [122–124].

6.19 Red Dwarf and Brown Dwarf Systems

Red dwarf systems orbit red dwarf stars, which may represent 70% to greater than 90% of the stars in the galaxy. Gliese 876 was the first such system discovered in 1998. Most planets of these systems are assumed to be tidally locked. This and high stellar variation are postulated to be major impediments to planetary habitability. Gliese 581 is another red dwarf system that is speculated to have multiple planets within the habitable zone. Figure 6.29 illustrates a young red dwarf surrounded by three planets.

2M1207, the first planetary system to be imaged, was also the first such system discovered. Some companions of brown dwarfs, such as those of 2M1207 and GQ Lupi, are assumed to have been formed by cloud collapse rather than accretion, and hence these are sometimes considered sub-brown dwarfs rather than planets. Disks around brown dwarfs have been observed to have many of the same features as disks around stars. It is expected that there will be accretion-formed planets around brown dwarfs [125]. Given the small mass of brown dwarf disks, most planets will be terrestrial planets rather than gas giants [126]. It is thought that many brown dwarfs may have planets and a short-lived habitable zone. Most planets of these systems are thought to be tidally locked. Figure 6.30 reveals infrared image of 2M1207 (light gray) and 2M1207b (dark gray) that was taken using the ESO's 8.2-m Yepun VLT.

FIGURE 6.29
Artist's concept illustrating a young red dwarf surrounded by three planets Brown dwarf planetary system orbit brown dwarfs.

FIGURE 6.30
Infrared image of 2M1207 (light gray) and 2M1207b (dark gray). (Courtesy of ESO's 8.2-m Yepun VLT.)

6.20 Pulsar Systems

A pulsar (portmanteau of pulsating star) is a highly magnetized, rotating neutron star that emits a beam of electromagnetic radiation. PSR B1257+12 was not only the first pulsar exoplanetary system but also the first planetary system discovered. PSR B1620-26 is confirmed; however, many other pulsar systems remain unconfirmed. Figure 6.31 reveals an artist's conception of PSR B1257+12's system of planets.

FIGURE 6.31
An artist's conception of PSR B1257+12's system of planets. The formation of pulsar systems is still the subject of research. It is thought that planets form from the debris of companion stars.

References

1. NASA. NASA—Planets Around Other Stars. http://science.nasa.gov/astrophysics/focus-areas/exoplanet-exploration/ (retrieved April 15, 2016).
2. *Wikipedia.* Exoplanet. http://en.wikipedia.org/wiki/Exoplanet (last modified on August 19, 2016).
3. Microlensing, Beyond our Cosmic Neighborhood. The Planetory Society. http://www.planetary.org/explore/space-topics/exoplanets/microlensing.html.
4. M. Williams (2016). The Planet Neptune. *University Today* (Space and Astronomy News). http://www.universetoday.com/21581/neptune/ (retrieved August 23, 2016).
5. New evidence for Solar-like planetary system around nearby star. Royal Observatory, Edinburgh. http://www.roe.ac.uk/roe/support/pr/pressreleases/vega.html (retrieved August 23, 2016).
6. M. D. Lemonick, A New Exoplanet — with Pictures to Prove It. *Time,* 30 June 2010. http://content.time.com/time/health/article/0,8599,2000787,00.html?iid=sr-link1 (retrieved August 19, 2016).
7. NASA. Kepler. http://kepler.nasa.gov/ (retrieved August 12, 2016).
8. NASA. Testing the Theory: Other Planetary Systems. http://www.astronomynotes.com/solfluf/s12.htm (retrieved July 2, 2014); Scarch for habitable planet. http://kepler.nasa.gov/ (retrieved August 12, 2016).
9. NASA. 1,091 New Kepler Planet Candidates. http://kepler.nasa.gov/news/index.cfm?FuseAction=ShowNews&NewsID=190 (retrieved Febuary 28, 2012).
10. NASA. Kepler-20 system: 5 planets including two that are Earth-size. http://kepler.nasa.gov/news/nasakeplernews/index.cfm?FuseAction=ShowNews&NewsID=172 (retrieved August 18, 2016).
11. J. O. Bennett et al. (2010). *The Cosmic Perspective.* 6th ed. San Francisco, CA: Pearson Addison-Wesley.
12. P. Armitage (2008). Planetary formation and migration. Scholarpedia 3(3): 4479.
13. M. Mayor, D. Queloz (1995). *Nature* 378 (6555): 355–359.
14. M. Mayor, D. Queloz (1995). A Jupiter-mass companion to a solar-type star. *Nature* 378: 355–359.
15. J. Schneider (2011). Interactive extra-solar planets catalog. The Extrasolar Planets Encyclopaedia. Paris Observatory.
16. Extrasolar Hot Jupiter: The Planet That 'Shouldn't Exist'. ScienceDaily. http://www.sciencedaily.com/releases/2009/08/090827132901.htm (retrieved August 20, 2016).
17. European Sapce Agency (ESA) Science & Technology. CoRoT. http://sci.esa.int/corot/ (retrieved September 28, 2016).
18. NASA Ames Research Center. Discoveries (2011). Kepler: Home Page.
19. J. O. Bennett et al. (2010). *The Cosmic Perspective.* 6th ed. San Francisco, CA: Pearson Addison-Wesley.
20. B. Mathiesen (2010). Hot Jupiter' systems may harbor Earth-like planets. PhysOrg.com. http://phys.org/news/2006-03-hot-jupiter-harbor-earth-like-planets.html (retrieved September 28, 2016).
21. The Origin of the Solar System (2011). Astronomy 161: The Solar System. Department of Physics and Astronomy University of Tennessee. http://csep10.phys.utk.edu/astr161/lect/solarsys/nebular.html (retrieved January 16, 2013).
22. V2: Accepted to ApJ on 2014April 11. Spitzer photometry and model fitting Matlab pipeline code is publicly available at: This http URL Earth and Planetary Astrophysics (astro-ph. EP) arXiv:1403.6831[astro-ph. EP] (or arXiv:1403.6831v2[astro-ph. EP] for this version).
23. A. Shporer et al. (2014). Atmospheric characterization of the Hot Jupiter Kepler-13Ab. *The Astrophysical Journal* 788(1): 92.
24. J. A. Johnson et al. (2012). Characterizing the cool KOIs II: The M Dwarf KOI-254 and its Hot Jupiter. *The Astronomical Journal* 143 (5): 111.
25. Laboratory for Atmospheric and Space Physics (LASP) - University of Colorado at Boulder (August 2007). The Outer planets - Extrasolar planets. http://lasp.colorado.edu/education/outerplanets/exoplanets.php#comparing (retrieved September 28, 2016).
26. W. M. Farrell et al. On the possibility of coherent cyclotron emission from extrasolar planets. *Journal of Geophysical Research: Planets (1991–2012)* 104 (E6): 14025–14032.

27. P. Zarka et al. (2004). Jupiter's low-frequency radio spectrum from Cassini/Radio and Plasma Wave Science (RPWS) absolute flux density measurements. *Journal of Geophysical Research: Space Physics (1978–2012)* 109 (A9): 1–15.
28. J. D. Nichols (2011). Magnetosphere-ionosphere coupling in Jupiter's middle magnetosphere: Computations including a self-consistent current sheet magnetic field model. *Journal of Geophysical Research: Space Physics (1978–2012)* 116 (A10).
29. R. Rosner et al. (1985). On stellar X-ray emission. *Annual Review of Astronomy and Astrophysics* 23: 413–452.
30. Royal Astronomical Society (RAS). Exoplanets Exposed to the Core (2009). https://www.ras.org.uk/news-and-press/newsarchive/68-news2009/1611-ras-pn0928-nam15-mass-loss-leaves-close-in-exoplanetsexposed-to-the-core.
31. G. E. Ballester et al. (2007). The signature of hot hydrogen in the atmosphere of the extrasolar planet HD 209458b. *Nature* 445: 511.
32. G. E. Ballester et al. (2003). Detection of an extended upper atmosphere in the extra-solar planet HD209458b. http://vega.lpl.arizona.edu/gilda/hothydrogen/ (retrieved September 28, 2016).
33. J. Walker et al. (2011). *Fundamentals of Physics.* 9th ed. Hoboken, NJ: Wiley.
34. Astrobites.org. http://astrobites.org/2012/08/02/lonely-no-more-do-other-planets-and-exomoonskeep-hot-jupiters-company/; Wikipedia, https://en.wikipedia.org/wiki/Methods_of_detecting_exoplanets (last modified September 08, 2016).
35. J. Chambers (2007). Planet formation with type I and type II migration. AAS/Division of Dynamical Astronomy Meeting 38.
36. J. M. Grießmeier et al. (2005). The potential of exoplanetary radio emissions as an observation method. *10th Anniversary of 51Peg-b: Status and Prospects for Hot Jupiter Studies.*
37. D. Lynden-Bell, J. P. O'Dwyer. (2001). One mass-radius relation for planets, white dwarfs and neutron stars. arXiv: astro-ph/0104450.
38. T. Guillot et al. (1996). Giant planets at small orbital distances. *The Astrophysical Journal Letters* 459 (1): L35.
39. D. Charbonneau et al. (2002). Detection of an extrasolar planet atmosphere. *The Astrophysical Journal* 568 (1): 377.
40. G. W. Henry et al. (2000). A transiting. *The Astrophysical Journal Letters* 529 (1): L41.
41. T. M. Brown et al. (2001). Hubble space telescope time-series photometry of the transiting planet of HD 209458. *The Astrophysical Journal* 552 (2): 699.
42. A. Burrows et al. (2000). On the radii of close-in giant planets. *The Astrophysical Journal Letters* 534 (1): L97.
43. P. Bodenheimer et al. (2001). On the tidal inflation of short-period extrasolar planets. *The Astrophysical Journal* 548 (1): 466.
44. T. Guillot, A. P. Showman (2002). Evolution of "51Pegasus b-like" planets. *Astronomy & Astrophysics* 385 (1): 156–165.
45. Geneva observatory-University of Geneva (January 07, 2005). HD 209458: A transiting hot Jupiter. http://obswww.unige.ch/~udry/planet/hd209458.html (retrieved September 28, 2016).
46. K. Rice. Planet formation via core accretion. http://faculty.ucr.edu/~krice/coreacc.html (retrieved September 28, 2016).
47. N. P. Gibson et al. (2011). A new look at NICMOS transmission spectroscopy of HD 189733, GJ-436and XO-1: No conclusive evidence for molecular features. *Monthly Notices of the Royal Astronomical Society* 411 (4): 2199–2213.
48. A. B. Bhattacharya et al. (2013). Is the stormy weather of Jupiter is comparable to earth? *International Journal of Application or Innovation in Engineering and Management* 2: 162–171.
49. D. Charbonneau et al. (2008). The broadband infrared emission spectrum of the exoplanet HD 189733b. *The Astrophysical Journal* 686 (2): 1341.
50. A. B. Bhattacharya et al. (2014). A comparison of four solar planetary systems from models consideration and reported data of visible clouds, metals and hydrogen compounds. *International Journal of Innovative Research in Science, Engineering and Technology* 3 (2): 54–58.

51. D. Sudarsky et al. (2000). Albedo and reflection spectra of extrasolar giant planets. *The Astrophysical Journal* 538 (2): 885.
52. M. S. Marley et al. (1999). Reflected spectra and albedos of extrasolar giant planets. I. Clear and cloudy atmospheres. *The Astrophysical Journal* 513 (2): 879.
53. S. Seager, D. D. Sasselov (2000). Theoretical transmission spectra during extrasolar giant planet transits. *The Astrophysical Journal* 537 (2): 916.
54. A. B. Bhattacharya, B. Raha (2014). "Hot Jupiters": The most astounding planets in the galaxy. *Science and Culture* 80: 309–313.
55. G. Sharp et al. (2004). Predicting the atmospheric composition of extrasolar giant planets. *35th Lunar and Planetary Science Conference* 35: 1152.
56. M. Fellman (2011). Research on extrasolar planets helps us better understand our solar system. *Flipping Hot Jupiters*. Northwestern University Newscenter.
57. S. Raymond et al. (2005). The formation and habitability of terrestrial planets in the presence of close-in giant planets. *Icarus* 177: 256–261.
58. B. Mathiesen (2006). "Hot Jupiter" systems may harbor earth-like planets. PhysOrg.com.
59. F. Pont (2011). The Hot Jupiter HD 189733b. ExoClimes.com. http://www.exoclimes.com/topics/the-hot-jupiter-hd-189733-b/ (retrieved August 12, 2016).
60. C. D. Impey (2011). *The Formation of the Solar System*.
61. J. Barnes, D. O'Brien (2002). Stability of satellites around close-in extrasolar giant planets. *The Astrophysical Journal* 575 (2): 1087–1093.
62. Royal Astronomical Society (2013). Hubble Space Telescope reveals variation between hot extra-solar planet atmospheres. http://www.ras.org.uk/search/article-archive/2309-hst-reveals-variation-between-hotjupiters. (retrieved September 28, 2016).
63. G. E. Ballester et al. (2007). The signature of hot hydrogen in the atmosphere of the extrasolar planet HD 209458b. *Nature* 445: 511–514.
64. J. Stoller-Conrad (2014). Detection of water vapor in the atmosphere of a hot Jupiter. Caltech News. https://www.caltech.edu/news/detection-water-vapor-atmosphere-hot-jupiter-42115 (retrieved September 28, 2016).
65. M. Wall (2013). Why "Hot Jupiter" exoplanets aren't eaten by their stars. Space.com. Lambert Academic Publishing, Germany.
66. A. B. Bhattacharya, B. Raha (2013). *Oriental Journal of Physics* 5: 52–62.
67. A. B. Bhattacharya, B. Raha (2013). *Science and Culture* 79 (1–2): 120–122.
68. J. Schneider (2011). Notes for Planet HD 178911Bb. *The Extrasolar Planets Encyclopaedia*. http://exoplanet.eu/catalog/hd_178911_b_b/ (retrieved September 26, 2016).
69. J. Schneider (2011). Notes for Planet HD 41004A b. *The Extrasolar Planets Encyclopaedia*. http://exoplanet.eu/catalog/hd_41004_a_b/ (retrieved September 26, 2016).
70. G. Anglada-Escudé et al. (2010). How eccentric orbital solutions can hide planetary systems in 2:1 resonant orbits. *The Astrophysical Journal* 709 (1): 168.
71. The Exoplanet Eccentricity Distribution from Kepler Planet Candidates, Stephen R. Kane, David R. Ciardi, Dawn M. Gelino, Kaspar von Braun (Submitted on March 7, 2012 (v1), last revised July 2, 2012 (this version, v2)).
72. J. Schneider (2011). Notes for planet Tau Boo b. The Extrasolar Planets Encyclopaedia. http://exoplanet.eu/catalog/tau_boo_b/ (retrieved September 26, 2016).
73. G. Marcy et al. (2005). Observed properties of exoplanets: Masses, orbits and metallicities. *Progress of Theoretical Physics Supplement* 158: 24–42.
74. J. T. Wright et al. (2012). The frequency of hot Jupiters orbiting nearby solar-type stars . The Astrophysical Journal 753 (160): 5pp.
75. A. Cassan et al. (2012). One or more bound planets per Milky Way star from microlensing observations. *Nature* 481 (7380): 167–169.
76. R. Sanders (2013). Astronomers answer key question: How common are habitable planets?. http://news.berkeley.edu/2013/11/04/astronomers-answer-key-question-how-common-are-habitable-planets/(retrieved August 23, 2016).

77. E. A. Petigura et al. (2013). Prevalence of Earth-size planets orbiting Sun-like stars. *Proceedings of the National Academy of Sciences* 110 (48): 19273.
78. A. Cumming et al. (2008). The Keck planet search: Detectability and the minimum mass and orbital period distribution of extrasolarplanets. *Publications of the Astronomical Society of the Pacific* 120 (867): 531–554.
79. A. W. Howard et al. (2012). Planet occurrence within 0.25AU of Solar-type Stars from Kepler . *The Astrophysical Journal Supplement Series* 201 (15): 20pp.
80. J. Amos (October 19, 2009). Scientists announce planet bounty. *BBC News*. http://news.bbc.co.uk/2/hi/science/nature/8314581.stm (retrieved September 28, 2016).
81. D. P. Bennett et al. (2006). Identification of the OGLE-2003-BLG-235/MOA-2003-BLG-53 planetary host star. *The Astrophysical Journal Letters* 647 (2): L171–L174.
82. X. Bonfils et al. (2005). The HARPS search for southern extra-solar planets: VI. A Neptune-mass planet around the nearby M dwarf Gl 581. *Astronomy & Astrophysics* 443 (3): L15–L18.
83. S. E. Thorsett, R. J. Dewey (September 16, 1993). Limits on planets orbiting massive stars from radio pulsar timing. http://hdl.handle.net/2014/35943 (retrieved September 28, 2016).
84. P. Robertson, S. Mahadevan (2014). Disentangling planets and stellar activity for Gliese 667C *The Astrophysical Journal Letters* 793: 24.
85. G. D. Mulders et al. (2014). A stellar-mass-dependent drop in planet occurrence rates (Submitted on June 28, 2014).
86. European Southern Observatory (ESO). Young stars paint spectacular stellar landscape. ESO Press Release. https://www.eso.org/public/news/eso1347/ (retrieved November 20, 2013).
87. H. Frommert, C. Kronberg (August 27, 2007). Open star clusters. SEDS. University of Arizona, Lunar and Planetary Lab. http://seds.org/messier/open.html (retrieved January 2, 2009).
88. H. Karttunen et al. (2003). *Fundamental Astronomy: Physics and Astronomy Online Library*. 4th ed. Springer. p. 321.
89. L. A. Buchhave et al. (2012). An abundance of small exoplanets around stars with a wide range of metallicities. *Nature* 486: 375–377. Bibcode:2012Natur.486.375B.
90. G. Israelian et al. (2009). Enhanced lithium depletion in Sun-like stars with orbiting planets. *Nature* 462 (7270): 189–191.
91. Binary Catalogue of Exoplanets. Maintained by Richard Schwarz. http://www.univie.ac.at/adg/schwarz/multiple.html (retrieved September 28, 2013).
92. P. D. Sackett (2010). Microlensing exoplanets. *Scholarpedia* 5(1):3991. http://www.scholarpedia.org/article/Microlensing_exoplanets (last modified December 03, 2014).http://www.scholarpedia.org/article/Microlensing_exoplanets.
93. NASA. Wide Field Infrared Survey Telescope (WFIRST). http://wfirst.gsfc.nasa.gov/ (retrieved September 28, 2016).
94. S. Hall (September 4, 2014). One planet, two stars: A system more common than previously thought. *Universe Today*. http://www.universetoday.com (retrieved August 12, 2016).
95. E. P. Horch et al. (September 3, 2014). Most sub-arcsecond companions of Kepler exoplanet candidate host stars are gravitationally bound. *The Astrophysical Journal* 795: 1.
96. W. F. Welsh, L. R. Doyle (2013). Worlds with two Suns. *Scientific American* 309 (5): 40.
97. J. L. van Saders, B. S. Gaudi (2011). Ensemble analysis of open cluster transit surveys: Upper limits on the frequency of short-period planets consistent with the field. *The Astrophysical Journal* 729 (1): 14pp.
98. A. Brucalassi et al. (2014). Three planetary companions around M67 stars. *Astronomy & Astrophysics* 561(L9): 7.
99. A. P. Boss et al. (2003). Nomenclature: Brown dwarfs, gas giant planets, and ?. *Brown Dwarfs* 211: 529.
100. S. Rappaport et al. (2013). The Roche limit for close-orbiting planets: Minimum density, composition constraints, and application to the 4.2-hour planet KOI 1843.03. arXiv:1307.4080.
101. M. Johnson, J. D. Harrington (February 26, 2014). NASA's Kepler mission announces a planet bonanza, 715 new worlds. NASA. http://www.nasa.gov/ames/kepler/nasas-kepler-mission-announces-a-planet-bonanza (retrieved February 26, 2014).

102. C. Scharf, K. Menou (2009). Long-period exoplanets from dynamical relaxation. *The Astrophysical Journal* 693 (2): L113.
103. The Extrasolar Planet Encyclopaedia. Catalog. http://exoplanet.eu/catalog/ (retrieved September 28, 2016).
104. T. Rodigas, Hinz (2009). Which radial velocity exoplanets have undetected outer companions? *The Astrophysics Journal* 702: 716–723.
105. J. Mason (2008). *Exoplanets: Detection, Formation, Properties, Habitability*. Springer. p. 2.
106. Tilting stars may explain backwards planets. *New Scientist*, September 1, 2010, issue 2776. https://www.newscientist.com/article/mg20727765.200-tilting-stars-may-explain-backwards-planets/ (retrieved September 28, 2016).
107. I. A. G. Snellen et al. (April 2014). The fast spin-rotation of a young extra-solar planet. *Nature* 509: 63–65.
108. A. C. M. Correia, J. Laskar (2010). Tidal evolution of exoplanets. In S. Seager (ed.). *Exoplanets*. University of Arizona Press.
109. R. Cowen and Nature magazine (April 30, 2014). Exoplanet rotation detected for the first time. https://www.scientificamerican.com/article/exoplanets-rotation-detected-for-the-first-time/ (retrieved September 28, 2016).
110. I. J. M. Crossfield (2014). Doppler imaging of exoplanets and brown dwarfs. *Astronomy & Astrophysics* 566 (2014): A130.
111. B. Tingley et al. Confirmation of an exoplanet using the transit color signature: Kepler-418b, a blended giant planet in a multiplanet system. *Astronomy & Astrophysics* 567(A14): 8.
112. M. C. Johnson (2013). Doppler tomographic observations of exoplanetary transits.
113. J. Horner et al. Dynamical constraints on multi-planet exoplanetary systems (Submitted on February 21, 2013).
114. International Astronomical Union (IAU) (July 2014). Press release, NameExoWorlds: An IAU Worldwide contest to name exoplanets and their host stars. http://www.iau.org/news/pressreleases/detail/iau1404/ (retrieved September 28, 2016).
115. International Astronomical Union (IAU). NameExoWorlds. http://nameexoworlds.iau.org/ (retrieved September 28, 2016).
116. J. Schneider et al. (2011). Defining and cataloging exoplanets: The exoplanet database. *Astronomy & Astrophysics* 532 (79): A79.
117. J. T. Wright *et al.* (2011). The exoplanet orbit database. *Publications of the Astronomical Society of the Pacific* 123: 412–422.
118. Exoplanet criteria for inclusion in the archive. NASA Exoplanet Archive.
119. Planetesimals to brown dwarfs: What is a planet? (2006). *Annual Review of Earth and Planetary Sciences* 34: 193–216.
120. T. J. J. See (1896). Researches on the orbit of F.70Ophiuchi, and on a periodic perturbation in the motion of the system arising from the action of an unseen body. *The Astronomical Journal* 16: 17.
121. T. J. Sherrill (1999). A career of controversy: The anomaly of T. J. J. See. *Journal for the History of Astronomy* 30 (98): 25–50.
122. NASA. NASA Kepler results usher in a new era of astronomy. http://www.nasa.gov/content/nasa-kepler-results-usher-in-a-new-era-of-astronomy (retrieved November 4, 2013).
123. J. D. Harrington, M. Johnson (November 4, 2013). NASA Kepler results usher in a new era of astronomy. http://www.nasa.gov/press/2013/november/nasa-kepler-results-usher-in-a-new-era-of-astronomy (retrieved August 23, 2016).
124. P. van de Kamp (1969). Alternate dynamical analysis of Barnard's star. *The Astronomical Journal* 74: 757–759.
125. I. Baraffe et al. (2008). Structure and evolution of super-Earth to super-Jupiter exoplanets. I. Heavy element enrichment in the interior. *Astronomy & Astrophysics* 482 (1): 315–332.
126. F. Bouchy et al. (2009). The Sophie search for northern extrasolar planets. I. A companion around HD 16760 with mass close to the planet–brown-dwarf transition. *Astronomy & Astrophysics* 505 (2): 853–858.

7

Detecting Exoplanets and Searching for Habitable Zones in Solar and Extrasolar Planetary Systems

7.1 Introduction

As we have pointed out in Chapter 6, the first confirmed detection of a planetary system was in 1992, with the discovery of few terrestrial-mass planets orbiting the pulsar PSR B1257+12. It was in 1995 that the detection of planetary system of a main-sequence star was first confirmed. Giant planet, 51 Pegasi b, was noted in a 4-day orbit around the nearby G-type star 51 Pegasi. The frequencies of detections of exoplanets have increased since then. In particular, tremendous success came through the advancements in methods of detecting extrasolar planets and dedicated planet finding programs such as the Kepler mission. It has been observed that certain characteristics are common in most known exoplanets, as well as the stars they are orbiting. Most of the stars that host planets are main-sequence stars and similar in spectral class to our own Sun. The majority of the known exoplanets are massive, which does not necessarily mean that Earth-sized exoplanets do not exist. In fact, the detection methods we have developed largely favor for finding massive planets orbiting close to stars. Some of them even orbit more closely than Mercury orbits the Sun. Dissimilar to the nearly circular orbits of the planets in our solar system, most exoplanets reveal largely eccentric orbits [1–5]. Major numbers of the known exoplanets are gaseous and similar to the giant planets in our solar system, whereas some of the smaller exoplanets found exhibit signs of rockier compositions. Until recently, there was no evidence that any planets existed around the 100 billion stars in our galaxy, except for around our Sun. However, over the last few decades, we have detected evidence of over a hundred planets orbiting other stars, even though they are far too small to be observed by any telescope. But we know their presence because of the effect they have on their host star. At the time when a planet orbits a star, its gravity causes the star to wobble. Changes in the wavelength of the light arriving from the star can be detected when the star wobbles in the direction of Earth. This variation of wavelength is known as the Doppler effect and is similar to the variation in pitch of the siren when a police car passes the listener [6–14].

7.2 Search of Suitable Detection Methods

There are different methods scientists use when trying to detect extrasolar planets [15–24]. The methods can be preliminarily classified under two categories:

1. Direct evidence:
   a. Images of the planet
   b. Spectra of the planet
2. Indirect evidence:
   a. Doppler technique
   b. Transits evidence of getting exoplanets directly is very difficult to obtain using photography. Image will look like a very small round blob. It is really hard to take a good-quality snap of something that is so far away. On average, Pluto orbits about 40 times farther from the Sun than Earth. We may now think about a planet which is 4 million times farther from our Sun than we are. Not only the distance factor is considerable but also the planets under consideration are really dim compared with their parent stars and hence they are really hard to see. For that reason, indirect observations like the Doppler technique, transits, and eclipses are much more commonly used when searching for exoplanets. Figure 7.1a shows the Doppler effect on a star. It is seen that the Doppler shifts permit to detect the slight motion of a star caused by an orbiting planet. Figure 7.1b, on the other hand, reveals a periodic Doppler shift in the spectrum of the star 51 Pegasi [25].

The Doppler technique is definitely a good method for discovering exoplanets. The technique is used to analyze the motion and properties of the star and planet. Both of them are orbiting a common center of mass. The star and the planet gravitationally attract one another, causing them to orbit around a point of mass central to both bodies. It is something like trying to perfectly balance a healthy child and a small child on a teeter-totter. If the teeter-totter is suspended in the center, the healthy child will be on the ground, but

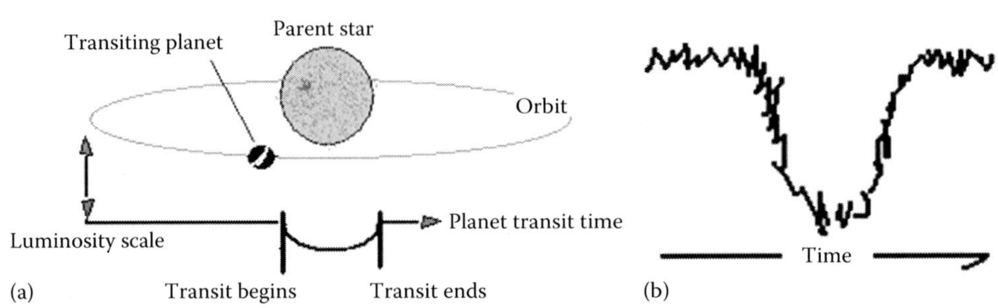

FIGURE 7.1
Illustrating the Doppler effect on a star: (a) Doppler shifts allow detecting the transiting of a planet in front of its parent star. (b) A simulated light curve. (Courtesy of Lichtman, Radio Astronomy Supplies, TX.)

both the children will be in perfect balance if the healthy child is moved very close to the center. If we could rotate the children around each other, both the healthy child and the small child orbit a common center of mass.

With the solar system, all bodies orbit a common center of mass, including the Sun. But the Sun being very large in comparison to the planets, the center of mass actually lies inside the Sun (Figure 7.2). This makes the Sun seem to wobble back and forth, this also effects the motion of the solar spectrum. We search for this spectral shift in other stars to find if there are one or more planets orbiting that star. When the star moves toward us, the light emitted has a shorter wavelength and we say its spectrum is blue shifted, whereas when it is moving away from us, the light has a longer wavelength and we say its spectrum is red shifted. The Doppler technique is most widely used for detecting extrasolar planets. It is best suited to find very massive planets orbiting close to their parent star as the parent star will wobble more with a large planet nearby. This scenario will thus produce a larger and more easily detectable spectral shift. Most of the planets discovered around other stars are very massive and orbit extremely close to their parent star. Another indirect observational technique is known as a transit. This happens when a planet crosses in front of its parent star (occulting), temporarily diminishing the brightness of the star by a small amount. This method permits to determine the density and size of the planet with respect to the size of the star. Although these methods provide us an idea of size, distance, and orbital period of the planets, they cannot give us

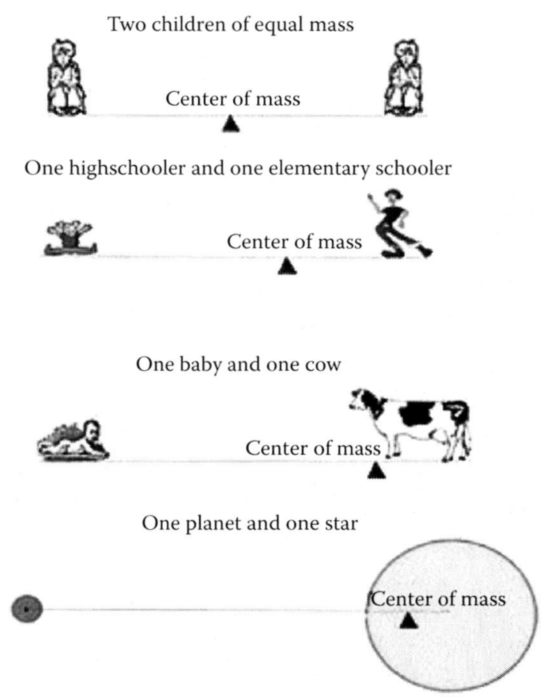

FIGURE 7.2
Changing the position of center of mass between two bodies.

any concrete information about the planet. Direct observation is a better method than indirect, but because these planets are far away and also hidden from view and owing to how small and dim they are compared to their parent stars, direct observation is almost impossible.

7.2.1 Typical Examples of Direct and Indirect Imaging

We have already mentioned that the planets are extremely faint compared with their parent stars. It is reported that at visible wavelengths, they generally have less than a millionth of the brightness of the parent star. It becomes complicated to detect such a faint light source. Further, the parent star causes a glare that tends to wash it out. The light from the parent star needs to be blocked for reducing the glare while leaving the light from the planet detectable; doing so is a major technical challenge [26]. Figure 7.3 reveals a directly imaged planet, Beta Pictoris b.

All directly imaged exoplanets are both large (more massive than Jupiter) and largely separated from their parent star. Most of those exoplanets are also very hot, so that they emit intense infrared radiation. The images have then been made at infrared wherein the planet is brighter than it is at visible wavelengths. During the gas-accretion phase of the giant planet formation, the star-planet contrast may be even better in H-alpha than it is in infrared [27]. Thus, direct imaging has used infrared wavelengths to find planets. At infrared wavelengths, a star like the Sun is only 100 times brighter than Jupiter, in comparison to a billion (10^9) times brighter at visual wavelengths. This is applicable for planets that are very far from their stars, so in that case, an orbit might take hundreds or thousands of years for a planet discovered by this method. Astronomers need to verify that the planet and the star move together through space to prove that the planet orbits the star because observing an entire orbit will perhaps take many years. This method does not permit astronomers to determine the mass of a planet directly, but by the use of spectral research, astronomers can determine surface temperature. Up until December 7, 2011, approximately 29 planets have been discovered by employing

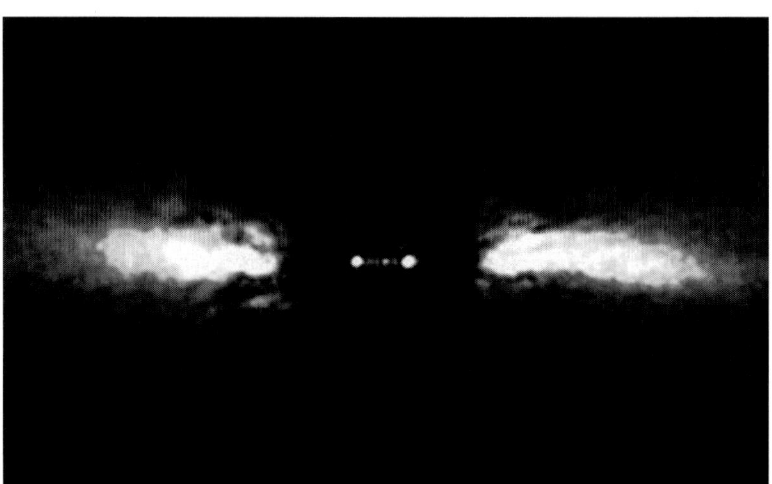

FIGURE 7.3
Directly imaged planet, Beta Pictoris b.

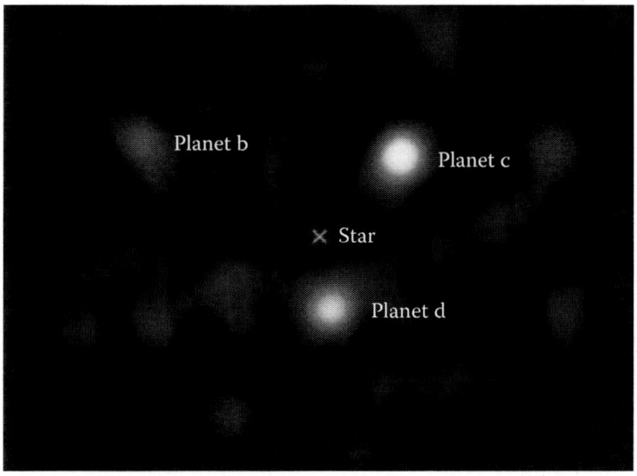

FIGURE 7.4
The parent star (center) and the planets HR 8799b (top left), HR 8799c (top right), and HR 8799d (bottom) as obtained by using direct-imaging instruments.

this technique. Figure 7.4 reveals the parent star and planets obtained by using direct-imaging instruments.

Specially designed direct-imaging instruments like the Gemini Planet Imager, VLTSPHERE, and SCExAO have taken images of dozens of gas giants. A vast majority of known extrasolar planets have been detected only through indirect methods. Indirect methods have been vigorously employed and have proven very useful. For example, an infrared image of the HR 8799 system was a direct result and provided very useful information. Figure 7.5 shows a central blob is noise left over after the light from the star has been largely removed. The three known planets can be observed by this method. They are HR 8799d (bottom), HR 8799c (upper right), and HR 8799b (upper left).

If a planet crosses (or transits) in front of the parent star, then the observed brightness of the star drops by some amount. The amount by which the star dims depends on the size of the star as well as on the size of the planet, among other factors. This technique suffers from a substantial rate of false positives and hence confirmation from other detection techniques is considered important. The transit method shows the radius of a planet

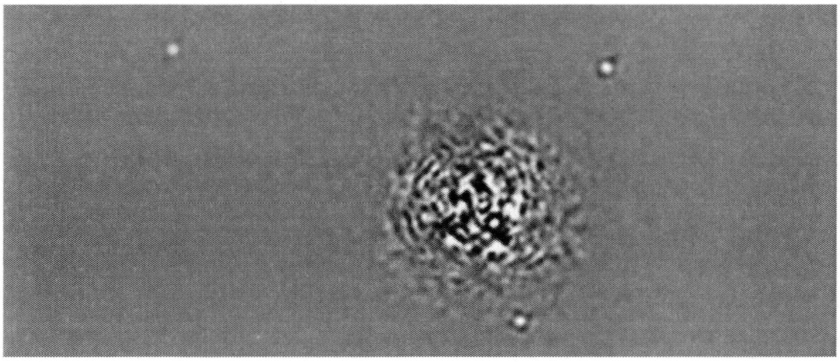

FIGURE 7.5
The three known planets: HR 8799b (top left), HR 8799c (top right), and HR 8799d (bottom).

and permits further investigation of the planets' atmosphere through spectroscopy. As the transit method needs that part of the planet's orbit to intersect a line-of-sight between the host star and Earth, the probability that an exoplanet in a randomly oriented orbit will be found to transit the star is to some extent small.

7.3 Methods of Detecting Extrasolar Planets

Initially, there were three main detection techniques developed for finding extrasolar planets. All of these methods rely on detecting a planet's effect on its parent star and to infer the planet's existence. All the three techniques are very simple in principle but difficult in practice because extreme precision is essential to register the effect of the planet on the much larger star. This is again difficult from Earth, as the atmosphere distorts our view of the stars, limiting the accuracy of the investigation. This is where space missions can help to solve this problem. The methods in question are the radial velocity method, the astrometry method, and the transit method. All these methods are referred to as *indirect* methods. An example of the indirect method is illustrated in Figure 7.6a, which reveals how, depending on the position of the planet and the star, the brightness level reduces and again regains when the planet comes out of the star's obstruction, whereas Figure 7.6b shows clearly the luminosity versus time (not to the scale) of the exoplanet transit.

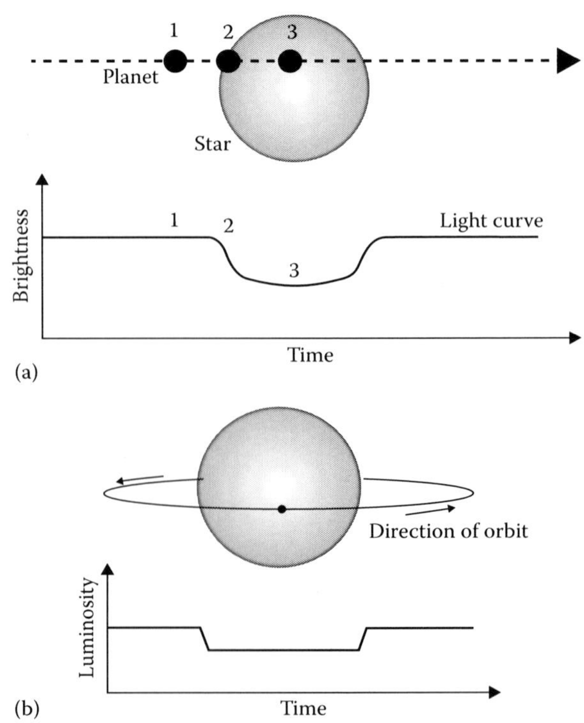

FIGURE 7.6
(a) How the brightness level varies with the position of the planet and the star due to the transit. (b) Luminosity versus time (not to the scale) of the exoplanet transit.

Initially developed, the three main detection techniques for finding extrasolar planets were improved and five well-accepted methods for the purpose were subsequently implemented. These methods will be discussed in the next subsections.

7.3.1 Radial Velocity or Doppler Method

When a planet orbits a star, the star also moves in its own small orbit around the system's center of mass. Variations in the radial velocity of the star, the speed with which it moves toward or away from Earth, can be detected from displacements in the spectral lines of the star due to the Doppler effect. Extremely small variations in the radial velocity can be found, up to 1 m/s or even less [27]. The method has the advantage of being applicable to stars with varying characteristics, but one of its disadvantages is that it cannot provide the true mass of a planet and can only set a lower limit on that mass. But if the radial velocity of the planet itself can be distinguished from the radial velocity of the star, then the true mass can be found out [28]. The method uses the fact that if a star has a planet (or planets) around it, then it is not strictly correct to mention that the planet orbits the star. Rather, the planet and the star orbit their common center of mass. As the star is more massive than the planets, the center of mass is within the star and the star appears to wobble slightly as the planet moves around it. Astronomers can determine this wobble by using spectroscopy. If a star is moving toward us, its light will appear blue shifted, whereas if it is moving away, the light will be red shifted. Spectroscopy reveals this change in color from a star as it moves toward and away from us, orbiting the center of mass of the star–planet system. Many planets discovered by this method are very large and very close to their stars. These planets are known as super-Jupiters. They are much larger than Jupiter and orbit their stars in a few days. These large planets cause a large wobble and the short orbit time makes them easier to find than ones that are smaller or farther from their stars. Up to December 2011, over 650 planets have been discovered by this method.

7.3.2 Transit Timing Variation

When multiple planets are present, each of them slightly perturbs the orbits of the other. Slight variations in the times of transit for one planet can indicate the presence of another planet, which itself may or may not transit. As for example, changes in the transits of the planet Kepler-19b indicate the existence of a second planet in the system, the nontransiting Kepler-19c [29,30]. If multiple transiting planets exist in one particular system, then this method can be utilized to assure their existence [31]. In another form, timing the eclipses in an eclipsing binary star can show an outer planet that orbits both stars. In August 2013, a few planets have been observed in that manner and many planets have been confirmed following this technique. It may be mentioned that this method only works for star–planet systems that have orbits aligned in such a way that as found from Earth, the planet moves between us and the star and blocks some of the light from the star. The exoplanet orbit orientations are illustrated in Figure 7.7.

It is interesting to note that a planet does not block enough light from a star (only 1% or less), but even this can be detected. Similar to the radial velocity method, this method has a bias toward discovering large planets orbiting near to their stars, which block more light and transit more frequently and so they are easier to detect. This technique will not work for all systems because only about 10% of super-Jupiters are aligned in such a way that we can observe the transit. Smaller planets in larger orbits are less likely to be aligned

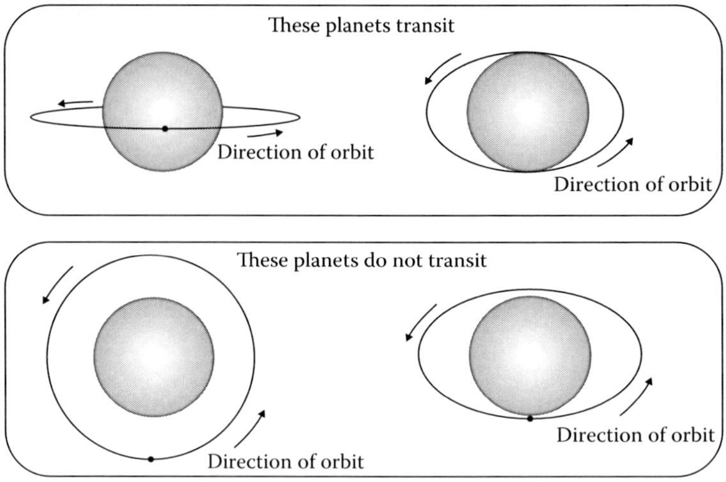

FIGURE 7.7
Exoplanet orbit orientations.

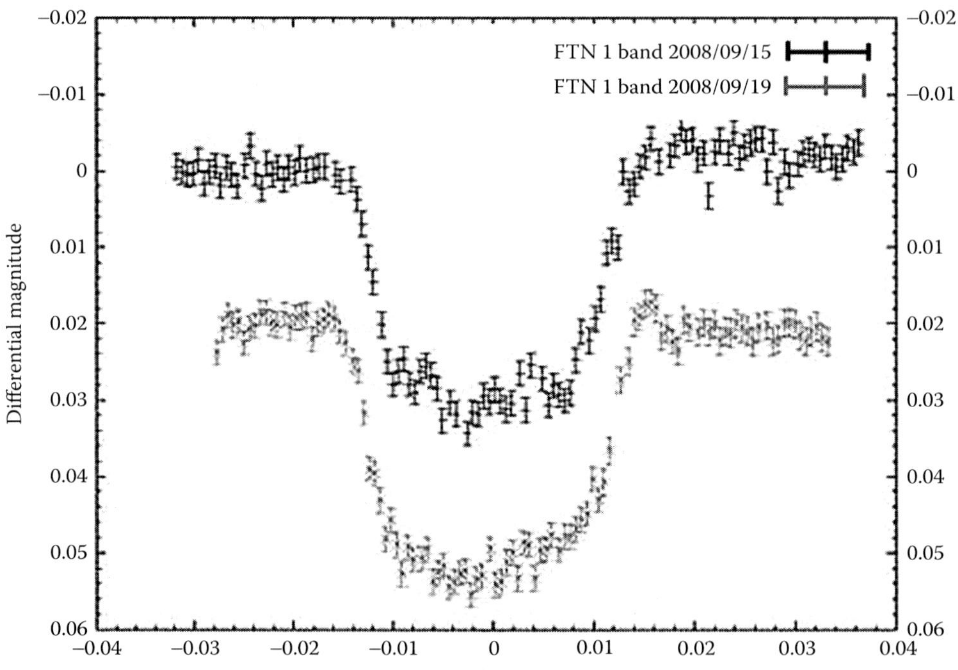

FIGURE 7.8
Variations of differential magnitude versus phase at two separate dates.

in such a manner that we can see transits. Astronomers get valuable information about the atmosphere of the planet that transits. Over 180 planets have been discovered by this method up to December 2011. We have illustrated in Figure 7.8 the variations of differential magnitude versus phase corresponding to two different dates indicating a similar nature of variation, but with magnitude differences.

7.3.3 Transit Duration Variation

When a planet orbits multiple stars or if the concerned planet has moons, its transit time can alter significantly per transit. Although, at this point, no new planets or moons have been discovered by applying this technique, it has been successfully utilized to confirm many transiting circumbinary planets [32]. Figure 7.9 represents an animation showing the difference between planet transit timing of one- and two-planet systems.

It should be pointed out that the transiting planetary systems are those with multiple planets in orbital planes along the line of sight between their star and the observer. The transit method is, in fact, one of the most successful observation methods. Kepler mission was designed with a view to discover systems of this type and has observed numerous candidate systems. A prominent example of a transiting system is Kepler-11, in which multiple simultaneous planetary transits have been found. Figure 7.10 shows an artist's conception of the simultaneous transit of three planets in the Kepler-11 system, found by NASA's Kepler spacecraft on August 26, 2010.

Transiting planetary systems are considered excellent candidates for accurately analyzing the properties of the systems, including the detection of extraterrestrial atmospheres, dust, or asteroid belts and possibly natural satellites through astronomical spectroscopy and transit timing effects.

7.3.4 Gravitational Microlensing

When the gravitational field of a star acts like a lens, microlensing occurs, which can be successfully utilized for magnifying the light of a distant background star. Planets orbiting the lensing star can cause detectable anomalies in the magnification as it changes over time. Unlike many other methods that have detection bias toward planets with small (or for resolved imaging, large) orbits, the microlensing technique is most sensitive for detecting planets around 1–10 AU away from Sun-like stars. Gravitational microlensing is based

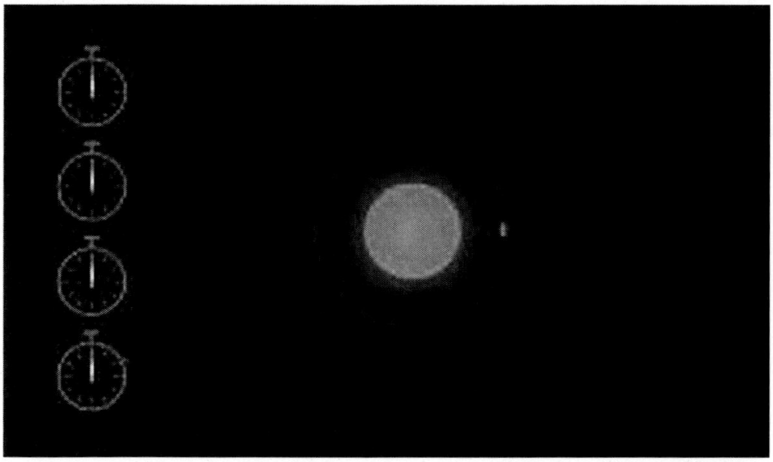

FIGURE 7.9
Animation indicating difference between planet transit timing of one- and two-planet systems. (Courtesy of Wikipedia, https://en.wikipedia.org/wiki/Methods_of_detecting_exoplanets [last modified September 08, 2016]; Courtesy of NASA/Kepler Mission, http://kepler.nasa.gov/ [retrieved August 17, 2016]; For full visualization of the process, refer to animated NASA video available at: https://commons.wikimedia.org/wiki/File:201008-2a_PlanetOrbits_16x9-_Transit_timing_of_1-planet_vs_2-planet_systems.ogv [last modified February 27, 2014].)

FIGURE 7.10
Artist's conception of a simultaneous transit of three planets before in the Kepler-11 system. (Courtesy of NASA, NASA's Kepler spacecraft discovers extraordinary new planetary system, http://www.nasa.gov/mission_pages/kepler/news/new_planetary_system.html [retrieved October 24, 2015].)

on chance events wherein, from our viewpoint, one star passes in front of another star. Usually, the farther star is a bright star and the near one is one we could not ordinarily find from Earth. When it passes in front of the farther star, its gravity causes the light from the farther star to bend and the star is magnified from our point of view. The principle of gravitational microlensing is illustrated in Figure 7.11.

The microlens event method can be utilized for detecting Jovian-mass and terrestrial-mass exoplanets near their parent stars that are distance from the Earth. Like the transit

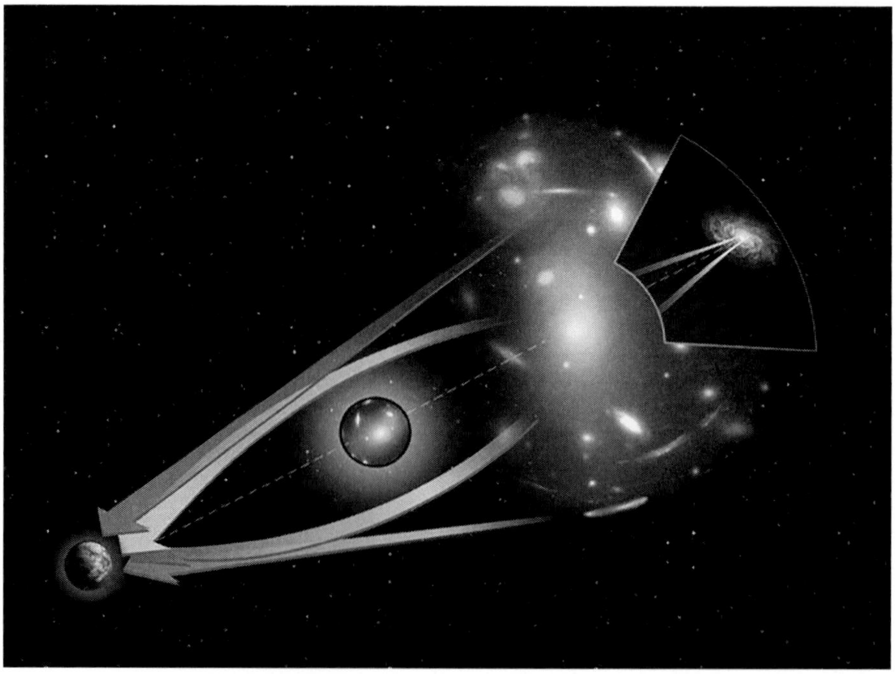

FIGURE 7.11
Gravitational microlensing observing principle.

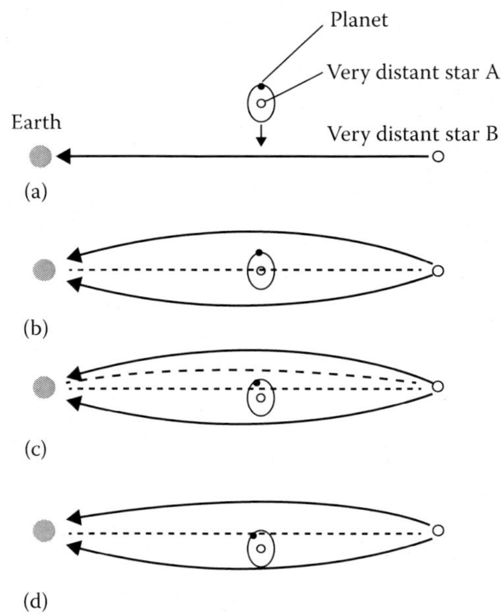

FIGURE 7.12
Different steps in the method of gravitational microlensing as marked by (a) to (d), respectively.

method, many stars must be monitored to pick up even a single stellar microlens event. The microlens events are due to chance alignments that are not repeatable. Twenty-nine exoplanets orbiting stars have been observed using the microlens technique as of June 2014. Figure 7.12 shows the different steps in the method of gravitational microlensing: (a) closer but still distant star A with planet in small orbit begins to move between us and very distant star B; (b) closer star A warps space so additional light from very distant star B reaches us and very distant star B appears to brighten; (c) the planet of the closer star A warps space less than closer star A, so a little extra light from very distant star B is added to already lensed light; and (d) the planet of the closer star A is now enough out of the way to not lens the very distant star B's light, but the gravity of the closer star is still enough to lens some of the light: we find extra short-lived increase or other significant deviation in the otherwise symmetric larger and longer-lived brightening then dimming in the very distant star B's brightness due to the closer star.

7.3.5 Astrometry

Astrometry consists of measuring the precise the position of a star in the sky and observing the changes in that position over time. The motion of a star owing to the gravitational influence of a planet may be observable. However, this method has not yet been very productive. It has produced only some disputed detections, although it has been used successfully to find the properties of planets observed in other ways. Until recently, telescopes did not have the required resolution for detecting this type of motion. Astrometry is very sensitive to planets with large orbits, an advantage over other methods. No extrasolar planet discoveries by this method have been confirmed by other methods. However, the field is highly promising.

7.4 Some Salient Features of Different Detection Methods

- *Pulsar timing*: A pulsar is the small, ultradense remnant of a star that has exploded as a supernova emitting radio waves regularly as it rotates. If planets orbit the pulsar, they will cause small anomalies in the timing of its observed radio pulses. The first confirmed discovery of an extrasolar planet was made using this method. As of 2011, it has not been very productive. Five planets have been detected in this manner, around three different pulsars.
- *Variable star timing (pulsation frequency)*: Similar to pulsars, there are some other types of stars that show periodic activity. Deviations from the periodicity can occasionally be caused by a planet orbiting the pulsar. As of 2013, some planets have been discovered with this method [33].
- *Reflection/emission modulations*: When a planet orbits very close to the star, it catches a certain amount of starlight. As the planet orbits around the star, the amount of light varies due to the planet phases, as seen from Earth's viewpoint or planet brightness (albedo) more from one side than the other owing to temperature differences [34].
- *Polarimetry*: With the polarimetry method, a polarized light reflected off the planet can be separated from unpolarized light emitted from the star. No new planets have been found with this method, although a few discovered planets have been detected by applying the technique [35,36].
- *Circumstellar disks*: Disks of space dust surround many stars that are believed to originate from collisions among asteroids and comets or other interstellar objects. The dust can be detected, as it absorbs starlight and reemits it as infrared radiation. Features in the disks may suggest the presence of planets, although this is not considered a definitive detection method.
- *Relativistic beaming*: Relativistic beaming determines the observed flux from the star owing to its motion. The brightness of the star varies as the planet moves closer or further away from its host star [37].
- *Ellipsoidal variations*: Massive planets close to their host stars can slightly deform in shape due to of the host stars tugging motion, thus varying the brightness [38].

7.5 Exoplanet Discoveries and Their Sizes

Figure 7.13 shows a histogram of exoplanet discoveries. In the figure, the gray bar displays new planets *verified by multiplicity*. The exoplanets reported are up to February 26, 2014.

Since 1998, over 1800 extrasolar planets have been discovered [39] either through radial velocity, direct imaging, transit, microlensing, and timing methods. The histogram shown in Figure 7.14 presents exoplanets by size. In the figure, the deep gray bars represent Kepler's latest newly verified exoplanets. The data used are up to February 26, 2014.

The majority of planetary detections so far have been obtained by using the radial velocity method from ground-based telescopes. The method needs the light from a star to be passed through a prism and split into a spectrum, rather like water droplets in the atmosphere splitting sunlight into a rainbow. Figure 7.15 illustrates radial velocity data from 51 Pegasi.

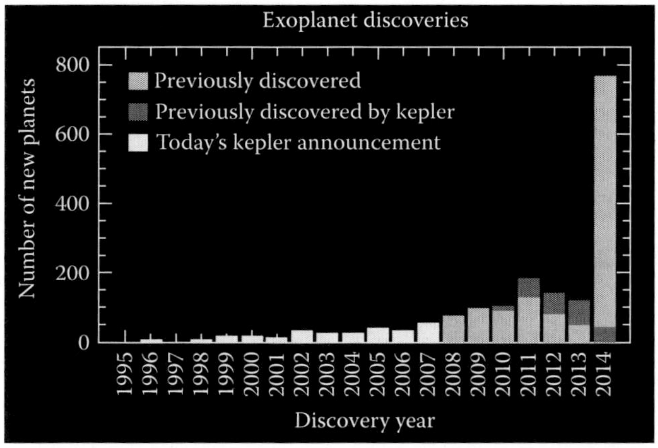

FIGURE 7.13
Histogram of exoplanet discoveries. (Courtesy of NASA, Exoplanet discoveries, http://www.nasa.gov/content/exoplanet-discoveries [retrieved July 31, 2015].)

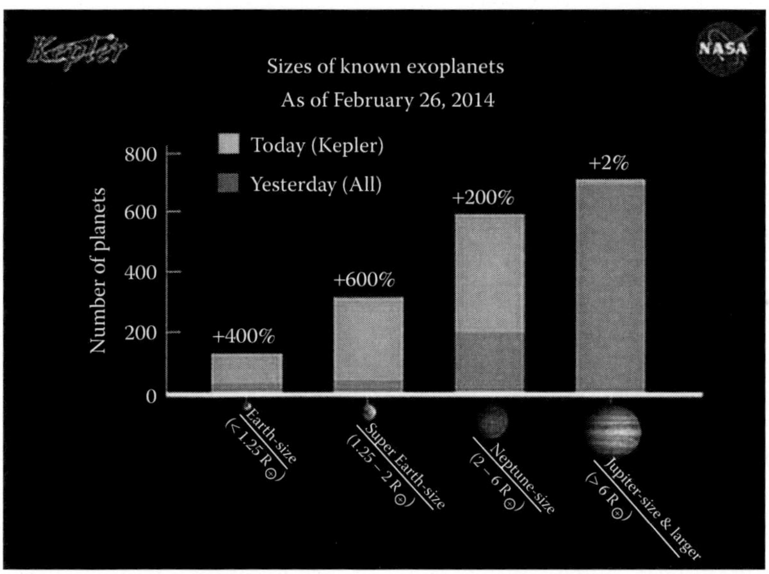

FIGURE 7.14
Histogram of exoplanets by size. (Courtesy of NASA, Sizes of known exoplanets, http://www.nasa.gov/content/sizes-of-known-exoplanets [retrieved July 31, 2015].)

If the spectrum is magnified, straight black lines can be found superimposed on the colors. These spectral lines correspond to the wavelengths of light that have been absorbed by chemicals on the surface of the star where the light is originated. Every element and molecule generates its own chemical fingerprint through unique *spectral lines* at various wavelengths, which provide an indication of the amount of that element present in the object. Studying these lines can reveal which stars have large planets around them. As the planet orbits the star, it pulls on it with its gravitational field. This causes forcing the star into a small orbit or wobble. The star will sometimes be spinning toward Earth, whereas at other times spinning away. When the star travels toward Earth, the wavelengths of the spectral lines in the light it emits move toward the blue end of the spectrum. When the

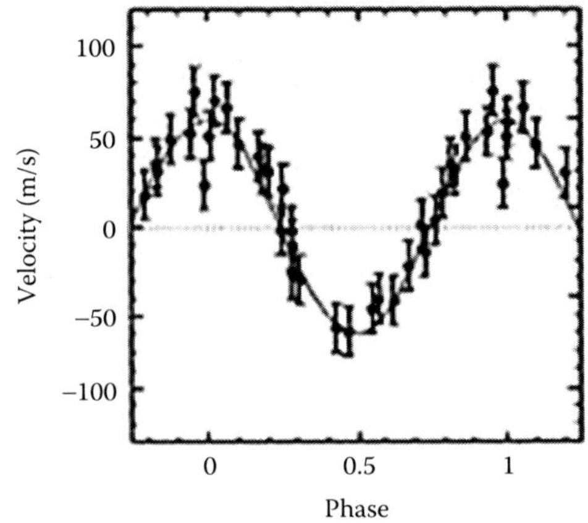

51 Peg
P = 4,7 jours, A = 0.05 u.a.
M = 0,47 MJupiter

FIGURE 7.15
Radial velocity measurements for the star 51 Pegasi taken at different times. (Courtesy of Wikipedia, Doppler spectroscopy, https://en.wikipedia.org/wiki/Doppler_spectroscopy#/media/File:Doppler_Shift_vs_Time.jpg [last modified June 19, 2016].)

star moves away from Earth, the opposite happens and the wavelengths are moved toward the red part of the spectrum. Measuring the amount of movement with time, the mass of the planet and its orbit can be found out. This technique is limited, as it will never be able to detect small, Earth-sized worlds. With the best spectroscopes, astronomers can detect motions of about 15 m/s.

7.6 Space Telescopes

Most confirmed extrasolar planets have been obtained utilizing ground-based telescopes. However, some methods can work more effectively with space-based telescopes, which avoid atmospheric haze and turbulence. COROT and Kepler were space missions used for searching extrasolar planets. Hubble Space Telescope and MOST have also observed or confirmed some planets. The Gaia mission [40], launched in December 2013, is using astrometry to find the true masses of 1000 nearby exoplanets. The CHEOPS and TESS missions, to be launched in 2017, as well as the PLATO mission in 2024 will use the transit method [41] (Table 7.1).

7.7 Kepler Mission

The majority of extrasolar planets detected so far are gas giants, like the outer planets in our solar system. Scientists are largely assured that life could not exist on the outer planets in our solar system. The challenge is mainly to find terrestrial exoplanets that may

TABLE 7.1

Primary and Secondary Detection

Method	Primary	Secondary
Transit	Primary eclipse. Planet passes in front of the star.	Secondary eclipse. Star passes in front of planet.
Radial velocity	Radial velocity of star.	Radial velocity of planet [42]. This has been used for Tau Boötis b.
Astrometry	Astrometry of star. Position of star moves more for large planets with large orbits.	Astrometry of planet. Color-differential astrometry [43]. Position of planet moves quicker for planets with small orbits. Theoretical method has been suggested for use for the SPICA (spacecraft).

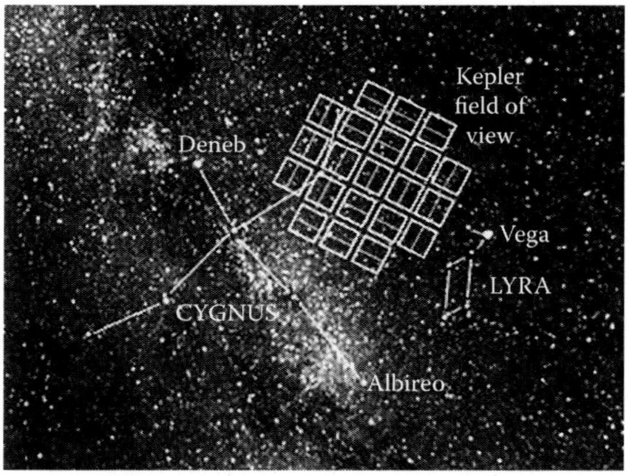

FIGURE 7.16
Kepler image of the Milky Way in the Cygnus region. (Courtesy of NASA, Kepler field of view, http://kepler.nasa.gov/multimedia/photos/?ImageID=10 [retrieved August 12, 2016].)

be capable of sustaining life. The Kepler Mission was designed to search a region of the Milky Way for smaller, Earth-sized exoplanets in or near the habitable zone of their parent star. Kepler concentrated at a single portion of the sky in the Cygnus region along the Orion arm for its entire mission with a view capable of detecting hundreds of exoplanets. Figure 7.16 exhibits image of the Milky Way in the Cygnus region.

There are some essential conditions to make a planet habitable and also there are certain key factors that influence these conditions. The size and temperature of the star and the orbit of the planet widely determine the condition of having liquid water on the surface. Planets situated very close to a star are so hot that any water on the surface would boil away. On the other hand, planets too far from a star are so cold that any liquid water freezes. The size and mass of the planet are responsible largely to ascertain whether or not it can sustain an atmosphere. Smaller planets may not be sufficiently massive to have the required surface gravity to retain an atmosphere. Other factors, such as the compositions of the atmosphere, have influence on the temperature of the planet and the type of protection from various forms of harmful radiation. For the purpose of detecting planets, Kepler followed the transit method. As we know, a transit is an event in which a planet passes in front of a star as found from Earth. In comparison to a larger planet, transits by terrestrial-sized planets create a small change in the brightness of the star. This change in brightness lasts only for a few hours depending on the distance between the planet and the star. We know a planet is causing the change in brightness if the change is periodic.

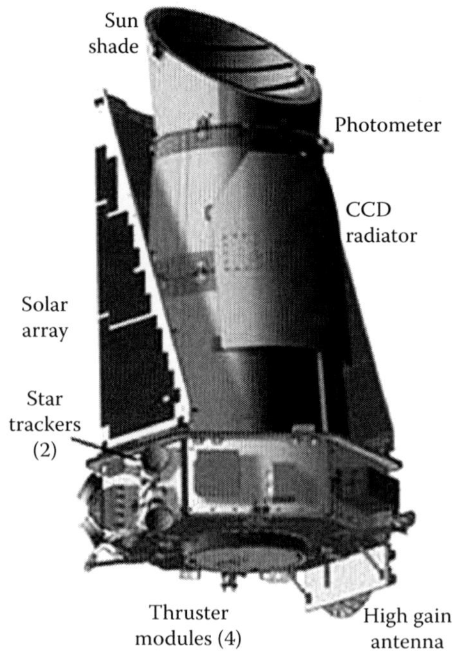

FIGURE 7.17
Image of the Kepler photometer and spacecraft. (Courtesy of NASA, http://kepler.nasa.gov/ [retrieved August 12, 2016].)

This means that it happens repeatedly on the same time period. The main Kepler instrument is a telescope, called a photometer or light meter with a very large field of view. It will stare at the same star field for the entire mission and continuously monitor the brightness of more than 100,000 stars. If Earth-size planets are common, Kepler has the capacity to detect hundreds of them. An image of the Kepler photometer and the spacecraft is shown in Figure 7.17.

7.8 Earth Similarity Index

Earth similarity index (ESI) (or the *easy scale*) is a multiparameter first assessment of Earth-likeness for solar and extrasolar planets as a number between zero (no similarity) and one (identical to Earth). Such similarity indices can be used in many fields and give a powerful tool for categorizing patterns from complex datasets. The ESI can be successfully utilized to prioritize exoplanets observations as well as to perform statistical assessments and develop planetary classifications. It can also be used as a way for sorting the objects of interest of the habitable exoplanets catalog. The basic ESI expression can be written as

$$\mathrm{ESI} = \prod_{i=1}^{n} \left(1 - \left|\frac{x_i - x_{io}}{x_i + x_{io}}\right|\right)^{\frac{w_i}{n}}$$

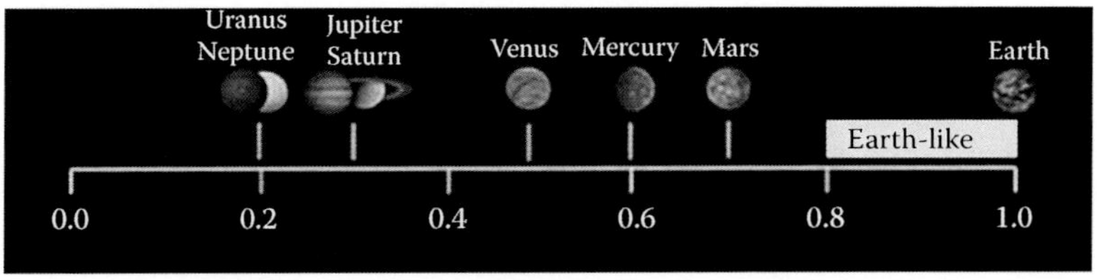

FIGURE 7.18
Earth similarity index (ESI). (Courtesy of Wikipedia, https://en.wikipedia.org/wiki/Earth_Similarity_Index [last modified August 05, 2016]; Courtesy of Professor Abel Mendez and the University of Puerto Rico, The Habitable ExoplanetsCatalog, Planetary Habitability Laboratory, http://www.phl.upr.edu [retrieved September 10, 2016].)

TABLE 7.2
Reference Values and Weight Exponents for the Four Planetary Properties Used to Define the Earth Similarity Index

Planetary Property	Reference Value (Eu = Earth's Units)	Weight Exponent
Mean radius	1.0Eu	0.57
Bulk density	1.0Eu	1.07
Escape velocity	1.0Eu	0.70
Surface temperature	288K	5.58

where ESI is the similarity measure, x_i is a planetary property (e.g., surface temperature), x_{io} is the corresponding terrestrial reference value (e.g., 288K), w_i is a weight exponent, and n is the number of planetary properties. The weighting exponents are utilized for adjusting the sensitivity of the scale and equalize its meaning between different properties. Earth-like planets are defined as any planetary body with a similar terrestrial composition and a temperate atmosphere. As a general rule, any planetary body with an ESI value over 0.8 can be taken as Earth-like planet, which means that the planet is rocky in composition (silicates) and has an atmosphere appropriate for most terrestrial vegetation including complex life. Planets with ESI value lie in the range 0.6–0.8 (e.g., Mars) might still be habitable but only by simple extremophilic life, as those are either too cold or too hot. Figure 7.18 reveals the ESI.

The ESI measures the Earth-likeness for exoplanets as a number between zero and one. The ESI scale is a function of the radius, bulk density, escape velocity, and surface temperature of an exoplanet, but it is much more sensitive to surface temperature. The parameters for the ESI equation for mean radius, bulk density, escape velocity, and surface temperature are shown in Table 7.2.

Referring to Figure 7.19, calculations of ESI for solar and extrasolar planets are shown. In the figure, ESI for 47 solar system bodies with radius greater than 100 km (small circles with black color) and 258 known extrasolar planets (small black squares) have been considered. Only some of the most notable bodies are labeled. The ESI scale makes a distinction between those rocky interior (light red area) and temperate surface (light blue area) planets. Only planets within these two categories can be taken as Earth-like planets (light green area). The dotted lines indicate constant ESI values. If confirmed, only Gliese 581g is in the Earth-like category together with Earth. They are further divided for convenience into an interior ESI based on the mean radius and bulk density and a surface ESI based on the escape velocity and surface temperature. Both the interior and surface ESI are then combined into a global ESI. The ESI is more a surface than a subsurface habitability indicator owing to its Earth-centric definition. However, a similar formulation can be made for other planetary bodies using different reference values (e.g., ocean-like planets).

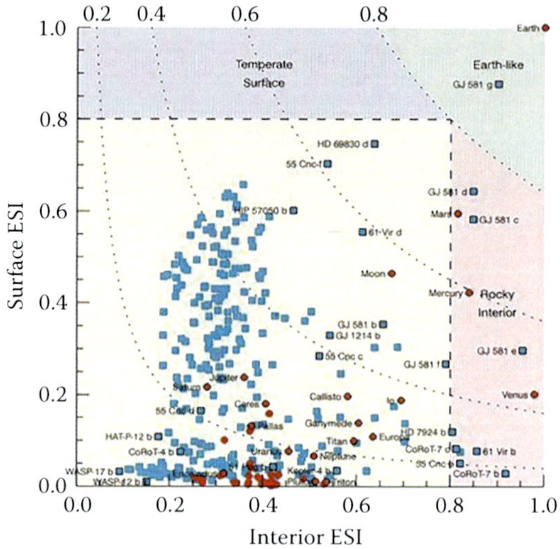

FIGURE 7.19
ESI for 47 solar system bodies with radius greater than 100 km (small circles with black color) and 258 known extra solar planets (small black squares). (From A. Mendez, Planetary Habitability Laboratory at University of Puerto Rico at Arecibo. Earth similarity index, 2011. http://phl.upr.edu/projects/earth-similarity-index-esi [retrieved September 10, 2016]. With permission.)

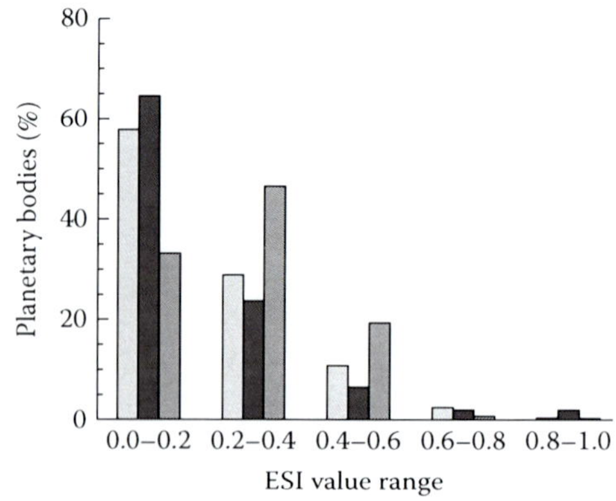

FIGURE 7.20
Distribution of ESI values based on a theoretical statistical prediction (fade ash color) for 47 solar system bodies with radius greater than 100 km (black color) and 258 known extrasolar planets (deep ash color).

One important practical application of the ESI is in studies about the distribution and diversity of Earth-like planets, as shown in Figure 7.20. The figure shows the distribution of ESI values based on a theoretical statistical prediction (fade ash color), for 47 solar system bodies with radius greater than 100 km (black color) and 258 known extrasolar planets (deep ash color). Our solar system matches the prediction, but the bars for the known extrasolar planets reveal the bias of current observational techniques toward large planetary bodies (ESI values between 0.2 and 0.4). This analysis with the ESI can be utilized to predict the number of expected Earth-like planets within a sample of stars.

The ESI values of the Kepler Mission of planets candidates is shown in Figure 7.21. The figure shows the ESI for 47 solar system bodies with radius greater than 100 km (small circles with black color), 258 known extrasolar planets (small black squares), and Kepler's 1235 planet

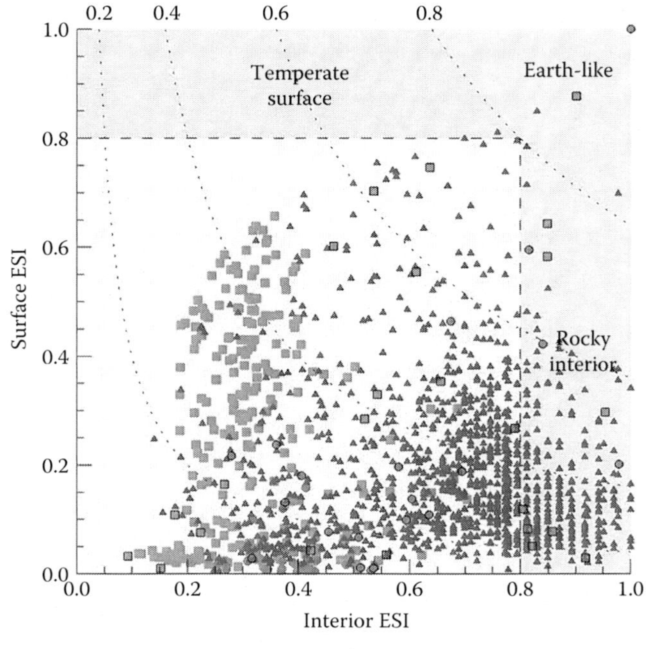

FIGURE 7.21
ESI for 47 solar system bodies with radius greater than 100 km (small circles with black color), 258 known extra solar planets (small black squares) and Kepler's 1235 planet candidates (small black color triangles). (From A. Mendez, Planetary Habitability Laboratory at University of Puerto Rico at Arecibo. Earth similarity index, 2011. http://phl.upr.edu/projects/earth-similarity-index-esi [retrieved September 10, 2016]. With permission.)

candidates (small black color triangles). The mass for the ESI calculations was estimated using generic mass–radius relationships for gas, ocean, and rocky planets. The interesting result in the Kepler data is the potential abundance of rocky bodies and the presence of two Earth-like planets candidates, plus a few more close to this category. The information on these planetary candidates is very limited and further observations will be necessary to confirm them.

7.9 Terrestrial Exoplanets Compared to Earth

Exoplanets are different from Earth. They are massive; the biggest one we know is 3180 times larger than Earth. Even the smallest exoplanet is nearly 30 times larger. Most of them have orbits about five times closer to their host star. This suggests that they must have incredibly high surface temperatures. The biggest known extrasolar planetary system found is around Upsilon Andromedae, a star 44 light years away that has three orbiting giant planets. This does not necessarily mean that there are no exoplanet systems like ours. Exoplanets that are large and close to their host star produce a larger and more noticeable wobble, which means that their Doppler effect can be more conveniently detected. Presently, we are not able to detect the Doppler effect for an exoplanet system similar to ours, but it is possible they are out there. Some exoplanets have been imaged by blocking the light from the brighter star with a device called a *coronograph* [44], so that the feeble light from the exoplanet can be detected. Use of a coronograph was essential to produce the first visible light image of an exoplanet. The Hubble Space Telescope has detected disks of gas and dust around 50% of the stars still forming in the Orion Nebula. It, therefore, appears that the formation of planet systems is a common process in the universe. In the context of exoplanet research, a *super-Earth* is defined by mass only: a world up to 10 times the mass of Earth. The term does not mean the planet is habitable. The upper limit of 10 times the mass of Earth is used, as it is assumed that planets larger than 10 times the mass of Earth have enough gravity to suck up the hydrogen and helium surrounding it, as it is forming and becoming a Jovian planet. However, the discovery of Kepler-10c announced in June 2014 with a mass 17.2 times the mass of the Earth but just 2.35 times the diameter of the Earth means it has a density 7.1 times that of water, so it is definitely a rocky world instead of something like Jupiter or Neptune. Astronomers are yet unable to find the diameters of most of the exoplanets and their densities. (Their composition is still unknown.) Only fractions of the exoplanets have been observed to move in front of their stars and cause an eclipse or dimming of the starlight. One exoplanet, HD 189733b, has water, methane, and carbon dioxide in its atmosphere. However, the exoplanet is extremely hot and massive to support life. Until January 2010, astronomers had not been able to take the spectrum of an exoplanet directly. This will be an important step in eventually being able to analyze the spectrum of a terrestrial exoplanet and to find if it is supporting life on it.

7.10 Understanding of Exoplanets in the Search of Life

A clear understanding of exoplanets will tell us about how our solar system formed, why it has small, rocky planets near the Sun, why it has gas giant planets far from the Sun, why the Earth has the chemicals and necessary conditions that can support life, and why conditions on other planets are hostile to life. Theories of planet formation and evolution are yet incomplete but provide some specific predictions. Detections of exoplanets are already testing, validating, and of course, in many cases, invalidating details of these theories. Perhaps the most interesting question and one of the most difficult to answer concerns the uniqueness of the Earth. Are there planets similar to the Earth around other stars and does life (as we know it or much different) exist on any other planet beyond our own Earth? Dates of some important events are given in Table 7.3.

7.11 Formation of the Protoplanetary Disk

Early theories of planetary systems were based on idea of the formation (gravitational collapse or molecular cloud) and the evolution of the solar system. Planetary systems are believed to form, in general, as part of the same process that results in star formation. Early theories involved another star passing extremely close to the star, drawing material out from it, which then coalesced for forming the planets. However, the probability of such a near collision is not yet very clear to make this a viable model. Accepted theories argue that a protoplanetary disk forms by gravitational collapse of a molecular cloud and thereby evolves into a planetary system by collisions and gravitational capture [45]. An artist's concept of a protoplanetary disk is shown in Figure 7.22.

Some planetary systems may form differently. Stars that emit periodic bursts of electromagnetic radiation have been discovered by the slight variations they cause in the timing

TABLE 7.3

Dates of Some Important Events

Date	Event
April 24, 2014	Astronomical Forensics Uncover Planetary Disks in Hubble Archive
April 17, 2014	Kepler Discovers First Earth-Size Planet in the 'Habitable Zone' of Another Star (Kepler-186f)
March 18, 2014	Herschel Completes Largest Survey of Dust in the Local Universe
March 6, 2014	Nearby Star's Icy Debris Suggest *Shepherd* Planet
March 6, 2014	Loops of Gas and Dust Rise from Planetary Disks
February 25, 2014	Bullying Black Holes Force Galaxies to Stay Red and Dead
February 4, 2014	Kepler Finds a Very Wobbly Planet (Kepler-413b)
January 22, 2014	Herschel Detects Water on Dwarf Planet (Ceres)
January 6, 2014	Kepler Provides Insight About Five New Rocky Planets (Kepler-99b, Kepler-406b...)
December 3, 2013	Subtle Signals of Water on Hazy Worlds
October 30, 2013	First Earth-Sized Rocky Planet (Kepler-78b)
October 18, 2013	Giant Misalignment in a Multiple Planet System
October 10, 2013	Water-rich Planetary Building Blocks Found Around White Dwarf
September 30, 2013	Partially Cloudy Skies on Kepler-7b
July 29, 2013	2013 Chandra Sees Eclipsing Planet in X-rays for the First Time (HD 189733)
July 17, 2013	How Scientists Search for Habitable Planets
July 11, 2013	Hubble Finds a True Blue Planet

FIGURE 7.22
An artist's concept of a protoplanetary disk.

of these bursts. Pulsars are formed in violent supernova explosions and a normal planetary system could not survive such a blast. One theory is that existing stellar companions were almost evaporated by the supernova blast, leaving behind planet-sized bodies [46].

7.12 Location of Habitable Zone around Types of Stars

The habitable zone may be considered as the region in the planetary system where an Earth-like planet (or moon) could maintain liquid water on its surface and potentially, therefore, life. The distance of this zone from the parent star can vary largely from system to system. The lack of surface water on Mars, Venus, the Moon, and Ceres does not exclude the possibility of extraterrestrial surface life in the solar system, but makes it highly unlikely, though according to some estimates, those regions lie within the habitable zone of the solar system. An example of a system to predict the location of the habitable zone around types of stars is illustrated in Figure 7.23. However, the abundance of life on Earth makes the habitable zones of exoplanetary systems of particular interest to exobiologists and the search for extraterrestrial intelligence.

HD 37124 has multiple well-known planets in the zone. Systems with gas giants have more commonly discovered within the habitable zone, though they could possibly harbor Earth-like natural satellites.

7.13 Desert Planets Having a Reduced Greenhouse Effect

The habitable zone around a star is the region where the temperature is just right to permit liquid water to exist on a planet. This is not too close to the star for the water to evaporate and not too far away from the star for the water to freeze. The heat produced

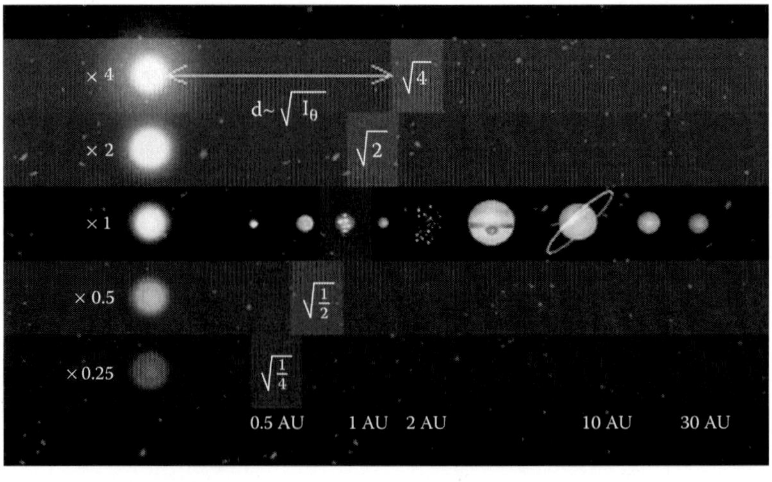

FIGURE 7.23
An example of a system to predict the location of the habitable zone around types of stars. Gliese 581 was the first exoplanetary system observed with super-Earths in the habitable zone.

by stars changes depending on the size and age of the star, so that the habitable zone can be at different distances. Further, the atmospheric conditions on the planet influence the ability of the planet to retain heat, so that the location of the habitable zone becomes specific to each type of planet. Desert planets, also known as dry planets with very little water, will have less water vapor in the atmosphere than Earth and hence have a reduced greenhouse effect, meaning that a desert planet could maintain oases of water closer to its star than Earth is to the Sun. The lack of water further means that there is less ice to reflect heat into space, so the outer edge of desert-planet habitable zones is further out [47]. Rocky planets with a thick hydrogen atmosphere could maintain surface water much further out than the Earth–Sun distance [48]. Habitable zones are generally defined in terms of surface temperature. However, over half of Earth's biomass is from subsurface microbes [49] and the temperature increases as we go deeper underground. Hence, the subsurface can be habitable when the surface is frozen and if this is taken into account, then the habitable zone extends much further from the star [50]. Even rogue planets (those without a star) could have liquid water at sufficient depths underground [51]. In earlier era of the universe, the temperature of the Cosmic Microwave Background would permit any rocky planets that existed to have liquid water on their surface regardless of their distance from a star [52]. Jupiter-like planets might not be habitable, but they could have habitable moons.

7.14 Ice Ages and Snowball States

The outer edge of the habitable zone is an environment where planets will be completely frozen, but even planets well inside the habitable zone can periodically become frozen. If orbital fluctuations or other causes create cooling, then this produces more ice. Ice reflects sunlight, causing even more cooling, creating a feedback loop until the planet is completely (or nearly) frozen. When the surface is frozen, carbon dioxide weathering stops, causing a build-up of carbon dioxide in the atmosphere possibly from volcanic emissions, which creates a greenhouse effect and unfreezes the planet again. Planets with a high axial tilt [53] are less likely to enter snowball states and can retain liquid water further from their star. Large fluctuations of axial tilt can have even more of a warming effect than a fixed large tilt [54,55]. Paradoxically, planets around cooler stars (e.g., red dwarfs) are less likely to enter snowball states as the infrared radiation emitted by cooler stars is largely at wavelengths that are absorbed by ice, which heats it up [56,57].

7.14.1 Tidal Heating

When a planet has an eccentric orbit, then tidal heating can provide another source of energy besides stellar irradiation. Eccentric planets in the radiative habitable zone can be too hot for liquid water (Tidal Venus) [58]. Eccentric planets further out than the radiative habitable zone would still have frozen surfaces, but the tidal heating could create a subsurface ocean similar to Europa's. For some of the planetary systems, such as in the Upsilon Andromedae system, the eccentricity of orbits can be maintained or even periodically varied by perturbations from other planets in the system. Tidal heating can cause outgassing from the mantle, thus helping to the formation and replenishment of an atmosphere [59].

7.15 Potentially Habitable Planets

Orbital motions of the planet candidates around Gliese 667C are considered as one important proposed orbital solution with seven planets [60]. Three of these planets are super-Earths mass objects orbiting in the habitable zone, where widespread surface liquid water may exist. The orbit of the planet Mercury in the solar system may be included for scale (Figure 7.24).

Confirmed planet discoveries in the habitable zone include the Kepler-22b, the first super-Earth located in the habitable zone of a Sun-like star [61]. In September 2012, the discovery of two planets orbiting the red dwarf [62], Gliese 163, was reported [63,64]. One of the planets, Gliese 163c, which is nearly 6.9 times the mass of Earth and somewhat hotter, was considered to be within the habitable zone. In 2013, three more potentially habitable planets, Kepler-62e, Kepler-62f, and Kepler-69c, orbiting Kepler-62 and Kepler-69, respectively, were identified [65,66]. All three planets were super-Earths [65] and may be covered by oceans thousands of kilometers deep. In June 2013, a dynamically packed planetary system around the nearby red dwarf, Gliese 667C, was reported. The system was observed to contain at least three super-Earths in its habitable zone (Gliese 667Cc, Gliese 667Ce, and Gliese 667Cf), creating the new record in the number of potentially habitable worlds around a single star [61]. The system contains two other planet candidates (Gliese 667Cd and Gliese 667Ch), which would lie in the cold/hot edges of the habitable zone of star. This result highlights the prevalence of low-mass stars as hosts of potentially habitable worlds.

7.16 Earth-Size Planets

In November 2013, it was announced that 22±8% of Sun-like stars have an Earth-size planet in the habitable zone [67,68]. Given that there are 200 billion stars in the Milky Way, it would be 11 billion potentially habitable Earths, rising to 40 billion if red dwarf

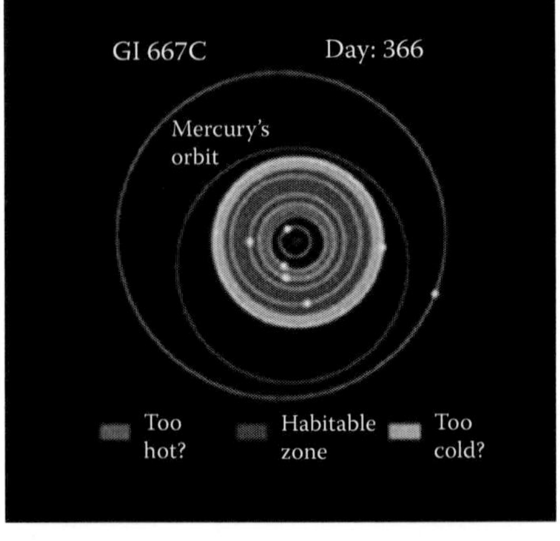

FIGURE 7.24
Proposed orbital solution in relation to potentially habitable planets.

stars are included [69]. Kepler-186f is the first Earth-size planet in a habitable zone to have been discovered. A 1.1 Earth-radius planet in the habitable zone of a red dwarf was announced in April 2014. In February 2013, researchers determined that up to 6% of small red dwarfs may have planets with Earth-like properties, which suggests that the closest *alien Earth* to the solar system could be 13 light years away. The estimated distance increases to 21 light years when a 95% confidence interval is used [70]. In March 2013, a revised estimate based on a more accurate consideration of the size of the habitable zone around red dwarfs provided an occurrence rate of 50% for Earth-size planets in the habitable zone of red dwarfs [71].

7.17 Orbital Dynamics of Planetary Systems

Planetary systems are categorized according to their orbital dynamics as resonant, nonresonant-interacting, and hierarchical, or some combination of these. In resonant systems, the orbital periods of the planets are in integer ratios. The Kepler-223 system contains four planets in an 8:6:4:3 orbital resonances [72]. Planet orbits in interacting systems are sufficiently close together that they perturb the orbital parameters. The solar system could be explained as weakly interacting. In strongly interacting systems, Kepler's laws does not apply [73]. In hierarchical systems, the planets are arranged so that the system can be gravitationally considered as a nested system of two bodies, for example, in a star with a close-in hot Jupiter with another gas giant much further out, the star and hot Jupiter form a pair that appears as a single object to another planet that is far enough out. A system can contain bodies of different dynamical types, for example, the Galilean moons of Jupiter, where Io, Europa, and Ganymede are in resonance but Callisto is too distant to be part of this resonance. Orbital dynamics of planetary systems are shown in Figure 7.25.

The spacing's between orbits change largely among the different systems discovered by the Kepler spacecraft.

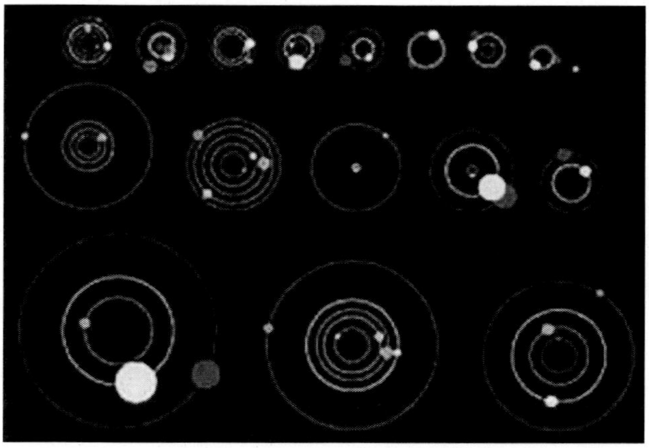

FIGURE 7.25
Orbital dynamics of planetary systems.

7.18 Thermal Planetary Habitability Classification

It is strongly believed that recent and future observations by ground and orbital missions will be able to locate habitable exoplanets [74,75]. As a first assessment, the surface temperature of Earth-like exoplanets can be utilized as a proxy for habitability. Surface temperature is indirectly obtained by the distance of the exoplanet to the parent star together with the luminosity of the star within the stellar habitable zone. So far, this is the simplest technique to assess the potential for life of exoplanets. Future observations will be required for better assessments, which include direct planetary temperature measurements together with atmospheric composition. Not all habitable exoplanets, however, are expected to be equally habitable. Even Earth's habitability has changed through time. The number of habitable exoplanets that will be detected by current (i.e., Kepler) and future missions (i.e., TPF) will need more detailed study in order for proper classification. To develop a planetary habitability classification based on temperature, we have to consider our understanding of the thermal requirements of life, in general. Microbial life such as bacteria has a wide thermal tolerance whose growth has been measured at temperatures from –15°C (5°F) to 121°C (249.8°F). Macrobial life like animals (metazoa) reveals a more restrictive tolerance usually between 0°C (32°F) and 50°C (122°F). This is true for many microbial (i.e., cyanobacteria) and macrobial (i.e., plants) primary producers to generate most of the energy for consumers in the trophic scale. Particularly, majority of the plants are efficient at temperature close to 25°C (77°F). At standard atmospheric pressures, water is liquid between 0°C (32°F) and 100°C (212°F) but is also known to stay liquid down to near –50°C (58°F) in supercooled states or in combination with antifreeze agents, salts. For simplicity, it is practical to divide the thermal tolerance of life in bins of 50°C (122°F) around the freezing point of water. There is a simple thermal classification for microbial life in the microbiology field. Mesophiles are those microorganisms that can grow best at moderate temperatures between 10°C (50°F) and 45°C (113°F). This is true for psychrophiles between –15°C (5°F) and 10°C (50°F), thermophiles between 45°C (113°F) and 80°C (176°F), and hyperthermophiles above 80°C (176°F). These limits were decided to some extent arbitrarily for historical reasons, and there is no strict standard between the exact values. This is also extendible to the classification of our planetary habitability using the same Greek prefixes for exoplanets as mesoplanets, psychroplanets, and thermoplanets. However, the limits may be revised to 50°C (122°F) increments, as they are easier to remember and also relevant to both simple and complex life. Proposed Thermal Planetary Habitability Classification (T-PHC) for exoplanets is presented in Figure 7.26. The classification concludes the use of the term mesoplanet or Class M planet to refer to *Earth-like exoplanets*, or those terrestrial planets with mean global surface temperatures between 0°C (32°F) and 50°C (122°F). The proposed names of our classification can be abbreviated as M-planets (mesoplanets), P-planets (psychroplanets), and T-planets (thermoplanets). All three classes represent potential habitable exoplanets based on their mean global surface temperature. Only mesoplanets (temperate) correspond to Earth-like planets, which may be potentially habitable by complex life as we know it, such as plants and animals. Psychroplanets (cold) and thermoplanets (hot) may only be habitable for microbial life. Even an extension of the classification as hypopsychroplanets or hyperthermoplanets might still be habitable, but these conditions are in the limits of our understanding of carbon-based life in aqueous environments.

The above classification matched the letters and general description of Class P and Class M planets. The name mesoplanet was also used before by Isaac Asimov to refer

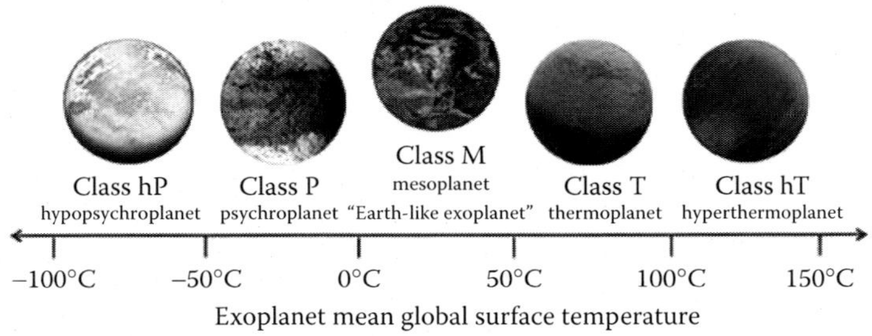

FIGURE 7.26
Proposed Thermal Planetary Habitability Classification (T-PHC) for exoplanets. (From A. Mendez, Planetary Habitability Laboratory at University of Puerto Rico at Arecibo. A thermal planetary habitability classification for exoplanets, 2011. http://phl.upr.edu/library/notes/athermalplanetaryhabitabilityclassificationforexoplanets [retrieved August 12, 2016]. With permission.)

to planets between the size of Ceres and Mercury. The proposed classification is easy to understand by the scientific community and also familiar to microbiologists. With the help of this classification, we can put the recent discoveries of potential habitable planets. In particular, the classification can be utilized for the exoplanets Gliese 581g (unconfirmed) and Gliese 581d and the Kepler's candidates KOI-701.03 and KOI-326.01. We can further include Earth and Mars for comparison, which is shown in Figure 7.27. A thermal classification of potential habitable exoplanets has made in the figure. Only Gliese 581g and KOI-701.03 fall in the category of mesoplanets or truly Earth-like exoplanets, if confirmed. Exoplanets like Gliese 581d and KOI-326.01 lie in the categories that only supports microbial life, if any. Mars is in one of the extremes on the categories where any surface microbial life is much less likely. Only further observations will be able to confirm these assessments of habitability. However, there is still much uncertainty in current observations and large shifts in the exoplanets classifications are probable. The other objects of the solar system will be out of the scale and hence nonhabitable by the proposed classification, at least at the surface. Planetary bodies like Mars, Europa, Titan, and Enceladus are more interesting and dominant at the subsurface levels. The evolution of Earth's habitability since the origin of life is highly interesting, as it probably changed

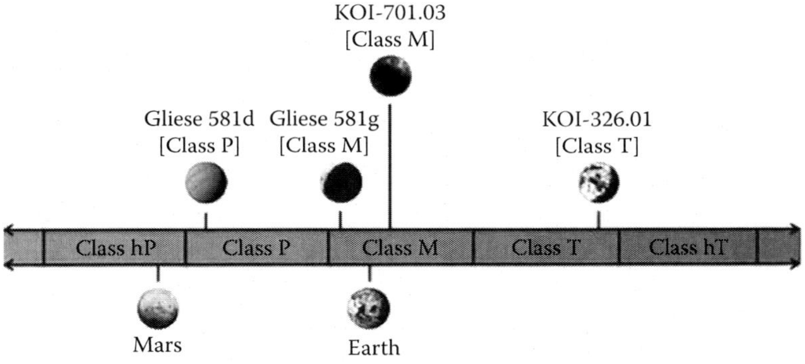

FIGURE 7.27
Thermal classifications of potential habitable exoplanets with Earth and Mars for comparison. (From A. Mendez, Planetary Habitability Laboratory at University of Puerto Rico at Arecibo. A thermal planetary habitability classification for exoplanets, 2011. http://phl.upr.edu/library/notes/athermalplanetaryhabitabilityclassificationforexoplanets [retrieved August 12, 2016]. With permission.)

from a Class hT to a Class M planet in 4 billion years. Mars perhaps began as a Class M planet, which rapidly changed to Class hP in less than 1 billion years.

The thermal planetary habitability classifications provide a simple method based on temperature for terrestrial exoplanets. It is compatible with our knowledge of the thermal limits for terrestrial life and the observations that are currently possible for exoplanets. The classification can be expanded with other planetary parameters preserving its functional meaning as more detailed information of terrestrial exoplanets are obtained. It may be noted that the mesoplanets (Class M) are the main goal of current efforts in exoplanets searches, but psychroplanets (Class P) and thermoplanets (Class T) are still interesting.

References

1. W. Claven (2013). Billions and billions of planets. NASA. http://www.nasa.gov/mission_pages/kepler/news/kepler20130103.html (retrieved January 3, 2013).
2. Staff (January 2, 2013). 100billion alien planets fill our Milky Way galaxy: Study. Space.com. http://www.space.com/19103-milky-way-100-billion-planets.html (retrieved January 3, 2013).
3. A. Cassan et al. (2012). One or more bound planets per Milky Way star from microlensing observations. *Nature* 481 (7380): 167–169. http://www.nature.com/nature/journal/v481/n7380/full/nature10684.html (retrieved August 27, 2016).
4. ESO. Planet population is plentiful (January 11, 2012). https://www.eso.org/public/news/eso1204/ (retrieved January 13, 2012).
5. J. Schneider (2010). Interactive extra-solar planets catalog. The Extrasolar Planets Encyclopedia. http://exoplanet.eu/catalog.php (retrieved May 18, 2010).
6. C. Beichman et al. (2014). WISE Y dwarfs as probes of the brown dwarf-exoplanet connection. *The Astrophysical Journal* 783 (2): 68. arXiv:1401.1194v2. Bibcode:2014ApJ...783...68B. doi:10.1088/0004-637X/783/2/68.
7. NASA—Kepler. http://www.nasa.gov/sites/default/files/kepler-nov-4-2013-media-briefing-audio.mp3 (retrieved November 4, 2013).
8. J. D. Harrington and M. Johnson (2013). NASA Kepler results usher in a new era of astronomy. NASA press release 13–323, 4 November 2013. http://www.nasa.gov/press/2013/november/nasa-kepler-results-usher-in-a-new-era-of-astronomy/ (retrieved August 03, 2016).
9. P. Tenenbaum et al. (2012). Detection of potential transit signals in the first twelve quarters of Kepler mission data. arXiv:1212.2915[astro-ph. EP].
10. My God, it's full of planets! They should have sent a poet. (Press release). Planetary Habitability Laboratory, University of Puerto Rico at Arecibo. January 3, 2012. http://phl.upr.edu/press-releases/mygoditsfullofplanetstheyshouldhavesentapoet (retrieved January 4, 2013).
11. A. Santerne et al. (2013). Astrophysical false positives in exoplanet transit surveys: Why do we need bright stars? arXiv:1310.2133[astro-ph. EP].
12. R. Sanders (November 4, 2013). Astronomers answer key question: How common are habitable planets? Berkeley News. http://news.berkeley.edu/2013/11/04/astronomers-answer-key-question-how-common-are-habitable-planets/ (retrieved August 12, 2016).
13. E. A. Petigura et al. (2013). Prevalence of Earth-size planets orbiting Sun-like stars. *Proceedings of the National Academy of Sciences* 110 (48): 19273. arXiv:1311.6806. Bibcode:2013PNAS..11019273P. doi:10.1073/pnas.1319909110.
14. A. Khan (November 4, 2013). Milky Way may host billions of Earth-size planets. *Los Angeles Times*. http://www.latimes.com/science/la-sci-earth-like-planets-20131105-story.html (retrieved November 5, 2013).

15. L. E. Strigari et al. (2012). Nomads of the galaxy. *Monthly Notices of the Royal Astronomical Society* 423 (2): 1856–1865. arXiv:1201.2687. Bibcode:2012MNRAS.423.1856S. doi:10.1111/j.1365-2966.2012.21009.x. estimates 700objects >10—6solar masses (roughly the mass of Mars) per main-sequence star between 0.08and 1Solar mass, of which there are billions in the Milky Way.
16. M. Johnson and J. D. Harrington (February 26, 2014). NASA's Kepler mission announces a planet bonanza, 715new worlds. NASA. http://www.nasa.gov/ames/kepler/nasas-kepler-mission-announces-a-planet-bonanza (retrieved August 12, 2016).
17. M. Wall. Population of known alien planets nearly doubles as NASA discovers 715new worlds. http://www.space.com/24824-alien-planets-population-doubles-nasa-kepler.html (retrieved February 26, 2014).
18. Kepler telescope bags huge haul of planets. BBC News. http://www.bbc.com/news/science-environment-26362433 (retrieved February 27, 2014).
19. NASA Exoplanet Archive. DENIS-P J082303.1—491201b. http://exoplanetarchive.ipac.caltech.edu/cgi-bin/DisplayOverview/nph-DisplayOverview?objname=DENIS-P+J082303.1-491201+b&type=CONFIRMED_PLANET (retrieved March 8, 2014).
20. J. Sahlmann et al. (2013). Astrometric orbit of a low-mass companion to an ultracool dwarf. *Astronomy & Astrophysics* 556 (2013): A133.
21. A. Wolszczan and D. A. Frail (1992). A planetary system around the millisecond pulsar PSR1257 + 12. *Nature* 355 (6356): 145. doi:10.1038/355145a0.
22. G. Giorbran (2007). Giordano Bruno. http://everythingforever.com/bruno.htm (retrieved September 28, 2016).
23. I. Newton et al. (1999[1713]). *The Principia: A New Translation and Guide*. University of California Press, Berkeley, CA. p. 940.
24. W. S. Jacob (1855). On certain anomalies presented by the binary star 70ophiuchi. *Monthly Notices of the Royal Astronomical Society* 15: 228. Bibcode:1855MNRAS..15..228J.
25. SolStation.com. 51 Pegasi. http://www.solstation.com/stars2/51pegasi.htm (retrieved September 28, 2016).
26. NASA's exoplanet archive KOI table. NASA. http://exoplanetarchive.ipac.caltech.edu/cgi-bin/TblView/nph-tblView?app=ExoTbls&config=cumulative (retrieved February 28, 2014).
27. M. Perryman (2011). *The Exoplanet Handbook*. Cambridge University Press, New York. p. 149.
28. F. Rodler et al. (2012). Weighing the non-transiting hot Jupiter Tau BOO b. *The Astrophysical Journal Letters* 753 (25): L25.
29. Scientific Computing (2010). Planet hunting: Finding Earth-like planets. http://www.scientific-computing.com/news/2010/07/planet-hunting-finding-earth-planets (retrieved August 12, 2016).
30. S. Ballard et al. (2011). The Kepler-19 system: A transiting 2.2 R_Earthplanet and a second planet detected via transit timing variations. *The Astrophysical Journal* 743 (2): 20pp. .
31. Y. Abe et al. (June 2011). Habitable zone limits for dry planets. *Astrobiology* 11 (5): 443–460.
32. A. Pál and B. Kocsis (2008). Periastron precession measurements in transiting extrasolar planetary systems at the level of general relativity. *Monthly Notices ofthe Royal Astronomical Society* 389: 191–198.
33. R. Silvotti et al. (2007). A giant planet orbiting the "extreme horizontal branch" star V391Pegasi. *Nature* 449 (7159): 189–191.
34. J. M. Jenkins and R. D. Laurance (September 20, 2003). Detecting reflected light from closein giant planets using space-based photometers. *The Astrophysical Journal* 1 (595): 429–445.
35. H. M. Schmid et al. (2006). Search and investigation of extra-solar planets with polarimetry. *Direct Imaging of Exoplanets: Science & Techniques. Proceedings of the IAU Colloquium #2001* (C200): 165–170. Bibcode:2006dies.conf.165S. doi:10.1017/S1743921306009252.
36. S. V. Berdyugina et al. (January 20, 2008). First detection of polarized scattered light from an exoplanetary atmosphere. *The Astrophysical Journal* 673: L83.
37. A. Loeb and B. S. Gaudi (2003). Periodic flux variability of stars due to the reflex doppler effect induced by planetary companions. *The Astrophysical Journal Letters* 588 (2): L117.

38. N. Atkinson (May 13, 2013). Using the theory of relativity and BEER to find exoplanets. *Universe Today*. http://www.universetoday.com/102112/using-the-theory-of-relativity-and-beer-to-find-exoplanets/# (retrieved August 12, 2016).
39. The Extrasolar Planet Encyclopaedia, Exoplanet.eu (1995–2015). Catalog. http://exoplanet.eu/catalog/ (retrieved August 20, 2016).
40. Gaia Science (2013). Gaia Science Homepage. http://www.gaiascience.com.sg/ (retrieved August 20, 2016).
41. P. Dunn (2014). Space eye with 34 telescopes will investigate one million stars. Warwick News & Events, http://www2.warwick.ac.uk/newsandevents/pressreleases/space_eye_with/ (last modified February 21, 2014).
42. H. Kawahara et al. (April 23, 2014). Spectroscopic coronagraphy for planetary radial velocimetry of exoplanets. *The Astrophysical Journal Supplement Series* 212: 27.
43. L. Abel et al. (2009). . Characterizing extra-solar planets with color differential astrometry on SPICA. SPICA Workshop 2009.
44. Wikipedia. Coronagraph. http://en.wikipedia.org/wiki/Coronagraph (last modified on May 07, 2016).
45. M. E. Fayad et al. (2007). Planetary systems, formation of, David Darling, entry. *The Internet Encyclopedia of Science*, http://www.daviddarling.info/encyclopedia/P/plansysform.html (accessed September 23, 2007).
46. P. Podsiadlowski (1993). Planet formation scenarios. In: *PlanetsAroundPulsars; Proceedings of the Conference* 36: 149. Bibcode:1993ASPC...36..149P.
47. Charles Q. Choi (September 1, 2011). Alien Life More Likely on "Dune" Planets. *Astrobiology Magazine*.
48. S. Seager (2013). Exoplanet habitability. *Science* 340: 577. doi: 10.1126/science.1232226.
49. J. P. Amend, A. Teske (2005). Expanding frontiers in deep subsurface microbiology. *Palaeogeography, Palaeoclimatology, Palaeoecology* 219 (1–2): 131–155.
50. Further away planets "can support life" say researchers. *BBC*, January 7, 2014. Last updated at 12:40.
51. D. S. Abbot, E. R. Switzer. The Steppenwolf: A proposal for a habitable planet in interstellar space. *The Astrophysical Journal Letters* 735 (L27): 4pp.
52. A. Loeb. The habitable epoch of the early Universe. *International Journal of Astrobiology* 13 (4): 337-339.
53. M. Linsenmeier, S. Pascale, V. Lucarini (January 21, 2014). Habitability of Earth-like planets with high obliquity and eccentric orbits: Results from a general circulation model. *Astronomy & Astrophysics*, 19pp. arXiv:1401.5323.
54. P. Kelley (April 15, 2014). Astronomers: "Tilt-a-worlds" could harbor life. *UW Today*. http://www.washington.edu/news/2014/04/15/astronomers-tilt-a-worlds-could-harbor-life/ (retrieved August 12, 2016).
55. J. C. Armstrong et al. (April 14, 2014). Effects of extreme obliquity variations on the habitability of exoplanets. *Astrobiology* 14(4): 277–291.
56. P. Kelley (July 18, 2013). A warmer planetary haven around cool stars, as ice warms rather than cools. *UW Today*. http://www.washington.edu/news/2013/07/18/a-warmer-planetary-haven-around-cool-stars-as-ice-warms-rather-than-cools/ (retrieved August 12, 2016).
57. A. L. Shields et al. (March 14, 2014). Spectrum-driven planetary deglaciation due to increases in stellar luminosity. *The Astrophysical Journal Letters* 785: L9. doi:1088/2041-8205/785/1/L9.
58. R. Barnes et al. (2013). Tidal Venuses: Triggering a climate catastrophe via tidal heating. *Astrobiology* 13 (3): 225–250..
59. B. Jackson, R. Barnes, R. Greenberg (2008). Tidal heating of terrestrial extra-solar planets and implications for their habitability. *Monthly Notices of the Royal Astronomical Society* 391 (1): 237–245.
60. G. Anglada-Escudé et al. (2013). A dynamically-packed planetary system around GJ 667C with three super-Earths in its habitable zone. *Astronomy & Astrophysics* 556: A126.

61. NASA Ames Research Center. Kepler. Kepler-22b, our first planet in the habitable zone of a Sun-like Star (December 05, 2011). http://kepler.nasa.gov/news/index.cfm?FuseAction=ShowNews&NewsID=165 (retrieved February 28, 2012).
62. Staff (September 20, 2012). LHS 188—High proper-motion star. Center de donnéesastronomiques de Strasbourg (Strasbourg astronomical Data Center). http://simbad.u-strasbg.fr/simbad/sim-id?Ident=HIP+19394 (retrieved September 20, 2012).
63. A. Méndez (August 29, 2012). A hot potential habitable exoplanet around gliese 163. University of Puerto Rico at Arecibo (Planetary Habitability Laboratory). http://phl.upr.edu/library/notes/athermalplanetaryhabitabilityclassificationforexoplanets (retrieved September 20, 2012).
64. N. T. Redd (September 20, 2012). Newfound alien planet a top contender to host life. Space.com. http://www.space.com/17684-alien-planet-gliese-163c-extraterrestrial-life.html (retrieved September 20, 2012).
65. W. J. Borucki et al. (2013). Kepler-62: A five-planet system with planets of 1.4and 1.6 Earth radii in the habitable zone. *Science* 340 (6132): 587.
66. NASA. NASA's Keplerdiscovers its smallest "habitable zone" planets to date. http://www.nasa.gov/mission_pages/kepler/news/kepler-62-kepler-69.html#.V_eWDvl9670 (retrieved April 18, 2013).
67. G. Israelian, E. D. Mena, N. C. Santos, S. G. Sousa, M. Mayor, S. Udry, C. D. Cerdeña, R. Rebolo, S. Randich (2009). Enhanced lithium depletion in Sun-like stars with orbiting planets. *Nature* 462 (7270): 189–191.
68. L. Vu (2006). Planets prefer safe neighborhoods. NASA Spitzer Space Telescope News. http://www.spitzer.caltech.edu/news/863-feature06-31-Planets-Prefer-Safe-Neighborhoods (retrieved August 20, 2016).
69. The Royal Observatory, Edinburgh. http://www.roe.ac.uk/roe/support/pr/rsexhibition/typesofsolarsystem.pdf (retrieved August 03, 2016).
70. E. Howell (February 6, 2013). Closest "Alien Earth" may be 13 light-years away. Space.com. TechMediaNetwork. http://www.space.com/19667-closest-alien-earth-exoplanets.html (retrieved February 7, 2013).
71. R. Kopparapu (March 2013). A revised estimate of the occurrence rate of terrestrial planets in the habitable zones around kepler m-dwarfs. *The Astrophysical Journal Letters* 767: L8. doi:10.1088/2041-8205/767/1/L8.
72. M. Lukac (2011). Keplerfinds bizarre systems. *International Business Times*. International Business Times Inc. http://www.ibtimes.com/kepler-finds-bizarre-systems-272649 (retrieved March 2, 2011).
73. D. C. Fabrycky (2010). Non-Keplerian dynamics.arXiv:1006.3834[astroph. EP].
74. A. B. Bhattacharya et al. (2009). A roadmap for detecting extraterrestrial intelligent life. *Proceedings for Progress in Electromagnetic Research* (Beijing, China). 1400–1403.
75. A. B. Bhattacharya et al. (2010). An alternative search strategy for interstellar communication. *Indian Journal of Physics* 84: 511–515.

8

Wonders of the Solar and Extrasolar Planetary Systems

8.1 Introduction

The universe, our galaxy, our solar system, and the Earth–Moon double planet system demonstrate clearly some remarkable evidence of highly intelligent design. If we consider them separately, each characteristic appears to be highly improbable due to random chance. When taken all of them together, the probability of random chance becomes as small as to be impossible. An alternative thought, designed by an intelligent creator is a more realistic explanation to many of the civilized people. In either way, we must admit that we are nothing but a product of a miracle—either a miracle of chance or a miracle of design. Let us look at some of the improbable highlights for the design of the wonders of our Solar and extrasolar planetary systems.

8.2 Incredible Design of the Solar System

Modern astronomers are gradually gaining knowledge about the motions of the heavens. Astronomers are observing and uncovering many astonishing examples of chaotic behavior of the heavens. Nonetheless, the long-term stability of the solar system remains a perplexing, unsolved issue [1].

8.2.1 Our Unique Location in the Galaxy

The Sun and our solar system are located in a stable orbit within our galaxy for the last 4.5 billion years. This orbit lies far from the center of our galaxy, in between the spiral arms. The stability of our position is possible as the Sun is one of the rare stars that lie within the *galactic corotation radius*. Typically, the stars in our galaxy orbit the center of the galaxy at a rate that differs from the rate of the trailing spiral arms. Most of the stars located between spiral arms will not remain there for long periods of time, but would eventually be swept inside a spiral arm. Our location in the universe gives us an unobstructed view of the debris and gases found in the spiral arms [2]. If we were within the spiral arms, our view would be significantly impaired. Second, being outside the spiral arms, we are essentially in a location that is safer than anywhere else in the universe. We are, in fact, removed from the more densely occupied areas, where stellar interactions can lead to disruption of planetary orbits. Moreover, we are farther from the deadly effects of supernovae explosions. The 4+-billion-year longevity of life on Earth, considering the time required to prepare the planet for human occupation, would not have been possible at most other locations in our galaxy [2].

8.3 Mercury–Venus Encounter

Investigation reveals some unusual design in our solar system. With the sophistication of the computer systems, scientists are gaining the ability to model the dynamics of the solar system. They are constantly asking *what-if* questions regarding the presence and size of planets. The presence of Jupiter is needed to permit advanced life to exist on the Earth. But the large mass of Jupiter, along with the other gas giants, has a profound destabilizing effect on the inner planets [3]. In the absence of the Earth–Moon system, the orbital period of Jupiter sets up resonance over the period of 8 million years. The effect of this resonance causes the orbits of Venus and Mercury to become highly eccentric. As a consequence, the orbits will eventually become close enough so that there would be a *strong Mercury–Venus encounter.* This encounter would lead to the ejection of Mercury from the solar system and will cause an alteration of the orbit of Venus. For simulations, the scientists implemented the stabilizing effect of the Earth–Moon presence, which requires a planet with at least the mass of Mars and within 10% of the distance of the Earth from the Sun [3].

8.4 Earth at the Edge of the Life Zone

The unique arrangement of large and small planetary bodies in the solar system may be considered essential to ascertain the 4+-billion-year stability of the system [4]. Moreover, it appears from the cycle of ice ages that the Earth is at the edge of the life zone for our star. It is not yet clear that the stable planetary systems, in which a small Earth-like planet resides in the habitable zone, exist in any other galaxy in our universe [4,5]. It is not certain that the other essential components are required for life to exist anywhere in the universe.

8.4.1 Axial Tilt and Eccentricity of Orbit

The Earth is titled on its axis at an angle of 23.5°, which accounts for the seasons. Two factors impact the progression of seasons, of which the most important is the location of land masses on the Earth. All of the continental land mass is partially located in the Northern Hemisphere [6,7]. As land has a higher capacity to absorb the Sun's energy, the Earth becomes warmer when the Northern Hemisphere points toward the Sun. This appears to be the point at which the Earth is farthest from the Sun: at the aphelion of its orbit. If the reverse were true, the seasons on the Earth would be much more severe (hotter summers and colder winters).

8.5 Presence of an *Impossibly* Large Moon

The best explanation for the Moon's existence is that a Mars-sized planet crashed into the Earth around 4.25 billion years ago (the age of the Moon). The probability of two planets colliding in the same solar system is very remote at this point in time but more likely to have happened in the early formation of our solar system. Any *normal* or severe collision would not have resulted in the formation of the Moon. The small planet must have collided with a glancing blow accounting the angular momentum of the Earth–Moon system. Further, the Moon is moving away from the Earth, presently at 2 inches per year. If we calculate backward, we find that the Moon must have formed just outside the Roche limit, the point where an object would be torn apart by the Earth's gravity. A collision ejecting material less than the Roche limit would have only formed rings around the Earth. Computer models reveal that a collision of a small planet with the Earth would have been very precise for any moon to have been formed at all.

8.6 Earth's Atmosphere

The role of Moon is considered important for life on Earth. The collision of the small planet with the Earth caused the ejection of the majority of the Earth's primordial atmosphere. In the absence of this collision, we would have an atmosphere similar to Venus, which is 80 times more than that of the Earth. Such a thick atmosphere on Venus produced a runaway greenhouse effect, leaving a dry planet with a surface temperature of 800°F. The Earth would have suffered similar fate if most of its primordial atmosphere had not been ejected into the outer space. It is important to note that the Earth is 20% more massive than Venus, as well as being a certain distance from the Sun. These two factors have led to a terrestrial atmosphere much thicker than that of Venus. Due to some surprising reason, we have a very thin atmosphere and also the right density for maintaining the presence of liquid, solid, and gaseous water essential for life.

8.7 Slowing Rotation Making Advanced Life Possible

Scientists explained that the Earth originally had a rapid rotational period of 8 hours, which would have resulted in surface wind velocities in excess of 500 miles/h. The gravitational tug of the Moon for the last 4+ billion years has diminished the rotation period of the Earth to 24 hours. Similarly, the gravitational attraction of the Earth on the Moon has reduced its rotational period to 29 days. The wind velocity of 500 miles/h would not be conducive to the existence of higher life forms.

8.7.1 Van Allen Radiation Shield: Unique to Earth

Another important result of the collision of the Mars-sized planet with the Earth is the presence of large and heavy metallic core of the Earth. Earth has the highest density of any of the planets in our solar system. The large nickel-iron core is responsible for our high magnetic field that produces the Van Allen radiation shield. This radiation shield protects the Earth from radiation bombardment. In the absence of this shield, life would not be possible on the Earth. The only other rocky planet to have magnetic field is Mercury. However, its field strength is 100 times less than that of the Earth. The Van Allen radiation shield may be considered as a unique design to Earth. Even our sister planet Venus has no magnetic field. The lack of a magnetic field on Venus is assumed to have caused the planet to lose virtually all of its water through stripping by the solar wind.

8.7.2 Unique Continental Crust and Tectonic Activity

Earth is unique in many ways compared to other rocky planets in our solar system. The Earth has a unique continental crust, different from any other planet in our solar system. The mechanisms responsible for this unique continental crust are not certain. Perhaps the greatest dilemma facing those interested in understanding or gathering the knowledge of how the continents formed is their composition [7,8]. However, the Earth's crust is much thinner (4 km) than that of Venus (30 km). Tectonic processes cannot occur with such thick plates. If most of the crust of our Earth had not been blown away during the formation of its moon, the Earth would have no continents but would be completely covered by water. The tectonic processes are extremely important to maintain life on our planet by recycling minerals and nutrients.

8.8 Other Earth-Size Planets: Either Deserts or Water Worlds

Planets like the Earth, with huge amounts of water and land, are virtually impossible to form. Large planets do not form continents as the increased gravity opposes significant mountain and continent formation. Earth-sized planets are completely flooded and any land formed is eroded by the seas in a very short period of time in the absence of tectonic activity. Smaller planets lack tectonic activity and hence would have no land masses but would be completely covered with water. In theory, within the countless galaxies and stars, there will be terrestrial planets that will only have small dull rocks in orbits and much different than Earth.

8.9 Reduction of Greenhouse Gases with Increasing Solar Luminosity

Another uniqueness of the Earth is the temperatures suitable for life over a period of billions of years. At the formation of the solar system, the Sun was nearly one-third less luminous than it is now. It is postulated that certain greenhouse gases must have been present at higher concentrations to prevent the Earth from becoming a frozen planet. Carbon dioxide levels could not have been sufficiently high to compensate for the lower solar luminosity. The presence of other greenhouse gases, for example, ammonia or methane, is also problematical, as it is known that the Earth possessed an oxidative atmosphere even 4 billion years ago [9]. Moreover, ammonia is highly sensitive to solar UV radiation, and ammonia at levels required to influence the Earth's temperature would have prevented photosynthetic organisms from fixing nitrogen (i.e., protein, DNA, and RNA synthesis would have been prevented). Fossil evidence suggests that photosynthetic organisms have been present on the Earth for at least 3.5 billion years. Methane also has an identical problem to ammonia; it is sensitive to solar UV radiation in an oxidative atmosphere. The problem is still unsolved, but some unique conditions have existed to prevent the Earth from becoming a planet frozen in solid ice in the early stages or a sweltering inferno at present. It seems that life itself may have been responsible for preventing the Earth from turning into a scorched planet similar to Venus. Scientists have reported a methane-metabolizing Archaea [10] in the extreme pressures of deep sea sediments. It has been estimated that these bacteria-like organisms consume 300 million tons of methane every year to prevent the Earth from turning into a furnace. If they had not been established at some point in the history of Earth, we probably would not have been here [11]. On early Earth, the microbes might have been even more significant. Atmospheric scientists pointed out that methane levels in the atmosphere may have been 1000 times greater than they are today, initially formed by volcanoes and then by methane-producing microbes. Initially, methane may have been beneficial, producing a greenhouse effect that prevented the planet from freezing. However, if the rise in methane levels had gone unchecked, Earth might have become too hot for life to exist, as Venus.

8.10 Need for Jupiter-Sized Planets at 5 AU from Its Star

We have mentioned the destabilizing effects of large planets in our solar system. But these large bodies are essential for life to exist on the Earth. The study shows that Jupiter is the indirect cause of oceans on the Earth. Investigations also concluded that comets brought water to the Earth. However, this theory has certain problems. The water on the Earth contains 150 ppm deuterium or heavy hydrogen, which is five to six times the deuterium-to-hydrogen ratio observed in the Sun and in the solar nebula gas. Further, it is only about a third of the deuterium-to-hydrogen ratio measured in the comets Halley, Hyakutake, and Hale-Bopp. However, the deuterium-to-hydrogen ratio in meteorites is similar to that found in the oceans of the Earth. It is hypothesized that the presence of Jupiter transferred huge amount of water-containing meteorites into the inner solar system soon after it formed. It is further possible that Jupiter was responsible for sending the Mars-sized planet that formed the Moon. Almost all large planets have been seen to be closer to their stars than the Earth is to the Sun. Despite having been responsible for the shower of meteors that pelted the early Earth, Jupiter is our great protector and is responsible for collecting and ejecting a large proportion of the comets that enter into the orbit around the Sun (e.g., comet Shoemaker-Levy). Without Jupiter, life on Earth at this time would be difficult or impossible due to the large number of cometary collisions (nearly 1000–10,000 times more collisions) with the Earth [12]. There are many large planets observed around other stars, but these planets are not far away from their stars to stabilize the orbits of planets in the zone that can support life or protect these inner planets from cometary bombardment. From that point of view, the presence of Jupiter-like planets in the universe is a rare event.

8.11 Life Support—Galaxy–Sun–Earth–Moon System

The uniqueness of the galaxy–Sun–Earth–Moon system for life support is based on the assumption that life is based on and also on boron or silicon [13]. But these elements do not form very long-chained compounds that would make any form of life based on these elements virtually impossible [14]. Life based on carbon needs water to exist in the liquid state. For all practical purposes, this range is even narrower. There are some bacteria that can exist near the boiling point but they are very specialized [15]. Nearly all other life forms must exist below a temperature of 50°C (122°F). This is the major constraint on the system that demands stabile galaxies, stabile stars, stabile planetary orbits, and exact rotational characteristics.

8.12 Are Planetary Systems Filled to Capacity?

Computer simulations indicate that the answer may be yes. However, observations of extrasolar systems may provide the ultimate test. In 1605, Kepler discovered that the orbits of the planets are ellipses rather than combinations of circles, as astronomers had considered since antiquity. Newton was then able to prove that the same force of gravity that pulls apples to the ground keeps planets in their elliptical orbits around the Sun. But he was worried that the accumulated effects of the weak gravitational tugs between neighboring planets would increase their orbital eccentricities until their paths eventually crossed, leading to collisions and ultimately to the destruction of the solar system [16]. By 1800, the mathematician Laplace had concluded that the solar system needs no such guiding but is, in fact, naturally self-correcting and stable. Laplace thought that, given the exact position and momentum of every object in the solar system at any one time, it would be possible to calculate from the laws of motion precisely where everything would be at any future instant, no matter how remote [16]. In the late nineteenth century, Henri Poincaré suggested that Laplace had simplified some of his equations by removing terms as superfluous, leading him to overlook the possibility of chaos in the solar system. Calculations using high-speed computers provided evidence that the solar system is only marginally stable and that its detailed behavior is unpredictable over long time periods [16]. Theorists have used powerful computer simulations to explore the formation of planetary systems under a wide range of starting conditions, while observers are rapidly discovering planetary systems around many other stars. The evidence ultimately suggests that such systems may be filled nearly to capacity. The abundance of observational data from the planetary systems will stimulate and test many ideas about the delicate balance between order and chaos among the worlds.

8.12.1 Gaps in Understanding

In 1866, the American astronomer Daniel Kirkwood gave the first real evidence for instability in the solar system in his studies of the asteroid belt [16]. This belt lies between the orbits of Mars and Jupiter. At the time, only about 90 asteroids were known but that meager population was sufficient for Kirkwood to notice several *gaps* in the distribution of their orbital periods. It was noticed that the orbital periods of planets, asteroids, and comets increase with orbital size in a well-defined manner. Kirkwood observed that no asteroid had a period near 3.9 years, which he noted to be one-third that of Jupiter.

An asteroid that orbits the Sun exactly three times, while Jupiter goes around just once, would make its closest approaches to the giant planet at the same point in its own orbit and in addition get a similar gravitational kick from its massive celestial neighbor each time. Astronomers refer to such an asteroid as being in a 3:1 mean-motion resonance. Other gaps in the asteroid belt correspond to places where the ratio of the orbital period of Jupiter to that of an asteroid would be 5:2 or 7:3. An asteroid in such a resonant orbit has its eccentricity increased until the body either collides with the Sun or a planet. Many of the meteorites that strike Earth are fragments that were ejected from the asteroid belt after straying into one of the resonant gaps. When the solar system was forming, the Kuiper belt contained hundreds of times more mass than it does now. The objects in the belt represent only the small fraction that managed to survive and it is equally true for the asteroid belt. Gravitational sculpting by the planets has severely depleted both populations, leaving the Kuiper and asteroid belts as remnants of the primordial planetesimal disk. Thousands of small worlds, known as Trojan asteroids, share Jupiter's orbit around the Sun (Figure 8.1a), leading or following the planet by about 60°. These bodies are trapped in a so-called 1:1 mean-motion resonance, the planet and asteroid having the same orbital period. This configuration inhibits very close approaches to Jupiter and is relatively stable. Similar families of co-orbital asteroids accompany both Neptune and Mars around the Sun. Eros (Figure 8.1b) gives a sense of what most asteroids probably look like. Eros is about 30 km long, much too small for its gravity to make it spherical.

Chaotic systems are basically deterministic. In that they follow the laws of classical physics precisely but they are fundamentally unpredictable. The nature of chaos was not well understood until recently, when developing computer power permitted mathematicians to explore it in sufficient detail. No one in Laplace's day imagined that the solar system, then taken as the paradigm of clockwork stability, is really vulnerable to chaos.

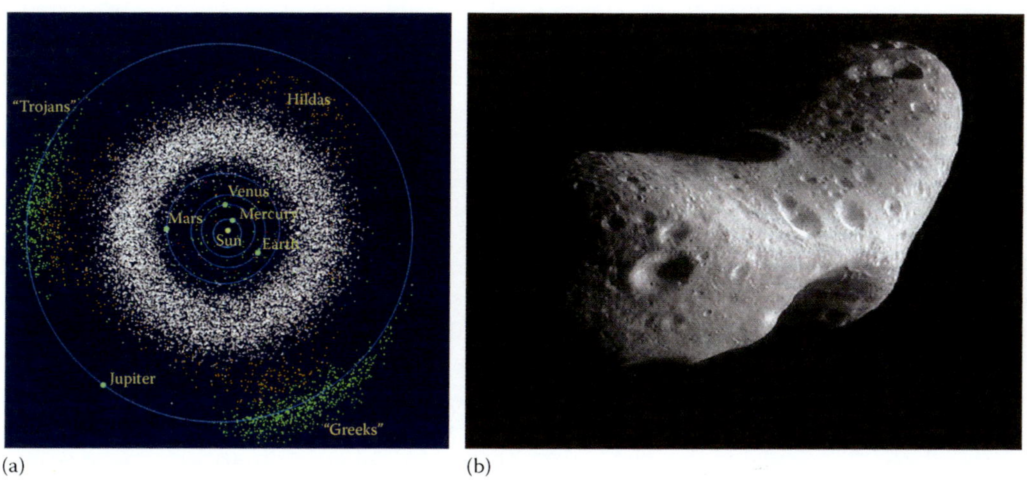

FIGURE 8.1
(a) The asteroids of the inner solar system and Jupiter: The donut-shaped asteroid belt is located between the orbits of Jupiter and Mars. ■ Sun ■ Jupiter trojans ■ Orbits of planets ■ Asteroid belt ■ Hilda asteroids (Hildas) ■ Near-Earth objects (selection) and (b) NASA photograph of EROS. (Courtesy of Wikimedia Commons, https://commons.wikimedia.org/w/index.php?curid=1951518 [last modified November 08, 2014]; Courtesy of NASA/Johns Hopkins University Applied Physics Laboratory, http://discovery.nasa.gov/SmallWorlds/cometNEXT.cfml [retrieved August 12, 2016].)

8.13 Cleaning Up the Solar System

Jacques Laskar [17] of the Bureau des longitudes in Paris carried out the most extensive calculations for investigating the long-term stability of the solar system. He simulated the gravitational interactions among all the eight planets over a period of 25 billion years or for a period five times the age of the solar system. He found that the eccentricities and other elements of the orbits undergo chaotic excursions. This simulation makes it impossible to predict the locations of the planets after a hundred million years. Does Laskar's result indicate that the Earth might eventually find itself in a highly elliptical orbit, taking it much closer to and farther from the Sun? In other words, will it mean the solar system could lose a planet? The immediate answer is perhaps, *no*. This is because of the fact that even the chaos has to operate within physical limits. As an example, meteorologists cannot predict the weather that is another chaotic system as far as a month in advance, but they can be quite confident that those conditions will fall within a certain range. This is because external constraints, for example, the Sun's brightness and the length of the day, set certain limits on the overall system. Interestingly, Laskar observed that despite the influence of chaos on the exact locations of the planets, their orbits remain relatively stable over billions of years. This means that the long-term configuration is highly unpredictable, and the orbits remain well behaved for preventing collisions between the neighboring planets. In this case, an external constraint is imposed by the conservation of angular momentum in the system, which limits the excursions of orbital eccentricity for bodies of planetary mass [17]. The orbits of the giant outer planets are most stable, while the smaller terrestrial planets (particularly, Mars and Mercury) are more vigorously tossed about. The simulations reveal that over millions of years the terrestrial planets undergo substantial excursions in their eccentricities to clear out debris from the intervening orbital space. But this is not sufficiently large enough to permit collisions between them [17]. Laskar noted one possible exception for Mercury that has a small but finite chance of colliding with Venus on a timescale of billions of years. From an elaborate analysis, he ultimately concluded that the solar system is *marginally stable*. To maintain its marginal stability, the solar system has been eliminating objects on a timescale comparable with its age at every epoch. It appears from the analysis that billions of years ago the solar system may have contained more planets that it does now. According to this concept as the solar system matured, it managed to remain stable against the breakout of large-scale chaos by reducing the number of planets and thereby increasing the spacing between them, which means that the present number must be about as large and their spacing about as small as allowed by the system's long-term stability. The solar system has increased its internal order by exporting disorder, called entropy, to the rest of the galaxy to receive the chaotically ejected objects. This process is known as dynamical relaxation that operates in star clusters and in entire galaxies and also in evolving planetary systems. As such systems expel their most unstable members, the orbits of the remaining objects become automatically more compact. Extensive computer simulations reveal that the eight planets largely disturb the motions of test particles placed on circular orbits at most of the locations in the solar system. These particles are sent into close encounters with the planets that remove them in only a few million years that is essentially a small fraction of the age of the solar system. These simulations also identify some regions where objects can survive for far longer times. One of the regions, according to the simulation, is between the orbits of Mars and Jupiter—and the asteroid belt. Computer simulations by Jack Lissauer and colleagues

[18] at NASA suggested that if an Earth-size planet had formed there, it could remain in a stable orbit for billions of years because the zone of the asteroid belt is well populated and therefore be relatively immune to disturbance. The same study, however, reported that a giant planet in the asteroid belt would soon become unstable. The Kuiper belt is another region of stability, because there are no other planets to stir up the neighborhood beyond the orbit of Neptune. Aside from such islands of stability, interplanetary space is considerably empty. The majority of the small objects orbiting between the planets, for example, Earth-crossing asteroids and short-period comets, are transient interlopers that leaked into the neighborhood from the asteroid and Kuiper belts.

8.14 Resonance and Predictability of Kuiper Belt Objects

Resonance occurs when two periods have a simple numerical ratio. The fundamental period for an object in the solar system is its orbital period and orbital resonances pervade the solar system. In 1867, the American astronomer Daniel Kirkwood observed that the asteroid belt is not distributed at random [19]; rather there were distinct gaps in the belt at locations that corresponded to resonances with Jupiter. Another very common form of resonance in the solar system is the spin–orbit resonance where the period of spin (i.e., the time it takes the planet or Moon to rotate once about its axis) has a numerical relationship with its orbital period. An example is our own Moon that is in a 1:1 spin–orbit resonance, which keeps the far side of the Moon away from the Earth. The histogram in Figure 8.2 shows the numbers of Kuiper belt objects for a given distance (in AU) from the Sun.

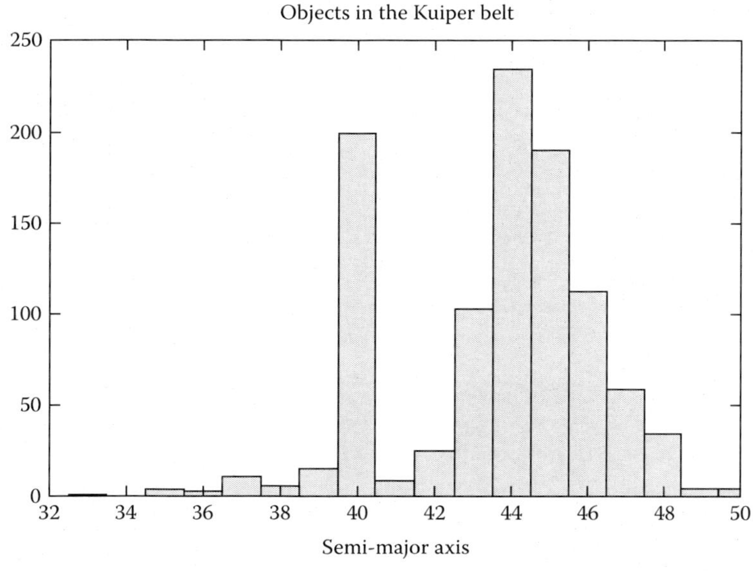

FIGURE 8.2
Histograms showing the numbers of Kuiper belt objects for a given distance (in AU) from the Sun.

The planets' orbits are chaotic over longer timescales [20]. This means that the position of a planet along its orbit becomes ultimately impossible to predict with any certainty. However, in some cases the orbits themselves may vary dramatically. In fact, the chaos manifests most strongly as changes in eccentricity, with some orbits of the planets becoming significantly elliptical [21].

8.15 Some Scenarios of Resonances

- ***Neptune–Pluto resonance***: The Neptune–Pluto system is in a 3:2 orbital resonance. Cohen and Hubbard at the Naval Surface Warfare Center Dahlgren Division discovered the resonance in 1965. Although the resonance remains stable in the short term, it becomes impossible to predict the position of Pluto with any degree of accuracy, as the uncertainty in the position develops by a factor e (mathematical constant) with each Lyapunov time, which for Pluto is 10–20 million years into the future [22]. Hence, on the timescale of hundreds of millions of years Pluto's orbital phase becomes impossible to find even if Pluto's orbit appears to be perfectly stable on 10 Gyr timescales [23].

NOTE 1: The number e is an important mathematical constant that is the base of the natural logarithm and is nearly equal to 2.71828. This is the limit of $(1 + 1/n)\ n$ as n approaches infinity, an expression that arises in the study of compound interest. It can also be calculated as the sum of the infinite series: The constant e can be defined as the unique positive number a so that the graph of the function $y = ax$ has unit slope at $x = 0$. The function $f(x) = ex$ is known as the exponential function, and its inverse is the natural logarithm, or logarithm to base e.

$$e = \sum_{n=0}^{\infty} \frac{1}{n!} = 1 + \frac{1}{1} + \frac{1}{1 \cdot 2} + \frac{1}{1 \cdot 2 \cdot 3} + \cdots$$

NOTE 2: Gyr is the abbreviation for gigayear (i.e., 109 years or a billion years).

- ***Jovian moon resonance***: Jupiter's moon Io has an orbital period of 1.769 days, approximately half that of the next satellite Europa (3.551 days). They are in a 2:1 orbit–orbit resonance, which has important consequences as Europa's gravity perturbs the orbit of Io. With the movement of Io closer to Jupiter and then further away in the course of an orbit, it gathers significant tidal stresses causing active volcanoes as reported and observed by Voyager. Europa is also in a 2:1 resonance with the Jovian satellite Ganymede.
- ***Mercury–Jupiter 1:1 resonance***: Mercury is susceptible to Jupiter's influence due to a small celestial coincidence: Mercury's perihelion, the position where it gets nearest to the Sun, precesses at a rate of about 1.5° every 1000 years and Jupiter's perihelion precesses only a little slower. At one point, the two may fall into sync when Jupiter's constant gravitational tugs could accumulate and pull Mercury. This could either eject it from the solar system [17] or send it on a collision course with Venus, the Sun, or Earth with 1%–2% probability, hundreds of millions of years into the future [24].

8.16 Chaos from Geological Processes

The Earth's axial tilt owing to friction raised within Earth's mantle by tidal interactions with the Moon will be rendered chaotic at some point between 1.5 and 4.5 billion years from now [25,26]. Another project involved use of a supercomputer to integrate the orbits of the outer planets over 845 million years, nearly 20% of the age of the solar system. In 1988, Sussman and Wisdom recorded data that revealed that the orbit of Pluto shows signs of chaos, for the part to its peculiar resonance with Neptune [22]. If Pluto's orbit becomes chaotic, then technically the whole solar system is chaotic as each body, even one as small as Pluto, affects the others to some extent through gravitational interactions.

8.17 Making Worlds Is a Messy Business

These concepts fit into the prevailing theory of solar system formation, initially proposed by the philosopher Immanuel Kant [27] in 1755. According to the nebular accretion theory, the solar system and other planetary systems are formed by the condensation and accumulation of dust and gas in flattened debris disks orbiting around young stars. The theory has confirmed strong support in modern observations and astronomers now routinely detect such debris disks around newborn stars. The dust-sized particles in such a disk first coagulate to produce trillions of rocky asteroids and icy comets a few kilometers in diameter, termed planetesimals. These objects in turn collide and grow to form scores to hundreds of Moon- to Mars-sized bodies, known as planetary embryos, orbiting amid the swarm of remaining planetesimals. Some embryos in the outer parts of the disk grow sufficiently large for their gravity to capture the abundant gas from the nebula, giving rise to giant planets. The observational evidence supports that the worlds produced in the young solar system were subjected to intense bombardment, as seen by their surfaces being saturated with craters. A Mars-sized planetary embryo collided with and entirely melted the proto-Earth, explosively throwing off a great abundance of debris. Some part of this reassembled to form the Moon. Theoreticians use computer models for simulating the late stages of planetary formation taking into account the dynamical evolution of such systems while using a range of starting conditions to represent different debris disks. Many simulations generate planets orbits and masses that resemble those in our solar system. Some of the simulations produce systems with giant planets in more eccentric orbits. In these simulations, collisions and ejections reduce the number of growing planets and increase the average spacing between them. The planets effectively compete for space, *elbowing* each other apart. Astronomers are trying to verify by using various means whether such simulations reflect reality. For a long time, observers have been charting the configuration of other planetary systems, which were long assumed to exist. Planet hunters have detected more than 240 worlds orbiting around other stars. Of these, more than 60 have two or more known planets. So far, the observing techniques have limitations for detecting giant planets, on a minimal scale of 10 times more massive than Earth. Smaller terrestrial planets exist around many of those stars but current measurements cannot yet reveal them. Astronomers were surprised that most of extrasolar planets have orbits much more eccentric than those of giant planets in our solar system. It was believed that the other systems would resemble our own, with planets in nearly circular orbits. Initially many astronomers argued that our solar system is exceptional and most planetary systems were formed in a different way. This now appears very unlikely. The largest worlds left behind continue to grow by sweeping up smaller objects that remain bound to the central star, thus making planet formation seemingly an extremely messy business. A growing planetary system may be taken as an overly energetic infant learning to eat cereal with a spoon: some portion of the food is consumed but a major part ends up on the floor, walls, and ceiling. The majority of the known extrasolar planets are more massive and have shorter periods and more eccentric orbits than the planets of our solar system. However, that does not necessarily indicate that our system is anomalous. Subsequently developed observational techniques strongly support the discovery of massive planets with orbital periods of only a few years or less. This would make the giant planets of our solar system (with their longer orbital periods) appear to be near the limits of detection, when found from the distance of a nearby star.

8.18 Worlds on the Edge

Barnes and Quinn [28] at the University of Washington used computer simulations to find the stability of extrasolar systems having two or more planets. They observed that almost all systems with planets that are close enough to affect one another gravitationally lie near the edge of instability. The simulations revealed that small alterations in the orbits of the planets in those systems would lead to catastrophic disruptions. This interesting result might seem surprising. Further investigations revealed that a large fraction of planetary systems, including our own, lie dangerously close to instability. Barnes and Raymond [29] identified apparently empty regions of stability around the central star. They predicted that those regions contain planets small enough to have evaded detection. For example, the star 55 Cancri has four known giant planets; three of them close in with short orbital periods and a more distant planet with a period of about 15 years. Between the inner three and the outermost planet lies a large area where one or more new planets will eventually be found. This region includes the *habitable zone*, suggesting that the planet's surface temperature would permit liquid water to exist. There is a fascinating new hypothesis, which mentions that our solar system and other mature planetary systems are filled nearly to capacity. The present configurations of such systems contain as many planets as they can hold, spaced about as closely together as stability permits. This is, in fact, the expected outcome of the chaotic process that makes planets. A family of planetary embryos develops by feeding on a vast swarm of smaller objects in a debris disk until the system loses its brakes. Global instability then erupts and the larger worlds consume or eject the more erratic ones. This is continued till the system settles down into the mature state of marginal stability. The process is one of self-organization, increasing order within the system by means of exporting disorder to the external environment, particularly for a galaxy. Similar to good scientific hypothesis, this makes testable predictions. Astronomers will look for new planets in the stable regions in other systems. This process will continue for a long time, as the smaller planets are difficult to detect. As observational methods continue to improve, we may find whether the idea of *packed planetary systems* stands up to critical scrutiny.

8.19 Set of Explanations for Large-Scale and Small-Scale Properties

The origins and the characteristics of all the planets are governed by the natural forces that created and shaped the Solar System. This appears as a continuous process and not a random one. Modeling or theory for the process can be developed from the consideration of a set of illustration for the large-scale and small-scale properties. As far as the large scale is concerned, the planets are separated periodically with nearly circular orbits in the same plane. Also they revolve prograde.

For the small scale, the prograde rotations are in abundance. The systems of moons for the small scale can be divided into direct spherical orbits. The terrestrial planets have higher densities with thin or merely no atmospheres with a slow rotation, whereas Jovian worlds have lower densities with thick atmospheres.

A great deal of evolution occurred in the Solar System after it formed at 4.6 billion year ago; for example, chemical processing happened to the secondary atmospheres of Venus (scorched), Mars (frozen), and Earth (presence of life). Other significant examples are orbital evolution, ring formation, capturing of moons, and tidal locking between worlds as in the Pluto–Charon system. As a matter of fact, we are always witnessing a nonstatic or dynamic Solar System.*

8.20 More Planetary Systems: Exciting Times Are Ahead!

It is likely that more planetary systems will be known using the methods that found 51 Pegasi, 70 Virginis, and 47 Ursae Majoris. Some basic points in addition to some exciting questions are mentioned below:

- The 51 Pegasi planetary systems are quite different from our solar system. But the 70 Virginis and 47 Ursae Majoris systems appear to be more *normal*. With several known planetary systems, it is now possible to make more general statements and to better test theories of planetary formation.
- None of the extrasolar planets discovered so far are similar to the Earth, as expected given our current methods.
- What is the origin of the solar system? It is generally agreed that it condensed from a nebula of dust and gas. But the details are yet far from clear.
- How common are planetary systems around other stars? There is now good evidence of Jupiter-sized objects orbiting several nearby stars. What conditions allow the formation of terrestrial planets? It seems unlikely that the Earth is totally unique but we still have no direct evidence one way or the other.

* M. K. Rulison, Solar system formation, http://abyss.uoregon.edu/js/ast121/lectures/lec24.html and http://sites.oglethorpe.edu/mrulison/astronomy-web-lecture-notes/solar-system-formation/ (retrieved September 28, 2016).

- Is there life elsewhere in the solar system? If not, why is Earth special? Is there life beyond the solar system? Is intelligent life a rare and unusual or even unique event in the evolution of the universe or is it adaptable, widespread, and common?

8.21 Second Planetary System Like Ours Discovered

A European team of astronomers discovered a second planetary system, the closest parallel to our own solar system yet found. The system includes seven exoplanets orbiting a star with small rocky planets close to their host star and the gas giant planets further away. Interestingly, the system was hidden within the wealth of data from the Kepler Space Telescope. KOI-351 was the first system with a good number of planets showing a clear hierarchy like the solar system with small rocky planets in the interior and gas giants in the exterior. Three of the seven planets orbiting KOI-351 were detected and have periods of 59, 210, and 331 days, which are similar to the periods of Mercury, Venus, and Earth. However, the orbital periods of these planets change by as much as 25.7 hours. This is the highest variation detected in an orbital period of exoplanets so far. The gravitational pull of nearby planets, in closely packed systems, can cause the acceleration or deceleration of a planet along its orbit. These *tugs* are responsible for the variations in orbital periods. They are also useful to provide indirect evidence of further planets. By making use of advanced computer algorithms, Cabrera and his group [29] successfully detected four new planets orbiting KOI-351. But these planets are very close to their host star than Mercury is to our Sun, with orbital periods of 7, 9, 92, and 125 days. The system is highly compact with the outermost planet having an orbital period smaller than the Earth's. Most of the exoplanets are considerably different from the planets in our own solar system. A comparison between our solar system and a second planetary system: KOI-351 is made in Figure 8.3.

Some theories including planet migration and planet–planet scattering have been suggested for interpreting these differences [30]. But the fact of that planet formation is still poorly understood, though it is believed to be a key system in understanding planetary formation in general and the formation of the solar system in particular. Follow-up observations in this direction will not only permit astronomers to find how this planetary system formed, it will also provide further insight about the role of our own solar system at the time of early formation. Figure 8.4 shows the eight planets and Pluto with approximately correct relative sizes. One method to visualize the relative sizes in the solar system is to imagine a model wherein everything is reduced in size by a factor of a billion. If it is imagined in that way, then the Earth would be about 1.3 cm in diameter, which is similar to the size of a grape while the Moon would be nearly at a distance of 30 cm (about a foot) from the Earth. The Sun would be approximately 1.5 m in diameter, or about the height of a man and 150 m from the Earth, while the giant planet Jupiter would be 15 cm in diameter. A man on this scale would be the size of an atom, but the nearest star would be over 40,000 km away.

In Figure 8.4, we have not shown the numerous smaller bodies that inhabit the solar system. These are the satellites of the planets; the large number of asteroids (small rocky bodies) orbiting the Sun, mainly between Mars and Jupiter but also elsewhere; the

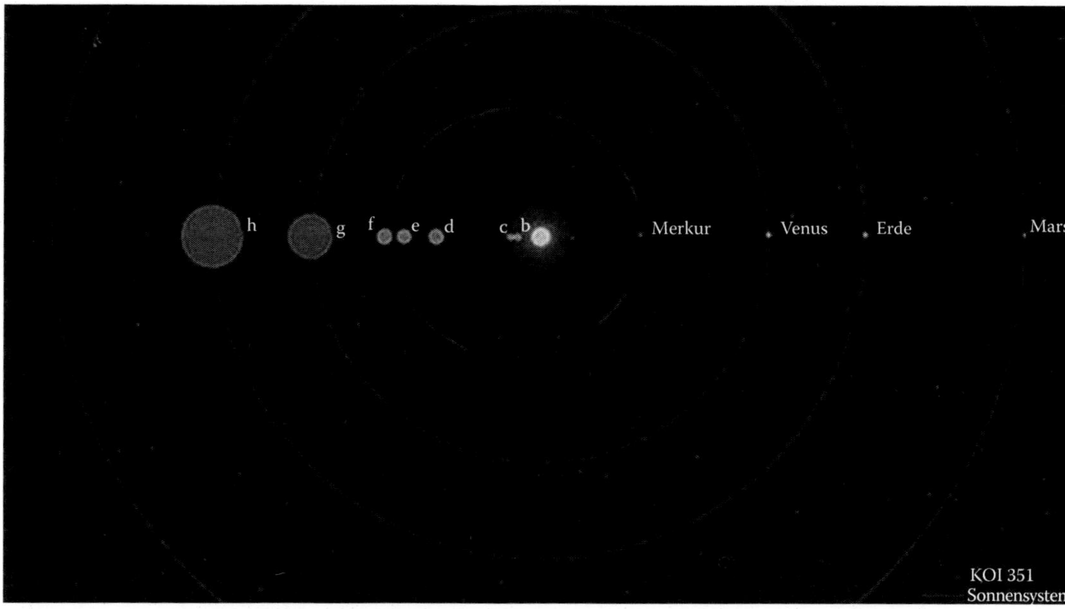

FIGURE 8.3
A comparison between our solar system and a second planetary system: KOI-351.

FIGURE 8.4
The eight planets and Pluto with approximately correct relative sizes.

comets (small icy bodies) that come and go from the inner parts of the solar system in highly elongated orbits and at random orientations to the ecliptic; and the many small icy bodies beyond Neptune in the Kuiper belt. The classification of these objects is a matter of some controversy. In general, the solar system is divided into planets (the big bodies orbiting the Sun), their satellites, popularly named as moons orbiting the planets, and asteroids, small dense objects orbiting the Sun and comets (small icy objects with highly eccentric orbits). However, in reality the solar system has been found to be much more complicated. The complicacy arises owing to many reasons, some of which are as follows:

- There are many small moons which probably started out as asteroids and were later captured by a planet.
- There are some moons larger than Pluto and two larger than Mercury.
- The Earth/Moon and Pluto/Charon systems are sometimes considered *double planets.*
- Sometimes comets fizzle out and become indistinguishable from asteroids.
- The Kuiper belt objects, including Pluto and others like Chiron, do not fit this scheme well. The eight bodies categorized as planets officially are also classified in several ways:
- By history:
  - Classical planets: Mercury, Venus, Mars, Jupiter, and Saturn
    - Known since pre-historical times
    - Visible to the unaided eye
    - In ancient times, this term also referred to the Sun and the Moon; the order was usually specified as Saturn, Jupiter, Mars, Sun, Venus, Mercury, and Moon, based on the time for them to go *all the way round* the sphere of the *fixed* stars
  - Modern planets: Uranus and Neptune
    - Discovered in modern times
    - Visible only with optical aid
  - Earth
- By size:
  - Small planets: Mercury, Venus, Earth, and Mars
    - The small planets have diameters less than 13,000 km
  - Giant planets: Jupiter, Saturn, Uranus, and Neptune
    - The giant planets have diameters greater than 48,000 km
  - The giant planets are sometimes also referred to as gas giants
- By position relative to the Sun:
  - Inner planets: Mercury, Venus, Earth, and Mars
  - Outer planets: Jupiter, Saturn, Uranus, and Neptune
  - The asteroid belt between Mars and Jupiter forms the boundary between the inner solar system and the outer solar system
- By composition:
  - Terrestrial or rocky planets: Mercury, Venus, Earth, and Mars
    - The terrestrial planets are mainly composed of rock and metal and have relatively high densities, slow rotation, solid surfaces, no rings, and few satellites.

  - Jovian or gas planets: Jupiter, Saturn, Uranus, and Neptune
    - The gas planets are mainly composed of hydrogen and helium and have low densities, rapid rotation, deep atmospheres, rings, and many satellites.
- By position relative to Earth:
  - Inferior planets: Mercury and Venus
    - Closer to the Sun than the Earth
    - The inferior planets show phases like the Moon's when seen from the Earth
  - Earth
  - Superior planets: Mars through Neptune
    - Farther from the Sun than the Earth
    - The superior planets always appear full or nearly so. It should be pointed out here that the International Astronomical Union (IAU) has decided that *classical* should refer to all eight planets but not Pluto. This makes some sense from a twenty-first century perspective

8.22 Speculation on Extrasolar Planetary Systems

In the sixteenth century, the Italian philosopher Bruno, an early supporter of the Copernican theory that the Earth and other planets orbit the Sun, gave the view that the fixed stars are similar to the Sun and are accompanied by planets [31]. In the eighteenth century, the same possibility was pointed out by Newton in his *Principia*. By a comparison to the Sun's planets, he wrote: "And if the fixed stars are the centers of similar systems, they will all be constructed according to a similar design and subject to the dominion of *one*" [32]. The theories he developed gained traction through the nineteenth and twentieth centuries despite a lack of supporting evidence. Long before their support by astronomers, conjecture on the nature of planetary systems had been a focus of search for extraterrestrial intelligence and has been a prevalent theme in science fiction. *De revolutionibus* of Nicolas Copernicus, (dated 1520–1541) is reproduced in Figure 8.5.

Prior to the discovery of extrasolar planets, astronomers believed that the configuration of our solar system was typical. Many of the astronomers used to believe that our solar system was representative of most planetary systems. An artistic impression of Kepler-186f is presented in Figure 8.6. Kepler-186f is an interesting exoplanet orbiting the red dwarf Kepler-186, about 500 light years from the Earth.

8.23 Multiplanet Systems

Presently, 462 systems, including the solar system, are known to consist of two or more planets. In terms of total confirmed planets now, the solar system with eight remains the largest planetary system, followed by HD 10180 with seven, the second largest [33].

NICOLAI COPERNICI

net, in quo terram cum orbe lunari tanquam epicyclo contineri diximus. Quinto loco Venus nono mense reducitur. Sextum denique locum Mercurius tenet, octuaginta dierum spacio circũ currens. In medio uero omnium residet Sol. Quis enim in hoc

pulcherrimo templo lampadem hanc in alio uel meliori loco poneret, quàm unde totum simul possit illuminare? Siquidem non inepte quidam lucernam mundi, alij mentem, alij rectorem uocant. Trimegistus uisibilem Deum, Sophoclis Electra intuentẽ omnia. Ita profecto tanquam in solio regali Sol residens circum agentem gubernat Astrorum familiam. Tellus quoque minime fraudatur lunari ministerio, sed ut Aristoteles de animalibus ait, maximam Luna cum terra cognationẽ habet. Cõcipit interea à Sole terra, & impregnatur annno partu. Inuenimus igitur sub hac

FIGURE 8.5
De revolutionibus of Nicolas Copernicus (p. 9 version, dated 1520–1541). (Courtesy of Wikipedia, De revolutionibus orbium coelestium, https://en.wikipedia.org/wiki/De_revolutionibus_orbium_coelestium [last modified August 18, 2016].)

FIGURE 8.6
Artistic impression of Kepler-186f (Courtesy of Wikipedia, https://commons.wikimedia.org/wiki/File:Kepler 186f-ArtistConcept-20140417.jpg [last modified January 29, 2015].)

For some time 55 Cancri was the largest known system [34]. Some systems, such as HD 10180 and Gliese 581, contain multiple unconfirmed planets pushing the total expected number of planets to be discovered in their systems higher. If sometimes HD 10180's two candidate planets are confirmed, it would be the largest by number of known planets with a total of 9 [34].

References

1. I. Peterson (1993). *Newton's Clock: Chaos in the Solar System*. W.H. Freeman and Co, New York.
2. M. V. Medvedev and A. L. Melott (2007). Do extragalactic cosmic rays induce cycles in fossil diversity? *The Astrophysical Journal* 664: 879–889.
3. K. Innanen et al. (1998). The earth-moon system and the dynamical stability of the inner solar system. *The Astronomical Journal* 116: 2055–2057.
4. R. Kerr (1999). Why the ice ages don't keep time. *Science* 285: 503–505.
5. J. A. Rial (1999). Pacemaking the ice ages by frequency modulation of Earth's orbital eccentricity. *Science* 285: 564–568.
6. J. E. Chambers (1998). How special is Earth's orbit? *American Astronomical Society*. DPS meeting #30, #21.07.
7. G. J. Taylor (1994). The scientific legacy of Apollo. *Scientific American* 271 (1): 40–47.
8. R. Rudnick (1995). Making continental crust. *Nature* 378: 571–578.
9. K. A. Maher and D. J. Stevenson (1988). Impact frustrations of the origin of life. *Nature* 331: 612–614.
10. Wikipedia. Archaea. http://en.wikipedia.org/wiki/Archaea (last modified on August 03, 2016).
11. C. Zimmer (2001). "Inconceivable" bugs eat methane on the ocean floor. *Science* 293: 418–419.
12. F. A. Rasio and E. B. Ford (1996). Dynamical instabilities and the formation of extrasolar planetary systems. *Science* 274: 954–956.
13. Wikipedia. Hypothetical types of biochemistry. http://en.wikipedia.org/wiki/Hypothetical_types_of_biochemistry (last modified on June 19, 2016).
    D. Darling. Silicon-based life. http://www.daviddarling.info/encyclopedia/S/siliconlife.html (accessed on August 06, 2016).
14. R. H. Dicke (1961). Dirac's cosmology and Mach's principle. *Nature* 192: 440.
15. Yellowstone National Park. Rare Microbes Flourish in Yellowstone National Park Geysers and Hot Springs. http://www.yellowstonepark.com/2011/06/some-like-it-hot-rare-microbes-flourishin-yellowstone-national-park-geysers-and-hot-springs/ (accessed on August 06, 2016).
16. S. Soter. Are planetary systems filled to capacity? *Astrobiology Magazine*, August 13, 2007. http://www.astrobio.net/topic/deep-space/cosmic-evolution/are-planetary-systemsfilled-to-capacity/ (accessed on August 06, 2016).
17. J. Laskar (1994). Large-scale chaos in the Solar System. *Astronomy & Astrophysics* 287: L9–L12.
18. J. J. Lissauer et al. (2014). Validation of Kepler's multiple planet candidates. II: Refined statistical framework and descriptions of systems of special interest. *The Astrophysical Journal* 784 (1): 21pp.
19. N. Hall (1994). *Exploring Chaos: A Guide to the New Science of Disorder*, p. 110. WW Norton & Company, New York.
20. W. B. Hayes (2007). Is the outer Solar System chaotic? *Nature Physics* 3 (10): 689–691.
21. I. Stewart (1997). *Does God Play Dice?* 2nd ed., pp. 246–249. Penguin Books, Westminster, UK.
22. G. J. Sussman and J. Wisdom (1988). Numerical evidence that the motion of Pluto is chaotic. *Science* 241 (4864): 433–437.
23. T. Ito and K. Tanikawa (2002). Long-term integrations and stability of planetary orbits in our solar system. *Monthly Notices of the Royal Astronomical Society* 336: 483–500.

24. D. Shiga. The Solar System could go haywire before the Sun dies. *New Scientist*, April 23, 2008. https://www.newscientist.com/article/dn13757-solar-system-could-go-haywire-before-the-sun-dies/ (accessed on August 06, 2016).
25. O. Neron de Surgy, J. Laskar (1997). On the long term evolution of the spin of the Earth. *Astronomy & Astrophysics* 318: 975–989.
26. I. Kant (1755). *Universal Natural History and Theory of the Heavens*. p. 367; translated by Stephen Palmquist in Kant's Critical Religion. Aldershot: Ashgate (2000), p. 320.
27. R. Barnes, T. Quinn (2001). A statistical examination of the short-termstability of the ν andromedae planetary system. *The Astrophysical Journal* 550: 884–889.
28. R. Barnes, S. N. Raymond (2004). Predicting planets in known extrasolar planetary systems. I. Test particle simulations. *The Astrophysical Journal* 617: 569–574.
29. J. Cabrera et al. (2013). The planetary system to KIC 11442793: A compact analogue to the Solar System. *The Astrophysical Journal* 781 (1): 13pp.
30. Cosmos (1991). *The New Encyclopædia Britannica*. 15th ed. Chicago 16: 787:2a. "For his advocacy of an infinity of suns and earths, he was burned at the stake in 1600."
31. I. Newton et al. (1999 [1713]). *The Principia: The Authoritative Translation and Guide*. p. 940. University of California Press, Oakland, CA.
32. F. vanLeeuwen (2007). Validation of the new Hipparcos reduction. *Astronomy & Astrophysics* 474 (2): 653–664.
33. R. Sanders (2002). Astronomers discover first planet in Jupiter-like orbit plus smallest planet outside solar system. University of California-Berkeley Press Release. http://www.berkeley.edu/news/media/releases/2002/06/13_planet.html (accessed September 28, 2016).
34. M. Tuomi (April 6, 2012). Evidence for 9 planets in the 10180 system. *Astronomy & Astrophysics*. e-print: 1204.1254v1.pdf.
35. M. K. Rulison, Solar system formation, http://abyss.uoregon.edu/js/ast121/lectures/lec24.html and http://sites.oglethorpe.edu/mrulison/astronomy-web-lecture-notes/solar-system-formation/ (retrieved September 28, 2016).

9

Solar Modulation and the Planetary Global Electric Circuit

9.1 Introduction

The subject of atmospheric electricity had its origins in the eighteenth century. The concept of the global atmospheric electric circuit in the early twentieth century by C. T. R. Wilson and matured considerably in the first decade of the twenty-first century [1–4]. Atmospheric electrical coupling occurs from near the Earth's surface up to the ionosphere at ~80 km altitude. This coupling takes place rapidly close to the speed of light (c), contrary to coupling mechanisms involving mechanical waves which propagate at speeds much lower than c [5]. The Earth's atmosphere is a layer of gases surrounding the Earth and is retained by Earth's gravity. On the basis of temperature distribution, the atmosphere can be classified into four layers consisting of the troposphere, stratosphere, mesosphere, and thermosphere. The temperature in the thermosphere remains nearly constant. The stratosphere and mesosphere regions are counted as the middle atmosphere above which the atmosphere is marked as the upper atmosphere. At the upper atmosphere the solar radiation and other sources ionize the neutral constituents producing plasma of ions and electrons. The region extending from the mesosphere to the thermosphere is known as the ionosphere where plasma dynamics is controlled by the collisions between the ionized particles and neutrals and also between the ionized particles themselves [6,7]. The region above the ionosphere is called the magnetosphere wherein the dynamics of charged particles is controlled largely by the Earth's magnetic field, as the density collision frequency is very low in this region. There is no sharp boundary between the upper ionosphere and the lower magnetosphere region. The study of electrical coupling between the troposphere and the ionosphere is an important assignment related to atmospheric electrodynamics [8]. Observations in mesosphere support strong electric fields of up to 10 V/m [9,10], indicating that the mesosphere should be treated as an active element in the atmospheric circuit. The formation of transient optical emissions in the mesosphere and lower ionosphere are named as sprites, which further supports the concept of strong mesospheric electric fields [11]. All this phenomena demand a critical searching of mechanisms related to electrodynamics caused by the effects of disturbances of tropospheric conductivity on lower ionosphere [12,13]. It is the purpose of this chapter to examine the present understanding of the link between the processes operative in the lower atmosphere and their electrodynamical coupling with the ionosphere. In spite of the existence of aerosol, ions, polar molecules, and convection, it has become difficult to apply the terrestrial model of a global circuit to outer planets of the solar system mainly because of the probable absence of a conducting surface. A different electrical model for the gas giant planetary atmospheres may be attempted but further theoretical work is necessary to enhance our understanding first. The same or at least similar atmospheric electrical processes may act across the solar system but a clear concept is initially essential to implement it successfully. Lightning has been detected on many other planets but without their elaborate knowledge of electrical and physical properties we are only able to approach so far as quasi-terrestrial global circuits. Ion-mediated nucleation has been noted on Earth but no successful attempt has yet been made to identify it in other planetary atmospheres. A comparative approach with all the available future information may provide new insight for successful development of electrical global circuits for other planets where the earth's electrical global circuit may be taken as a ready reference.

9.2 Essential Conditions for a Planetary Global Electric Circuit

The essential conditions for a dc global atmospheric electric circuit may be largely governed by (i) charge separation and generation by the method of convection to form the dipole structure causing electrostatic discharges and when no discharges are present, the role of precipitation to carry charges to the ground; (ii) mobile charged particles, to provide current flow and polar atmospheric molecules, to sustain charges; and (iii) evidence for conducting upper and lower layers. Table 9.1 provides a comparative statement of magnetic fields of Earth and Mars, which may contribute widely in the development for dc global electric circuit (GEC).

9.2.1 Planets: Earth and Mars

With a consideration that atmospheres of both Earth and Mars have some charge separation due to ionization, the following three aspects have been identified for characterization of the electrical systems: (a) If convective processes or any other processes cause sufficient charge separation for electrical discharges, the meteorological variations will be responsible to produce a global modulation of the electrical system; (b) If ions have dominant roles in the formation or removal of aerosol particles, there is a potential link between cosmic ray ionization and the atmospheric radiative balance of the concerned planet; (c) If any aerosol is present, it may have some link to local atmospheric electrification due to ion–aerosol interactions. In addition to the results of approximate surface magnetic field (in T) as presented in Table 9.1 some other available information related to the development of global circuit criteria met by planetary atmosphere are shown in Table 9.2.

TABLE 9.1

Comparison of Magnetic Fields of Earth and Mars

Planet	Earth	Mars
Approximate surface magnetic field (in T)	$<10^{-8}$	10^{-9}–10^{-4}

Sources: Smoluchowski, R., *Physics of Earth and Planetary Interiors*, 20, 247–254, 1979; Ip, W. H. et al., *Advances in Space Research*, 26, 1559–1563, 2000; Stevenson, J., *Earth and Planetary Science Letters*, 208, 1–11, 2003.

TABLE 9.2

Available Information of Global Circuit Criteria for Earth and Mars

Planet	Earth	Mars
Polar molecules	Yes	Yes
Charge separation	Yes	Has possibility
Conducting upper/lower layer	Yes	Upper: Yes Lower: Doubtful
Mobile charged particle in lower atmosphere	Yes	Yes

TABLE 9.3

Planetary Atmospheres of Earth and Mars, Showing the Principal Atmospheric Constituents and the Electrical Processes

Planet	Principal Atmospheric Constituents (% in Volume)	Lightning	Charged Aerosol	Ion Nucleation	References
Earth	N2 78.0 O2 20.9	C-G, I-C & C-C	Yes	No *Wilson* nucleation but may be ion-mediated nucleation	Rycroft et al. [17], Harrison and Aplin [18]
Mars	CO_2 95.3 N_2 2.7 Ar 1.6 O_2 0.13 CO 0.08	Electrostatic dust discharges are likely	Triboelectrically charged dust	Unlikely; no super saturation or condensable species	Fillingim [19], Farrell and Desch [20]

Table 9.3 gives an overview of the atmospheres of Earth and Mars including their electrical systems, in terms of the three principal aspects listed above. It provides information regarding theoretical predictions of ionization and other electrical processes and thereby the likelihood of a global atmospheric electric circuit.

9.3 Processes Linking the Global Atmospheric Electrical Circuit

Atmospheric electric currents can be transported by two different means, such as mobile charged particles and electrical discharges. This can be termed as an *atmospheric GEC*. If the conductive ionosphere of the earth and its surface are considered as spherical equipotentials, a potential difference of 250–300 kV exists between them, with the ionosphere as the positive spherical plate. The spherical conductor produced in this manner with the ionosphere surface system has a finite capacitance with an *RC* time constant of a few minutes. This suggests that there must be a constant current source to maintain the electric fields observed in the atmosphere [21]. The GEC is maintained due to almost constant generation of thunderstorms in the troposphere and thereby making a link with the upper atmosphere to the surface of the earth. The variations in the electrical potential of the ionosphere are communicated directly by the air–earth current flowing through cloud layers

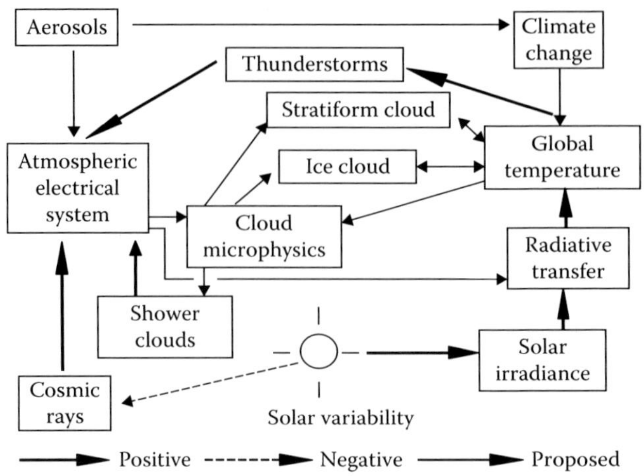

FIGURE 9.1
Flowchart indicates the processes linking the global atmospheric electrical circuit with global climate. (Data from Bhattacharya, A. B. et al., *International Journal of Engineering Science and Technology*, 3, 3221–3228, 2011.)

to the surface of the earth [22–25]. In fact, thunderstorms generated in the troposphere supply a net current by transferring positive charge to the conductive upper atmosphere. As a matter of fact, in a reverse way currents also flow from the surface up to the thundercloud [26]. Interchange of charges from particle interactions within a cumulonimbus cloud is followed by charge separation from gravitational settling of negatively charged particles and thus forming a dipole within the cloud. In course of time when the electric field exceeds the breakdown voltage of air, the cloud is discharged by lightning from cloud to ground, within the cloud, or cloud to other clouds [27]. It has also been noticed that under certain atmospheric conditions electrical discharges directed upward from the thundercloud to the ionosphere in the form of sprites, elves, or blue jets [27–31]. Electrostatic discharges by various means and associated precipitation from electrified shower clouds comprise the disturbed weather part of the electrical global circuit [32]. Figure 9.1 [33] summarizes the different processes that can result from the solar modulation of cosmic rays and link global atmospheric electrical circuit with global climate. The thick lines in the figure indicate established links while the thin lines indicate suggested links.

9.3.1 Electrification and Meteorological Processes in the Martian Atmosphere

Mars is unique in the solar system owing to the presence of dust in its climate. The fine dust grains on the surface can be suspended by wind to form dust storms and can even cover large areas of the planet. On the basis of the deductions from the terrestrial analogues in desert sandstorms, Mars is reasonably assumed to have substantial atmospheric electrification. Convective vortices which are very similar in structure and size to terrestrial dust devils play a very important role in lofting dust into the atmosphere. Meteorological models developed from terrestrial general circulation models have been utilized to find the role of dust in Martian global climate [34,35]. It appears that in many ways they resemble the electrically active terrestrial storm. When the electron density values are plotted against altitude for both terrestrial and Martian atmosphere during day as well as during night times separately (Figure 9.2), it appears that the characteristic variations are similar to some extent [34–36].

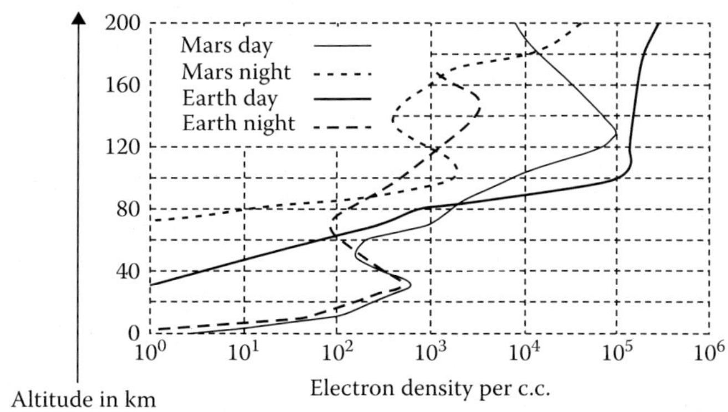

FIGURE 9.2
Variation of electron density in Earth and Mars. (Data from Bhattacharya, A. B. et al., *International Journal of Engineering Science and Technology*, 3, 3221–3228, 2011.)

The atmosphere of Mars consists of five levels, these are the troposphere, isothermal mesosphere, thermosphere, and exosphere with a mean surface temperature of about 200 K. So far there are no direct observations of electrical activity on Mars but it is reasonable to assume that the frequent small dust devils ($z < 100$ m) and larger storms ($z \sim 10$ km) would be electrified due to triboelectric charging process, with a possibility of resulting in some form of electrical discharges [37]. Numerical simulation suggested that the charging process is highly efficient to generate local fields within the dust devils (~20 kVm–1) which are enough to lead to generate breakdown [38]. It was further pointed out that larger storms are unavailable to sustain huge amounts of charge and will only initiate a corona and glow discharge. Using the existing knowledge of the Martian atmosphere and the outputs derived from various model calculations, we have predicted the appearance of lightning on Mars and probability of branching of streamers. As we have read earlier, convective vortices play an important role in lofting dust into the atmosphere. Photoelectrons dominate the Martian daytime atmospheric conductivity due to their high concentration and electrical mobility [39], given as 200 $m^2V^{-1}s^{-1}$ in CO_2 at 1.33 hPa. Martian atmospheric CO_2^-ions are orders of magnitude less mobile than the electron. During the night, no photoelectron emission occurs and charged particles are produced only by cosmic ray ionization, causing a significant diurnal air conductivity variation [40]. Charge exchange from particle interactions within a dust storm is followed by charge separation from gravitational settling of positively charged particles, creating a dipole within the cloud. At the time when the electric field becomes greater than the breakdown voltage, the cloud is discharged by lightning. These electrical discharges are directed downward from the dust storm, unlike terrestrial electrical discharges usually being directed upward from the thundercloud to the ionosphere thus generating sprites [41–44]. The principal atmospheric constituents of the planet Mars in percent (by volume) are reported [45] as CO_2 (95.3), N_2 (2.7), Ar (1.6), O_2 (0.13), and CO (0.08). The absorption of solar radiation by the atmospheric gases is responsible for producing photoelectrons which in turn loose their energy in the atmosphere through excitation and ionization. Charging experienced in the Martian dust storms were simulated in laboratory experiments as well as modeled extensively. The results showed conclusively that significant charges are generated due to triboelectric effect [46,47]. This effect is a type of contact electrification where certain materials become electrically charged. After coming into contact with a different material they separate from each other. During this triboelectric charging within the disturbed weather, the polarity and the strength of the charges produced differ according to their properties. Calculations [37] established that even for the 10 C (coulomb) amount of charge or for larger charges at 10 km above the surface, the conventional breakdown is surpassed within the dust cloud and not above it. This suggests the possibility of forming high current, high-temperature lightning channels on Mars in spite of low-current low-temperature sprite lightning discharges above the charge center. Once the critical breakdown field in the dust cloud had been locally surpassed, electron avalanches can evolve into streamers. This condition happens when the pre-breakdown (ambient) electron density is appropriate. From the measurements of pure nitrogen and pure argon, it has been reported that streamer properties like inception, length, width, and branching largely depend on gas composition. The presence of CO_2, N_2, and Ar as principal constituents of Mars have a vital role in atmospheric electrical processes causing triboelectrically charged dust and eventually produced lightning as electrostatic dust discharges under favorable condition [48]. Assuming sufficient atmospheric dynamics in the Martian atmosphere we would expect the occurrence of vigorous charge separation and dendritic structure of lightning [42]. According to our proposed hypothesis, highly dynamic electrical discharges on Mars continues to grow into new regions of space charges, thereby transferring the electric charges and maintaining the current. The hypothesis relies on the integrating effect of a large number of branch tips advancing into the space charge regions within the dust storm. The total number of tips of first order can be obtained from $N = rbs–1$; where *rb* is the branching ratio and *s* is the order of network. As a simple consideration, taking $rb = 3$ and $s = 5$ we have $N = 35–1 = 81$. The resistance (102–103 ohm) in the air column above the dust storm can be estimated from electron density models. Voltage in the dust cloud is assumed to be 1 kV m^{-1} multiplied by the cloud height [49]. As a conservative calculation, if the effective part of the dust cloud height is assumed to vary from 10 to 100 m, then the voltage in the dust cloud will be in the range 10–100 kV. Applying Ohm's law, we can calculate the corresponding current as 100 A (ampere) (upper limit) and 10 A (lower limit). To achieve a minimum current of 10 A would require individual branch tip currents of 10/81 ~0.12 A. The probability of charge rearrangement during the inter stroke processes corresponding to current of 10 A is illustrated in Figure 9.3.

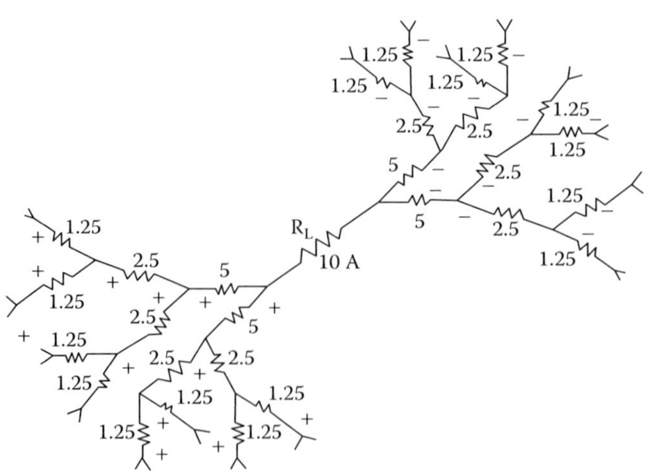

FIGURE 9.3
Probable charge rearrangement during the inter stroke processes. (Data from Bhattacharya, A. B. et al., *International Journal of Engineering Science and Technology*, 3, 3221–3228, 2011.)

A quasi-steady arc-like current of 10 A in a simple bifurcating tree (*rb* = 2) needs a halving of current at each bifurcation. In this scheme, we can proceed over 5–6 orders before the branch current drops to a value when the current no longer maintains an over dense plasma response. For Mars, the possibility that corona or other type of discharges occur inside dust storms may be optically verified by sensitive cameras like the lightning and airglow camera (LAC) [48,49]. These can be less bright by a factor of 100 compared to terrestrial flashes. By implementing orbiting spacecraft, such discharges may be remotely sensed by planning oblique observations of the dust column allowing for a view of escaping photons to reach the sensors.

9.3.2 Luminous Events in the Martian Atmosphere

Transient phenomena on planetary bodies are luminous events of varying intensities, which occur in their atmospheres. They essentially consist of lightning, meteors, impact flashes on solid surfaces, and so on. Atmospheric electrification could be important throughout the solar system. Atmospheric electrification of the earth is a highly prominent transient phenomena producing cloud-to-cloud; intra-cloud; and cloud-to-ground lightning, sprites, elves, and various other luminous events as reported from time to time [50]. Here, the essential conditions for a planetary atmospheric electric circuit similar to that of our Earth are first focused. A comparative approach is then made to re-examine the role of electrification in the atmospheres of earth and Mars with the possibility of a GEC [33]. Figure 9.4 [51] illustrates the comparisons of optical spectra.

Liszka [52] suggested the generation of infrasound waves by sprites, whose signatures were detected by network sensors in Sweden [53]. The shape of the chirp signature in the spectrograms of infrasound can be explained by the horizontal size of the sprite [54]. Neubert [55] have concluded that sprite detection by infrared is an attractive alternative to optical detection, because it is not limited by clear viewing condition or by the absence of daylight. Elves are concentric rings of optical emissions propagating horizontally outward at the bottom edge of ionosphere, between 75 and 95 km altitudes. The cause of this type of emission is by EMP being radiated by the cloud to ground lightning discharge current of either plus or minus polarity. Elves produce average electron density enhancements of 210–460 electrons cm^{-3} over large (165–220 km diameter) circular regions and having an assumed 10 km altitude extent [56,57]. Elves produce perturbation in electrical conductivity of the mesosphere. This can be evaluated by measuring changes in the amplitude of very low frequency (VLF) waves [58,59] from transmitters propagating in the Earth–ionosphere waveguide (discussed in Section 9.18). Mika [60] has discussed experimental data in which the incident VLF transmitter signal seems to be scattered from horizontally extended diffused regions of electron density enhancements. This is most likely associated with halos or diffuse regions of the upper part of carrot sprites, rather than small-scale streamers observed at lower altitudes. The electromagnetic pulse (EMP) generating elves also create ionization [61,62], which depends on its intensity. The individual contributions of different components of constituent gas to the optical emissions of sprite/elves are not known. An attempt should be made to investigate lightning chemistry and energetic in the region of sprites/elves. Lightning-generated whistler mode waves propagate along geomagnetic field lines without appreciable attenuation and thus connect the troposphere with the ionosphere and the magnetosphere. Whistler mode waves while propagating through the equatorial region of the magnetosphere interact with the Earth's radiation belt. Trapped electrons cause their (whistler mode waves) precipitation in the form of plasma resulting in homogeneities in the high-latitude lower ionosphere [63–73].

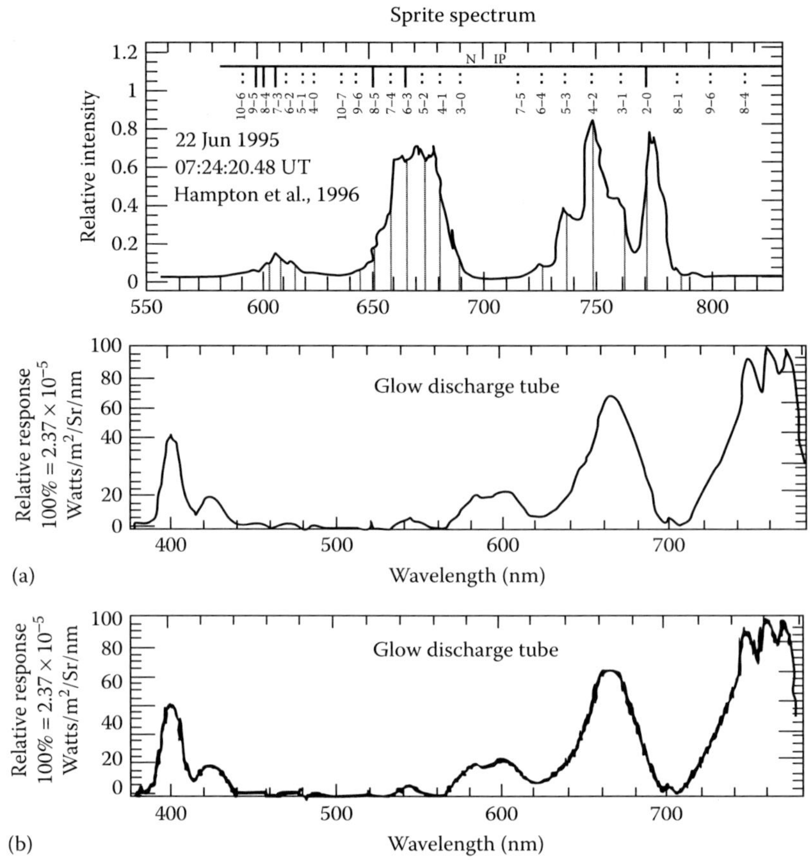

FIGURE 9.4
Comparison of optical spectra for (a) sprite in the mesosphere and for (b) the positive column of large glow discharge tube. (Data from Williams, E. et al. *Calibrated Radiance Measurements with an Air-Field Glow Discharge Tube: Application to Sprites in the Mesosphere*, Springer, 2006.)

Such a type of coupling has been confirmed in ground-based observations [74], balloon measurements [75], rocket experiments [76,77], and satellite observations [78,79].

9.4 Thoughts about the GEC

The role of thunderstorms is to couple the atmosphere and the ionosphere. Potential difference between the ionosphere and the Earth is maintained due to thunderstorms' pumping action of lightning discharges. Intense atmospheric disturbances owing to earthquakes, volcanoes, tropical storms, and typhoons affect electrical properties of the lower ionosphere. Thunderstorms and lightning discharges produce major current source in GEC [80]. Basically, the GEC is based on the concept that the quantity of electric charge has to be conserved in the Earth's atmosphere–ionosphere–magnetosphere regions. The essential features of the global atmospheric electric circuit are shown in Figure 9.5.

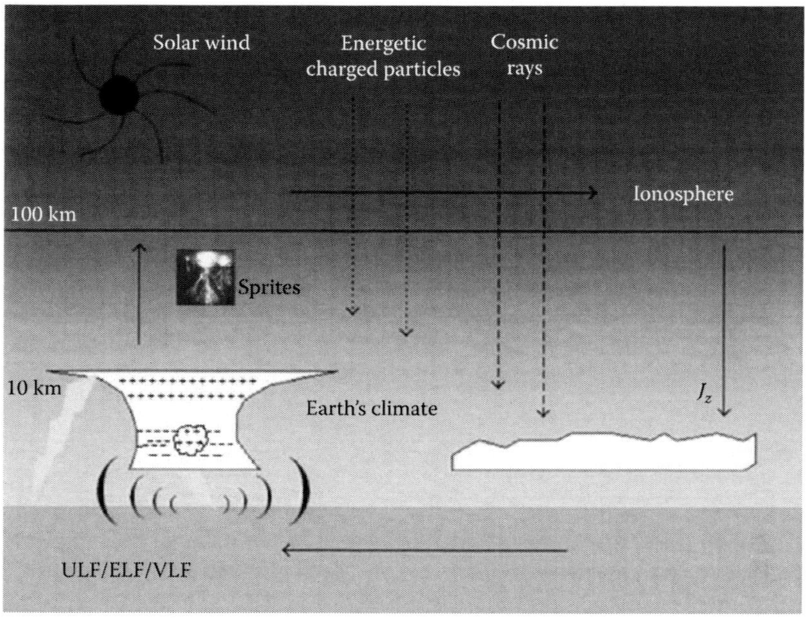

FIGURE 9.5
Features of the GEC.

Solar wind interactions with the Earth's magnetic field generate additional current [81]; potential gradient modulation may arise owing to coupling of geomagnetically induced variations in the magnetospheric dynamo through the global circuit. Precipitations from electrified clouds also act as current drivers [82]. In addition, the schematic diagram (Figure 9.6) can be applied for computing external currents and enhancement of electrical conductivity in the lower atmosphere, which is responsible for producing an increase

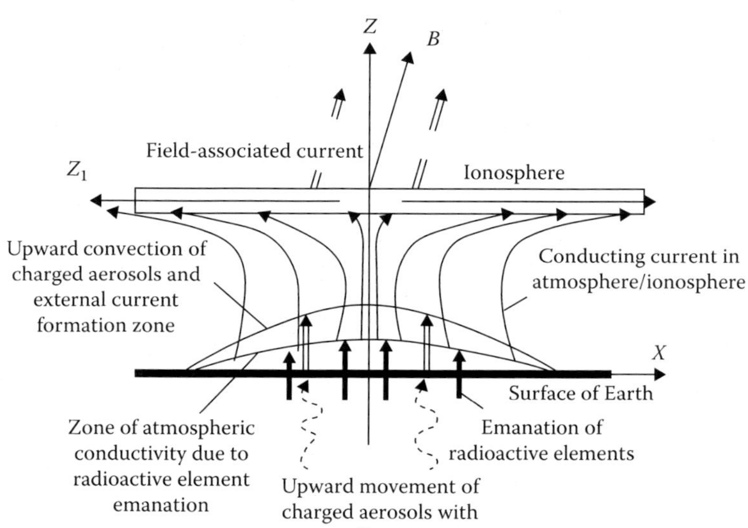

FIGURE 9.6
Schematic model used for the calculation of GEC.

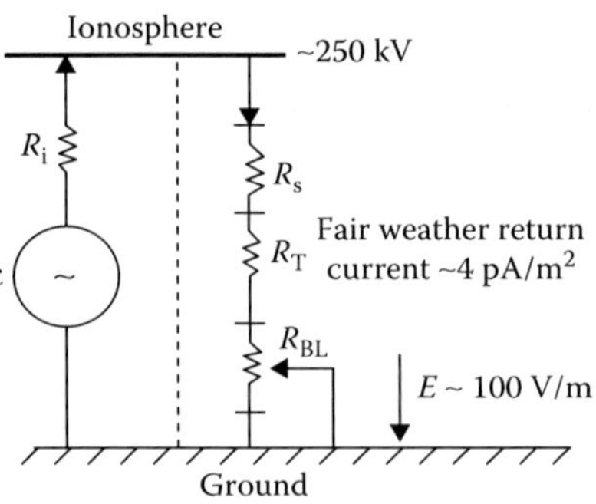

FIGURE 9.7
Schematic presentation of the GEC.

in the ionospheric electric field due to the injection of charged aerosols and radioactive elements [83]. Considering the realistic model, an equivalent circuit with capacitors, resistors, and switches is shown in Figure 9.7 [51]. To operate the receiver, the switch drawn in the figure is closed for a short time when a certain type of discharge occurs. For example, switch $S1$ closes for a few ms when a sprite occurs above a particular type of thunderstorm and fair-weather time constant, $r\,C$ ~ 2 min (r ~ 200 ohms, C ~ 0.7 F). Generators act over <1% of Earth's surface and rest 99% of the Earth's surface region behaves as a load on the circuit. The total current (I) flowing in the circuit becomes ~1 kA [51,84].

Continuing the process discussion, both radioactive substances and charged aerosols injected into the atmosphere modify the altitude profile of conductivity, perturbation of electric field, generation of external currents, and current in the ionospheric layer. Consequently, the Joule heating of the ionosphere and instability of acoustic gravity waves take place to manifest in the formation of horizontal inhomogeneities of ionospheric conductivity. Again the excitation of plasma density fluctuations and ULF/ELF (ultralow frequency/extremely low frequency) emissions in the ionosphere, upward plasma transport, and generation of field-aligned currents and plasma layers, modification of F2-layer, and change in the ion composition of the upper ionosphere take place [85,86]. All these changes affect the GEC and the Earth's climate. The schematic model shown in Figure 9.7 can be used to calculate electric field which is due to the injection of charged aerosols in the atmosphere–ionosphere circuit.

It should be pointed out that there are some other complex processes participating in multiple scales of variability. Turbulent mixing and also eddy diffusion are other ways of transport from bottom. In the lowest part of the atmosphere over continental surfaces, ionization is produced from the escape of radon isotopes [87] as well as by galactic cosmic rays arriving from beyond the solar system [88–90].

9.5 Equivalent GEC

Irregularities of the order of 10° in latitude and longitude [91] create an essential difference in vertical current of the GEC. In Figure 9.7, the GEC has been presented in the form of equivalent electric circuit. Considering that the planetary boundary layer (PBL) contributes 95% of columnar resistance [92] as indicated by the variable resistor in Figure 9.8, it appears that the variation of the PBL resistance will cause a modulation of the ionospheric potential over the area. In the GEC presented in Figure 9.7, RBL, RT, RS, and Ri represent

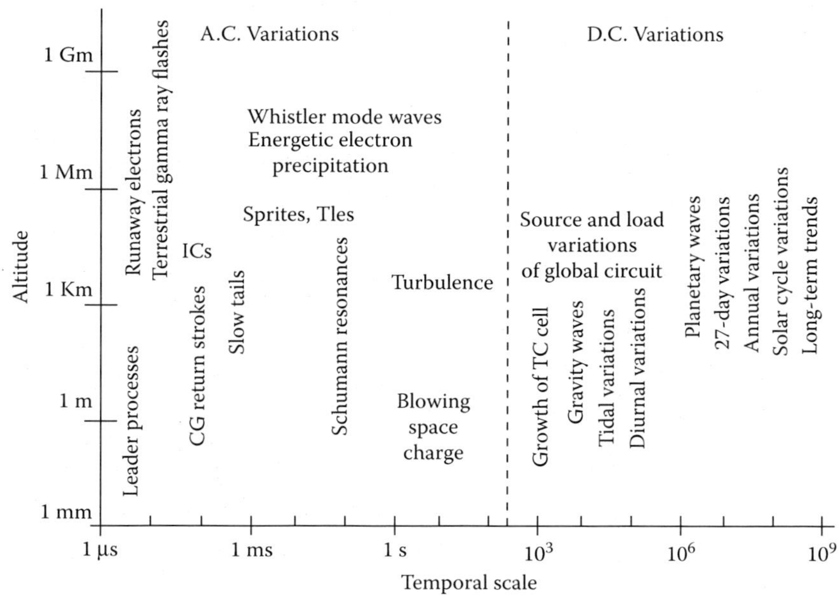

FIGURE 9.8
AC and DC variations on timescale from seconds to hundreds of seconds corresponding to different altitudes.

resistances of the boundary layer (BL), troposphere, stratosphere and internal resistance of the thunderstorm electric generator, respectively.

Some sources like dust storms, thunderstorms, volcanic eruptions, earthquakes, and radioactive contamination may be considered responsible for significant electric field variations at the ground level. At the time of the formation of a dust storm, a maximum deviation of the potential gradient, has been noted from +15 kV/m up to −10 kV/m [93]. A spatial type of electrical charge appears as a result of dispersion and friction of dust particles including small ions. Volcanic activity can be the reason of variability of an electrical field when huge amount of ashes and other particles are ejected into the atmosphere. It has been reported that an uncompensated charge from the dust cloud can cause a thunderstorm [94]. Perturbations of the ground-level electrostatic field on the order of 1000 V/m are frequently found before earthquakes in the regions of tectonic faults [95]. All these processes contribute on the level of ionization rate in the atmosphere and hence it is possible to consider the sources of ionization as a generator of a local electric field. The atmospheric layer situated at about 60–100 km altitude can be defined as a transition region, where different fundamental physical mechanisms coexist, showing complicated atmospheric characteristics. It may be pointed out that the lower ionosphere exhibits not only a solar but also a strong nonsolar control which is partly of meteorological in nature.

9.6 Atmospheric Electrification

Above large thunderstorms transient luminous events (TLEs) such as sprites, elves, and blue jets [96] may occur which may extend up to the ionosphere. It has been reported that the lower ionosphere responds to activity from above in the form of wave–particle interactions between whistler mode waves from lightning and energetic electrons trapped in the magnetosphere [97]. This produces extra ionization in the lowest ionosphere [98]. Figure 9.8 reveals the broad range of temporal scales involving many important phenomena. These electrical discharge phenomena events are on the shortest timescales in the order of microseconds [99]. Lightning radiates at all radio frequencies where the spectrum shows peaks at 10 kHz [100,101] to *slow tails* at frequency ~100 Hz [102] and to the longest wavelength electromagnetic waves at frequency ~10 Hz. These latter waves are responsible for exciting Schumann resonances of the spherical shell cavity between the good conducting Earth and ionosphere [103–109]. Some stations around the world are recording the radiation at different frequencies generated by lightning and sprites. Williams et al. [110] recorded this for radio signals produced over Africa while Whitely et al. [111] have noted that with four stations around the world; sources can be identified to an accuracy of ~10 km. Shvets et al. [112,113], Nakamura et al. [114], Shvets and Hayakawa [115], and Yamashita et al. [116] have also reported interesting results on these aspects. Figure 9.8 clearly shows that on the timescale from seconds to hundreds of seconds, a blowing space charge related to turbulence in the BL from the Earth's surface up to ~2 km altitude generates electrical fluctuations. In order to illustrate the different components associated with both AC and DC variations, we have plotted the altitude against the temporal scales. Primarily, the space charge can be ionic in clean air or particulate in polluted air. These particulate space charges are subject to turbulent motions that results in electrical fluctuations. The processes shown in the figure contributes widely to the generation of AC variations in the global circuit.

The EMP from a large ground-to-cloud (GC) discharge deposits energy in the mesosphere to create an expanding ring of light at ~90 km altitude, which is called elves. The consequent large electrostatic field above the thundercloud produces sprites from about 80 km at the base of the ionosphere down to around 55 km [117–121], while blue jets appear at the top of the thundercloud. The DC variations, in the scale of a fraction of an hour, are linked to the development and evolution of thunderstorm cells. Mechanical waves produce a form of gravity waves, tidal variations (e.g., diurnal, semi-diurnal, and two-day waves), and long period planetary waves. At longer timescales there are variations associated with the changing energy input over the 27-day solar rotation period, annual variations, variations over the 11-year solar cycle, and long-term trends. In the above figure, the boundary between DC and AC phenomena is shown at 200 s which is the electrical time constant of the global atmospheric electric circuit. In addition, we see that the spherical capacitor is formed between the Earth and the ionosphere [122].

9.7 Ionospheric Potential

Integrating the vertical electric field profile derived from aircraft or balloon measurements up to the troposphere, one can determine the ionospheric potential. As reported at 12 UT, this has the value 230 kV and it rises to 310 kV at 16 and 20 UT, which then falls to 200 kV at 24 UT. The curve closely resembles the Carnegie curve [123] and is considered as one of the *confirming ideas* supporting the behavior of the Earth–atmosphere global circuit system [124]. A significant generator of the conduction currents up to the ionosphere, also known as Wilson currents, is owing to the action of thunderstorms [124] which serve as batteries. Figure 9.9 is taken from Harrison and Usoskin [125] wherein the

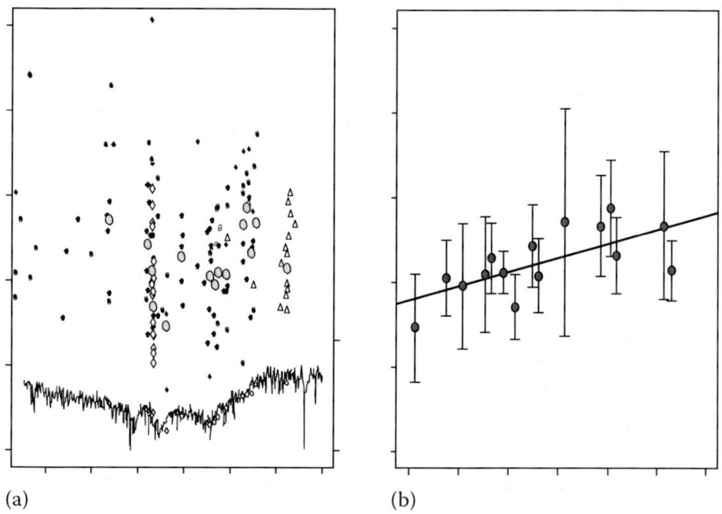

FIGURE 9.9
(a) The ionospheric potential (*Vi*) from various data sets and (b) plots of the monthly averaged *Vi* values against the monthly average neutron count rate. (Data from Harrison, R. G. and Usoskin, I., *Journal of Atmospheric and Solar-Terrestrial Physics*, 72, 176–182, 2010.)

left panel shows the ionospheric potential (*Vi*) from various data sets. In the figure, the larger circles show monthly averages for months having four or more *Vi* values, and the neutron count rate at Climax, Colorado, as the gray line, observed from 1966 to 1972. The right-hand side figure, on the other hand, shows a plot of the monthly averaged *Vi* values against the monthly average neutron count rate. In the figure the error bars show ± two standard errors.

9.8 Electric Field and the Molecular Number Density

The electron mobility μe is a complicated function of the electric field, which may be determined by many loss processes, each having different cross sections [126]. One may apply a functional dependence of μe on the magnitude of the electric field *E* which is derived from experimental data [127,128]. This function can be expressed in analytical form

$$\log(\mu_e N) = \sum_{i=0}^{2} a_i x^i, EN_0/N \geq 1.62 \times 10^3 \text{ V/m} \quad (9.1)$$

$$\mu_e N = 1.36\text{N}_0, EN_0/N \geq 1.62 \times 10^3 \text{V/m} \quad (9.2)$$

where:

$x = \log(E/N)$
$a_0 = 50.970$
$a_1 = 3.0260$
$a_2 = 8.4733 \times 10^{-2}$

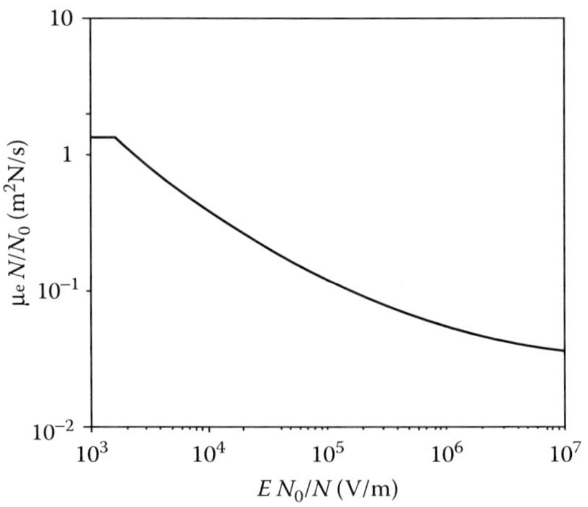

FIGURE 9.10
Electron mobility μe as a function of the electric field E and the molecular number density N of atmospheric gas.

The analytical expression can be utilized for computationally efficient calculation of the electron mobility for any combination of electric field and number density of air molecules (i.e., altitude). This method provides results that can be compared well with those obtained from kinetic formulations. The electron mobility μe when plotted as a function of the electric field E and the molecular number density N of atmospheric gas, as shown in Figure 9.10, interesting variations have been noted [127].

9.9 The Thunderstorm Process

Thunderstorms and lightning discharges are the main sources of electromagnetic and electrostatic energy in the Earth's atmosphere. Thunderstorms consist of a cluster of clouds. The process of building a thunderstorm starts with the right conditions. Vigorous vertical air currents and thunderstorms are a consequence of excessive warmth and moisture at low altitudes. When the right conditions are present, the following processes start. One of the first processes is the buildup of electrostatics. This is a build-up of static charges on the surface of objects (clouds, particles, etc.) due to the interaction with other surfaces. The effects of charge exchange are usually only noticed when at least one of the surfaces has a high resistive electrical flow. In thunderstorms, a rapid vertical rearrangement of deep air layers takes place. Large processes promote vertical and horizontal air motions, while other processes within the storms control the development of rain and strong local winds. Electrical charges generate and separate during the developmental stage of thunderstorm and are neutralized during lightning discharges. Numerous processes operating within the environment of a mature convective cloud with varying effectiveness and time dependencies affect cloud electrification [129]. The charging of thunderstorms can be discussed as inductive charging or noninductive charging. An inductive process requires pre-existing electric fields to induce charges on a particle. This interaction between the re-bounding particles separates the charge and enhances the field. In the atmosphere, the fair-weather electric field resulting from positive charges in the atmosphere and negative charges on the ground could be considered as the pre-existing field. However, there are experimental results from airborne instruments that require some other processes of charging. Noninductive processes are independent of the presence of an external electric field. In an ordinary thundercloud, the smaller ice crystals are charged positively and move upward, whereas larger graupel. (They are the precipitation that forms when supercooled droplets of water are collected and freeze on a falling snowflake, forming a 2–5 mm (0.079–0.197 in) ball of rime. The term graupel comes from the German language.) Particles charged negatively descend relative to the smaller particles under the action of gravity. This is the normal situation, depending on the prevailing conditions of temperature, liquid water content, and mixing in the thunderstorm. Different movements and situations will vary and the charging process may not remain as discussed above. The charge buildup during the active period of a thunderstorm ultimately results into lightning discharges either through the processes of dielectric breakdown or runaway breakdown [130–133]. The electric field computation favors the runaway breakdown mechanism [134]. However, discharge processes are not fully understood, and further observations and works are required. Lightning discharge creates a charge imbalance in the cloud/thunderstorm, and as a result an upward propagating quasi-electrostatic (QE) field initiates from the upper charge layer of thunderstorm. This QE field connects troposphere with the stratosphere and the lower ionosphere for a very short duration. There are about 40–100 lightning discharges occurring every second. Apart from the upward propagating QE fields, lightning discharge current generates intense EMP of ~20 GW peak powers [135], which may cause significant ionospheric disturbances due to the heating and acceleration of electrons, production of ionization, optical emissions, and so forth.

9.10 Large QE Field

It is estimated, that there are around 2000 active thunderstorms at any time over the Earth's surface and on average lightning strikes the Earth 100 times per second [136]. About 1000 coulombs of bound space charge are separated and maintained during the active periods of a thunderstorm [137] with GC lightning discharges involving the transfer to the ground of up to 300 coulombs of charge in several ms [138]. This creates a large QE field which exists in the mesosphere and lower ionosphere regions over millisecond timescales. These fields combine with intense EMP [139] generated by lightning currents, heat the ambient electrons [139] to accelerate energetic runaway electrons [140,141], producing ionization and optical emissions. Experimental evidence suggests that strong upward electrodynamical coupling of thunderstorms to the mesosphere and lower [142–147] as well as optical emissions in clear air above thunderstorms. This coupling is associated with sprites [148–154] and blue jets [155] as well as airglow enhancements [156]. The latter are thought to be associated with rapid (<1 ms) optical emissions at 80–95 km altitudes and are referred to as *elves* [157–159]. Sprites are also uniquely associated with positive CG lightning discharges [140,153,160]. The observations of gamma ray bursts of terrestrial origin [161,162], associated with positive CG lightning discharges [163] and intense VHF bursts [164] constitute further phenomena of upward coupling of energy associated with thunderstorms.

9.11 QE Heating Model

The input parameters which are specifically required for QE heating for each model calculation are (i) the altitude profiles of ambient ion conductivity and electron density, (ii) the magnitude (Q_0) of the charge transferred and the time constant (τ_s) which determines the rate at which the charge is transferred, and (iii) the altitude of the charge removed. None of these parameters can be obtained directly and also, some restrictions apply, these are the magnitude of the removed charge Q and the timescale of the charge removal τ_s. For the QE heating model [165], the thundercloud charging and discharging process is considered as two stages. During the pre-discharge stage with duration τ_f, the charge Q_0 slowly accumulates in the thundercloud. The second stage is the lightning discharge when the thundercloud charge is completely removed very fast over a time τ_s, which is generally of the order of several ms. The continuous thundercloud charge dynamics can be expressed mathematically in the form

$$Q(t) = Q_0 \frac{\tanh\left(\dfrac{t}{\tau_f}\right)}{\tanh(1)}, \quad 0 < t < \tau \tag{9.3}$$

$$Q(t) = Q_0 \left[1 - \frac{\tanh\left(\dfrac{t}{\tau_s}\right)}{\tanh(1)}\right], \quad \tau_f < t < \tau_f + \tau_s \tag{9.4}$$

$$Q(t) = 0, \quad \tau_f + \tau_s < t \tag{9.5}$$

It may be noted that the functional variation as tanh (·) is somewhat arbitrary. For the functional dependence of $Q(t)$ the amplitude of the peak current at $t = \tau_f$ can be obtained from

$$I_p = \frac{dQ}{dt} = \frac{Q_0}{\tau \tanh(1)}, \quad t = \tau_f \tag{9.6}$$

From knowledge of I_p, the peak electric field at a distance of 100 km from the lightning discharge can be calculated as [166]

$$E_{100} = \frac{I_p v}{2\pi D \varepsilon_0 c^2} \tag{9.7}$$

Here, c is the speed of light in free space, ε_0 is dielectric permittivity of free space, $D = 105$ m and $v = 1.5 \times 10^8$ m/s. It may be noted that $E_{100} = 75$ V/m corresponds to a peak current of

I_p = 270 kA and a total charge removed in lightning of Q_0 = 200 C, if we take into account τ_s = 1 ms which is a reasonable data based on experimental result [152].

9.12 Luminous Events

We briefly survey the information available concerning the recently discovered phenomena of optical flashes above thunderstorms. This phenomena spans the altitude range from the cloud tops and up to the lower edge of the ionosphere (90 km), and referred to as *sprites* and *blue jets*. Sprites are luminous glows occurring at altitudes typically ranging from 50 to 90 km. In video, they exhibit a red color at their top which gradually changes to blue at lower altitudes. Sprites may occur singly or in clusters of two or more [167]. The lateral extent of *unit* sprites is typically 5–10 km and they endure for several milliseconds. Blue jets or jets in short are upward moving (100 km/s) highly collimated beams of luminosity, emanating from the tops of thunderclouds, extending up to 50 km altitude [168]. Figure 9.11a and b show actual photos of these phenomena.

The upward propagating QE charge causes TELs (sprites, blue jets/gigantic jets, elves, etc.) in the stratosphere and the mesosphere. These events influence the lower ionosphere [169] with additional ionization and formation of ionospheric homogeneities. Our understanding of the electrical processes and properties of the atmosphere are being studied by those doing research in this area [170–177]. Figure 9.12 shows the variety of TELs and the typical altitude at which they occur. Sprites may occur in a cluster of 2 or 3, or more short-lived (~50 ms) carrot-shaped emissions (of ~1 km thickness) over a horizontal distance of 50–100 km, with the separation between sprite elements of ~10 km [178,179]. The upper portion of the sprites is red, with wispy, faint blue tendrils extending to 40 km or at times as low as the cloud top, and occurring at an altitude of 65–85 km [178]. They typically last for 5–50 ms and may take the form of one or more vertical columns measuring in a few hundred meter radius for the smaller column sprites or, large jellyfish-shaped structures measuring in the tens of kilometers of radius and extending from the ionosphere D-region and almost down to the thunderstorm cloud tops [178,179].

(a) (b)

FIGURE 9.11
(a) Red sprites and (b) blue jets. (Courtesy of Jason Ahrns, University of Alaska-Fairbanks, AK. http://elf.gi.alaska.edu/#chrjet and https://www.flickr.com/photos/musubk/9496938627/ [retrieved September 10, 2016].)

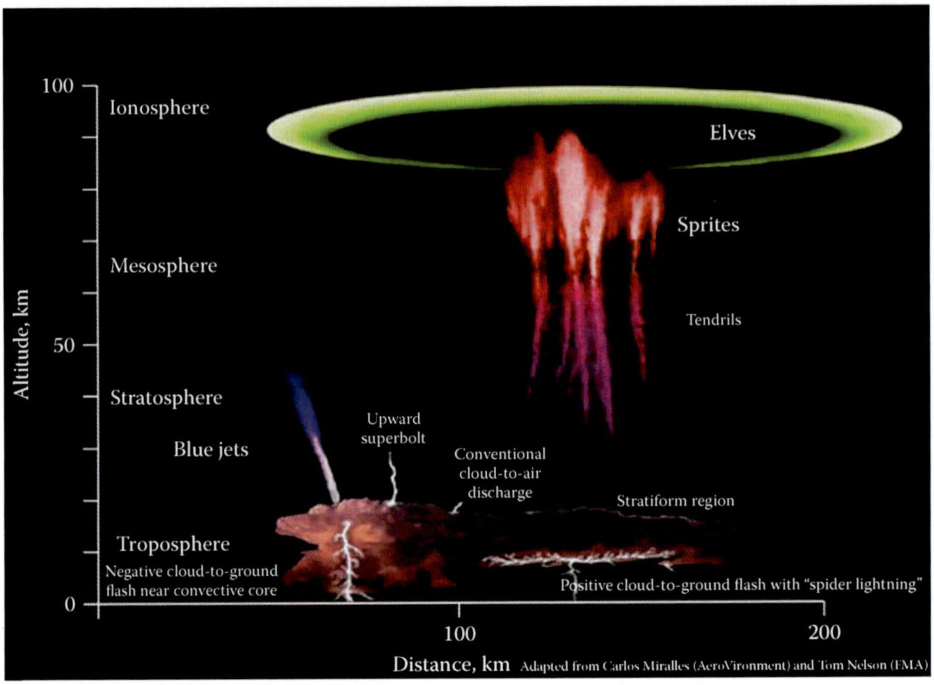

FIGURE 9.12
Depiction of various optical events in the atmosphere and the altitude at which they occur. (Data from Qin, J. et al., *Geophysical Research Letters*, 39, L22803, 2012.)

The knowledge gained from laboratory experiments of gas discharges at subatmospheric pressure has been used to understand sprite spectroscopy and associated phenomena [179]. Figure 9.3 shows broad identical spectral characteristics of light of positive column of the laboratory tube and sprites [179]. The difference in spectral characteristics may be due to difference in the applied electric field, gas pressure, and gas composition in the mesosphere and gas tube discharges. Physical processes associated with sprites and other optical events are also associated with thunderstorm activity in the troposphere and are thought to result in the gradual buildup of conductivity changes in the lower ionosphere [180]. The observed properties of the phenomenon of sprites and jets constitute an important component of the complex of physical processes. The electrodynamic coupling of energy originating in tropospheric thunderstorms to the mesosphere and lower ionosphere as well as the general understanding of these are not yet fully understood.

9.13 Heating and Related Processes Contributing to Luminous Events

The mechanisms related to TELs are based on heating (discussed in Section 9.16) of the ambient electrons by EMP generated due to lightning discharges [52,181–185] or due to large QE thundercloud fields [47,52,186–189] and on runaway electron processes [47,180,190,191]. It is suggested that the QE thundercloud fields are capable

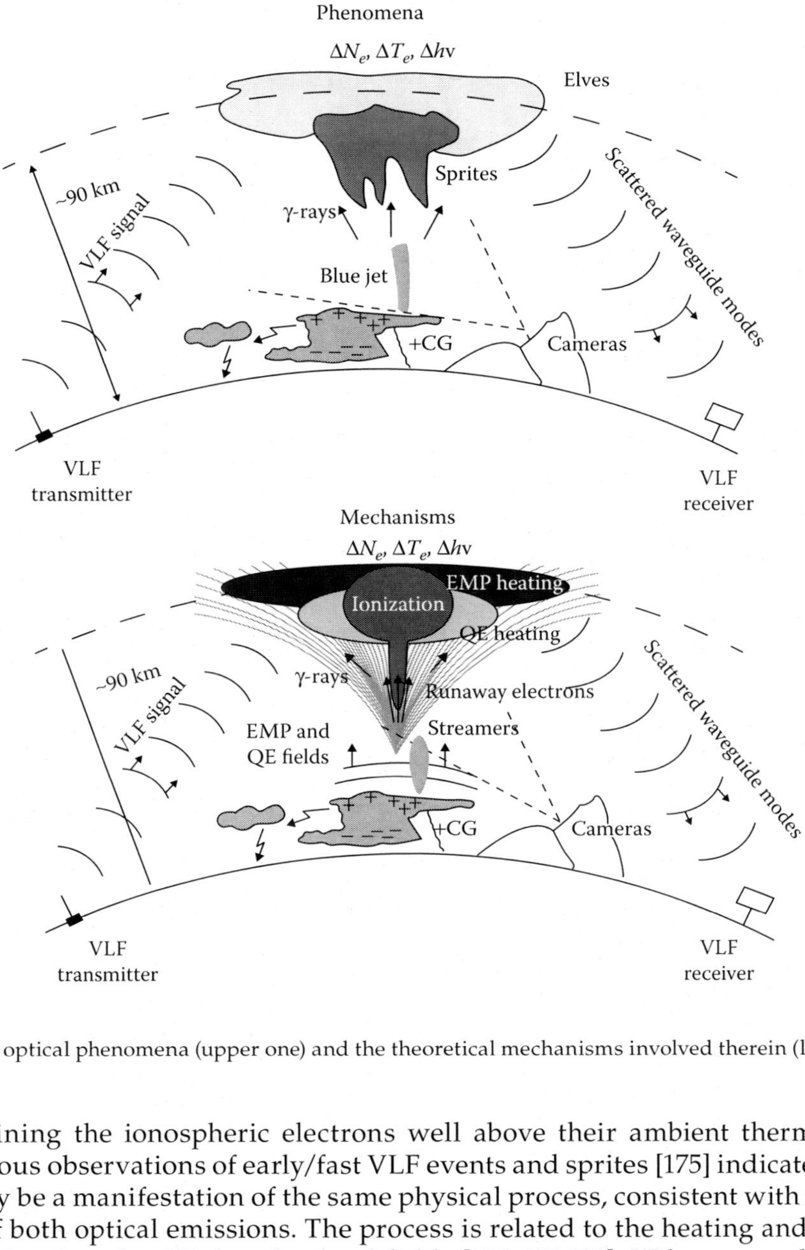

FIGURE 9.13
The different optical phenomena (upper one) and the theoretical mechanisms involved therein (lower one).

of maintaining the ionospheric electrons well above their ambient thermal energy. Simultaneous observations of early/fast VLF events and sprites [175] indicate that these effects may be a manifestation of the same physical process, consistent with model predictions of both optical emissions. The process is related to the heating and ionization changes related to the QE thundercloud fields [172–174,192]. Wilson speculated that a discharge between the top of a cloud and the ionosphere might be the normal accompaniment of a lightning discharge to ground [193]. Figure 9.13 illustrates the different phenomena (upper one) and the theoretical mechanisms involved therein (lower one) of lightning–ionosphere interactions operating at different altitudes. The figure clearly shows the optical emissions ($\Delta\nu$) observed as sprites, blue jets and elves, as well as heating (ΔT) and ionization changes (ΔN_e) detected as VLF signal changes.

As illustrated in Figure 9.14, a cylindrical coordinate system (r, ϕ, z) is taken with the z-axis representing the altitude. The ground and ionospheric boundaries, taken at z = 90 km, are assumed to be perfectly conducting and the whole system is considered to

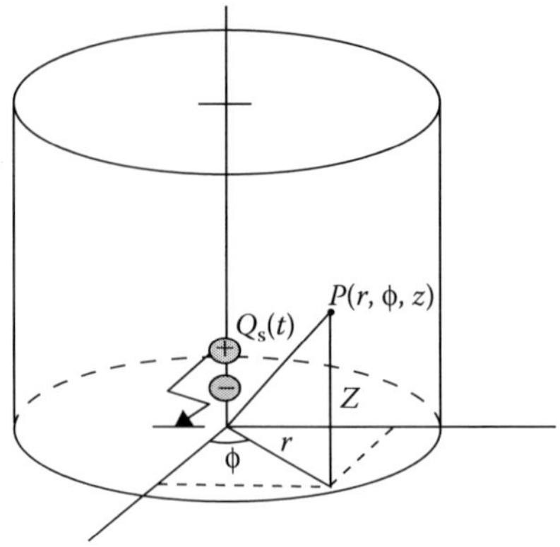

FIGURE 9.14
Coordinate system used in calculations.

be cylindrically symmetric about the z-axis. Two types of boundary conditions can be used at the cylindrical boundary. If we imagine a perfect conductor at $r = 60$ km, we have $\partial/\partial z = 0$ (or $\psi = 0$), where ψ is the electrostatic potential. This condition was applied by Illingworth [194] for considering the solution of the electrostatic fields in a *tin can*. Alternatively, one can assume that $\partial\psi/\partial r = 0$, following Tzur and Roble [195]. Both the boundary conditions taken in this fashion are artificial and lead to unphysical modification of the electric fields near the boundary. The temporal and spatial evolution of the electrostatic field system does not depend on these edge fields if the boundaries are taken at a distance far from the charge sources. The presence of the conducting boundary at $r = 60$ km leads to an error of less than 10% on the magnitude of the electric field at $r = 50$ km and $z = 10$ km.

The electrostatic field $\vec{E} = \vec{\nabla}\psi$, the charge density ρ, and the conduction current $J = \sigma E$ can be calculated using the following equations:

$$\frac{\partial(\rho + \rho_s)}{\partial t} + \nabla \cdot \left(\vec{J} + \vec{J}_s\right) = 0 \tag{9.8}$$

$$\nabla \cdot \vec{E} = \frac{(\rho + \rho_s)}{\varepsilon_0} \tag{9.9}$$

where ρ_s and $\vec{J}_s$ are the thundercloud source charge and current which must satisfy the equation

$$\frac{\partial \rho_s}{\partial t} = -\nabla \cdot \vec{J}_s' \tag{9.10}$$

where

$$-\nabla \cdot \vec{J}_s' = \nabla \cdot \vec{J}_s + \frac{\rho_s \sigma}{\varepsilon_0}$$

Here, the effective source current J_s' is introduced so as to compensate any change in the thundercloud charges due to the conduction current.

The equation can then be written as

$$\frac{\partial \rho}{\partial t} = \nabla \sigma \cdot \vec{E} + \frac{\rho \sigma}{\varepsilon_0} = 0 \tag{9.11}$$

If one simply assumes,

$$\frac{\partial \rho_s}{\partial t} = -\nabla \cdot \vec{J}_s \quad (\text{i.e., } \vec{J}_s = \vec{J}_s') \tag{9.12}$$

It is necessary that $\rho\sigma/\varepsilon_0$ be substituted with $[(\rho + \rho_s)\,\sigma]/\varepsilon_0$. In this case if one avoids the effects of the conductivity gradients, the quasi-stationary solution of Equation 9.10 becomes simply so that the source charge can be totally compensated by charge induced in the conducting medium over a timescale $/\sigma$. In fact, being a function of time, the quasi-stationary solution has a definite physical meaning when the characteristic timescale of (t) is sufficiently large. Equations 9.11 and 9.12 clearly indicate that the introduction of $\vec{J}_s'$ permits us to specify the charge dynamics inside the cloud as an externally noted function of time.

9.14 Effects of the Solar Wind Forces on the Ionosphere

The terrestrial ionosphere, a part of the upper atmosphere, is a cold magnetized plasma environment enveloping the Earth whose behavior can be described as ionospheric aeronomy. Due to the pervasive influence of gravity, the atmosphere and ionosphere are conventionally divided into layers based on the vertical structure of different parameters. The atmospheric structure can be clearly organized by a representative temperature profile, while the ionosphere is more sensibly organized by the number density of plasma. The concept of the GEC provides the existence of a vertical atmospheric electric field and coupling between the ground and ionosphere. The GEC is a system of quasi-stationary electric current between the ground of the earth and the ionosphere driven by global thunderstorm activity which is considered as an electric generator providing the negative Earth potential in relation to the ionosphere and thus creating the potential difference between the ground and ionosphere (150–600 kV). The downward return current closes the circuit in the areas of fair weather [160]. By convention, this downward electric current in fair weather regions of the globe is designated positive. Estimates of electric field penetration from thunderstorm clouds into the ionosphere [196,197] clearly indicate (a) that the effectiveness of the electric field penetration into the ionosphere from the cloud altitude level is very low and (b) that the penetration is most effective during nighttime.

9.15 Electrodynamic Coupling

The electrodynamic troposphere–ionosphere coupling is treated using the model of a troposphere–mesosphere electric circuit. This consists of a localized or powerful global-scale source of the mesospheric current 10^{-8}–10^{-9} A/m^2, a local ground level (or troposphere–stratosphere) resistance, a local mesospheric load resistance for the mesospheric source, and the resistance (approximately 200 Ohms) of the global atmospheric layer between the ground and the lower boundary of the ionosphere. The electric current discharge density in the global capacitor under undisturbed conditions can be neglected. Also during undisturbed conditions, the tropospheric resistance is much greater than the mesospheric resistance. During undisturbed conditions, the integral mesospheric source loading is equal to the mesospheric resistance, or the electrical troposphere–mesosphere coupling does not appear. During disturbed conditions, the tropospheric resistance can decrease by an order of magnitude or more due to an increase in the level of near-earth radiation presiding in the vicinity of strong earthquakes or during accidents at nuclear plants, with the possibility of tragic discharge of radioactive materials. Consequently, the ratio between the tropospheric resistance and the mesospheric resistance changes, and this leads to a change in the integral mesospheric source loading. The intensity of the strong electric field becomes dependant on the tropospheric resistance. The result of an increase in the tropospheric conductivity, the decrease in the integral mesospheric source loading and the tropospheric resistance, in turn, results in a corresponding decrease in the electric field intensity and the electron temperature in the mesosphere. On the other hand, the ionospheric electron cooling law goes under the influence of disturbances in tropospheric conductivity when strong mesospheric electric fields occur. Thus, strong mesospheric electric fields result in an additional electrodynamic troposphere–mesosphere coupling during disturbed conditions to estimate the effect of the decrease in the mesospheric electric field intensity due to changes in lower ionosphere parameters. To accomplish this, we have used the well-known system of equations related to the energy balance equation (in terms of the electron temperature), the two continuity equations in the electron density and the positive-ion density in the stratified inhomogeneous weakly ionized plasma, and the condition of quasi-neutrality. Numerical simulations show that the resulting decrease in the effective electron collision frequency plays a key role. The lowering of the lower boundary of the ionosphere was observed in VLF signals propagated during nuclear plants accidents with the discharge of radioactive materials into the atmosphere.

9.16 Contributing Sources in the Coupling Processes

There are at least three independent ways by which the solar wind modulates the current density (J_z) flowing in the GEC. These are (i) changes in the galactic cosmic ray energy spectrum, (ii) changes in the precipitation of relativistic electrons from the magnetosphere, and (iii) changes in the ionospheric potential distribution in the polar caps due to magnetosphere–ionosphere coupling. The current density (J_z) that is flowing between the ionosphere and the surface generates space charge concentrations and increases the conductivity gradient. The gradients are large at the cloud surfaces and space charge concentrations on the order of 1,000–10,000 elementary charges per cm^3 are generated at cloud tops [198]. It has been suggested that ionization produced by cosmic rays increases the production of condensation nuclei at upper tropospheric levels and so affects the concentration and size distributions of cloud particles [198]. Sources that play a role in tropospheric–ionospheric coupling processes are shown in Figures 9.15 and 9.16. The majority of the more persuasive existing observations fit within one of these frameworks are summarized in Table 9.4. However, there has been no confirmed theory yet which convincingly explains the linkage of pre-seismic surface changes to the ionosphere.

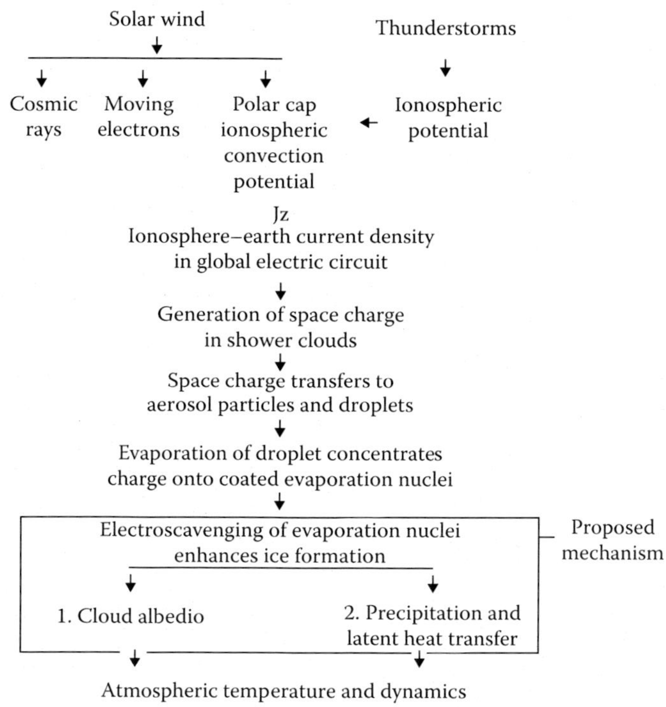

FIGURE 9.15
Sources responsible in tropospheric–ionospheric coupling processes.

The meridional section of the GEC with about 25 times vertical exaggeration is illustrated in Figure 9.17a and b. The arrows in the figure represent the flow of current density for which the main generator is tropical thunderstorms. The ionosphere–earth current density J_z of 1–4 pA m^{-2} is driven by the ionosphere–earth potential difference V_i of about 250 kV acting across the vertical column resistance at each location. A review of the GEC is given by [209] following an earlier comprehensive reviews by Israël [209] and NAS [210]. The global integral of J_z constitutes the return current and is equal to the charging current. At any location on the globe J_z is given by the product of the overhead ionospheric potential V_I and the resistance of the vertical earth–ionosphere column. This is expressed as

$$J_z = V_I \int \frac{dz}{\sigma_z} \tag{9.13}$$

where z is altitude and $\sigma(z)$ is the conductivity, which increases strongly with altitude in clear air, except within the lowest one or two km over land layer of mixed air. A simplified equivalent circuit with the only horizontal variation due to the magnetic latitude is shown in Figure 9.8. In this equivalent circuit, the ionosphere–earth return path is divided into a low latitude branch with relatively small variations due to solar wind inputs, including a varying high latitude branch with the inputs of at least several times greater.

The solar wind-modulated inputs can affect J_z either by varying $\sigma(z)$ in the troposphere and lower stratosphere, or by changing V_i, especially in the polar regions. Both V_i and

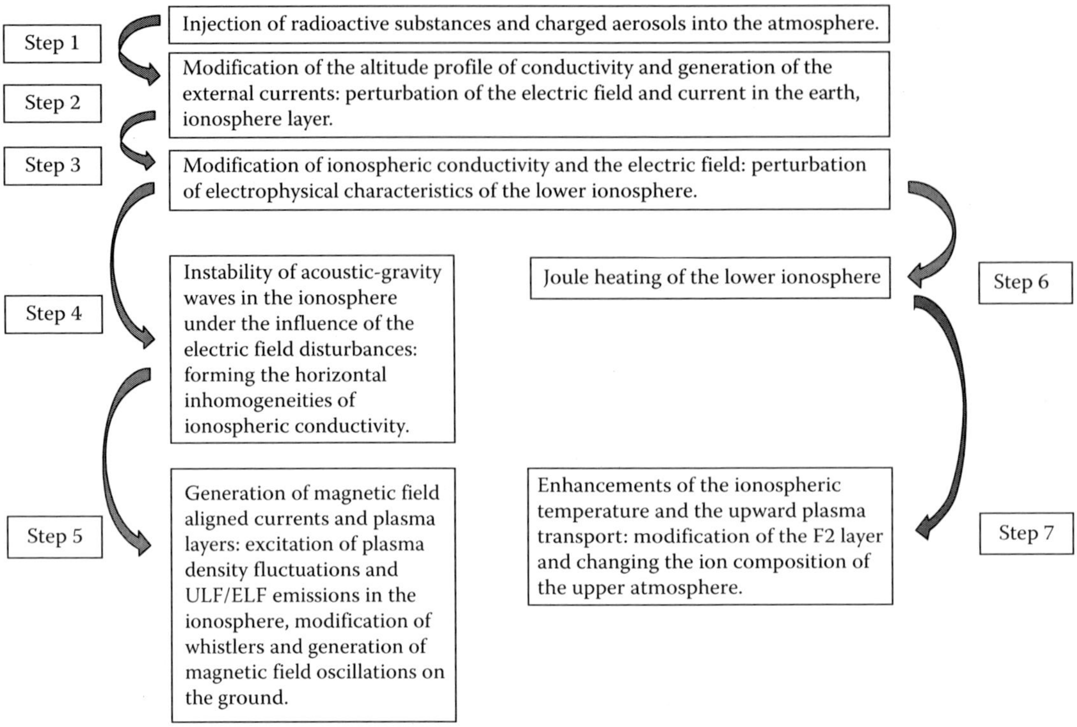

FIGURE 9.16
Diagram of the processes forming the atmosphere–ionosphere coupling.

TABLE 9.4
Summary of Proposed Surface–Ionosphere Coupling Mechanisms

Mechanism	Supporting Observation	Example References
Electromagnetic: Before the earthquake, direct emission of radio waves occur.	VLF radio anomalies	Fujiwara et al. [199] Ohta et al. [200] Schekotov et al. [201]
	UHF/VHF telemetry disturbances	Nagamoto et al. [202]
Acoustic: Excitations of atmospheric oscillations propagating upward to the ionosphere	F-region ionospheric effects attributed to gravity waves	Hegai et al. [203]
Pre-seismic emissions: Emissions due to rock motion stimulate the total electron density in the ionosphere	Atmospheric electric field variations	Kondo [204] Kamogawa et al. [205]
	Changing infra-red emissions from surface, suggesting release of gas	Surkov et al. [167]
	Over the horizon VHF emissions, suggesting changed properties of air	Hayakawa [168] Yonaiguchi et al. [206]
	Total ionospheric content variations	Naman et al. [207]
	Emission of ions from rock stresses	Freund et al. [208]

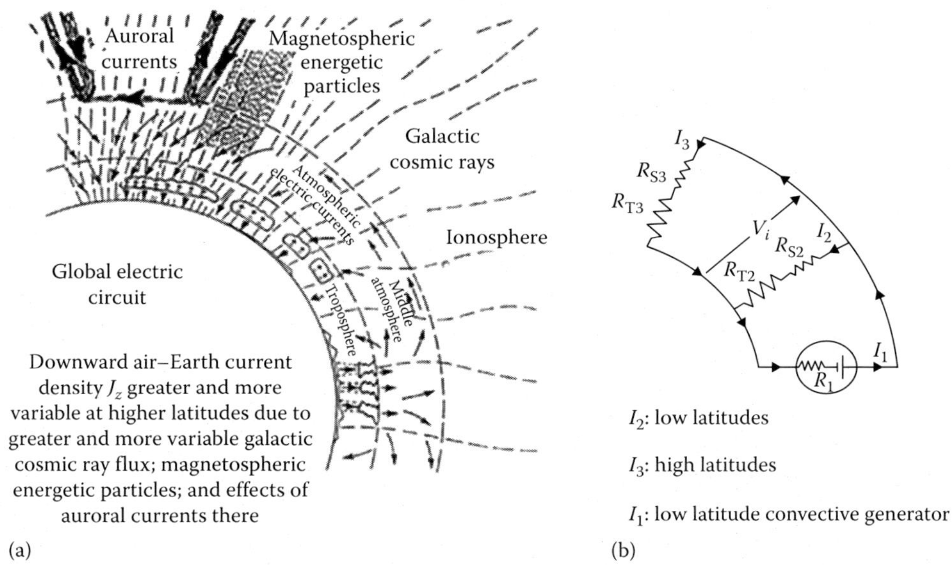

FIGURE 9.17
(a) Illustration of meridional section of the global electric circuit and (b) equivalent circuit with the only horizontal variation due to magnetic latitude.

$\sigma(z)$ have systematic and irregular natural variations and the small-scale chaotic variations of $\sigma(z)$ in the mixing layer make it complicated to measure at low altitude sites. It is possible also that the thunderstorm charging current and V_i are affected by the small (3%–6%) changes in the cosmic ray flux in the equatorial regions by solar wind variations. In Figure 9.17b, we have shown an equivalent global circuit with the only horizontal variation due to magnetic latitude and with the return path divided into one representing the high latitude regions. In addition, another representing the low latitude regions. We have divided the vertical column resistance integrated over either part into tropospheric (*RT*) and stratospheric (*RS*) components. The high latitude path is subject to the influences of solar wind at least several times stronger than the low latitude path. The latitude gradient in the modulation of vertical column conductivity redistributes J_z between the high and low latitude return paths. The cosmic ray flux variations modulate $\sigma(z)$ between about 2 and 60 km altitude. The relativistic electron flux and associated bremsstrahlung appear to affect J_z by modulating $\sigma(z)$ between about 20 and 60 km during episodes of high stratospheric resistivity *RS*. We have suggested a simple mechanism for coupling emanations to the electron density in the lower ionosphere. The coupling occurs through changes of the downward-directed dense conduction current flowing in the global atmospheric electrical circuit in fair weather regions [120]. Local modulation of the dense conduction current can arise through changes in the surface layer air conductivity. The continuous nature of the conduction current allows coupling to the ionosphere of surface radioactivity changes, thus producing changes in the local rate of charge transfer to the ionosphere. Figure 9.18 depicts the proposed coupling concept which distinctly indicates that a local current density J_c flows between the ionosphere and the surface in fair weather regions. The local current flow depends on the local columnar resistance R_c, which has significant contributions from the lowest atmosphere and the free troposphere (FT). A small (~7%) contribution from the stratosphere adds to the process [106]. Changes in the resistance of the atmospheric BL,

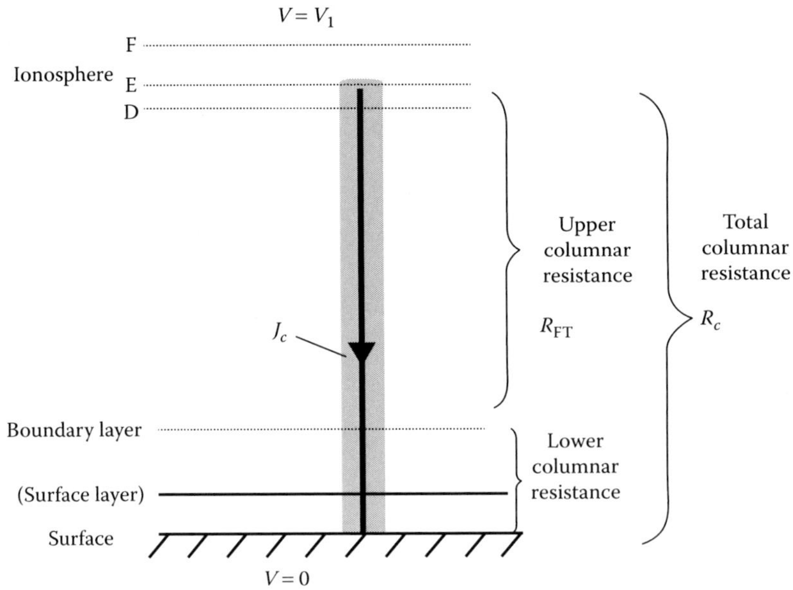

FIGURE 9.18
Schematic diagrams showing the current flowing between the ionosphere and the Earth's surface.

through the release of radioactive gases into the surface layer, modify the total columnar resistance and hence modify the flow of conduction current. This modifies the charge transferred to and from the ionosphere.

Surface layer conductivity significantly contributes to the production of atmospheric electricity. The global atmospheric electrical circuit links the charge separation in disturbed weather regions with current flow in fair weather regions [120,148]. This occurs as a result of current flow through the ionosphere and the Earth's surface [211]. The ionosphere at ~80 km altitude is typically maintained at a positive potential of ~250 kV with respect to the surface by the integrated effect of disturbed global weather activity, such as thunderstorms and shower clouds. Finite conductivity of atmospheric air leads to a small *conduction* current density (J_c) flowing between the surface and the ionosphere in fair weather conditions, which is typically ~2 pAm^{-2}. The flow of locally dense current is determined by the ionospheric potential V_I and the electrical resistance of the vertical column of atmosphere between the surface and the ionosphere. The resistance of a unit area atmospheric column is known as the columnar resistance, R_c, which is ~100–170 $p\Omega m^2$ [211].

9.17 Waveguide Propagation

Figure 9.19 reveals daily fair weather measurements made at Kew over 14 years, of the conduction current density J_c and air conductivity σ_s. These were derived from independent Wilson apparatus measurements of J_c and the potential gradient, using Ohm's law.

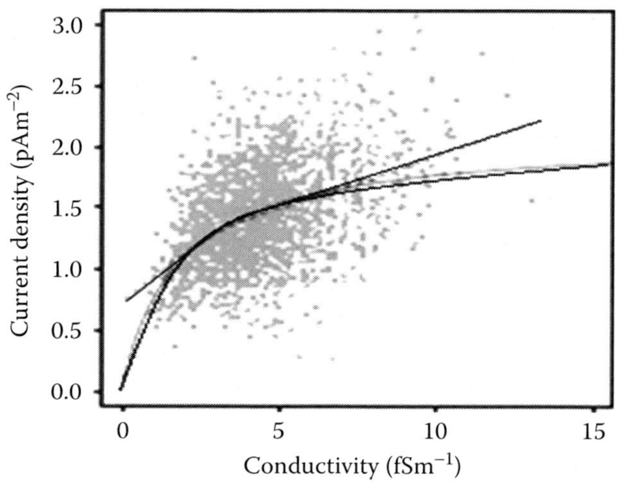

FIGURE 9.19
Daily fair weather measurements of current density versus conductivity.

In the figure, a locally weighted fit line has been added (dashed line) and thereafter the expression of Harrison and Bennett [212] is fitted (solid line).

The conduction current density J_c is constant with height between the surface and the ionosphere [213]. In addition, changes in J_c, due to surface layer air conductivity effects, are directly communicated to the ionosphere. In the ionosphere, the conduction current density is primarily due to the motion of electrons, which are highly mobile in comparison to ions. J_c is related to the electron concentration n_e, and drift speed v, by

$$J_c = n_e v e \tag{9.14}$$

For nighttime conditions in which no photoionization occurs, the electron concentration at 70–80 km altitude is $n_e \sim 1 \times 10^7$ m^{-3} [214,215]; and assuming that $J_c = 2$ pAm^{-2}, v is therefore 4.5 km h^{-1}. By drift speed effects alone, or neglecting any other production and loss terms in the continuity equation, a 20% regional change in J_c would cause a ~10 km variation between disturbed and fair weather regions after 12 h. The detection of such height changes could be made using remote radio frequency observations of the lower ionosphere. If an upward moving charge passes the effective position of the lower boundary of the ionosphere, the waveguide properties would be changed. This would modulate the waveguide cut-off frequency f_c, given by Budden [216] as

$$f_c = c/2h \tag{9.15}$$

where h is the effective height and c the speed of light. This relation is plotted in Figure 9.20. The perturbations to h of ~10 km per 12 h represent a ~13% change in f_c ~ 2 kHz, which would be readily detectable.

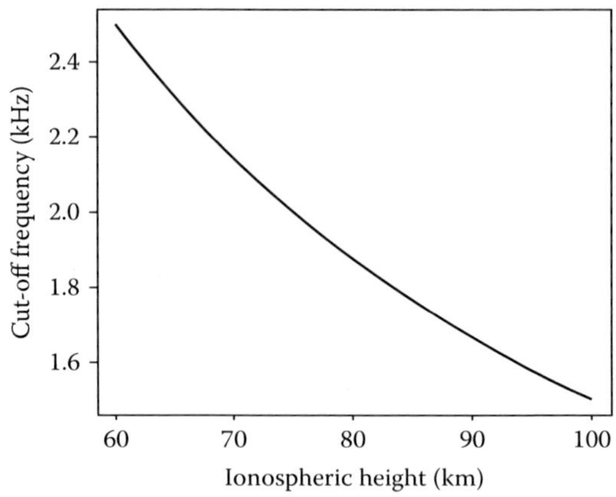

FIGURE 9.20
Variation of the cut-off frequency for Earth–ionosphere waveguide propagation (Equation 9.15), as a function of the height of the ionosphere.

References

1. C. T. R. Wilson (1921). Investigations on lightning discharges and on the electric field of thunderstorms. *Philosophical Transactions of the Royal Society of London* A221: 73–155.
2. C. T. R. Wilson (1929). Some thundercloud problems. *Journal of the Franklin Institute* 208: 1–12.
3. C. T. R. Wilson (1956). A theory of thundercloud electricity. *Proceedings of the Royal Society A* 236: 297–317.
4. K. L. Aplin et al. (2008). Balloon-borne disposable radiometer. *Space Science Reviews* 137: 11–27.
5. M. J. Rycroft (2006). Electrical processes coupling the atmosphere and ionosphere: An overview. *Journal of Atmospheric and Solar-Terrestrial Physics* 68: 445–456.
6. A. B. Bhattacharya and T. Das (2013). Tropospheric-Ionospheric coupling by electrical processes of the atmosphere. *International Journal of Application or Innovation in Engineering & Management (IJAIEM)* 2: 324–336.
7. A. B. Bhattacharya et al. (2014). A possible mechanism for solar wind forcing of tropospheric-ionospheric coupling driven by global thunderstorm activity. *International Journal on Electronics & Communication Technology (IJECT)* 5: 75–79.
8. D. K. Singh et al. (2004). The electrical environment of the Earth's atmosphere: A review. *Space Science Reviews* 113: 375–408.
9. T. Neubert (2003). Atmospheric science: On sprites and their exotic kin. *Science* 300: 747–749.
10. V. M. Sorokin and V. M. Chmyrev (2010). Atmosphere-ionosphere electrodynamic coupling. In L. Alfonsi (ed.). *The Atmosphere and Ionosphere, Physics of Earth and Space Environments*. New York: Springer. pp. 97–104.
11. E. S. Kazimirovsky (2002). Coupling from below as a source of ionospheric variability: A review. *Annals of Geophysics* 45: 1–30.
12. T. B. Jones et al. (1972). Observations of D-region modifications at low and very low frequencies. *Nature* 238: 33–34.
13. R. J. Gamble et al. (2008). Radiation belt electron precipitation by man-made VLF transmissions. *Journal of Geophysical Research* 113: A10211.
14. R. Smoluchowski (1979). Origin of the magnetic fields in the giant planets. *Physics of Earth and Planetary Interiors* 20: 247–254.
15. W. H. Ip et al. (2000). Plutos ionospheric models and solar wind interaction. *Advances in Space Research* 26: 1559–1563.

16. J. Stevenson (2003). Planetary magnetic fields. *Earth and Planetary Science Letters* 208: 1–11.
17. M. J. Rycroft et al. (2000). The global atmospheric electric circuit, solar activity and climate change. *Journal of Atmospheric and Sol-Terrestrial Physics* 62: 1563–1576.
18. R. G. Harrison and K. L. Aplin (2003). Nineteenth century parisian smoke variations inferred from atmospheric electrical observations. *Atmospheric Environment* 37(38): 5319–5324.
19. M. Fillingim (1998). Global Electric Circuit of Mars, Unpublished Report, Florida Institute of Technology.
20. W. M. Farrell and M. D. Desch (2001). Is there a Martian atmospheric electric circuit. *Journal of Geophysical Research* E4: 7591–7595.
21. R. G. Harrison and K. S. Carslaw (2003). Ion-aerosol-cloud processes in the lower atmosphere. *Reviews of Geophysics* 41: 3.
22. C. Mondal et al. (1996). Statistical analysis of heavy rainfall over two tropical stations in India. *Indian Journal of Physics* 70B: 183–195.
23. S. K. Sarkar et al. (1998). Some studies on attenuation and atmospheric water vapor measurements in India. *International Journal of Remote Sensing* 19: 473–480.
24. T. Kamilya et al. (2009). Effect of salt on the formation of alcohol-dehydrogenease monolayer: A study by Langmuir-Blodgett technique. *Journal of Physical Chemistry B* 113: 5128–5135.
25. J. Johny et al. (2009). Atmospheric phenomena deduced from radiosonde and GPS occultation measurements for various applications related studies. *Journal of Earth System Science* 118: 49–59.
26. R. MacGorman and W. D. Rust (2003). *The Electrical Nature of Storms*. Oxford University Press, Canada.
27. A. B. Bhattacharya et al. (2010). Earth's transient luminous events and their electromagnetic environment. *International Journal of Physics* 3: 87–94.
28. R. G. Harrison (2004). Long term measurements of the global atmospheric electric circuit at Eskdalemuir, Scotland, 1911–1981. *Atmospheric Research* 70: 1–19.
29. D. Mukherjee et al. (1999). Predictable electrical breakdown in composites. *Physical Review Letters* 83: 1215–1218.
30. A. L. Sharma and A. K. Thakur (2010). Improvement in voltage, thermal, mechanical stability and ion transport properties in polymer-clay nanocomposites. *Journal of Applied Polymer Science* 118: 2743–2753.
31. P. Mali et al. (2001). Effect of temperature on attenuation due to rain in terrestrial and satellite microwave communication. *International Journal of Infrared and Millimeter Waves* 22: 153–171.
32. A. Gurnett et al. (2001). Non-detection at Venus of high-frequency radio signals characteristic of terrestrial lightning. *Nature* 409: 313–315.
33. A. B. Bhattacharya et al. (2011). Electrification in the earth and martian atmospheres with the possibility of global electric circuits. *International Journal of Engineering Science and Technology* 3: 3221–3228.
34. S. Basu and M. I. Richardson (2004). Simulation of the Martian dust cycle with the GFDL mars GCME. *Journal of Geophysical Research* 109 (E11): 11006.
35. S. J. Desch et al. (2002). Progress in planetary lightning. *Reports on Progress in Physics* 65: 955–997.
36. S. Joardar and A. B. Bhattacharya (2006). Simultaneous resolving of frequency separated narrow band terrestrial radio sources by multi antenna spectrum monitoring systems assisting radio astronomy. *Journal of Electromagnetic Waves and Applications* 20: 1195–1209.
37. O. Melnik and M. Parrot (1998). Electrostatic discharging in Martian dust storms. *Journal of Geophysical Research* 103: 29107–29117.
38. S. Joardar and A. B. Bhattacharya (2007). Algorithms for categoric analysis of interference in low frequency radio astronomy. *Journal of Electromagnetic Waves and Applications* 21: 441–456.
39. R. Grard (1995). Solar photon interaction with the Martian surface and related electrical and chemical phenomena. *Icarus* 114: 130–138.

40. S. Joardar and A. B. Bhattacharya (2007). Design and analysis of a low frequency radio telescope for Jovian radio emission. *Progress in Electromagnetic Research* 72: 127–143.
41. R. Williams et al. (1989). A radar study of the plasma and geometry of lightning. *Journal of the Atmospheric Sciences* 46: 1173–1185.
42. R. Markson (1978). Solar modulation of atmospheric electrification and possible implications for the sun-weather relationship. *Nature* 273: 103–109.
43. J. Berthelier et al. (2000). ARES, atmospheric relaxation and electric field sensor, the electric field experiment on Netlander. *Planetary and Space Science* 48: 1193–1200.
44. L. Aplin (2006). Atmospheric electrification in the solar system. *Surveys in Geophysics* 27: 63–108.
45. C. E. Krauss et al. (2003). Experimental evidence for electrostatic discharging of dust near the surface of mars. *New Journal of Physics* 5: 70–79.
46. Z. Sternovsky et al. (2002). Contact charging of lunar and Martian dust simulants. *Journal of Geophysical Research* 107: 5105–5113.
47. V. A. Krasnopolsky (2006). A sensitive search for nitric oxide in the lower atmospheres of Venus and Mars: Detection on Venus and upper limit for Mars. *Icarus* 182: 80–91.
48. Y. J. Takahashi et al. (2008). Lightning detection by LAC onboard the Japanese Venus climate orbiter, Planet-C. *Space Science Reviews* 137: 317–334.
49. A. B. Bhattacharya et al. (2009). A roadmap for detecting extraterrestrial intelligent life. In *PIERS Proceedings*. EMW Publishing, Cambridge, MA. p. 1400.
50. A. B. Bhattacharya et al. (2008). *Astronomy & Astrophysics*. Hingham, Massachusetts: Infinity Science Press.
51. E. Williams et al. (2006). Calibrated radiance measurements with an air-field glow discharge tube: Application to sprites in the mesosphere. In M. Fullekrug, et al. (eds). Springer, New York. pp. 237–247.
52. L. Liszka (2004). On the possible infrasound generation by sprites. *Journal of Low Frequency Noise Vibration and Active Control* 23: 85–93.
53. L. Liszka and Y. Hobara (2006). Sprite-attributed infrasonic chirps their detection, occurrence and properties between 1994 and 2004. *Journal of Atmospheric and Solar-Terrestrial Physics* 68: 1179–1188.
54. T. Farges et al. (2005). Identification of infrasound produced by sprites during the sprite 2003 campaign. *Geophysical Research Letters* 32: 1–4.
55. T. Neubert et al. (2008). Recent results from studies of electric discharges in the mesosphere. *Surveys in Geophysics* 29: 71–137.
56. S. B. Mende et al. (2005). D region ionization by lightning-induced electromagnetic pulses. *Journal of Geophysical Research A* 110: A11312.
57. Z. Cheng et al. (2007). Broadband very low frequency measurement of D region ionospheric perturbations caused by lightning electromagnetic pulses. *Journal of Geophysical Research A* 112: A06318.
58. C. Haldoupis et al. (2010). More evidence for a one-to-one correlation between sprites and early VLF perturbations. *Journal of Geophysical Research* 115: A07304.
59. C. Haldoupis et al. (2009). Modeling the relaxation of early VLF perturbations associated with transient luminous events. *Journal of Geophysical Research A* 114: A00E04.
60. A. Mika (2007). Very low frequency EM wave studies of transient luminous events in the lower ionosphere. PhD Thesis. University of Crete, Greece.
61. Y. N. Taranenko et al. (1993). Interaction with the lower ionosphere of electromagnetic pulses from lightning: Heating, attachment and ionization. *Geophysical Research Letters* 20: 1539–1542.
62. L. Rowland (1998). Theories and simulations of elves, sprites and blue jets. *Journal of Atmospheric and Solar-Terrestrial Physics* 60: 831–844.
63. S. Celestin and V. P. Pasko (2013). Production of very high potential differences by intracloud lightning discharges in connection with terrestrial gamma ray flashes. *Journal of Geophysical Research* 118: 912–918.
64. R. J. Blakeslee et al. (2012). Seasonal variations in the lightning diurnal cycle and implications for the global electric circuit. *Atmospheric Research* 72: 376–381.

65. Q. Qin et al. (2012). Minimum charge moment change in positive and negative cloud to ground lightning discharges producing sprites. *Geophysical Research Letters* 39: L22801.
66. C. L. da Silva and V. P. Pasko (2012). Simulation of leader speeds at gigantic jet altitudes. *Geophysical Research Letters* 39: L13805.
67. S. Celestin et al. (2012). Terrestrial gamma ray flashes with energies up to 100 MeV produced by non-equilibrium acceleration of electrons in lightning. *Journal of Geophysical Research* 117: A05315.
68. W. Xu et al. (2012). Source altitudes of terrestrial gamma-ray flashes produced by lightning leaders. *Geophysical Research Letters* 39: L08801.
69. Q. Qin et al. (2012). Formation of single and double-headed streamers in sprite-halo events. *Geophysical Research Letters* 39: L05810.
70. S. Celestin and V. P. Pasko (2012). Compton scattering effects on the duration of terrestrial gamma-ray flashes. *Geophysical Research Letters* 39: L02802.
71. A. J. G. Baumgaertner et al. (2013). Toward a comprehensive global electric circuit model: Atmospheric conductivity and its variability in CESM1 (WACCM) model simulations. *Journal of Geophysical Research* 118: 9221–9232.
72. V. P. Pasko (2012). Infrasonic waves generated by supersonic auroral arcs. *Geophysical Research Letters* 39: L19105.
73. S. A. Mallios and V. P. Pasko (2012). Charge transfer to the ionosphere and to the ground during thunderstorms. *Journal of Geophysical Research* 117: A08303.
74. W. C. Burgess and U. S. Inan (1993). The role of ducted whistlers in the precipitation loss and equilibrium flux of radiation belt electrons. *Journal of Geophysical Research* 98: 15643–15650.
75. T. J. Rosenberg et al. (1971). Electron precipitation associated with discrete very low frequency emissions. *Journal of Geophysical Research* 76: 8445–8452.
76. J. Rycroft (1973). Enhanced energetic electron intensities at 100 km altitude and a whistler propagating through the plasmasphere. *Planetary and Space Science* 21: 239–251.
77. R. J. Goldberg et al. (1987). Detailed spectral structure of magnetospheric electron bursts precipitated by lightning. *Journal of Geophysical Research* 92: 2505–2512.
78. H. D. Voss et al. (1984). Lightning-induced electron precipitation. *Nature* 312: 740–742.
79. H. D. Voss (1998). Satellite observations of lightning-induced electron precipitation. *Journal of Geophysical Research A* 103: 11725–11744.
80. R. Williams and S. J. Heckman (1993). The local diurnal variation of cloud electrification and the global diurnal variation of negative charge on the earth. *Journal of Geophysical Research* 98 (3): 5221–5234.
81. M. J. Rycroft (2006). Electrical processes coupling the atmosphere and ionosphere: An overview. *Journal of Atmospheric and Solar-Terrestrial Physics* 68: 445–456.
82. M. J. Rycroft and R. G. Harrison (2011). Electromagnetic atmosphere—Plasma coupling: The global electric circuit. *Space Science Reviews* 168: 363–384.
83. M. Parrot (1994). Statistical study of ELF/VLF emissions recorded by a low-altitude satellite during seismic events. *Journal of Geophysical Research* 99: 23339–23347.
84. V. A. Liperovsky et al. (2008). Physical models of coupling in the lithosphere-atmosphere-ionosphere system before earthquakes. *Geomagnetism and Aeronomy* 48: 795–806.
85. R. G. Harrison and M. H. P. Ambaum (2010). Observing Forbush decreases in cloud at Shetland. *Journal of Atmospheric and Solar-Terrestrial Physics* 72: 1408–1414.
86. A. Bazilevskaya et al. (2000). Effects of cosmic rays on Earth's environment. *Journal of Atmospheric and Solar-Terrestrial Physics* 62: 1577–1586.
87. A. Bazilevskaya et al. (2008). Cosmic ray induced ion production in the atmosphere. *Space Science Reviews* 137: 149–173.
88. P. I. Y. Velinov et al. (2009). Model for induced ionization by galactic cosmic rays in the Earth atmosphere and ionosphere. *Advanced Space Research* 44: 1002–1007.
89. R. Markson (1986). Tropical convection ionospheric potentials and global circuit variation. *Nature* 320: 588–594.
90. K. L. Aplin et al. (2008). Balloon-borne disposable radiometer. *Space Science Reviews* 137: 11–27.

91. S. A. Pulinets and V. H. Depuev (2003). Global scale ionospheric irregularities associated with thunderstorm activity. *Proceeding of the IRI Task Force Activity* 3: 71–80.
92. R. G. Roble and I. Tzur (1986). *The Global Atmospheric-Electrical Circuit, the Earth's Electrical Environment.* National Academies Press, Washington, DC. pp. 206–231.
93. A. K. Kamra (1972). Measurements of the electrical properties of dust storms. *Journal of Geophysical Research* 77: 5856–5869.
94. E. Kolokolov and N. A. Shalagina (1978). Effects of explosion of Karymski volcano in the bottom area of ionosphere, researches on geomagnetism and problems of aeronomy in high latitudes. In *Academy of Science USSR.* Far-Eastern Scientific Center, Magadan. pp. 76–79.
95. F. Vershinin et al. (1997). Correlations of seismic activity with electromagnetic emissions and variations in Kamchatka Region. In *Proceedings of International Workshop on Seismo Electromagnetics.* Chofu, Tokyo, Japan: University of Electro-Communications, p. 23.
96. M. Fullekrug et al. (eds) (2006). *Sprites, Elves and Intense Lightning Discharges.* Dordrecht, the Netherlands: Springer.
97. M. J. Rycroft (2010). The plasma and radiation environment in Earth orbit. In R. Blockley, W. Shyy (eds). *Encyclopedia of Aerospace Engineering.* Chichester, UK: Wiley.
98. J. Rodger et al. (2001). Lower ionospheric modification by lightning-EMP: Simulation of the nighttime ionosphere over the United States. *Geophysical Research Letters* 28: 199–202.
99. V. A. Rakov and M. A. Uman (2003). *Lightning: Physics and Effects.* Cambridge University Press, Cambridge, UK.
100. A. B. Bhattacharya (1994). Multitechnique studies of Norwester using meteorological and electrical parameters. *Annales Geophysicae* 12: 232–239.
101. J. Smith et al. (2010). The statistics of natural ELF/VLF waves derived from a long continuous set of ground-based observations at high latitude. *Journal of Atmospheric and Solar-Terrestrial Physics* 72: 463–475.
102. V. A. Mullayarov et al. (2010). Patterns of spatial distribution of positive thunderstorm discharges in eastern Siberia. *Journal of Atmospheric and Solar-Terrestrial Physics* 72: 409–425.
103. E. R. Williams et al. (1989). A radar study of the plasma and geometry of lightning. *Journal of the Atmospheric Sciences* 46: 1173–1185.
104. E. Greenberg et al. (2007). ELF transients associated with sprites and elves in eastern Mediterranean winter thunderstorms. *Journal of Atmospheric and Solar-Terrestrial Physics* 69: 1569–1586.
105. F. Simões et al. (2008). Schumann resonances as a means of investigating the electromagnetic environment in the solar system. *Space Science Reviews* 13: 455–471.
106. H. Yeng et al. (2009). Seasonal variations of global lightning activity extracted from Schumann resonances using a genetic algorithm method. *Journal of Geophysical Research* 114: D01103.
107. A. V. Shvets et al. (2010). Variations of the global lightning distribution revealed from three-station Schumann resonance measurements. *Journal of Geophysical Research* 115: A12316.
108. P. Nickolaenko et al. (2010). Q-bursts: Natural ELF radio transients. *Surveys in Geophysics* 31: 409–425.
109. M. Golkowski et al. (2010). Amplitude and phase of nonlinear magnetospheric wave growth excited by the HAARP HF heater. *Journal of Geophysical Research* 115: A00F04.
110. E. R. Williams et al. (2010). Ground based detection of sprites and their parent lightning flashes over Africa during the 2006 AMMA campaign. *Quarterly Journal of the Royal Meteorological Society* 136: 257–271.
111. T. Whitely et al. (2011). Worldwide extremely low frequency magnetic field sensor network for sprite studies. *Radio Science* 46: RS4007.
112. A. V. Shvets et al. (2009). Reconstruction of the global lightning distribution from ELF electromagnetic background signals. *Journal of Atmospheric and Solar-Terrestrial Physics* 71: 1405–1412.
113. A. V. Shvets et al. (2010). Variations of the global lightning distribution revealed from three-station Schumann resonance measurements. *Journal of Geophysical Research* 115: A12316.
114. K. Nakamura; Particle Data Group (2010). Review of particle physics. *Journal of Physics G: Nuclear and Particle Physics* 37: 075021.

115. A. V. Shvets and M. Hayakawa (2011). Variations of the global lightning distribution revealed from three-station Schumann resonance measurements. *Surveys in Geophysics* 32: A12316.
116. K. Yamashita et al. (2011). Improvement in lightning geolocation by time-of-arrival method using global ELF network data. *Journal of Geophysical Research* 116: A00E61.
117. M. Fullekrug et al. (eds) (2006). *Sprites, Elves and Intense Lightning Discharges*. Dordrecht, the Netherlands: Springer.
118. V. P. Pasko (2010). Air-density-dependent model for analysis of air heating associated with streamers, leaders, and transient luminous events. *Journal of Geophysical Research* 115: A00E35.
119. V. P. Pasko et al. (2011). On the inception of streamers from sprite halo events produced by lightning discharges with positive and negative polarity. *Journal of Geophysical Research* 116: A06305.
120. M. J. Rycroft and A. Odzimek (2009). The impact of lightning flashes and sprites on the Earth's global electric circuit: An overview of recent modelling results. *AIP Conference Proceedings* 1118: 124–135.
121. M. J. Rycroft and A. Odzimek (2010). Effects of lightning and sprites on the ionospheric potential, and threshold effects on sprite initiation, obtained using an analog model of the global atmospheric electric circuit. *Journal of Geophysical Research* 115: A00E37.
122. M. J. Rycroft et al. (2000). The global atmospheric circuit, solar activity and climate change. *Journal of Atmospheric and Solar-Terrestrial Physics* 62: 1563–1576.
123. D. M. Mach et al. (2011). Global electric circuit implications of combined aircraft storm electric current measurements and satellite based diurnal lightning statistics. *Journal of Geophysical Research: Atmospheres* 116: 1984–2012.
124. Liu et al. (2010). Diurnal variations of global thunderstorms and electrified shower clouds and their contribution to the global electrical circuit. *Journal of Atmospheric Science* 67 (2): 309–323.
125. R. G. Harrison and I. Usoskin (2010). Solar modulation in surface atmospheric electricity. *Journal of Atmospheric and Solar-Terrestrial Physics* 72: 176–182.
126. R. Hegerberg and I. D. Reid (1980). Electron drift velocities in air. *Australian Journal of Physics* 33: 227–230.
127. M. J. Rycroft and M. Fullekrug (2004). The initiation and evolution of special. *Journal of Atmospheric and Solar-Terrestrial Physics* 66: 1103–1113.
128. M. J. Rycroft et al. (2007). New model simulations of the global atmospheric electric circuit driven by thunderstorms and electrified shower clouds: The roles of lightning and sprites. *Journal of Atmospheric and Solar-Terrestrial Physics* 69: 2485–2509.
129. H. Israel (1973). *Atmospheric Electricity: Fields, Charges, Currents*. Israel Program for Scientific Translations. Jerusalem, Israel. Vol. 2, p. 796.
130. T. C. Marshall et al. (2005). Observed electric fields associated with lightning initiation. *Geophysical Research Letters* 32: L03813.
131. M. Stolzenburg et al. (2007). Electric field values observed near lightning flash initiations. *Geophysical Research Letters* 34: L04804.
132. V. Gurevich and K. P. Zybin (2001). Runaway breakdown and electric discharges in thunderstorms. *Physics-Uspekhi* 44: 1119–1140.
133. V. Gurevich and K. P. Zybin (2005). Runaway breakdown and the mysteries of lightning. *Physics Today* 58: 37–43.
134. V. Gurevich et al. (2009). An intracloud discharge caused by extensive atmospheric shower. *Physics Letters Section A* 373: 3550–3553.
135. M. A. Uman (1987). *The Lightning Discharge*. Orlando, FL: Academic Press.
136. H. Volland (1984). *Atmospheric Electrodynamics*. New York: Springer-Verlag.
137. T. C. Marshall et al. (1996). Electric field measurements above Mesoscale convective systems. *Journal of Geophysical Research* 101: 6979–6996.
138. M. Brook et al. (1982). The electrical structure of the Hokuriku winter thunderstorms. *Journal of Geophysical Research* 87: 1207–1215.
139. M. A. Uman (1987). *The Lightning Discharge*. Orlando, FL: Academic Press.

140. R. A. Roussel-Dupre et al. (1994). Kinetic theory of runaway air breakdown. *Physics Reviews* 49: 2257–2271.
141. T. F. Bell et al. (1995). Runaway electrons as a source of red sprites in the mesosphere. *Geophysical Research Letters* 22: 2127–2130.
142. W. C. Armstrong (1983). Recent advances from studies of the Trimpi effect. *Antarctic Journal* 18: 281–283.
143. U. S. Inan et al. (1988). Subionospheric VLF signatures of night-time D-region perturbations in the vicinity of lightning discharges. *Journal of Geophysical Research* 93: 11455–11472.
144. U. S. Inan et al. (1993). VLF signatures of lightning-induced heating and ionization of the night time D-region. *Geophysical Research Letters* 20: 2355–2358.
145. U. S. Inan et al. (1995). VLF signatures of ionospheric disturbances associated with sprites. *Geophysical Research Letters* 22: 3461–3464.
146. U. S. Inan et al. (1996). VLF signatures of mesospheric/lower ionospheric response to lightning discharges. *Journal of Geophysical Research* 101: 5219–5238.
147. R. L. Dowden et al. (1994). Rapid onset, rapid decay (RORD), phase and amplitude pertubations of VLF subionospheric transmissions. *Journal of Atmospheric and Solar-Terrestrial Physics* 56: 1513–1527.
148. D. Sentman and E. M. Wescott (1995). Red sprites and blue jets: Thunderstorm-excited optical emissions in the stratosphere, mesosphere and ionosphere. *Physics Plasmas* 2: 2514–2522.
149. W. A. Lyons (1994). Characteristics of luminous structures in the stratosphere above thunderstorms as imaged by low-light video. *Geophysical Research Letters* 21: 875–878.
150. W. A. Lyons (1995). Low-light video observations of frequent luminous structures in the stratosphere above thunderstorms. *Monthly Weather Review* 122: 1940–1946.
151. W. A. Lyons (1996). Sprite observations above the U. S. High Plains in relation to their parent thunderstorm systems. *Journal of Geophysical Research* 101: 29641–29652.
152. W. L. Boeck et al. (1995). Observations of lightning in the stratosphere. *Journal of Geophysical Research* 100: 1465–1475.
153. R. L. Rairden and S. B. Mende (1995). Time resolved sprite imagery. *Geophysical Research Letters* 22: 3465–3468.
154. J. R. Winckler et al. (1996). New high-resolution ground based studies of sprites. *Journal of Geophysical Research* 101: 6997–7004.
155. M. Wescott et al. (1995). Preliminary results from the Sprites 94 aircraft campaign: 2 Blue jets. *Geophysical Research Letters* 22: 1209–1212.
156. W. L. Boeck et al. (1992). Lightning induced brightening in the airglow layer. *Geophysical Research Letters* 19: 99–102.
157. H. Fukunishi et al. (1996). Elves: Lightning induced transient luminous events in the lower ionosphere. *Geophysical Research Letters* 23: 2157–2160.
158. U. S. Inan et al. (1996). Space-time structure of lower ionospheric optical flashes and ionization changes produced by lightning EMP. *Geophysical Research Letters* 23: 133–136.
159. J. Boccippio et al. (1995). Sprites, ELF transients and positive ground strokes. *Science* 269: 1088–1091.
160. J. R. Winckler (1995). Further observations of cloud-ionosphere electrical discharges above thunderstorms. *Journal of Geophysical Research* 100: 14335–14345.
161. J. Fishman et al. (1994). Discovery of intense gamma-ray flashes of atmospheric origin. *Science* 264: 1313–1316.
162. A. B. Bhattacharya et al. (2010). On the relationship between lightning and terrestrial gamma ray flashes. *Science and Culture* 76: 566–568.
163. U. S. Inan et al. (1996). On the association of terrestrial gamma-ray bursts with lightning discharges and sprites. *Geophysical Research Letters* 23: 1017–1020.
164. N. Holden et al. (1995). Satellite observations of transionospheric pulse pairs. *Geophysical Research Letters* 22: 889–892.

165. M. E. Baginski and A. S. Hodel (1994). A case study comparing the lossy wave equation to the continuity equation in modeling late-time fields associated with lightning. *Applied Computational Electromagnetic Society Journal* 9: 98–110.
166. R. E. Orville (1991). Calibration of a magnetic direction finding network using measured triggered lightning return stroke peak currents. *Journal of Geophysical Research* 96: 17135–17142.
167. V. V. Surkov et al. (2006). On the origin stable IR anomalies detected by satellites above seismo-active regions. *Physics and Chemistry of the Earth Journal* 31: 164–171.
168. M. Hayakawa (2006). Electromagnetic Phenomena Associated with Earthquakes. *IEEJ Transactions on Fundamentals and Materials* 126: 211–214.
169. V. P. Pasko et al. (2012). Lightning related transient luminous events at high altitude in the Earth's atmosphere: Phenomenology, mechanisms and effects. *Space Science Reviews* 168: 475–516.
170. A. B. Bhattacharya et al. (2011). Cloud microphysical properties from Sferics measurements. *International Journal of Physics* 4: 73–83.
171. V. P. Pasko (2006). *Theoretical Modeling of Sprites and Jets, in Sprites, Elves and Intense Lightning Discharges*. NATO Science Series II: Mathematics, Physics and Chemistry. Heidleberg, Germany: Springer. Vol. 225, pp. 253–293.
172. V. P. Pasko et al. (2013). Toward better understanding of sprite streamers: Initiation, morphology and polarity asymmetry. *Surveys in Geophysics* 34: 797–830.
173. J. Qin et al. (2013). Mechanism of column and carrot sprites derived from optical and radio observations. *Geophysical Research Letters* 40: 4777.
174. L. da Silva and V. P. Pasko (2013). Vertical structuring of gigantic jets. *Geophysical Research Letters* 40: 3315–3319.
175. J. Qin et al. (2013). Dependence of positive and negative sprite morphology on lightning characteristics and upper atmospheric ambient conditions. *Journal of Geophysical Research* 118: 2623–2638.
176. R. Roussel-Dupre et al. (2008). Physical processes related to discharges in planetary atmospheres. *Space Science Reviews* 137: 51–82.
177. J. Qin et al. (2012). Low frequency electromagnetic radiation from sprite streamers. *Geophysical Research Letters* 39: L22803.
178. T. Neubert (2003). Atmospheric science: On sprites and their exotic kin. *Science* 300: 747–749.
179. D. Sentman and E. M. Wescott (1993). Observations of upper atmospheric optical flashes recorded from an aircraft. *Geophysical Research Letters* 20: 2857–2860.
180. R. Haldoupis et al. (2006). "Early/slow" events: A new category of VLF perturbations observed in relation with sprites. *Journal of Geophysical Research* 111: A11321.
181. U. S. Inan et al. (1991). Heating and ionization of the lower ionosphere by lightning. *Geophysical Research Letters* 18: 705–708.
182. Y. N. Taranenko et al. (1993). Interaction with the lower ionosphere of electromagnetic pulses from lightning: heating, attachment, and ionization. *Geophysical Research Letters* 20: 1539–1542.
183. Y. N. Taranenko et al. (1993). The interaction with the lower ionosphere of electromagnetic pulses form lightning: Excitation of optical emissions. *Geophysical Research Letters* 20: 2675–2678.
184. M. Milikh et al. (1995). On the physics of high altitude lightning. *Geophysical Research Letters* 22: 85–88.
185. L. Rowland et al. (1995). Lightning driven EMP in the upper atmosphere. *Geophysical Research Letters* 22: 361–364.
186. V. P. Pasko et al. (1995). Heating, ionization and upward discharges in the mesosphere due to intense quasi-electrostatic thundercloud fields. *Geophysical Research Letters* 22: 365–368.
187. V. P. Pasko et al. (1996). Sprites as luminous columns of ionization produced by quasi-electrostatic thundercloud fields. *Geophysical Research Letters* 23: 649–652.
188. V. P. Pasko et al. (1996). Blue jets produced by quasi-electrostatic pre-discharge thundercloud fields. *Geophysical Research Letters* 23: 301–304.

189. V. P. Pasko et al. (1996). Sprites produced by quasi-electrostatic heating and ionization in the lower ionosphere. *Journal of Geophysical Research* 23: 4529–4561.
190. R. A. Roussel-Dupre and A. V. Gurevich (1996). On runaway breakdown and upward propagating discharges. *Journal of Geophysical Research* 101: 2297–2311.
191. Y. N. Taranenko and R. A. Roussel-Dupre (1996). High altitude discharges and gamma-ray flashes: A manifestation of runaway air breakdown. *Geophysical Research Letters* 23: 571–575.
192. U. S. Inan et al. (1996). Early/fast VLF events as evidence of sustained heating of the ionosphere above thunderclouds. *Geophysical Research Letters* 23: 1067–1070.
193. T. Wilson (1956). A theory of thundercloud electricity. *Proceedings of the Royal Society of London Series A* 236: 297–317.
194. A. J. Illingworth (1972). Electric field recovery after lightning as the response of the conducting atmosphere to a field change. *Quarterly Journal of the Royal Meteorological Society* 98: 604–616.
195. I. Tzur and R. G. Roble (1985). The interaction of a polar thunderstorm with its global electrical environment. *Journal of Geophysical Research* 90: 5989–5999.
196. G. Park and M. Dejnakarintra (1973). Penetration of thundercloud electric fields into the ionosphere and magnetosphere: 1. Middle and subauroral latitudes. *Journal of Geophysical Research* 78: 6623–6633.
197. V. V. Hegai et al. (1990). The formation of a cavity in the night-time midlatitude ionospheric E-region above a thundercloud. *Planetary and Space Science* 38: 703–707.
198. H. Svensmark and E. Friis-Christensen (1997). Variation of cosmic ray flux and global cloud coverage—A missing link in solar-climate relationships. *Journal of Atmospheric and Solar-Terrestrial Physics* 59: 1225–1232.
199. H. Fujiwara et al. (2004). Atmospheric anomalies observed during earthquake occurrences. *Geophysical Research Letters* 31: L17110.
200. K. Ohta et al. (2002). Relationship between ELF magnetic fields and Taiwan earthquake. In M. Hayakawa, O. A. Molchanov (eds). *Seismo Electromagnetics: Lithosphere-Atmosphere-Ionosphere Coupling*. Tokyo, Japan: Terrapub. pp. 233–237.
201. A. Y. Schekotov et al. (2007). ULF/ELF magnetic field variations from atmosphere induced by seismicity. *Radio Science* 42: RS6S90.
202. H. Nagamoto et al. (2008). Disturbances in VHF/UHF telemetry links as a possible effect of the 2003 Hokkaido Tokachi-oki earthquake. *Natural Hazards Earth System Science* 8: 813–817.
203. V. V. Hegai et al. (2006). The ionospheric effect of atmospheric gravity waves excited prior to strong earthquake. *Advances in Space Research* 37: 653–659.
204. G. Kondo (1968). The variation of the atmospheric field at the time of earthquake. *Memoirs Kakioka Magnetic Observatory* 13: 11–23.
205. M. Kamogawa et al. (2004). Atmospheric field variations before the March 31, 2002 *M*6.8 earthquake in Taiwan. *Terrestrial, Atmospheric and Oceanic Sciences* 15: 397–412.
206. N. Yonaiguchi et al. (2007). On the statistical correlation of over horizon VHF signals with meteorological radio ducting and seismicity. *Journal of Atmospheric and Solar-Terrestrial Physics* 69: 661–674.
207. S. Naman et al. (2002). Comparison of simultaneous observations of the ionospheric total electron content and geomagnetic field associated with strong earthquakes. In M. Hayakawa, O. A. Molchanov (eds). *Seismo Electromagnetics: Lithosphere Atmosphere Ionosphere Coupling*. Tokyo, Japan: Terrapub. pp. 303–308.
208. T. Freund et al. (2006). Air ionization at rock surfaces and pre-earthquake signals. *Journal of Atmospheric and Solar-Terrestrial Physics* 71: 1824–1834.
209. A. Bering et al. (1998). The global electric circuit. *Physics Today* 51: 24–30.
210. NAS (National Academies Press) (1986). *The Earth's Electrical Environment*. Washington, DC: Geophysics Research Board, National Academies Press.
211. M. J. Rycroft et al. (2008). An overview of Earth's global electric circuit and atmospheric conductivity. *Space Science Reviews* 137: 83–105.

212. R. G. Harrison and A. J. Bennett (2007). Multi-station synthesis of early twentieth century surface atmospheric electricity measurements for upper tropospheric properties. *Advanced Geoscience* 13: 17–23.
213. A. Chalmers (1967). *Atmospheric Electricity*. 2nd ed. Oxford: Pergamon Press.
214. S. A. Cummer et al. (1988). Ionospheric D region remote sensing using VLF radio atmospherics. *Radio Science* 33: 1781–1792.
215. M. Friedrich and M. Rapp (2009). News from the lower ionosphere: A review of recent developments. *Surveys of Geophysics* 30: 525–559.
216. G. Budden (1962). *The Waveguide Mode Theory of Wave Propagation*. London: Logos Press.

10

Dynamic Atmosphere of the Sun

10.1 Introduction

Plasma is a high-energy gas that is responsible for the ionization of a significant number of its atoms or molecules, thus forming equal numbers of positive and negative ions (electrons). Plasmas exhibit some properties of gases, but differ from gases being good conductors of electricity and being affected by magnetic fields. Although there are free charges and ambipolar pairs in plasmas, overall the negative and positive charges compensate each other. Therefore, plasmas are electrically neutral, better known as quasi-neutrality [1]. Within the known universe, more than 99% of all observable matter is plasma, a state often highly dynamic and far from thermal and mechanical equilibrium. In particular, for our own solar–terrestrial system, various plasma active phenomena frequently occur such as solar flares, solar wind, and coronal mass ejections (CMEs) in the solar atmosphere; interplanetary magnetic clouds and anti-collision shock waves in interplanetary space; and geomagnetic storms and Earth's aurora occurring in Earth's magnetosphere and ionosphere. Such phenomena are not only the most important events resulting in changes in the space environment around our anthrosphere, but also provide natural laboratories for a detailed study of the basic plasma processes encountered in astrophysics. A large part of the energy flux released in the solar atmosphere travels into interplanetary space and impacts the Earth and is known as bow shock. The bow shock energizes the magnetosphere, resulting in changes in the composition, energy balance, and the ionospheric dynamics of the plasma sphere and plasma pause [2–4]. The activities of the plasma process of the solar atmosphere have a direct impact over the ionosphere down to troposphere, thus affecting the radio communication in the different frequency bands [5]. The layers of the atmosphere all react in different ways from the solar particles and plasma influx. The D layer [6] is the innermost layer and is one of the most active during solar activity. The D Layer is 60 km (37 mi) to 90 km (56 mi) above the surface of the Earth. Ionization here is due to Lyman series–alpha hydrogen radiation at a wavelength of 121.5 nanometer (nm) ionizing nitric oxide (NO). In addition, with high solar activity hard X-rays (wavelength <1 nm) may ionize (N_2, O_2). During the night, cosmic rays produce a residual amount of ionization. Recombination is high in the D layer, the net ionization effect is low, but loss of wave energy is great due to frequent collisions of the electrons (about 10 collisions every msec). As a result, high-frequency (HF) radio waves are not reflected by the D layer but suffer loss of energy therein. This is the main reason for absorption of HF radio waves, particularly at 10 MHz and below, with progressively smaller absorption as the frequency gets higher. The absorption is small at night and greatest about midday. The layer reduces greatly after sunset; a small part remains due to galactic cosmic rays. A common example of the D layer in action is the disappearance of distant AM broadcast band stations in the daytime. During solar proton events (SPEs), ionization can reach unusually high levels in the D-region over high and polar latitudes. Such very rare events are known as polar cap absorption (or PCA) events, because the increased ionization significantly enhances the absorption of radio signals passing through the region. In fact, absorption levels can increase by many tens of dB during intense events, which is enough to absorb most (if not all) transpolar HF radio signal transmissions. Such events typically last less than 24–48 h.

10.2 Evolution and Structure of the Sun

The Sun originated 4.57 Giga-annum (10^9 years) ago; its expected future lifetime is some 6 Ga. The Sun is a gaseous sphere with a radius of 432,450 miles and has no actual surface. At the time of its origin the total radiation flux, and the irradiance was lower by 30% than the present value. If the terrestrial albedo had been the same then, as it is at present, the Earth's average temperature would have been well below 0°C, which would have hampered or made it impossible for the origin of life. There is some evidence that a thick terrestrial CO_2 atmosphere existed at that time and has increased the Earth's troposphere's temperature. This is known as the greenhouse effect [7]. The photosphere is the part of the outer layers, which emits the continuous radiation in the visual spectral range. It has an effective thickness of about 250 km and a density scale height of 110 km (varying with depth). The effective (*surface*) temperature, defined as the temperature of a black body radiation [8] with the same wavelength-integrated radiation flux, is 5800 K. Upward, the layer temperature passes through a minimum value of 4200 K and it rises higher up. The regions immediately above the photosphere are called the chromosphere. It is defined by the property that, when viewed at the solar limb, it is transparent for continuous light and opaque in wavelengths of the strongest spectral lines like the Hα hydrogen line. This explains its colorful appearance witnessed during solar eclipses. The chromospheric temperature ranges between 5,000 and 10,000 K. It has a thickness of about 1800 km. At a height of 2100 km above the Sun's limb is already 500,000 K and further upward it assumes values of the order of 2 million K. This outer region, which extends to distances of many millions of kilometers from the Sun, is the corona. The corona, with its high temperature and low density is fully ionized: it is perfect plasma. Its characteristic of being very irregular and changing shape is defined by the magnetic properties of the underlying chromosphere and its features, notably by the centers of activity, and by the location of coronal holes [9].

10.3 Internal Variation of Temperature

With the compression of the gas, the atoms condense, thus increasing the density and temperature, producing a variation of the pressure with depth. The central temperature is 15.7×10^6 K; the density there is 160 kg/l. Nuclear fusion reactions in the core region are responsible for the solar radiation. In the reaction chains, four protons combine to form one helium nucleus thus causing nuclear energy to be radiated. Hence, the core's chemical composition changes with time and this applies primarily to the number ratio of helium to hydrogen atoms which has increased during the Sun's lifetime. Also, the numerical densities of some of the associated elements involved in the fusion reaction chains do change with time. The radiation flux, originating in the core is transported outward, driven by the temperature gradient. The radiative equilibrium is characterized by up- and downward moving gas 500,000 km from the center, while the uppermost 200,000 km is convective. The granulation cells, visible in the photosphere, are manifestations of these convective motions. They are short lived and moving; upward with velocities of about 1 km/s and downward with higher velocities because of the lower temperature of the downward streaming gas [10,11].

10.4 Solar Wind and the Origin of Heliosphere

The solar wind is a stream of plasma released from the upper atmosphere of the Sun. It consists of mostly electrons and protons with energy levels usually between 1.5 and 10 keV. The majority of the wind emanates from the coronal holes. These are regions of the corona, generally situated at high latitudes, where the magnetic fields are open and do not inhibit the escape of coronal plasma from the Sun. These fields connect as loops at great heights above the Sun's surface. The average particle velocity and density of the solar wind at earth distance are respectively 470 km/s and nine protons per cm^3. As the gas expands adiabatically, the wind carries frozen-in magnetic fields along into space. Since the Sun rotates around its axis, with a synodic rotation period of 27 days at the equator, these field lines are wound up in an Archimedean spiral [12,13]. The mean magnetic fields at the solar surface along the equator alternate in sign, roughly every 90° in longitude, and therefore the interplanetary magnetic field measured near the Earth changes sign roughly once a week. Clouds of particles that are emitted from the Sun, for example, those resulting from a CME and those related to the ephemeral polar regions and the polar facular regions will move along these interplanetary field lines. A consequence is that at Earth's distance we normally receive gas emitted from sources at about 50° western solar longitude. The Sun moves through interstellar space with a velocity (with respect to the near stellar and gaseous environment) of 20 km/s toward the apex, a point in the constellation of Hercules (right ascension 18 h, declination 30°). In the collision with the interstellar gas, a shock forms in front and around the moving Sun, at a distance where the dynamic pressure of the carried along solar wind equals the interstellar gas pressure. Thus, the heliosphere originates; it is the drop-shaped volume inside the bow shock where the motion of the gas is dominated by the solar magnetic fields and the solar wind. Its largest diameter, measured at right angles to the direction of motion, varies between about 80 and 100 AU. In the direction of motion, it is much larger because of its long tail [14]. The solar wind is represented by a stream of charged particles released from the Sun's upper atmosphere. It mainly consists of electrons and protons with energies between 1.5 and 10 keV, in general. Over time, the stream of particles varies in temperature and speed. These particles can escape the Sun's gravity because of their high kinetic energy and the high temperature of the corona [15]. At the orbit of the Earth, the solar wind flows with a speed of about 400 km-s and a density of 5 cm^{-3}. It carries a magnetic field of about 5 nT that lies near the ecliptic plane in an Archimedean spiral pattern and it is highly variable [16,17]. In this chapter, we report the behavior of this magnetized solar wind plasma and examine the properties of the various parameters that constitute the solar wind and their variation with solar activity cycle.

10.4.1 Formation of the Solar Wind

The formation of the solar wind is due to interaction between magnetic field lines which emit energized ions [18]. Magnetic field lines are *frozen in* the solar wind due to high electrical conductivity of plasma wind. The field lines are bent to form a spiral shape in the Sun's equatorial plane. Reflected ions can play a vital role in dissipating energy from all shock-related events. The reflected termination shock ions gain most of the energy while the reflected thermal ions gain less [19]. Due to rotation of the sun (27 day period) the spiral field lines get distorted. The number density of the charged particles such as wind flow speed, temperature of plasma, and so on is modulated by the interaction of interstellar medium in the outer heliosphere [20]. These quantities change with distance from the Sun. But apart from this, various parameters of the solar wind change due to solar activity cycle, CME, sunspot cycle, and the interplanetary magnetic field. This chapter critically analyses the solar wind plasma parameters and its modulation in the time slot of the rising phase of the present sunspot cycle 24. Solar wind parameters like wind flow speed and related solar wind parameters are greatly influenced at a distance greater than 90 AU from the Sun. In this region solar wind slows down and fluctuates rapidly. We elaborately discuss the principal features of solar wind and extract the correlation between them. Voyager 1 and 2 flew through the termination shock related to solar wind where the solar wind flow damped, deviated, and heated.

10.4.2 High- and Low-Speed Solar Wind

During the declining and minimum phases of the solar cycle, the solar wind is dominated by high-speed (500–800 km s^{-1}) flows emanating from coronal holes. In the ascending phase of the solar activity cycle and at solar maximum, the average solar wind speed diminishes as the polar coronal holes shrink and the high-speed flows narrow and weaken. At the same time, the ambient solar wind flow is largely perturbed by CMEs. In Figure 10.1, we have plotted the average variation of the 27-day of solar wind proton density and solar wind plasma speed for a period from 1970 to 2012. In general, the solar wind speed is found to exhibit an inverse correlation with the solar wind proton density [21].

10.4.3 Solar Wind and the Temperature at the Source

The temperature of the solar wind can be determined by a number of competing effects—the significant temperature at the source and high-speed wind generally having higher proton temperatures. Many processes like adiabatic cooling, heating due to stream interactions and shocks, dissipation of waves, and transfer of energy from pickup protons to the thermal protons affect the temperature of the outward moving solar wind. Figure 10.2 plots the average variation of the 27-day solar wind plasma temperature and the solar wind plasma speed for the same period. It appears from Figure 10.2 that the solar wind plasma speed and solar wind plasma temperature reveal the same nature of variation exhibiting a lower trend at solar maximum. Solar wind parameters show strong solar cycle variations in plasma flux, plasma pressure, and magnetic field strength. These variations produce variations in the geomagnetic activity also [22–24]. Spacecraft measurements of the solar wind have shown that the solar wind magnetic field strength and plasma speed and dynamic pressure also varied with solar cycle [25]. In Figure 10.3, we have plotted the average 27-day of solar wind flow pressure and a_p index with the sunspot number (SN).

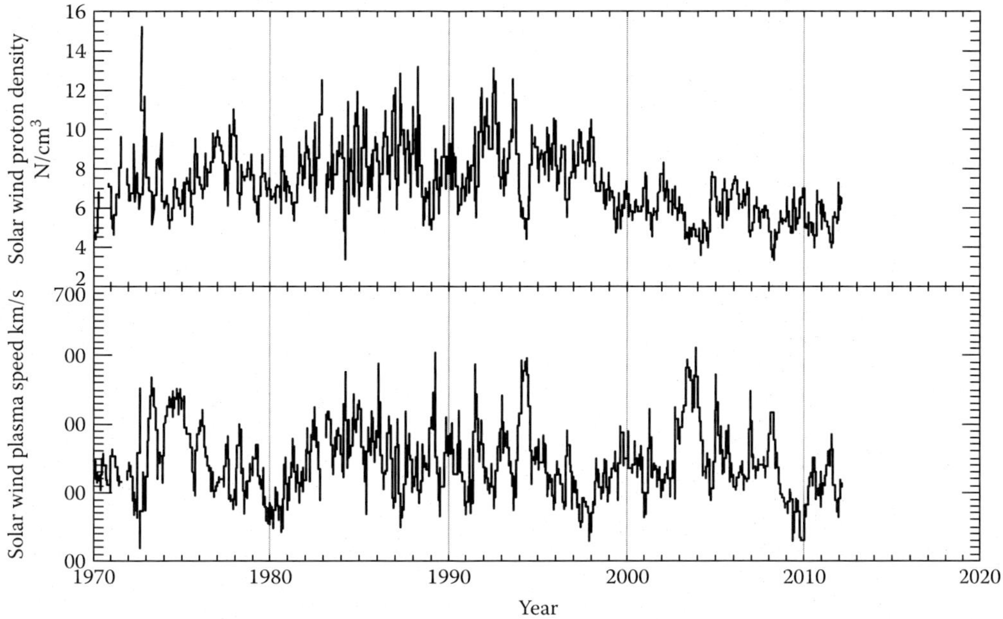

FIGURE 10.1
Variation of the 27-day average solar wind proton density and solar wind plasma speed.

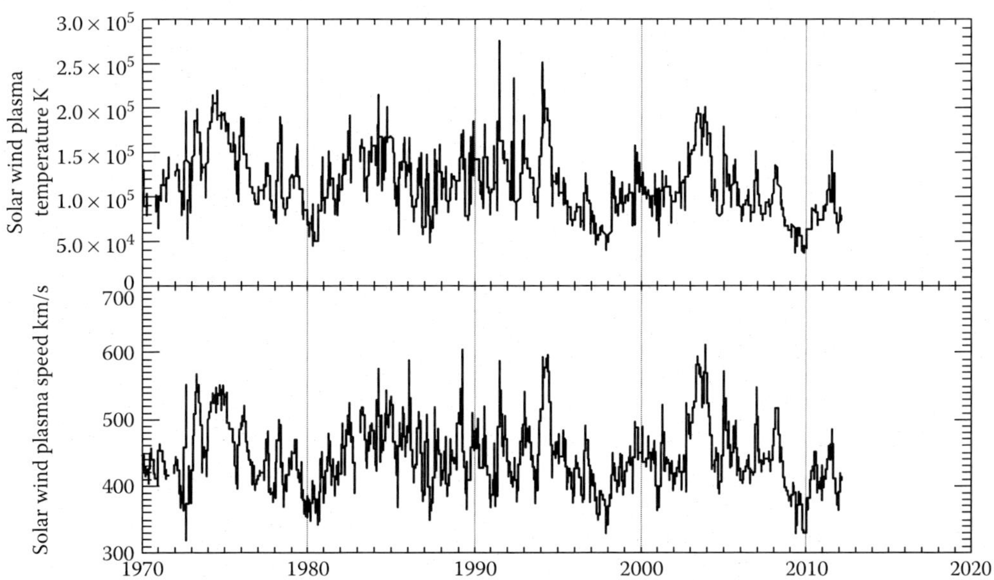

FIGURE 10.2
The variation of the 27-day average of solar wind plasma temperature and the solar wind plasma speed.

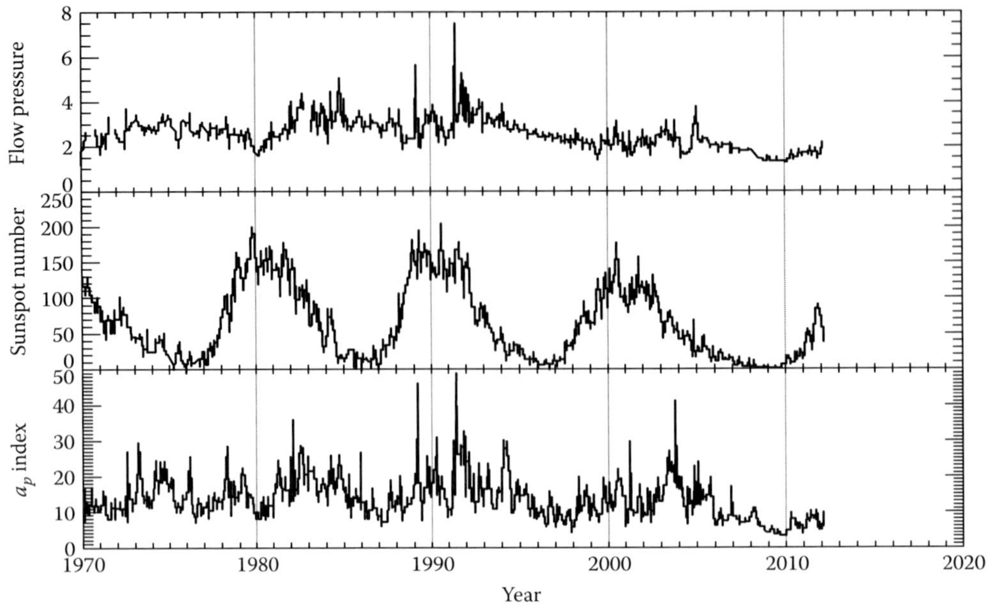

FIGURE 10.3
The 27-day average of solar wind flow pressure, the sunspot number, and the variation of a_p index.

From the graph (Figure 10.3), it is clear that the solar wind flow pressure varies in a reverse way with solar activity cycle. It increases rapidly after solar maximum and then decreases slowly toward the next solar maximum. But the geomagnetic activity index a_p falls at solar minimum, although the flow pressure is still high.

10.4.4 Variation of Proton Density and K_p Index

The proton density, K_p index, has been plotted for the period from 2008 to March 2012, which is the rising phase of solar cycle 24. It is clear from the figure that plasma proton density and K_p index are perfectly anti-correlated with each other. But fluctuation of proton density is greater than that of K_p index. But dependence of long-term fluctuation of proton number density upon sunspot activity has not been clearly examined. It appears from Figure 10.4 that the variation of proton density rises rapidly but falls slowly.

In Figure 10.5, we have a plotted histogram of magnitude of the magnetic field and latitude angle of the magnetic field with sunspot activity cycle for the same period. The magnetic field is asymmetric over this period. The asymmetry increases when sunspot activity increases rapidly. In the solar minimum of rising phase of cycle 24, magnetic fields in the solar wind is not at a minimum [26].

In Figure 10.6, we show a plot of the average moving wind flow pressure versus plasma temperature. From the figure, it is clear that plasma temperature increases with plasma flow pressure. A small change of plasma pressure causes rapid change in plasma temperature. Here the correlation coefficient 0.562 has been observed.

Amplitudes of high-speed solar wind fluctuate highly as shown in Figure 10.7. For sunspot minimum (in 2008) plasma temperature, speed and flow pressure were sufficiently high. It is clear from the figure that helium ion to proton ratio for flowing plasma is anticorrelated with plasma temperature. In addition, it is observed from the figure that though

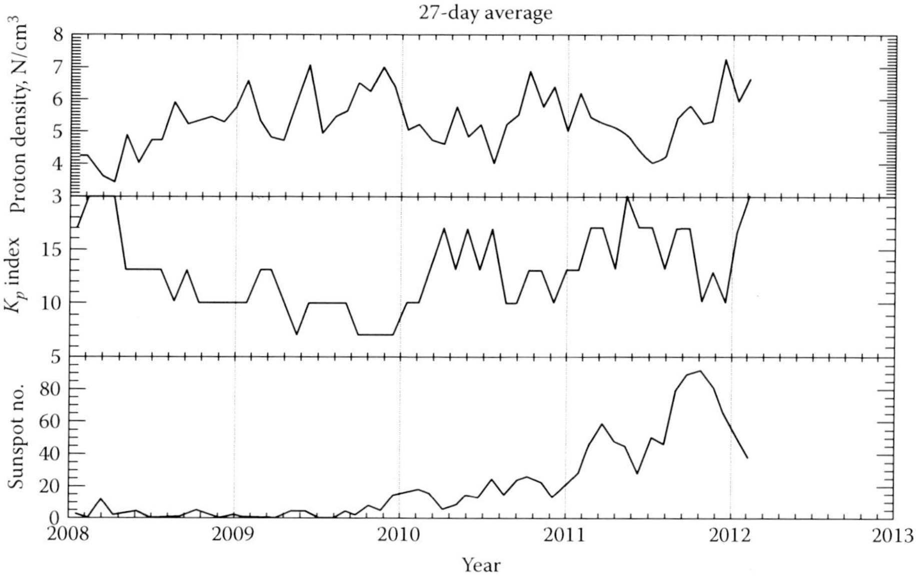

FIGURE 10.4
The 27-day averages of proton number density, K_p index, and SN.

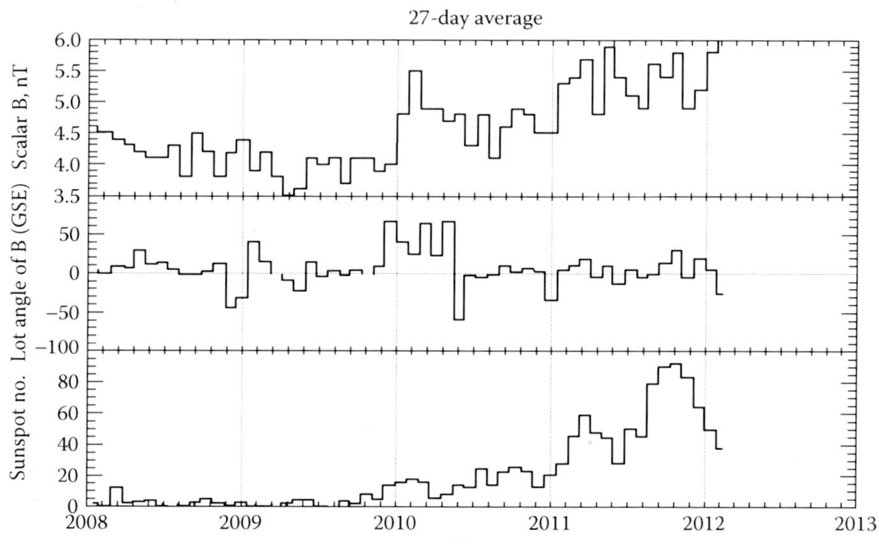

FIGURE 10.5
The 27-day averages of scalar magnetic field in nT, latitude variation of B, and SN.

the nature of fluctuation of plasma speed and plasma temperature is similar but there is a phase difference of approximately 20 days between the two. So the long-term variation of solar wind temperature clearly depends on the flow speed.

The results obtained from the present analysis may be summarized as follows: (i) the solar wind speed reveal an inverse correlation with the solar wind proton density. (ii) The

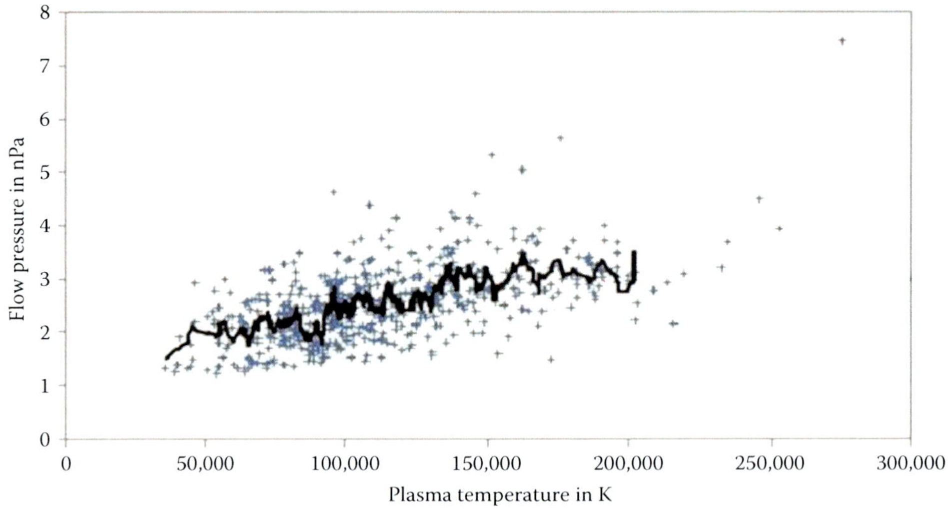

FIGURE 10.6
Variation of solar wind flow pressure with plasma temperature.

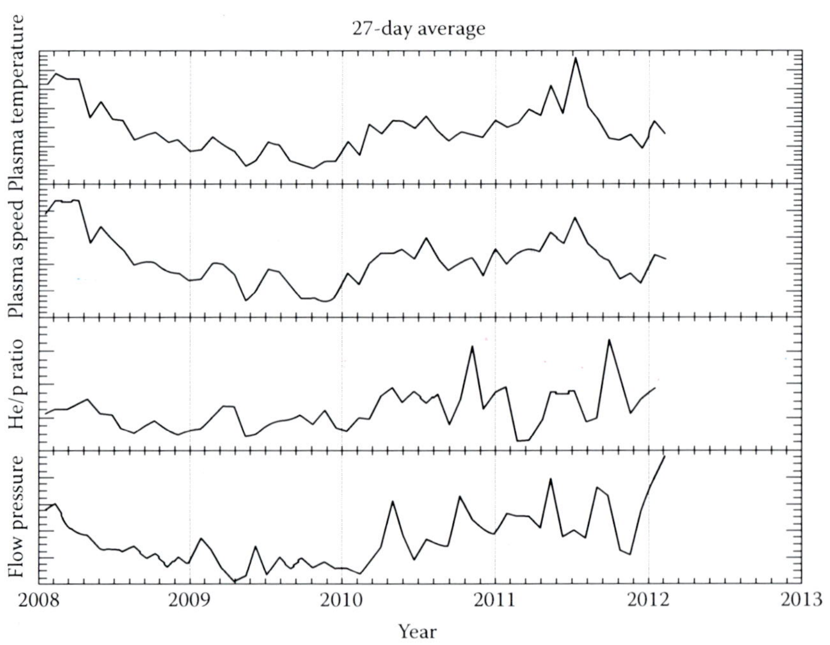

FIGURE 10.7
Variation of plasma temperature, plasma flow speed, helium ion to proton abundance in solar wind as well as the plasma flow pressure.

solar wind plasma speeds as well as solar wind plasma temperature show the same nature of variation producing a lower trend at solar maximum. (iii) The solar wind flow pressure changes in a reverse way with the solar activity cycle increasing rapidly after solar maximum and decreases slowly toward the next solar maximum. However, the geomagnetic activity index a_p is reduced at solar minimum, although the flow pressure remains high.

Again, we have discussed the 27-day averages of proton density data, K_p index, magnetic field, wind flow pressure and temperature, and helium ion and proton ratio in the rising phase of the ongoing solar cycle. Plasma mechanisms cannot explain clearly the variation of these solar wind parameters. Proton density and the K_p index are anti-correlated. The latitude magnetic field is asymmetric with respect to the equatorial plane of sun. Plasma proton temperature largely depends upon wind flow pressure. Plasma speed, plasma temperature, and wind flow speed are weakly modulated by the sunspot activity cycle. But the helium ion to proton ratio contained in the solar wind depends upon solar activity.

10.5 Active Region of the Sun

A center of activity (CA), also called an active region (AR) is a region where the larger part of photospheric activity manifests itself. The average size is 40,000 km (longitude) to 20,000 km (latitude), with large variations. The main feature is a weak—bipolar and fractionized—magnetic field, which shows a small tilt, with respect to the solar equator of the bipolar axis. Consequently, the AR is slightly inclined in latitude with the leading end closer to the solar equator. Its principal structures are transient phenomena: sunspots, faculae, prominences, and flares. CAs are also responsible for the CMEs. Akasofu [27] after having defined a new solar coordinate system, found that during the past few cycles main solar activity was chiefly concentrated in two quadrants, one around 90° in longitude in the northern hemisphere and the other around 270° in longitude in the southern hemisphere. The polar faculae and polar prominences are another sign of solar activity; these do not occur in the ARs, but are restricted to high latitudes, in the polar prominence zones.

10.6 Sudden Release of Energy

A flare is caused by a sudden release of energy in a CA. Flares seldom occur on or in a spot, but practically always in the vicinity of the spot. A typical flare starts with an impulsive phase. This phase, generally lasting for a few seconds or at most a few minutes, is the period in which energy is released. The heated plasma can then reach temperatures in the order of 50 million K. In one case, the flare of May 21, 1984, temperatures of 400–500 million K were determined [28]. The period during which the heated plasma cools down is called the gradual phase. That phase lasts longer, a typical duration is 10 min, but there are large variations from one flare to the other. In long-duration two-ribbon flares the energy release and plasma heating can continue far into the gradual phase. In the chromosphere, in light of the Balmer Hα line, flares are often quite impressive, and that was the way they were actually discovered. The chromospheric aspect of flares is a sudden increase of the density, by factors well above 100. A typical flare density in chromospheric levels is more than 10^{13} particles per cm^3. Reconnection or coalescence of flux tubes explains the origin (the *ignition*) of flares [29,30]. The process of reconnection transforms electromagnetic energy into heat and creates the environment for particle acceleration. After a flare in the magnetic fields, it was noted that they have changed very little, which shows that only part of the two reconnecting flux tubes interact, but that only a few threads reconnect.

This is confirmed by the finding that after a flare, the helicity is significantly reduced, but not wholly annihilated [31]. Flares are complicated and diverse, and so are the models proposed for the process of reconnection. At latitudes above ~40° activity is manifested by polar faculae [32–35], ephemeral ARs [36], and unipolar ARs [37,38]. The latter appearing as ~10^3 km spots with kilogauss magnetic fields. When the filament zones have reached the polar regions, these polar activity structures begin to show themselves.

10.7 Coronal Mass Ejections

The CMEs are magnetized clouds of plasma emitted from a CA. A large CME can contain 10^{16} g of matter. During the CMEs way outward in interplanetary space, plasma accelerated by the bow shock, driven by the CME, follows the Archimedean spiral magnetic field carried in the solar wind. Eventually, the shock waves associated with the many emitted CMEs can fill a significant part of the heliosphere. The magnetized heliospheric plasma slowly neutralizes or diffuses away through the heliospheric boundary. The consequence is that the heliosphere always contains a fair amount of magnetic field, even at solar minimum when only a few CMEs occur. To this, we have to add that corotating interaction regions (i.e., interplanetary shocks formed at the collision of fast solar winds with earlier emitted slower winds [39]) have their highest frequency during solar minimum. Hence, at no instant is the heliosphere without interplanetary magnetized shocks.

Note: 10 kG = 1 T (tesla), and 1 G = 10^{-4} T. 1,000,000,000 (or 10^9) γ (gammas), used in geophysics. Thus, 1 γ = 1 nT (nanotesla).

10.7.1 CME Bow Shock

Frequently, the Sun emits a beam of solar energetic particles called a CME. These particles are ejected from the central axis region of the CME bow shock. Since hydrogen is the major constituent of the solar plasma, these are often called SPEs. A beam of SPEs is usually defined as a cloud of particles with an energy >1 MeV. Others advocate higher limiting values, for example, El-Bone [39] assumes energy of 60 MeV. Observed energies range up to above 1000 MeV. The composition is coronal: hydrogen and helium. The first well-studied case was the burst of solar cosmic rays of February 23, 1956. The difference between solar CRs and SPEs is not always well defined and there is some confusion in literature about the distinction between them. All three classes of particles produce ground level enhancements (GLEs). A list of the major SPEs during the period 1977–2001 was published by Lario and Simnett [40]. CMEs, and SPEs are emitted from the CAs. The mechanism for their emission is not yet well understood, although it must be related to reconnecting flux tubes [41,42]. There is a certain correlation with solar flares. CMEs associated with flares are generally more important than those that are not. Also, the CME velocity correlates with the flare strength. The more energy that is stored in the relevant part of the active center, the better the correlation between the flares and CMEs [43]. The hypotheses that a flare would cause a CME, or the reverse, have not been confirmed. Rather, we should assume that a flare and the associated CME are both due to a *grand instability* in the AR [40]. In that line, Lin [43] assumes a *catastrophic loss of magnetohydrodynamics (MHD) equilibrium,* which may cause an energy release in a disrupted magnetic field in the CA. Green [44] seeks the origin in a magnetic polarity imbalance.

10.8 The 11-Year Solar Cycle

Solar activity rises and falls with a 11-year cycle (11 years of increase and 11 years of decrease, 22 years for a full cycle) that affects us in many ways. Increased solar activity includes increases in extreme ultraviolet and X-ray emissions from the Sun which produce dramatic effects in the Earth's upper atmosphere [45]. The associated atmospheric heating increases both the temperature and density of the atmosphere at many spacecraft altitudes. The increase in atmospheric drag on satellites in low Earth orbit can dramatically shorten the lifetime of these valuable assets [46]. Increases in the number of solar flares and CMEs raise the likelihood that sensitive instruments in space will be damaged by energetic particles accelerated in these events. These SPE particles can also threaten the health of both astronauts in space and airline travelers in high altitude, polar routes. Solar activity apparently affects terrestrial climate as well. Although the change in the total solar irradiance seems too small to produce significant climatic effects, there is good evidence that, to some extent, the Earth's climate heats and cools as solar activity rises and falls [47]. There is little doubt that the solar cycle is magnetic in nature and produced by dynamo processes within the Sun [48]. The solar cycle commonly known as the sunspot cycle is the periodic change in the Sun's activity and changes in the number of sunspots, flares, and other visible manifestations. Sunspots are regions on the solar surface that appear dark because they are cooler than the surrounding photosphere, typically by about 1500 K (thus, they are still at a temperature of about 4500 K, still cooler than the rest of the photosphere). They are only dark in a relative sense; a sunspot removed from the bright background of the Sun would glow quite brightly. The largest sunspots observed have had diameters of about 50,000 km, which makes them large enough to be seen with the naked eye. Sunspots often come in groups with as many as 100 in a group, though sunspot groups with more than about 10 are relatively rare. Sunspots were almost certainly seen by prehistoric humans viewing the Sun through hazy skies. The earliest actual recordings of sunspot observations were from China [49,50] over 2000 years ago. Yet, the existence of spots on the Sun came as a surprise to westerners when telescopes were first used to observe the Sun in the early seventeenth century. This is usually attributed to western philosophy in which the heavens and the Sun were thought to be perfect and unblemished [51,52].

10.9 Sunspot Observations

The largest sunspots and sunspot groups can be seen with unaided eye at sunrise and sunset or through smoke and haze. These sunspots were observed during the pre-telescopic age by ancient Oriental (particularly Chinese) [53] astronomers and now the data on ancient sunspot observations made by naked eye (SONE) is the longest nonproxy record of sunspot activity in the past. The most complete catalogue of SONE, covering more than 18 centuries and including more than 200 events, was collected by Wittmann and Xu [48]. Despite the first accurate description of sunspots [53] made by the English astronomers Thomas Harriot and Frisian astronomers Johannes and David Fabricius [53], systematic and qualified telescopic observations of sunspots begin only in the middle of the nineteenth century. But strong efforts by many scientists [54] particularly that of Rudolf Wolf [51], make it possible to extend the sunspot records back to the beginning of the eighteenth century [52].

The Wolf number datasets differ for some periods, however, is different: the reliability of the 1818–1848 data is reliable although, questionable from 1749 to 1817 and considered poor during 1700–1748 [55]. Recently, Hoyt and Schatten [56] have finished the reconstruction of group SN (GSSN) in which they used many historical sources missed by previous investigators. Their dataset covers the time interval 1610–1995 and authors consider GSSN data as quite reliable during 1653–1730, 1750–1788, and after 1795 [56]. Wolf numbers and GSSN show some differences in the eighteenth century; the values of Wolf numbers during this period are a bit higher. These two series are the longest direct instrumental records of South America. The reliability of the information on solar activity, extracted from historical chronicles, was analyzed in many works [49,54,57] and it was shown that the Oriental historic data really reflects such features of sunspot activity as 11-year and century-scale cycles, butterfly diagrams, and deep Maunder-type minima of sunspot activity. Qualification of ancient Chinese astronomers was high, observations of many lunar and solar eclipses, novae, supernovae, comets, meteor showers, and even a possible observation of Jupiter's satellite Ganymede (364 BC) confirm their professionalism. However, the SONE record has substantial shortcomings: (1) The ancient astronomers often mixed real sunspots with other celestial or meteorological phenomena. For example, Wittmann and Xu [48] compared more than 20 Oriental sunspot sightings with the data of European astronomers since 1848 and they found that only one-third of the naked-eye observations are confirmed by western telescopic records; (2) Ancient sunspot observations were not systematic and so are nonuniform in time. For example, very often sunspots were detected near the day of the new moon. This happened because the observation periods were more frequent and intensive determination of the new moon date had an important calendar purpose [49]. Dating of observations is also not always quite accurate; and (3) Some climatic effects may be present in the different SONE series. Nevertheless, SONE data is rather valuable for the investigation of solar variability over a long timescale. The Wittmann and Xu catalogue was vigorously used in the research work. Although, Christian Horrebow mentions that possible periodic variation were observed in 1776, the sunspot cycle was not truly discovered until 1844. In that year, Heinrich Schwabe reported [58] that his observations of the numbers of sunspot groups and spotless days over the previous 18 years indicated the presence of a cycle of activity with a period of about 10 years. Figure 10.8 shows his data for the number of sunspot groups observed yearly from 1826 to 1843.

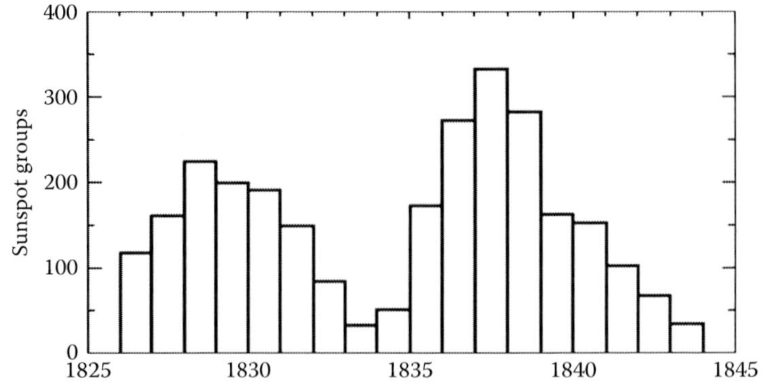

FIGURE 10.8
Sunspot groups observed from 1826 to 1843 by Heinrich Schwabe. (Data from Schwabe, H., *Astronomishe Nachrichten*, 21, 233–236, 1844.)

10.9.1 Wolf's SN

The modern era of sunspot counting began in the mid-1800s with the research of Bern Observatory director Rudolf Wolf, who introduced what he called the *universal SN* as an estimate of the Sun's activity [51]. This investigator, motivated by H. H. Schwabe's discovery of an apparent 10-year periodicity—*sunspot cycle*—in the frequency with which spots seem to appear over the years, sought to develop an index by which long-term trends could be monitored and periodicity verified and studied. It is said that Wolf would have preferred to measure the areas covered by the sunspots rather than their number but the methods and equipment of the day were not adequate for this task. As an alternative, he developed an index based on the number of spots and spot groups (clusters of related spots). His *relative* SN, R, thus emphasized sunspot groups with

$$R = k(10g + n) \tag{10.1}$$

where:

k is a correction factor for the observer
g is the number of identified sunspot groups
n is the number of individual sunspots

These Wolf, Zurich, or International Sunspot Numbers have been obtained daily since 1849. Wolf himself was the primary observer from 1848 to 1893 and had a personal correction factor $k = 1.0$ [51]. Wolf extended the data back to 1749 using the primary observers along with many secondary observers but much of that earlier data is incomplete. Wolf often filled in gaps in the sunspot observations using geomagnetic activity measurements as proxies for the SN. It is well recognized that the SNs are quite reliable since Wolf's time but that earlier numbers are far less reliable.

10.9.2 Sunspot Groups and Areas

The International Sunspot Number is the key indicator of solar activity. This is not because everyone agrees that it is the best indicator but rather because of the length of the available record [58,59]. Traditionally, SNs are given as daily numbers, monthly averages, yearly averages, and smoothed numbers. The standard smoothing is a 13-month running mean cantered on the month in question and using half weights for the months at the start and end. Solar cycle maxima and minima are usually given in terms of these smoothed numbers. Additional SNs do exist. The Boulder Sunspot Number is derived from the daily solar region summary produced by the US Air Force and National Oceanic and Atmospheric Administration (USAF/NOAA) from sunspot drawings obtained from the Solar Optical Observing Network (SOON) sites since 1977. These summaries identify each sunspot group and list the number of spots in each group. A fourth SN is the Group Sunspot Number, *RG*, devised by Hoyt and Schatten [55]. This index counts only the number of sunspot groups, averages together the observations from multiple observers [60] (rather than using the primary/secondary/tertiary observer system), and normalizes the numbers to the International Sunspot Numbers using

$$R_G = \frac{12.08}{N}\sum_{i=1}^{N} k_i G_i \tag{10.2}$$

where:

N is the number of observers
k_i is the ith observer's correction factor
G_i is the number of sunspot groups observed by observer i
12.08 normalizes the number to the International Sunspot Number

Hathaway et al. [60] found that the Group Sunspot Number follows the International Number fairly closely but not to the extent that it should supplant the International Number. In fact, the Group Sunspot Numbers are not readily available after 1995. The primary utility of the Group Sunspot Number is in extending the SN observations back to the earliest telescopic observations in 1610. These SNs are available from NOAA. The International Number can be obtained monthly, directly from SIDC. Sunspot areas (SAs) are thought to be more physical measures of solar activity. SAs and positions were diligently recorded by the Royal Observatory, Greenwich (RGO) from May of 1874 to the end of 1976 using measurements of photographic plates obtained from RGO itself and its sister observatories in Cape Town, South Africa; Kodaikanal, India; and Mauritius. Both umbral areas and whole spot areas were measured and corrected for foreshortening on the visible disc. SAs were given in units of millionths of a solar hemisphere (μHem). In 1977, NOAA began reporting much of the same SA and position information in its Solar Region Summary reports. These reports are derived from measurements taken from sunspot drawings done at the USAF SOON sites. The SAs were initially estimated by overlaying a grid and counting the number of cells that a sunspot covered. In late 1981 this procedure was changed to employ an overlay with a number of circles and ellipses with different areas. The SAs reported by USAF/NOAA are significantly smaller than those from RGO [61–64]. SAs are also available from a number of solar observatories. While individual observatories have data gaps, their data is very useful for helping to maintain consistency over the full interval from 1874 to the present. The combined RGO USAF/NOAA datasets are available online (RGO). These datasets have additional information that is not reflected in SNs—positional information—both latitude and longitude. The distribution of SA with latitude (Figure 10.9) shows that sunspots appear in two bands on either side of the Sun's equator. The average daily SA for each solar rotation since May 1874 is plotted as a function of time in the lower panel. The relative area in equal area latitude strips is illustrated with a color code in the upper panel. Sunspots form in two bands, one in each hemisphere, that start at about 25° from the equator at the start of a cycle and migrate toward the equator as the cycle progresses. At the start of each cycle spots appear at latitudes above about 20°–25°. As the cycle progresses the range of latitudes with sunspots broadens and the central latitude slowly drifts toward the equator, but with a zone of avoidance near the equator. This behavior is referred to as *Sporer's law of zones* by Maunder [65] and was famously illustrated by his *butterfly diagram* [66].

10.10 Solar Flux and the Magnetic Field

The 11.7 cm solar flux is the disk-integrated emission from the Sun at the radio wavelength of 11.7 cm (2800 MHz) [67,68]. This measure of solar activity has advantages over SNs and areas in that it is completely objective and can be made under virtually all weather

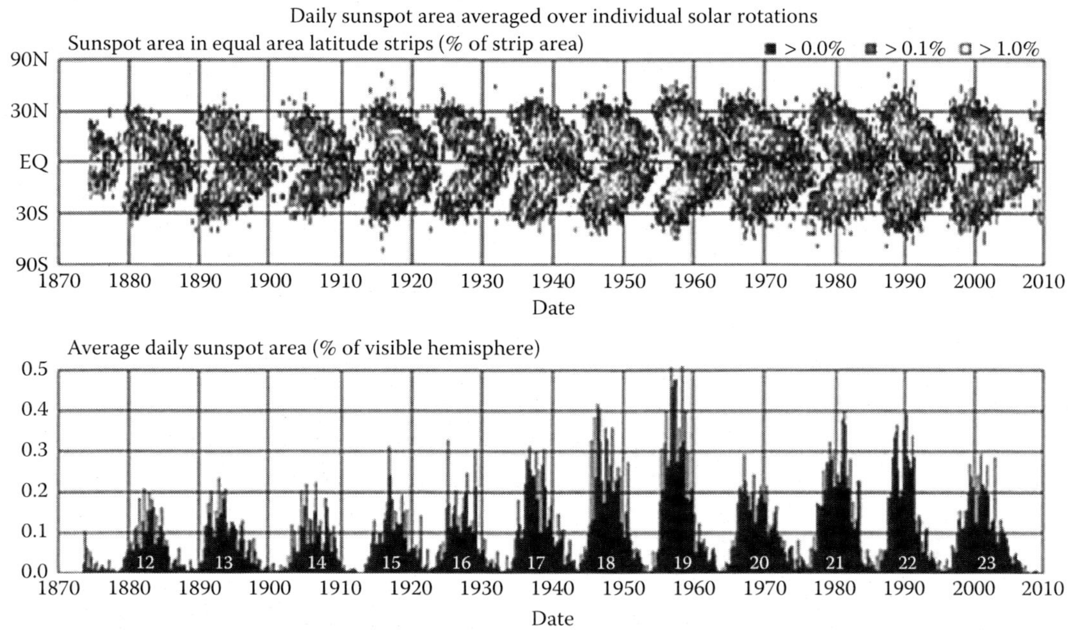

FIGURE 10.9
SAs as a function of latitude and time.

conditions. This index is a measure of the noise level generated by the sun at a wavelength of 11.7 cm at the earth's orbit. The global daily value of this index is measured at local noon at the Penticton Radio Observatory in Canada. Historically, this index has been used as an input to ionospheric models as a surrogate for the solar output in wavelengths that produce photo ionization in the earth's ionosphere (in the ultraviolet bands). Solar magnetic fields in sunspots were first measured by Hale [68,69]. The magnetic nature of the solar cycle became apparent once these observations extended over more than a single cycle [70]. A magnetogram from sunspot cycle 22 (August 2, 1989) is shown in Figure 10.10.

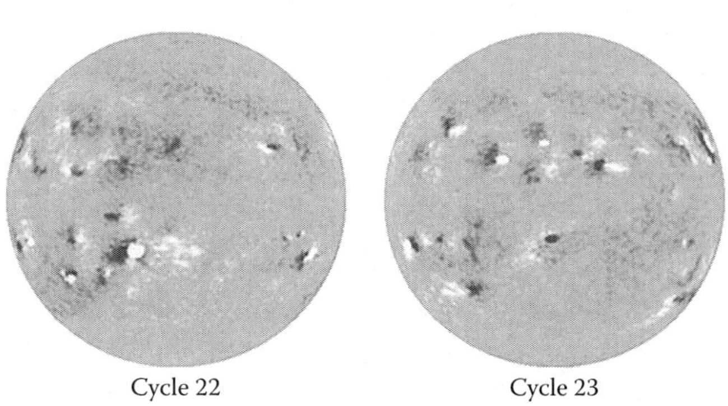

FIGURE 10.10
Magnetogram from sunspot cycles 22 and 23.

Leading spots in one hemisphere have opposite magnetic polarity to those in the other hemisphere and the polarities flip from one cycle to the next.

In addition to Hale's polarity law for sunspots, it was found that the Sun's polar fields reverse as well. Babcock [70] noted that the polar fields reversed at about the time of sunspot cycle maximum. The Sun's south polar field reversed in mid-1957 while its north polar field reversed in late 1958. The maximum for cycle 19 occurred in late 1957. The polar fields are thus out of phase with the sunspot cycle and polar fields are at their peak near sunspot minimum. This is also indicated by the presence of polar faculae small bright round patches seen in the polar regions in white light observations. It is noted that the numbers also peak at about the time of sunspot minimum [71,72].

10.11 Prediction of Mean SNs

Solar cycle predictions are needed to plan long-term space missions; just like weather predictions are needed to plan the launch. Fleets of satellites are responsible for collecting many types of science data and relaying solar information. Predictions of the SNs have been made since the cycle was discovered. A few prediction methods have been developed based on the precursor technique, which is found to be successful for forecasting the solar activity. Considering the geomagnetic activity *aa*-indices during the descending phase of the preceding solar cycle as the precursor, we predict the maximum amplitude of annual mean SN in cycle 24.

Note: Information about the *aa*-indices—The *aa*-index is a simple global geomagnetic activity index. It is derived from the *K* indices from two approximately antipodal observatories and has units of 1 nT. Considering the geomagnetic activity of the *aa*-index as the precursor, the maximum amplitude of the solar cycle 24, is predicted using the method employed by Jain [72]. The temporal behavior of observed annual mean SNs and annual mean *aa*-index for solar cycles 11–23 are considered in our investigation as shown in Figure 10.11. It is observed from the figure that the annual mean *aa*-index ranges from 5.7 (in 1901) to 36.6 (in 2003) which is an indicator of minimum and maximum geomagnetic activity, respectively, during the period of 1868–2008. Whereas, the annual mean SN varies between 1.4 (in 1913) and 190.2 (in 1957). SNs rise steadily to maximum and then fall steadily to a low level during each sunspot cycle, whereas geomagnetic indices (A_p or *aa*) show two or more maxima per cycle, one near or before the sunspot maximum and others in the declining phase, and the gap between the two primary maxima (the Gnevyshev gap) results in the quasi-biennial and quasi-triennial periodicities observed in the geomagnetic indices [73,74].

As the annual mean SN for the year 2008 is 2.86, which is within the range of sunspot minimum value, we have considered the sunspot minimum year for solar cycle 23 to be 2008 in the present study. The annual mean *aa*-index and annual mean SN are obtained by averaging the monthly mean of geomagnetic activity index *aa* and monthly mean of SN, respectively, for the period 1868–2008. For *n*th cycle, we determined $(aa^*_n)_{dsc}$, an average of the geomagnetic *aa*-index, of the year in which observed sunspot is minimum and four years preceding to it (i.e., total 5 years). Then we compared $(aa^*_n)_{dsc}$ of the *n*th cycle with the observed maximum annual mean SN $(R_{n+1})_{max}$ of $(n+1)$th cycle and obtained a relationship between $(aa^*_n)_{dsc}$ and $(R_{n+1})_{max}$ which is shown in Figure 10.12.

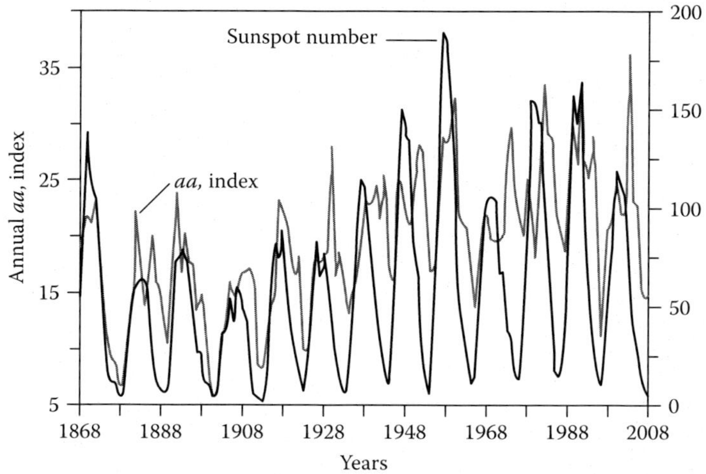

FIGURE 10.11
The observed annual mean *aa*-index and annual mean SN for the period of 1868–2008 indicating that the annual mean SN for 2008 is 2.86.

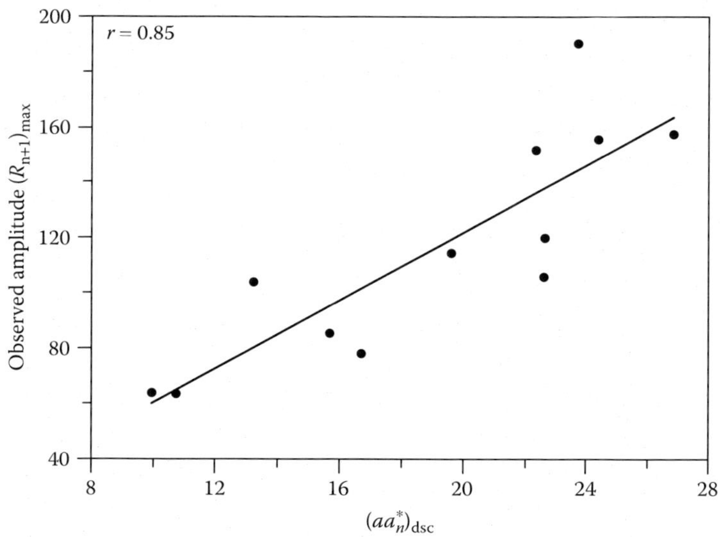

FIGURE 10.12
Observed amplitude $(R_{n+1})_{max}$ is plotted against $(aa^*_n)_{dsc}$.

Correlation coefficient is found to be $r = 0.85$. The best linear fit to the data with the correlation coefficient of 0.85 led us to derive an asymptotic relation as follows:

$$(R_{n+1})_{max} = 6.138(aa^*_n)_{dsc} - 1.1 \tag{10.3}$$

Using relation (10.3), we have obtained the maximum annual mean SN for cycles 12–23, which are almost in agreement with the observed values. The standard deviation $\sigma = \pm 21$ is found from the difference between the calculated and observed values. The relation (10.3) enabled us to predict the maximum annual mean SN for cycle 24 $(R_{23+1})_{max}$ to be

111 ± 21. This suggests that the maximum amplitude will be less than that of cycles 21–22. Our prediction of the maximum amplitude is in good agreement with the predictions made by a few earlier investigators [75–79] while in contrast to Hathaway and Wilson [79].

10.11.1 N–S Asymmetry of Solar Activity

The distribution of various solar activity phenomena with respect to heliographic latitudes as a function of time has been investigated in different studies. These activity features include flares, magnetic flux, SNs, SA, and so on. These studies indicate that a solar cycle is not symmetric considering the distribution of solar activity separately in the northern and southern hemisphere [80,81]. This intrinsic feature (N–S asymmetry) poses a challenge for dynamo model calculations. To study the spatial distribution of solar activity (e.g., solar flare) with respect to heliographic latitudes, we have calculated the number of flares in the interval of 10° latitude for northern and southern hemispheres. In this case those events have been excluded which occurred at 0° latitude. Since the number of flares above 50° latitude is very small in both the hemispheres, the number of flares occurring above 50° latitude is merged in one group. It is customary to describe N–S asymmetry by an asymmetry index where N and S are the yearly number of flares in the northern and southern hemisphere of the Sun, respectively.

$$A = \frac{N - S}{N + S} \tag{10.4}$$

The statistical significance of the flare dominance in northern and southern hemispheres has been assessed by using the binomial probability distribution. Let us consider a distribution of n objects in two classes. The binomial formula gives us the probability $P(k)$ of getting k objects in class 1 and $(n-k)$ objects in class 2, such that

$$P(k) = \frac{n!}{k!(n-k)!} p^k (1-p)^{n-k} \tag{10.5}$$

and the probability to get more than d objects in class 1 is given by

$$P(\geq d) = \sum_{k=d}^{n} P(k) \tag{10.6}$$

In general, when $P\ (\geq d) > 10\%$, it implies a statistically insignificant result (flare activity should be regarded as being equivalent for the two hemispheres); when $5\% < P(\geq d) < 10\%$ it is marginally significant; and when $P\ (\geq d) < 5\%$ we have a statistically significant result (flare occurrence is not due to random fluctuations) [80,81].

10.12 Solar Radio Bursts

It became apparent in the early 1960s that certain space weather events might interfere with the manned mission to the moon. In particular, the Sun emits continuous electromagnetic energy and electrically charged particles, which can cause disturbances in the near-Earth environment and disrupt satellite communications. Among these concerns there was the possibility of a geomagnetic storm of solar origin. Metric Type II radio bursts, signatures of coronal shock waves or CMEs, were known to be commonly associated with solar flares. The USAF Research Laboratory (AFRL) was thus assigned the task of developing and validating a network of ground-based solar observatories. AFRL established a worldwide network of sweep frequency recorders from which one can estimate the shock speed in the corona. This network, called the Radio Solar Telescope Network (RSTN), uses a bandwidth from 25 to 85 MHz. The prototype was assembled and operated at the Sagamore Hill Solar Radio Observatory. This observatory began operating solar patrols in 1966. The USAF Geophysics Laboratory (AFGL, currently Phillips Lab) transferred operation of the observatory to Detachment 2 of the 2nd Weather Group of the USAF Weather Agency in October 1978. However, Phillips Lab continues to work in an advisory capacity to the observatory. Solar radio bursts were among the first phenomena identified as targets for radio astronomy. Solar radio bursts originate in the same layers of the solar atmosphere where energy is released in solar flares, energetic particles are accelerated and CMEs are launched [53,68]. Characteristic studies of solar radio bursts have great importance for determining the solar flare and CME phenomenon. Solar radio bursts at frequencies below a few hundred MHz were classified into five types:

Type I: Short, narrow band events that usually occur in great numbers together with a broader band continuum. May last for hours or days.

Type II: Slow drift from high to low frequencies. Often show fundamental and second harmonic frequency structure.

Type III: Rapidly drift from high to low frequencies. May exhibit harmonics. Often accompany the flash phase of large flares.

Type IV: Flare-related broadband continual.

Type V: Broadband continua which may appear with III bursts. Lasts 1–2 min, with duration increasing as frequency decreases. Figures 10.13 and 10.14 show different types of solar radio bursts. The figure clearly shows the flare in minutes and the corresponding frequencies in MHz. Solar flares are also detected throughout the radio spectrum.

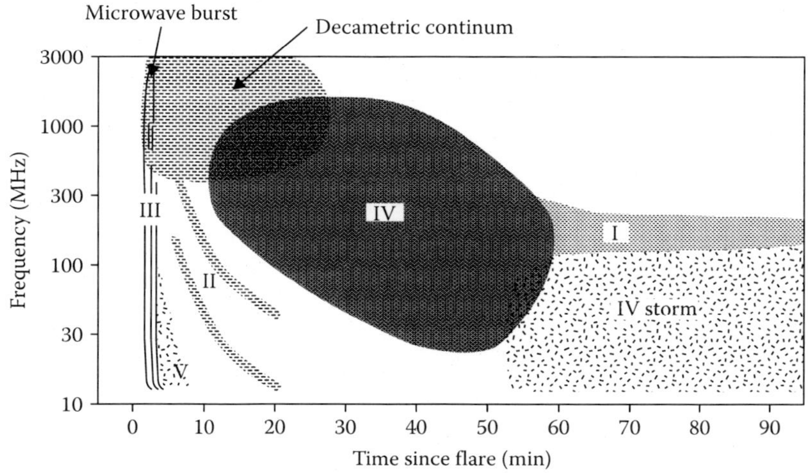

FIGURE 10.13
Schematic diagram of different bursts.

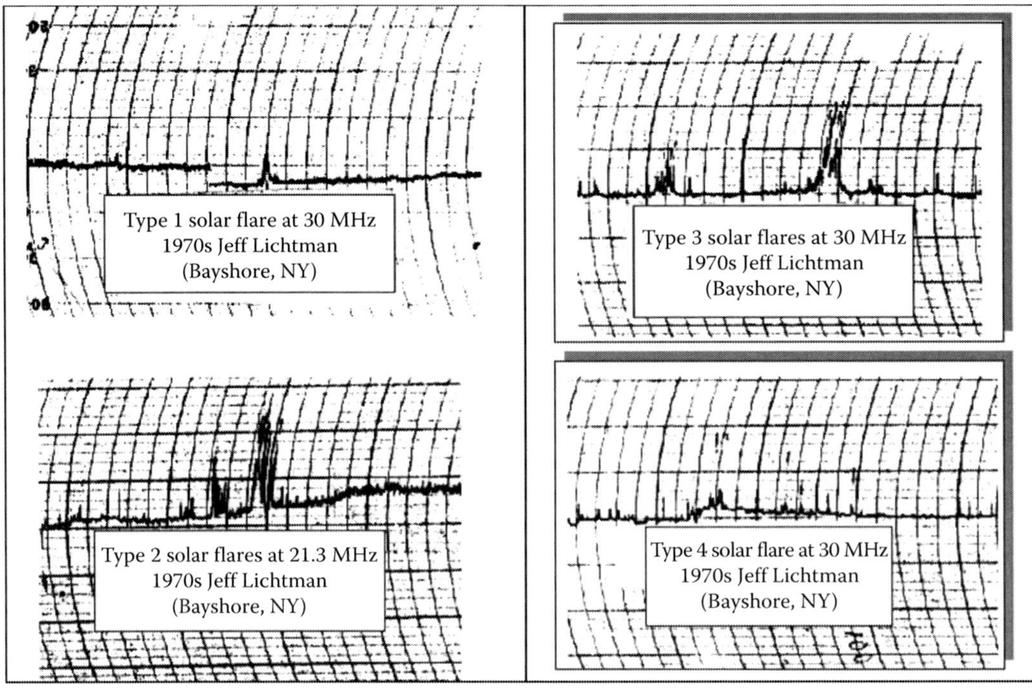

FIGURE 10.14
Actual solar flare chart recordings.

10.13 Sunspot Cycle and Hale Cycle

In 1932 Jansky first observed radio emissions from extraterrestrial origin [15]. But the first successful detection and interpretation of solar radio emission was in 1942, during World War II by British Radar units.

Note: Microwave emissions from the Sun were first reported by Southworth (1945). Radio bursts at the time of solar activity were discovered to be the source of interference in World War II radar systems [82]. From then various solar activity events were being reported of which the solar flare is the most gigantic solar activity releasing large amounts of electromagnetic energy. Basically, this transitory energy phenomenon releases energy from every window of electromagnetic spectrum. It develops in the magnetically active region and releasing the energy at the top of the magnetic arcade. They are not uniform but vary considerably from one another. The magnetic field forms short circuits through which intense and irregular currents flow [15]. The variation of the magnetic field is due to diffusion and advection. In the current sheet, due to diffusion of magnetic field, its energy is converted into heat energy.

Note: A current sheet is an electric current confined to a surface, rather than being spread through a volume of space. Current sheets feature in MHD, the study of the behavior of electrically conductive fluids. If there is an electric current through part of the volume of such a fluid, magnetic forces tend to expel it from the fluid, compressing the current into thin layers which pass through the volume. Due to decay and the emergence of solar ARs, the Sun's field changes as the shape of the current sheet evolving slowly over the months [18,19]. During most of the solar cycles the current sheet forms a tilted dipole, with varying quadruple distortion near the solar maximum; it then decays by forming a complicated structure. Solar activity has been measured by satellites during recent decades. Scientists are interested in understanding the effects of variations in the total and spectral solar irradiance on the Earth and its climate. In climate research, particularly concerning global warming, it is a very big question of how strongly the solar magnetic activities will affect the temperature of the Earth. Total solar irradiance and cosmic ray fluxes are phenomena that are responsible for causing variation in solar magnetic activity and which in turn can affect the global climate [20–22]. Owing to the perturbed condition and the continuously varying solar core and the solar neutrino flux, changes happen in the amount of total solar radiation and its spectral distribution over multimillennial timescales [18,23]. There are periodic and aperiodic components to these variations. Solar radiation reaches the earth surface at about 8.3 minutes traveling at the speed of light. The amount of average solar radiation received at the outer surface of Earth's atmosphere is about 1366 W/m^2. Total solar irradiance variation can well explain the internal solar structure and the solar–terrestrial relationship as well. The radiation environment of earth is also biased by it. From NASA observations, it has been observed that total solar irradiance varies about 0.1% over a complete solar cycle [18]. Out of the various solar activity events [20], the principal one is the 11-year solar or sunspot cycle. Without taking into consideration the magnetic variability and the temperature of the solar core, the 11-year solar activity cycle can't be explained. Activity of Sun gradually increases when the number of sunspots increases and then it decreases approximately at a 11-year cycle. Basically, sunspots are the dark spots on the solar disk with greater magnetic field. Already mentioned? The last solar maximum occurred in the year 2001 and currently we are in the rising phase of solar cycle 24. The sunspot cycle when analyzed from 1880 to 2010 exhibits an interesting variation. The result is shown in Figure 10.15. At the solar maximum the magnetically active sunspots produce energetic solar flares causing geomagnetic disturbances to a large extent [24].

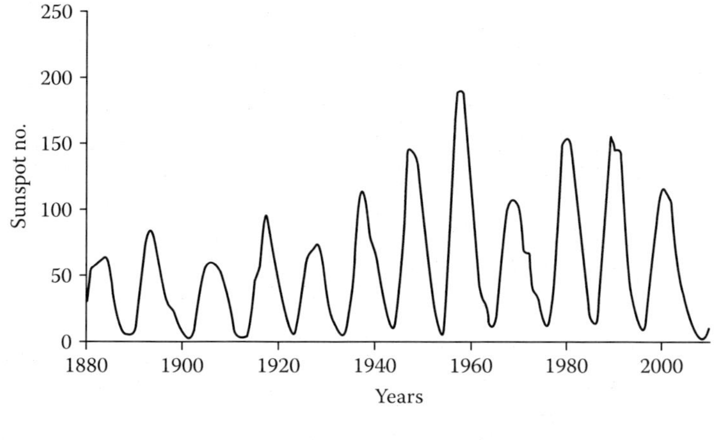

FIGURE 10.15
Variation of sunspot for the last 130 years.

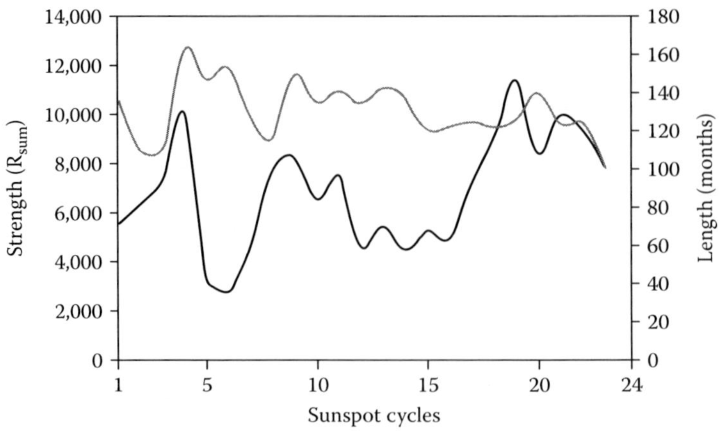

FIGURE 10.16
Variation of strength (dark) and length (light) of the sunspot cycles.

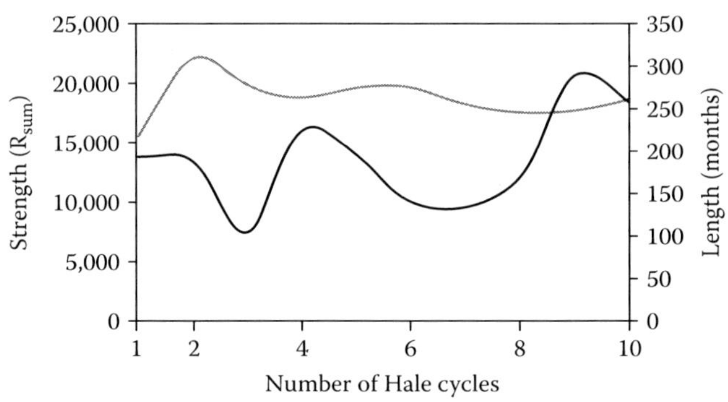

FIGURE 10.17
Variation of strength (dark) and length (light) of the Hale cycle (i.e., double solar cycle).

Figure 10.16 shows the variations of strength and length of the sunspot cycles while Figure 10.17 reveals the double sunspot cycles, technically named as Hale cycles [25]. It appears from the figure that the strength of the sunspot cycles rapidly falls from the fifth cycle and continued to an average value up to the 16th cycle. Thereafter, it increases showing a peak around 18th cycle. On the other hand, the length of the sunspot cycle exhibits a reduced value during cycle number 2 and 8, but for the remaining cycle the length remains more or less constant.

When the Hale cycle [83] (which essentially covers double solar cycles) is taken into account the variation in the respective parameter are rather smoothed out, particularly when the length is taken into consideration. However, the strength of the Hale cycle exhibited a minimum corresponding to cycle number 3. The two peaks are observed for cycle numbers 4 and 9. It is interesting to note that out of the two peaks mentioned the peak related to cycle number 9 is more prominent. Figure 10.18 depicts the normalized SN variation with the sunspot cycles. We have drawn histograms in Figure 10.18 showing the occurrences of solar cycles with the maximum number of sunspots in each cycle. From Figure 10.19, it is clear that the maximum SN for a sunspot cycle gives a peak around 100–150 counts and the frequency for that count is nearly 50% when we consider all the cycles.

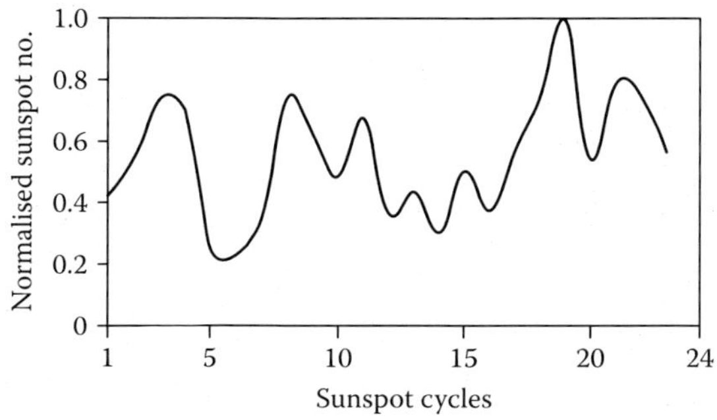

FIGURE 10.18
Variation of the normalized sunspot number with the sunspot cycles.

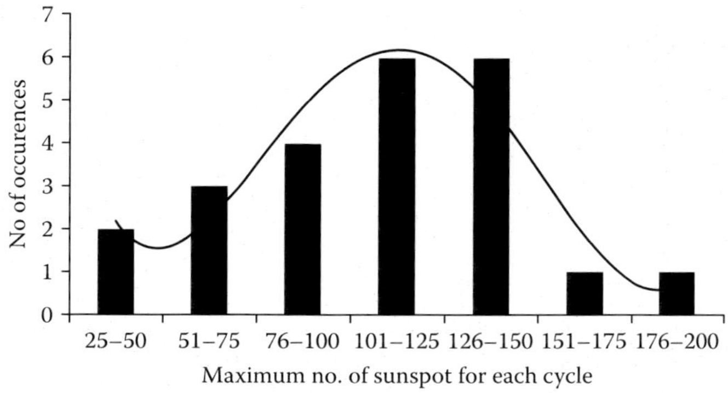

FIGURE 10.19
Histogram showing the occurrence of solar cycles with the maximum number of sunspots in each cycle.

10.13.1 Periodicities of the Solar Cycle

Sunspots occur when a concentrated portion of the solar magnetic field pokes through its surface. CMEs and solar flares are large explosions on the Sun's surface that occur close to the sunspot zones. Some statistical consideration has shown typical variations of a solar cycle and also focused on its periodicities. In reality, the Sun is one of the main climate drivers. The climate for the present century related to solar activity forecasts equivalent to a new Ice Age, with the minimum temperature achieved by the middle of this century. This points to a possible global cooling occurring in the future. We next note the deep minimum of solar cycle 23, the deepest during the past 140 years. This cycle started in May 1996 and had its peak in April 2000. The decline phase of this period extended from 2002 until December 2008, which is the longest decline phase in the past 23 solar cycles. On the other hand, solar cycle 24 started in December 2008; it started about three and half years later than the average of the strong cycles in the last twentieth century and almost three years later than the weak cycles of the late nineteenth century [26,27]. Figure 10.20 shows the variation of total SNs of the years 1950–2010 and Figure 10.21 shows the longest decline phase of solar cycle 23.

In Figure 10.22, we have compared solar cycles 1–4 with solar cycles 20–23. Solar cycles 5 and 6 of the Dalton minimum [84] are also presented in the same figure. It appears that solar cycles 22 and 23 are very similar to solar cycles 3 and 4, which preceded the Dalton minimum.

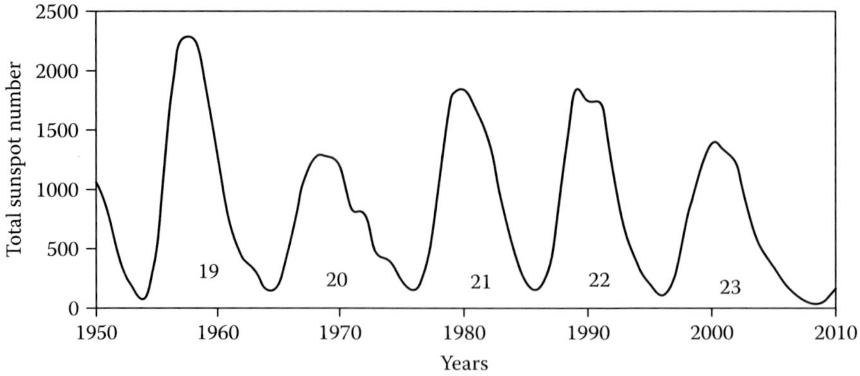

FIGURE 10.20
Sunspot cycles: Cycles 19–23.

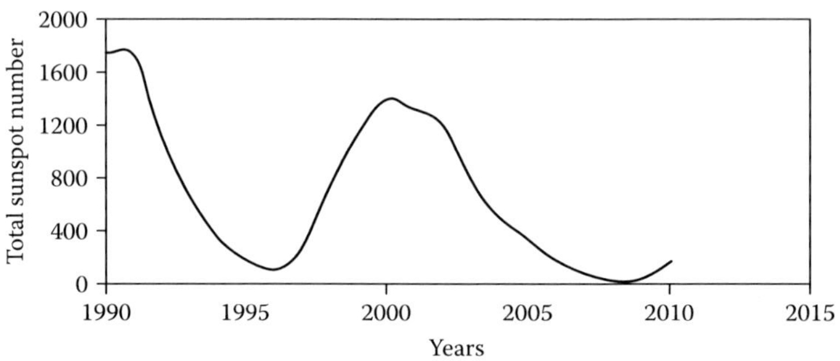

FIGURE 10.21
Solar cycle 23 revealing the longest decline phase.

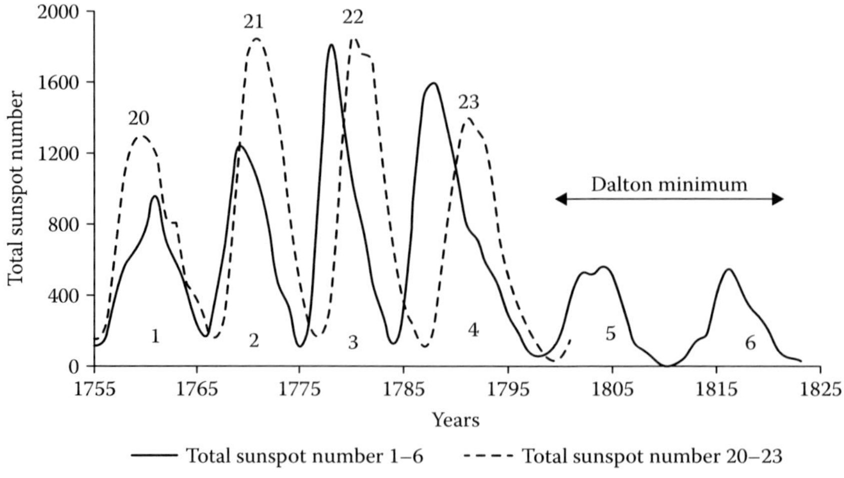

FIGURE 10.22
Comparison of solar cycles 1–4 with solar cycles 20–23.

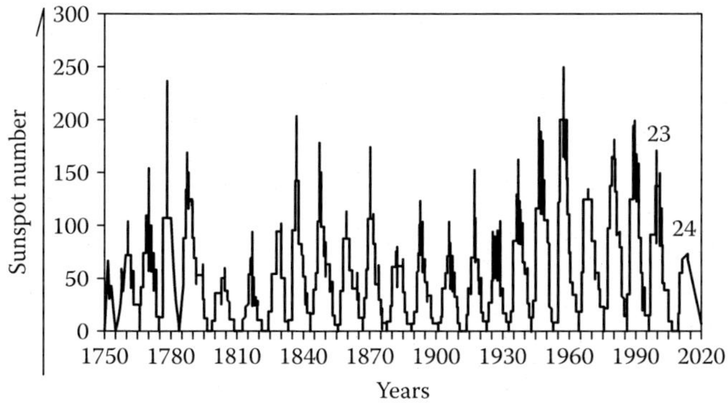

FIGURE 10.23
Monthly sunspot number observed and predicted.

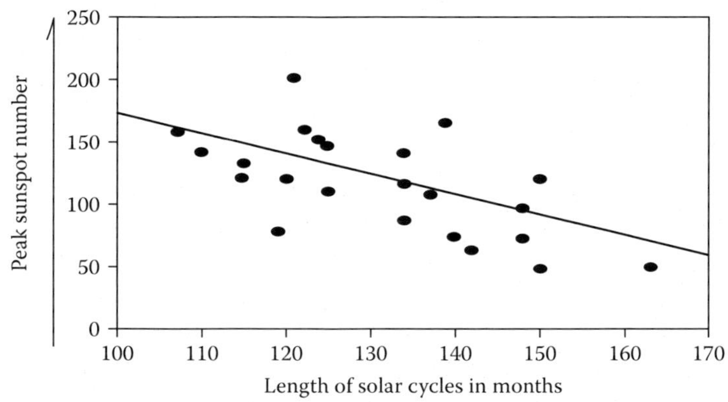

FIGURE 10.24
Plot showing correlation between solar-cycle length and peak sunspot number.

Figure 10.23 shows the variation of SNs over the years 1990–2020, which predicts some important information regarding the solar cycle 24. From the figure, it is clear that the peak amplitude of monthly SNs for cycle 24 will be less than 80. Also, cycle 24 may resemble solar cycles 5 or 6, as evident from Figure 10.21.

10.13.2 Solar Cycle Duration

In Figure 10.24, we have plotted the length of previous solar cycles with the peak SNs. It appears from the figure that the length of previous solar cycles increases when the peak SN decreases. This indicates a characteristic variation of solar activity with the change in length of the solar cycle. Atmospheric temperature solely depends on solar activity and so, if we can estimate the duration of upcoming cycles, then solar-based temperature variations can be predicted. Solar cycle 23 was the longest solar cycle since the reported cycle of 1798–1810. The length of the cycle was approximately 150 months and it was longer than the twentieth century average length of cycles, which is about 10.5 years. In Figure 10.25, we have considered SNs since 1750. The dotted portion in the figure exhibits the predicted

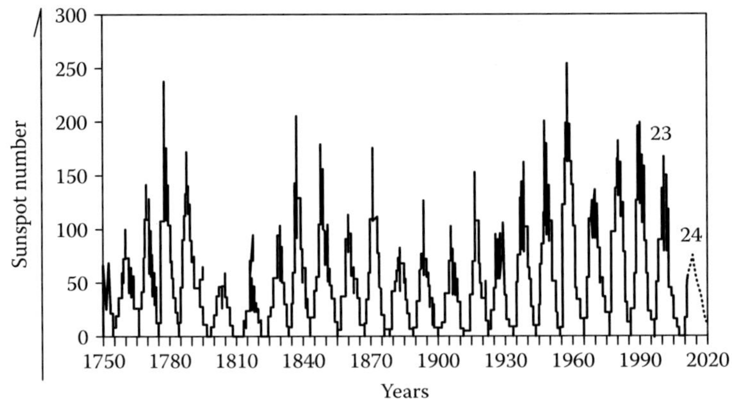

FIGURE 10.25
Plot showing sunspot number versus year. The dotted portion indicates the predicted sunspot number.

SN. From this analysis, one can predict that the present solar cycle 24 could be the weakest one. It is evident from the figure that the monthly mean SN is largely influenced by the noise which is superimposed on an apparent quasi-regular 11-year periodicity. In order to apply the SN traditionally, a running mean is taken to avoid monthly variation. It appears from the figure that the magnitude of the maximum SN changes in an oscillatory fashion.

The histogram drawn in Figure 10.26 demonstrates the number of different length occurrences of the solar cycle. It is seen from the figure that the most common cycle length is around 10.5 years. However, there are a few cycles in the range 11.5–12.5 years. Interestingly, only one cycle falls close to the average cycle length of 11.1 years.

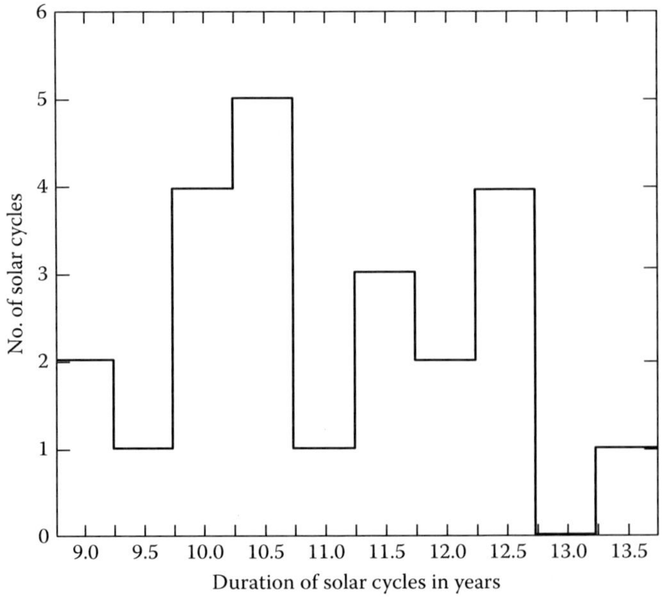

FIGURE 10.26
Number of solar cycles of different duration.

10.13.3 Solar Irradiance

Total solar irradiance becomes slightly lower during the periods of reduced sunspot activity and higher during periods of increased sunspot activity [85–87]. The total effect of long solar cycles and low SNs offers cumulatively more months on a decadal timescale with low average total solar irradiance, as indicated in Figure 10.27. The outcome is that weak and long solar cycles lead to cooling while intense and short cycles give rise to warming on a global basis. This can be re-confirmed in Figure 10.28. The behavior between solar activity and global basis temperature can be explained from the concept that sunspots are strong concentrations of magnetic flux at the surface of the sun. If the number of sunspots decreases, there is less solar wind allowing more galactic cosmic rays to enter into the earth's atmosphere. This causes more cloud formation at low level and sunlight reflected by the clouds back into the space, resulting in less heating to the earth's atmosphere.

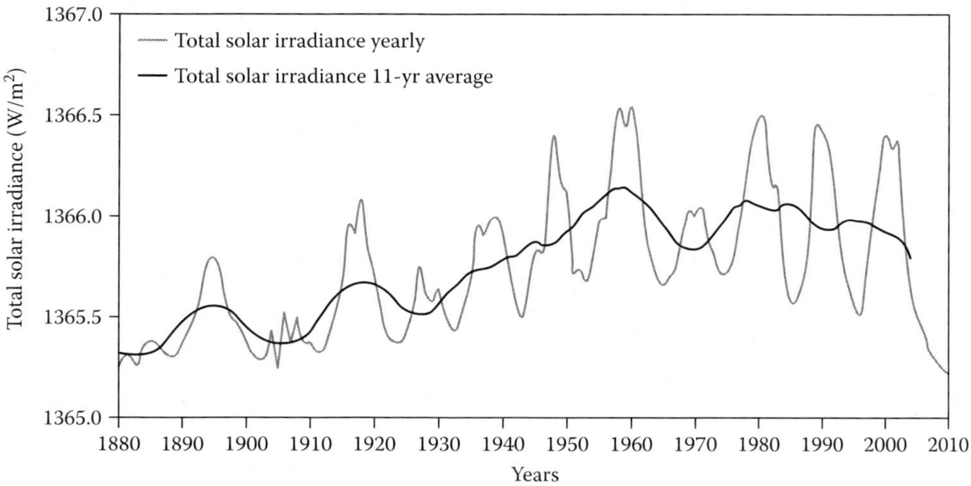

FIGURE 10.27
Total solar irradiance (TSI) from 1880 to 1978 from Solanki and TSI from 1979 to 2009 from software PMOD (Data from Physikalisch-Meteorologisches Observatorium Davos).

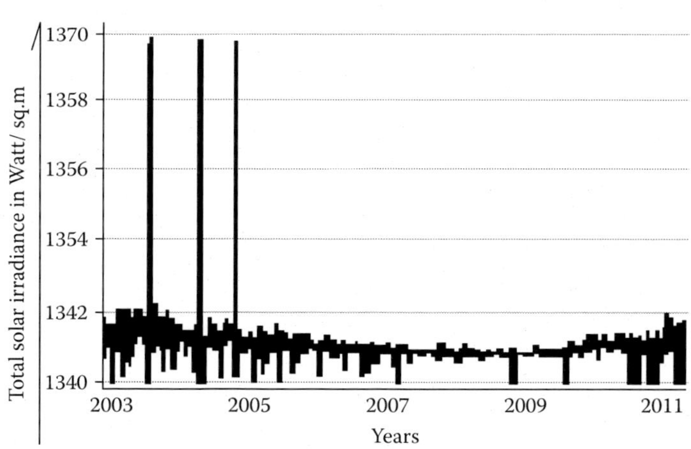

FIGURE 10.28
Plot of total solar irradiance versus year for the recent years.

Based on the evolutionary duration of the Schwabe cycle and the satellite measurements of the total solar irradiance during the past 30 years, astronomers predicted that the amplitudes of the next three Schwabe cycles are of about 70, 50, and 35 sunspots at the maximum as compared to 110 of the last complete cycle [88–90]. Present solar maximum and the Maunder and Dalton minima, detailing the causes of Little Ice Ages occurring once in about 200 years provides a strong probability of a global cooling effect [24,91]. Different spectral types of radio emissions are usually not used to define the solar cycles in relation to other solar parameters on the long-term basis. Prediction of the solar activity is a complicated problem, resembling in some ways to terrestrial weather forecasting [92]. The current solar cycle 24 started in January 2008 and showed peak in May 2013. There is a certain probability of the cycle to be a Dalton minimum as there is a certain resemblance to cycles 1, 2, 3, and 4 with cycles 20, 21, 22, and 23. As cycles 5 and 6 have low activity so we can say that cycles 24 and 25 will be of very low sunspot activity [93].

10.14 Global Solar Field Indices

Time variation of cosmic ray fluxes is very important to the spatial environment. The long-term variation of galactic cosmic rays has been compared with the heliospheric parameters and also with the index of solar activity from time to time. The Sun's magnetic field parameters calculated from solar wind source surface are used to determine the properties of the heliomagnetosphere. The tilt of the heliomagnetic current sheet α, and its variation control the long-term variation of cosmic ray fluxes, while the intensity of the average field B_{ss} dominates it during the maximum solar cycles. α, B_{ss}, and the polar magnetic field h_{pol} can't be explained except using global magnetic solar field indices on the solar wind source surface. Figure 10.29 shows the variation of the components of the solar magnetic fields [94,95].

Solar magnetic field characteristics are very useful to describe the long-term changes of cosmic ray density. For an adequate description of short-term variation of cosmic rays seen over the years, the modulation model is very useful for additional geophysical features. Figure 10.30 shows the global surface temperature variation for the period of 1860–2010.

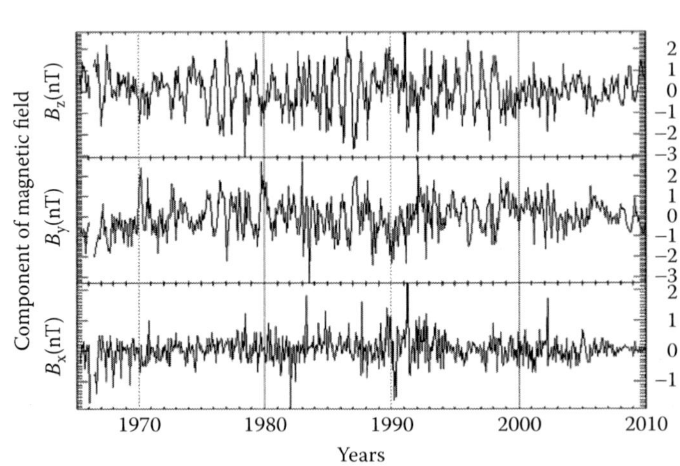

FIGURE 10.29
Variation of the components of the magnetic field.

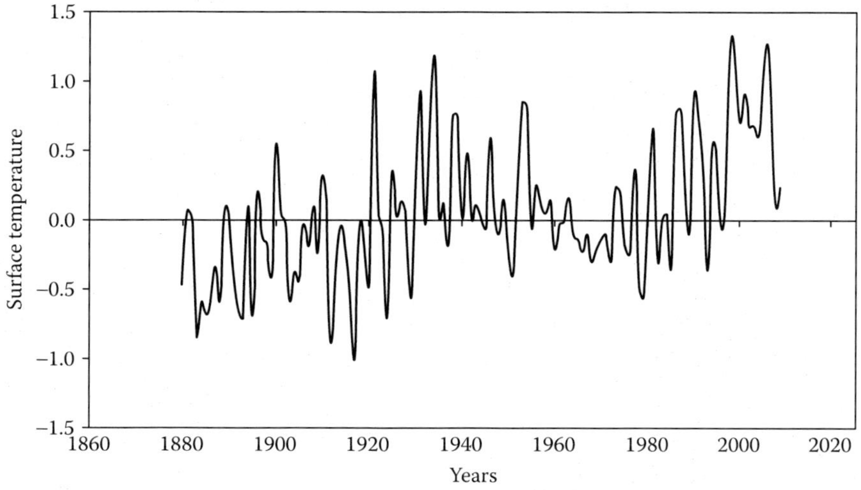

FIGURE 10.30
Global surface temperature variations.

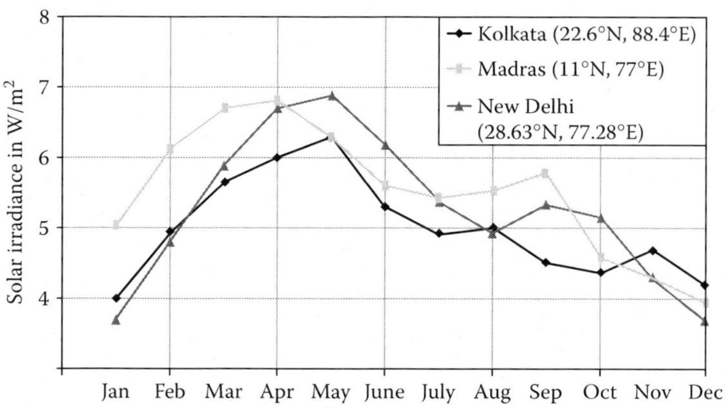

FIGURE 10.31
Graphical representation of monthly solar irradiance variation over three metropolitan cities in India.

Figure 10.30 clearly illustrates that the surface temperature keeps on rising from 1975 to till date, where also the SNs are greater.

Figure 10.31 represents the monthly variation of solar irradiance over three metropolitan cities in India. It reveals that the solar irradiance is maximum around April–May and minimum around November–December.

10.15 Variation of North–South Asymmetry of the Solar Flare Index and Associated Solar Activity

The images of the Sun showed that solar flares are one of the most powerful and explosive forms of solar activity [96]. Many investigations in the solar–terrestrial field classified solar flares as one of the most important solar events affecting the earth. Kleczek [97] introduced the quantity flare index (FI) as the product of i and t to quantify the daily flare activity over 24 hours per day, where i represents the intensity scale of importance and t the duration of the flare in minutes. This relationship provides an approximate amount of the total energy emitted by the flare and hence named it *flare index*. Solar activity is usually represented by either SNs or SAs. These two direct solar activity indices SN and SA are historically used to describe long-term solar variability [49,97]. When the smoothed monthly SN and SA in the northern and southern hemispheres over long time intervals were used, it was seen that sunspot activity never exhibits peak values at the same time in northern and southern hemispheres, and the hemispheric SN (or SA) maximum of a cycle usually differs from the maximum of the sunspot cycle [98,99]. A close scrutiny of the Schwabe cycle exhibits that the period for sunspot activity in the southern hemisphere was longer than in the northern hemisphere in the years from 1965 up to 1980. Since then and up to 1990, the activity was longer in the northern hemisphere. This section deals with two major aspects: (i) identify the changing characteristics for reported data from 1992 to February 2012 for the asymmetric variation of SNs and SAs in the northern and southern hemispheres and (ii) the behavior of the solar FI for a period from 1966 to 2010. In the analysis, priority is given to the unequal distribution of various solar activities in the two hemispheres of the Sun. Variations in the monthly total SAs over the whole solar disk and separately over the northern and southern hemispheres have been analyzed. Statistical properties of sunspots and their magnetic fields are extracted from the SOHO full disk, white light images, and magneto grams for the searchable Solar Feature Catalogue (SFC). To study asymmetry of FI (AFI), it is customary to use the following formula

$$\mathrm{AFI} = \frac{\mathrm{N} - \mathrm{S}}{\mathrm{N} + \mathrm{S}} \tag{10.7}$$

where N and S stand for the monthly north and south FI values, respectively. The asymmetry of the SN (ASN) has been considered according to their hemispherical position. The smoothed monthly mean SN and SA in the northern and southern hemispheres have been taken into consideration to investigate the characteristics of long-term hemispheric sunspot activity. SNs are computed according to the Wolf SN $R = k(10g + s)$, where g is the number of sunspot group (regions) and s is the total number of individual spots in all the groups; k is a variable scale factor which is usually less than one [100–102]. The total SAs in the Northern and Southern hemispheres measured from a single solar image and their cumulative areas during the whole cycle are shown to have a strong north–south asymmetry with the northern hemisphere prevailing. During and after the maximum while the southern hemisphere takes over moving to the solar minimum of 2008. It is clear from our study that a significant part of these variations can be accounted for by the fact that the two hemispheres are not exactly in phase. When the activity of the northern hemisphere leads that in the southern hemisphere, the north dominates early in the cycle while the south dominates in the declining phase [103]. In Figure 10.32, we have plotted the data of northern hemispherical FI and that occurring in the southern hemisphere during the period 1966–2010. It is observed from the plot that these two types of FI are not in phase. Also, we find that at the solar minimum in the southern hemisphere dominates for FI. The plot further shows that the northern hemisphere dominates at the beginning of cycle 21, but the southern hemisphere is more prominent at the end of this solar cycle. It is also observed that high values of FI sustain for a long time for both the hemispheres and there is no clear deep valley between two maxima for both hemispheres for cycle 21. It is very interesting to note that these two behaviors are also observed in cycle 23. For cycle 22, the figure shows

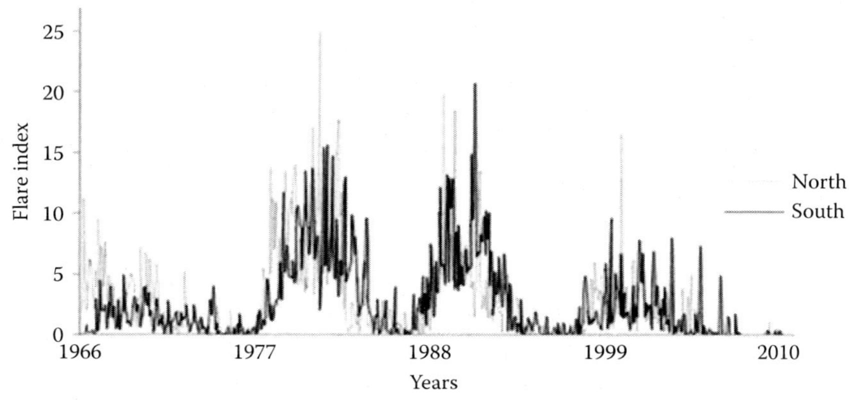

FIGURE 10.32
Monthly variation of North and South hemispherical FI during 1966–2010.

that there is a sharp fall of FI for the northern hemisphere at the end of the cycle. The plot shows that there is a deep valley between two maxima for cycle 22.

The yearly average values of solar flare indices both for the northern and southern hemispheres are used to investigate the characteristics of long-term hemispheric solar activity. Figure 10.33 reveals that flare indices exhibit peak at different times in the two hemispheres—earlier in the northern hemisphere and later in the southern hemisphere. It is interesting to note that the FI values show the two subsequent peaks in all solar activity cycles. The most prominent two peaks are in 21st solar cycle in northern hemisphere but in 22nd solar cycle in southern hemisphere. For cycles 20 and 23 the FI were of medium values. It is clear that the northern hemisphere leads in phase by 2 years when two adjacent peaks are observed.

It is evident from Figures 10.34 and 10.35 that cycle 23 is relatively weak compared to cycles 21 and 22. Figure 10.34 clearly shows the variation of total FI, while in Figure 10.35

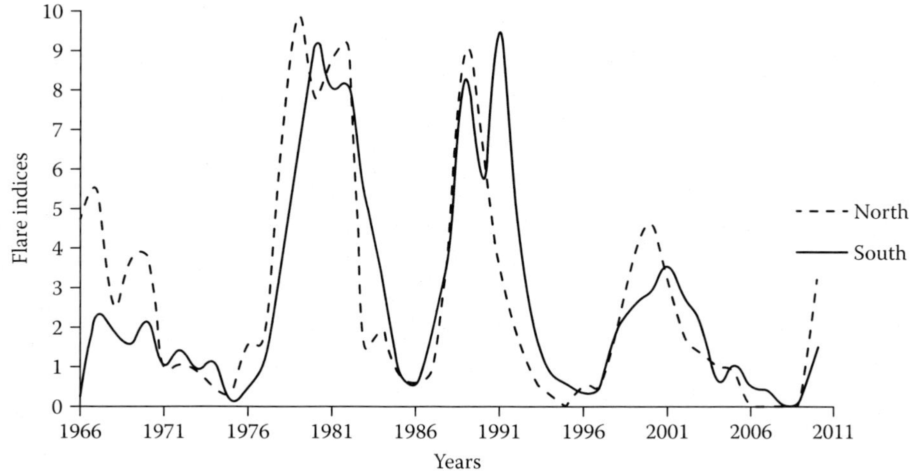

FIGURE 10.33
Yearly variation of flare indices with time (years 1966–2010).

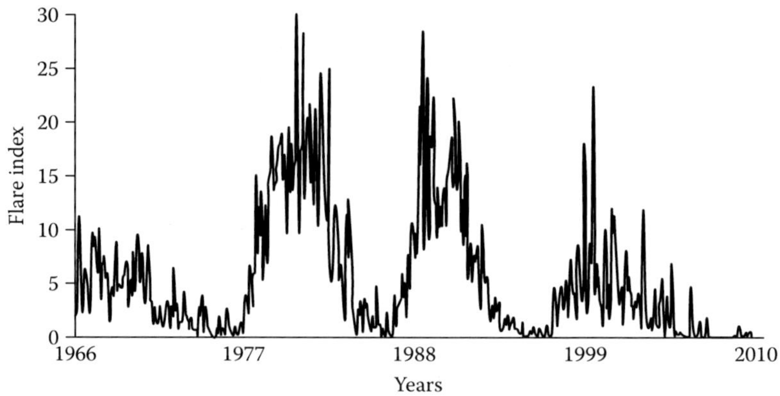

FIGURE 10.34
Variation of total FI for the period from January 1966 to December 2010.

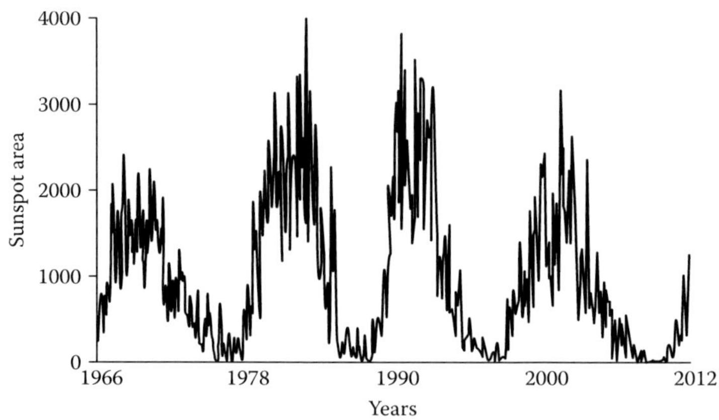

FIGURE 10.35
Variation of sunspot area from January 1966 to February 2012.

we have plotted SA during the period January 1966–February 2012. During the rising phase of cycle 23, the solar activity indices (SN and SA) are equal for the northern and southern hemispheres.

In Figure 10.36, we have plotted the average number of sunspots against the solar FI during 1992–2010, where monthly averages over each year are taken into account. It is observed that these two dominant features of solar activity vary linearly with respect to each other. A calculation shows that the correlation coefficient between them is 0.93.

Based on the derived hemispheric SNs, SAs, and solar flare indices we have studied the significance of N–S asymmetries. The asymmetry in solar features is found to be significantly high in statistical consideration. Sunspot activity never peaks at the same time in two hemispheres. Figure 10.37 shows the variation of SN from 1992 to February 2012. Figure 10.38 show the monthly variation of SA for both northern and southern hemisphere for the years 1966–February 2012. It is clearly evident from figure that SN is asymmetrically distributed in the northern and southern hemispheres.

The 11-year solar cycle appears in hemispheric sunspot activity whose periodic length appears to slightly vary during the period of consideration. In addition, it seems to be

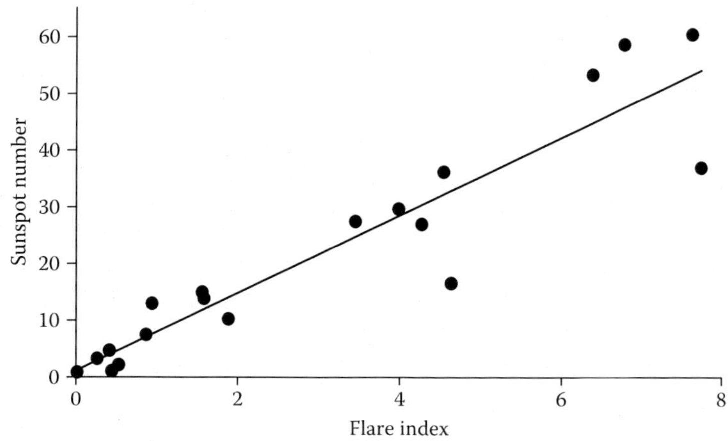

FIGURE 10.36
Sunspot number versus solar FI during 1992–2010.

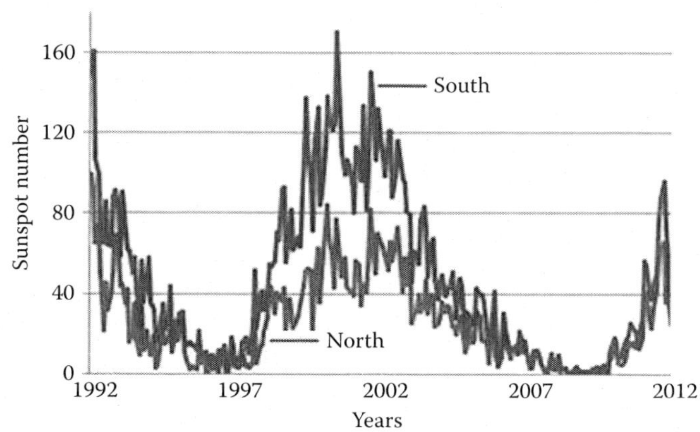

FIGURE 10.37
Variation of sunspot number January 1992–February 2012.

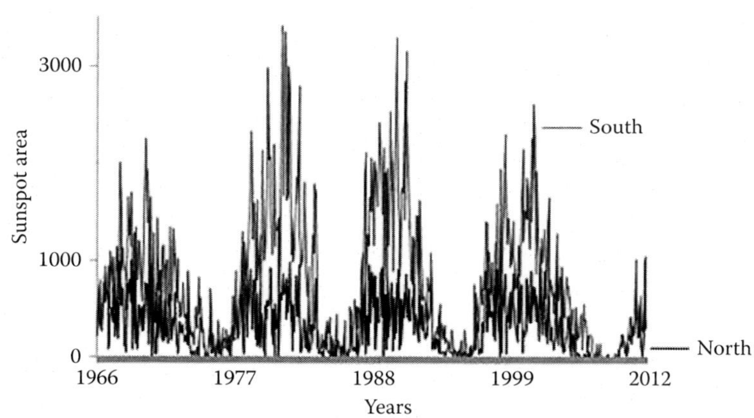

FIGURE 10.38
Monthly variation of sunspot area during January 1966–February 2012.

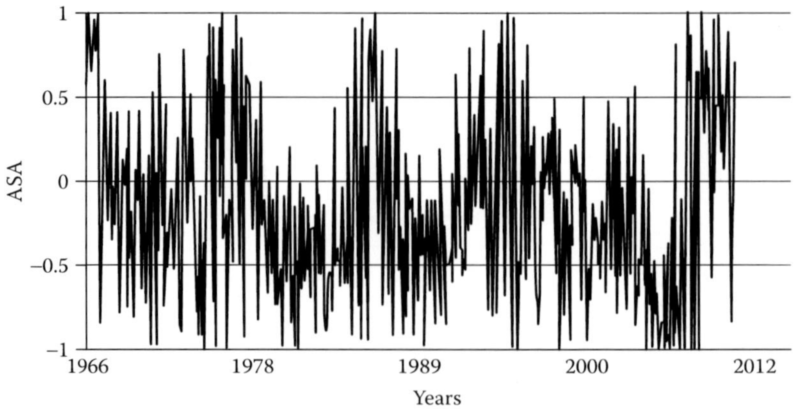

FIGURE 10.39
Asymmetry in sunspot area for the period 1966–February 2012.

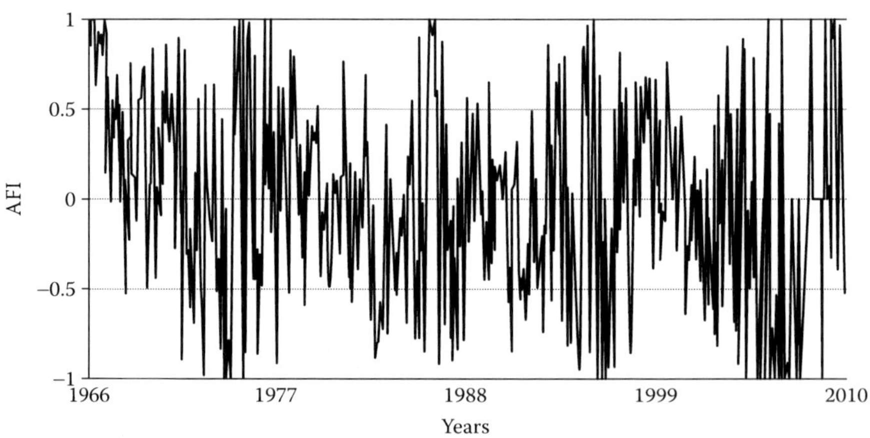

FIGURE 10.40
AFI during the year January 1992–December 2010.

longer in the southern hemisphere than in the northern hemisphere on the average. The sunspot activity is thus asymmetrically distributed in the hemispheres. The asymmetry in SA from 1966 to 2012 is plotted in Figure 10.39. It is observed that southern hemisphere dominates over northern hemisphere during cycles 20, 21, 22, and 23. But solar activities tend to shift from northern hemisphere to southern hemisphere (Figure 10.40). This is also seen in Figure 10.41, whereas SNs have been considered instead of ASA.

We have plotted AFI and ASN in the same diagram (Figure 10.42) when we observe that ASN and AFI are not in phase though total FI varies linearly with total sunspot numbers as evident from the figure. We find from the figure that AFI asymmetry peaks during solar activity minimum. During cycle 23 northern hemisphere dominates according to sunspot number. It is interesting to note that the flare activity dominates in southern hemisphere during the ending of cycle 22 and dominance increases during cycle 23.

We have plotted in Figure 10.43 the absolute north–south asymmetry of SA without any normalization: the north–south hemispheric difference of SA, which is defined as

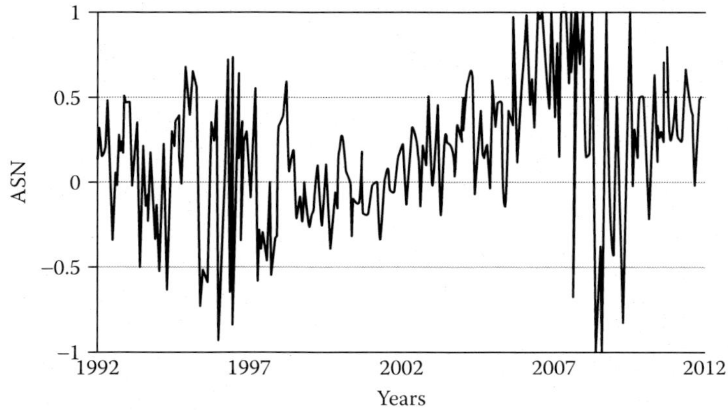

FIGURE 10.41
ASN during the year January 1992–February 2012.

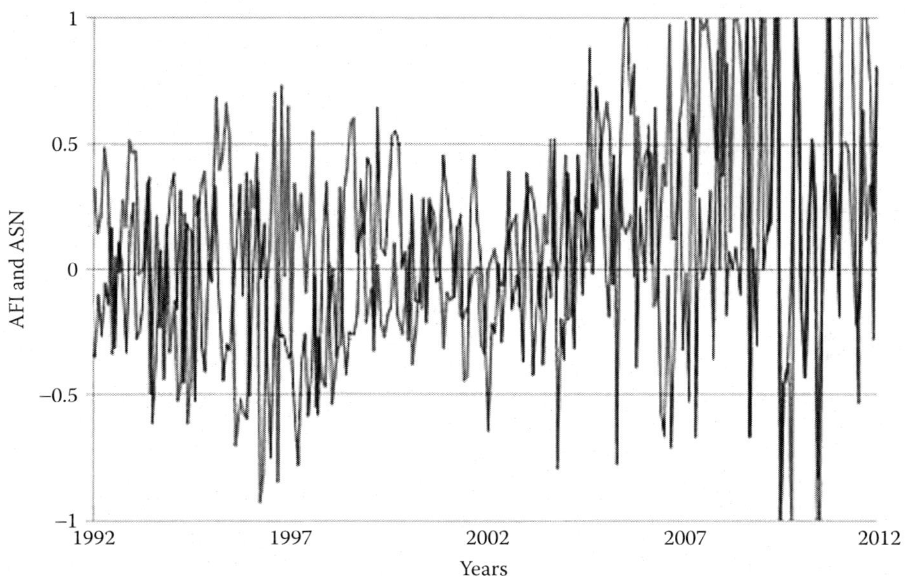

FIGURE 10.42
ASN and AFI during 1992–2010.

$\delta(area) = (SA)North - (SA)South$ [104]. Using sunspot groups and SAs we find that solar activity for cycle 23 (1996–2008) is dominant in the southern hemisphere, and our results enhance the inferred but uncertain conclusions obtained before. For the beginning part of cycle 24, northern hemispheres is stronger than southern hemisphere.

It is clear that though the southern hemisphere leads (with respect to SA) during 1966–2008, but during the beginning of new solar cycle 24 (considered up to February 2012) the northern hemisphere leads in phase. It is assumed that the north–south asymmetry of SA is due to phase asynchrony between the northern and southern hemispherical activities. This assumption is further supported by the cross-recurrence plot method to examine the synchronization of SA in the northern and southern hemispheres [103,105]. Long-term

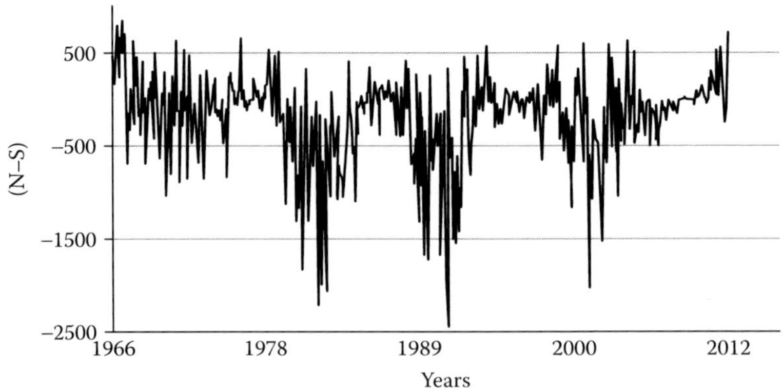

FIGURE 10.43
Difference in sunspot area 1966–February 2012.

observations of solar activity indicate that the behavior of solar activity is asymmetric in nature. For a long time it was believed that occurrences of different features on the northern and southern part of the solar disk are not uniform and that more features occur in one or the other part of the disk in different intervals. From the analysis of the N–S asymmetry of different features of solar activity, it is noted that the N–S asymmetry is statistically significant and not a random fluctuation and the asymmetry are increased during solar minimum. We have seen that one of the two hemispheres is always slightly advanced in the progress of its cyclic activity compared to the other hemisphere during the past 300 years. Interestingly, after several cycles, the hemispheres change their roles. Other aspects of solar activity reflect these uneven phenomena, like magnetic fields of the photosphere, faculae (literally a *bright spot*), monochromatic and K-corona, prominences, and possibility of solar wind. These problems are needed to be considered in a greater detail [106,107].

References

1. M. B. Kallenrode (1998). *Space Physics: An Introduction to Plasmas and Particles in the Heliosphere and Magnetospheres*. Berlin, Germany: Springer-Verlag.
2. V. Pierrard and M. Voiculescu (2011). The 3D model of the plasmasphere coupled to the ionosphere. *Geophysical Research Letters* 38: L12104.
3. J. M. Grebowsky et al. (2009). Altitude variation of the plasmapause signature in the main ionospheric trough. *Journal of Atmospheric and Solar Terrestrial Physics* 71: 1669–1676.
4. C. E. Rasmussen and R. W. Schunk (1990). A three-dimensional time-dependent model of the plasmasphere. *Journal of Geophysical Research: Space Physics* 95: 6133–6144.
5. R. P. Kane (2005). Sun–Earth relation: Historical development and present status—A brief review. *Advances in Space Research* 35: 866–881.
6. Y. K. Tulunay and P. A. Bradley (2004). The impact of space weather on communication. *Annals of Geophysics* 47: 929–944.
7. C. de Jager (2005). Solar forcing of climate. 1: Solar variability. *Space Science Reviews* 120: 197–241.
8. R. M. Goody and Y. L. Yung (1989). *Atmospheric Radiation: Theoretical Basis* (2nd ed.). Oxford University Press, Cambridge, UK.

9. G. Newkirk (1967). Structure of the solar corona. *Annual Review of Astronomy and Astrophysics* 5: 213–266.
10. S. C. Dubey et al. (2013). Effect of climate change on global surface temperature. *Journal of Engineering, Computers and Applied Sciences* 2: 2319–5606.
11. C. de Jager and H. Nieuwenhuijzen (2013). Terrestrial ground temperature variations in relation to solar magnetic variability, including the present schwabe cycle. *Natural Science* 5: 1112–1120.
12. R. J. Forsyth et al. (2002). The underlying direction of the heliospheric magnetic field through the ulysses first orbit. *Journal of Geophysical Research: Space Physics* 107: 1–11.
13. M. Lockwood and M. J. Owens (2013). Comment on "What causes the flux excess in the heliospheric magnetic field?" by E. J. Smith. *Journal of Geophysical Research: Space Physics* 118: 1880–1887.
14. G. Poletto and S. T. Suess (2004). *The Sun and the Heliosphere as an Integrated System*. Dordrecht, the Netherlands: Kluwer Academic Publishers. p. 422.
15. K. G. Jansky (1933). Radio waves from outside the solar system. *Nature* 132: 66–66.
16. S. I. Syrovatsky (1981). Pinch sheets and reconnection in Astrophysics. *Annual Review of Astronomy and Astrophysics* 19: 163.
17. A. B. Bhattacharya et al. (2012). Statistical studies of solar cycle and its periodicities. In *Proceedings of National Conference on Materials, Devices and Circuits in Communication Technology*. IETE Burdwan Sub-Center and Department of Physics, University of Burdwan, Bardhaman, India, pp. 50–52.
18. G. I. Usoskin et al. (2007). Grand minima and maxima of solar activity: New observational constraints. *Astronomy & Astrophysics* 471: 301–309.
19. J. Javaraiah et al. (2005). Long term variations in solar differential radiation and sunspot activity. *Solar Physics* 232: 25–40.
20. V. Belov et al. (2005). The relation of global magnetic solar field indices and solar wind characteristics with long term variations of galactic cosmic rays. In *29th International Cosmic Ray Conference*, Pune, India, vol. 2, pp. 235–238.
21. A. B. Bhattacharya et al. (2011). Variation of sunspot cycle and hale cycle over the century in relation to their peak activity. In *National Conference on Engineering Education In New Century*. Sir J. C. Bose School of Engineering, Sygnus Publication, Hooghly, Kolkata, India, pp. 183–186.
22. A. B. Bhattacharya et al. (2011). Anatomy of solar cycle length and sunspot number: Dependence of average global temperature on solar activity. *International Journal of Engineering Science and Technology* 3: 8012–8018.
23. A. B. Bhattacharya et al. (2012). Variation of North-South asymmetry of solar flare index and associated solar activity. *International Journal of Engineering Science and Technology* 4: 3268–3276.
24. A. B. Bhattacharya et al. (2012). A re-look on the solar activity and prediction of global temperature. In *Proceedings of National Conference on Materials, Devices and Circuits in Communication Technology*. IETE Burdwan SubCenter and Department of Physics, University of Burdwan, Bardhaman, India, pp. 47–49.
25. R. C. Willson and A. V. Mordvinov (2003). Secular total solar irradiance trend during solar cycles 21–23. *Geophysical Research Letters* 30: 1199.
26. M. L. Goelzer et al. (2013). An analysis of heliospheric magnetic field flux based on sunspot number from 1749 to today and prediction for the coming solar minimum. *Journal of Geophysical Research: Space Physics* 118: 7525–7531.
27. S. I. Akasofu et al. A new morphology of solar activity and recurrent geomagnetic disturbances: The late-declining phase of the sunspot cycle. *Space Science Reviews* 120: 27–65.
28. C. de Jager et al. (1987). A high-energy solar flare burst complex and the physical properties of its source region. *Solar Physics* 110: 317–326.
29. J. I. Sakai and C. de Jager (1996). Solar flares and collisions between current-carrying loops types and mechanisms of solar flares and coronal loop heating. *Space Science Reviews* 77: 1–192.
30. K. Hartkorn et al. (2004). Magnetic helicity change rate associated with X-class and M-class flares. In *Proceedings of the American Astronomical Society Meeting*. Bulletin of the American Astronomical Society, Washington D. C. p. 204.
31. T. Homann et al. (1997). Spectro-polarimetry of polar faculae. *Solar Physics* 175: 81–92.

32. M. Zhang and H. Q. Zhang (1999). Different magnetic features between solar polar and equatorial magnetic fields. *Astronomy & Astrophysic* 352: 317–320.
33. V. I. Makarov and K. R. Sivarama (1989). Evolution of latitude zonal structure of the large scale magnetic field in solar cycles. *Solar Physics* 119: 35–44.
34. V. I. Makarov and V. U. Makarova (1996). Polar faculae and sunspot cycles. *Solar Physics* 163: 267–289.
35. K. L. Harvey and S. F. Martin (1973). Ephemeral active regions. *Solar Physics* 32: 389–402.
36. D. K. Callebaut and V. I. Makarov (1992). Latitude time distribution of 3 types of magnetic field activity in the global solar cycle. *Solar Physics* 141: 381–390.
37. E. R. Priest (1982). *Solar Magneto-Hydrodynamics*. the Netherlands: Springer.
38. I. G. Richardson (2004). Energetic particles and corotating interaction regions in the solar wind. *Space Science Reviews* 111: 267–376.
39. M. A. El-Bone (2003). Major solar-energetic particle fluxes: I. Comparison with the associated ground level enhancements of cosmic rays. *Astroparticle Physics* 19: 549–558.
40. D. Lario et al. (2004). Solar variability and its effects on climate. *American Geophysical Union* 141: 195–216.
41. B. V. Somov et al. (2002). Magnetic reconnection scenario of the bastille day 2000 flare. *The Astrophysical Journal* 579: 863–873.
42. J. Lin et al. (2003). Theories of solar eruptions: A review. *New Astronomy Reviews* 47: 53–84.
43. J. Lin (2004). CME-flare association deduced from catastrophic model of CMEs. *Solar Physics* 219: 169–196.
44. L. M. Green et al. (2003). How are emerging flux, flares and CMEs related to the magnetic polarity imbalance in MDI data? *Solar Physics* 215: 307–325.
45. J. D. Haigh et al. (2010). An influence of solar spectral variations on radiative forcing of climate. *Nature* 467: 696–699.
46. P. Charbonneau (2005). Dynamo models of the solar cycle. *Living Reviews in Solar Physics* 2.
47. D. H. Clark and F. R. Stephenson (1978). An interpretation of pre-telescopic records from the orient. *Quarterly Journal of the Royal Astronomical Society* 19: 387–410.
48. A. D. Wittmann and Z. T. Xu (1987). A catalogue of sunspot observations from 165 BC to AD 1684. *Astronomy and Astrophysics Supplement* 70: 83–94.
49. R. J. Bray and R. E. Loughhead (1965). *Sunspots*. New York: John Wiley & Sons.
50. R. W. Noyes (1982). *The Sun, Our Star, The Harvard Books on Astronomy*. Cambridge: Harvard University Press.
51. R. Wolf (1861). Abstract of his latest results. *Monthly Notices of the Royal Astronomical Society* 21: 77–78.
52. A. Wittmann (1978). The sunspot cycle before the maunder minimum. *Astronomy & Astrophysics* 66: 93–97.
53. D. M. Willis et al. (1980). Seasonal variation of oriental sunspot sightings. *Nature* 287: 617–619.
54. J. A. Eddy (1976). The Maunder minimum. *Science* 192: 1189–1202.
55. D. V. Hoyt, K. H. Schatten (1988). Group sunspot numbers: A new solar activity reconstruction. *Solar Physics* 179: 189–219.
56. Y. A. Nagovitsyn (2001). Solar activity during the last two millennia: Solar patrol in ancient and medieval China. *International Journal of Geomagnitism and Aeronomy* 41: 711.
57. H. Schwabe (1844). Sonnen-Beobachtungen im Jahre 1843. *Astronomishe Nachrichten* 21: 233–236.
58. A. Sen Gupta (1999). Accumulation of scientific knowledge is the password for success in developing scientific attitude in society—A policy approach. *XXII Indian Social Science Congress*, pp. 13–18.
59. D. V. Hoyt and K. H. Schatten (1988). Group sunspot numbers: A new solar activity reconstruction. *Solar Physics* 179: 189–219.
60. D. H. Hathaway et al. (2002). Group sunspot numbers: Sunspot cycle characteristics. *Solar Physics* 211: 357–370.

61. M. Fligge and S. K. Solanki (1997). Inter-cycle variations of solar irradiance: Sunspot areas as a pointer. *Solar Physics* 173: 427–439.
62. T. Baranyi et al. (2001). Comparison of sunspot area data bases. *Monthly Notices of the Royal Astronomical Society* 323: 223–230.
63. L. A. Balmaceda et al. (2009). A homogeneous data base of sunspot areas covering more than 130 years. *Journal of Geophysical Research* 114: A07104.
64. E. W. Maunder (1903). Spoerer's law of zones. *Observatory* 26: 329–330.
65. E. W. Maunder (1904). Note on the distribution of Sun-spots in heliographic latitude, 1874 to 1902. *Monthly Notices of the Royal Astronomical Society* 64: 747–761.
66. K. F. Tapping and D. P. Charrois (1994). Limits to the accuracy of the 10.7 cm flux. *Solar Physics* 150: 305–315.
67. A. B. Bhattacharya et al. (1998). Observation of the microwave radio sun using L, S and C-band radio telescopes. *Indian Journal of Radio & Space Physics* 27: 124–129.
68. G. E. Hale (1908). On the probable existence of a magnetic field in sun-spots. *The Astrophysical Journal* 28: 315–343.
69. G. E. Hale et al. (1919). The magnetic polarity of sun-spots. *The Astrophysical Journal* 49: 153–178.
70. H. D. Babcock (1959). The sun's polar magnetic field. *The Astrophysical Journal* 130: 364–365.
71. N. R. Sheeley Jr. (1991). Polar faculae, 1906–1990. *The Astrophysical Journal* 374: 386–389.
72. R. Jain (1997). Prediction of the amplitude in sunspot cycle 23. *Solar Physics* 176 (2): 431–437.
73. R. P. Kane (1998). Correction to "A preliminary estimate of the size of the coming solar cycle 23, based on Ohl's precursor method" by R. P. Kane. *Geophysical Research Letters* 25: 3121.
74. J. L. Wang et al. (2002). The prediction of maximum amplitudes of solar cycles and the maximum amplitude of solar cycle 24. *Chinese Journal of Astronomy and Astrophysics* 2: 557–562.
75. E. Echer et al. (2004). Prediction of solar activity on the basis of spectral characteristics of sunspot number. *Annals of Geophysics* 22: 2239–2243.
76. R. S. Dabas et al. (2008). A prediction of solar cycle 24 using a modified precursor method. *Solar Physics* 250: 171–181.
77. K. M. Hiramath (2008). Solar forcing on the changing climate. *Astrophysics and Space Science* 314: 45–49.
78. J. Javaraiah (2008). Predicting the amplitude of a solar cycle using the north-south asymmetry in the previous cycle: II. An improved prediction for solar cycle 24. *Solar Physics* 252: 419–439.
79. D. H. Hathaway and R. M. Wilson (2006). Geomagnetic activity indicates large amplitude for sunspot cycle 24. *Geophysical Research Letters* 33: L18101.
80. B. Joshi et al. (2006). North-south distribution of solar flares during cycle 23. *Journal of Astrophysics and Astronomy* 27: 151–157.
81. M. Temmer et al. (2001). Statistical analysis of solar H-alpha flares. *Astronomy & Astrophysics* 375: 1049.
82. M. Temmer (2004). Solar activity patterns—A hemisphere related study. PhD Thesis, University at Graz, Austria.
83. D. E. Gary et al. (2005). Effects of solar radio bursts on wireless systems. In Ioannis A. Daglis (ed.). *Effects of Space Weather on Technology Infrastructure*. Springer, the Netherlands. pp 203–213.
84. G. I. Usoskin et al. (2007). Grand minima and maxima of solar activity: new observational constraints. *Astronomy & Astrophysics* 471: 301–309.
85. E. Zorita et al. (2005). The influence of volcanic, solar and CO2 forcing on the temperatures in the Dalton Minimum (1790–1830): a model study. *Climate Dynamics* 25: 205–218.
86. T. Landscheidt (2003). New little ice age instead of global warming? *Energy and Environment* 14: 327–350.
87. R. Reichel et al. (2001). The cause-and-effect relationship of solar cycle length and the Northern hemisphere air surface temperature. *Journal of Geophysical Research—Space Physics* 106: 15635–15641.

88. G. C. Reid (1991). Solar total irradiance variations and the global sea surface temperature record. *Journal of Geophysical Research* 96: 2835–2844.
89. J. Javaraiah (2000). 22-year periodicity in the solar differential rotation. *Journal of Astrophysics and Astronomy* 21: 167–170.
90. J. Javaraiah et al. (2009). Search for short-term periodicities in the Sun's surface rotation: A revisit. *Solar Physics* 257: 61–69.
91. J. Javaraiah (2013). A comparison of solar cycle variations in the, equatorial rotation rates of the sun's subsurface, surface, corona, and sunspot groups. *Solar Physics* 287: 197–214.
92. I. G. Usoskin et al. (2011). Grand minima of solar activity during the last millennia. *Proceedings of the IAU Symposium* No. 286.
93. J. E. Beckman and T. J. Mahoney (1988). The Maunder minimum and climate change: Have historical records aided current research? *Library and Information Services in Astronomy III: ASP Conference Series* 153: 212–217.
94. A. B. Bhattacharya et al. (2008). *Astronomy & Astrophysics*. Hingham, Massachusetts: Infinity Science Press.
95. E. Friis-Christensen and K. Lassen (1991). Length of the solar cycle: An indicator of solar activity closely associated with climate. *Science* 254: 698–700.
96. D. C. Archibald (2006). Solar cycles 24 and 25 and predicted climate response. *Energy and Environment* 17: 29–35.
97. J. Kleczek (1952). *Publications of Institute Central Astronomy of Prague*, Prague. 22.
98. K. J. Li et al. (2005). Sunspot unit area: A new parameter to describe long-term solar variability. *The Astrophysical Journal* 621: L81.
99. K. J. Li et al. The long–term hemispheric sunspot activity. *The Astrophysical Journal* 691: 75–82.
100. V. S. Nair and S. R. P. Nayer (2008). North south asymmetry in solar wind activity and solar cycle evolution. *Indian Journal of Radio and Space Physics* 37: 391.
101. O. C. Wilson et al. (1981). The activity cycles of stars. *Scientific American* 244: 104–119.
102. Y. M. Wang and E. Robbrecht (2011). Asymmetric sunspot activity and the southward displacement of the heliospheric current sheet. *The Astrophysical Journal* 736: 136–147.
103. M. Temmer et al. (2006). Hemispheric sunspot numbers Rn and Rs from 1945 to 2004: Catalogue and N-S asymmetry analysis for solar cycles. *Astronomy & Astrophysics* 447: 735–743.
104. N. V. Zolotova and D. I. Ponyavin (2006). Phase asynchrony of the north-south sunspot activity. *Astronomy & Astrophysics* 449: L1–L4.
105. J. L. Ballester et al. (2005). The periodic behaviour of the north-south asymmetry of sunspot areas revisited. *Astronomy & Astrophysics* 431: L5–L8.
106. R. Donner and M. Thiel (2007). Scale-resolved phase coherence analysis of hemispheric sunspot activity: A new look at the north-south asymmetry. *Astronomy & Astrophysics* 475: L33–L36.
107. A. B. Bhattacharya et al. (2011). Identification of solar bursts in the tropics. *International Journal of Advances in Science and Technology* 3: 57–60.

11

Plasma Processes of the Sun

11.1 Introduction

For quite some time, solar physicists have been wondering about the temperature difference between the Sun's apparent surface photosphere of the relatively low temperature of about 6000° K and the CORONAS temperature of a few million degrees. The laws of thermodynamics appear to rule out that the photosphere should be able to heat the corona to a temperature of several hundred times its own. Various plasma processes have been proposed from time to time, which would enable charged particles in the photosphere to accelerate causing high temperatures. However, all these processes cannot explain how unordered thermal energy particles should be transformed into ordered high-energy particles. In the coronal heating, apparently it has not been recognized that a temperature of few million degrees is the *natural* temperature of the solar plasma, whereas the photospheric temperature is the *abnormal* one. The coronal heating can be interpreted in a straightforward way by basic plasma kinetic and atomic processes. Plasma was first detected and described by William Crookes in 1879 [1]. The Crookes glass tube [2] was further identified and significantly improved by Thomson in 1897 [3]. Subsequently, the term *plasma* was used by Langmuir in 1928 [4], owing to the glowing discharge that molds itself to the shape of the Crookes tube [5].

11.2 The Fourth State of Matter

There are three well-known classic states of matter: solid, liquid, and gas. Besides all these three states, plasma is considered by scientists to be the fourth state of matter. The term has been used in physics since the 1920s for representing an ionized gas. Matter changes state as it is exposed to different physical conditions. Ice is a solid with hydrogen (H_2) and oxygen (O) molecules arranged in regular patterns. But if the ice melts, the H_2O enters a new state: liquid water. As the water molecules are warmed, they separate further to form steam, which is a gas. In these classic states, the positive charge of each atomic nucleus equals the total charge of all the electrons orbiting around it, so that the net charge becomes zero. Each entire atom is essentially electrically neutral. When further heat is applied, the steam may be ionized. An electron will gather sufficient energy to escape its atom, which is left one electron short and has a net positive charge called ion. In a highly heated gas, ionization occurs many times, producing clouds of free electrons and ions. But not all the atoms are necessarily ionized; some may remain completely intact with no net charge.

FIGURE 11.1
Lightning observed as a form of plasma.

This ionized gas mixture, consisting of ions, electrons, and neutral atoms, is termed as plasma, which must have sufficient numbers of charged particles so that the gas reveals a collective response to electric and magnetic fields. Hence, the plasma density refers to the density of the charged particles. Although plasma includes electrons and ions and conducts electricity, it is macroscopically neutral: in measurable quantities, the number of electrons and ions are equal. The charged particles are affected by electric and magnetic fields applied to the plasma. The motions of the particles in the plasma create fields and electric currents. This complex set of interactions makes plasma a fascinating, unique, and complex state of matter. Space plasma physics became an important aspect in the early 1950s with the discovery of the Van Allen radiation belts. Lightning is commonly observed as a form of plasma (Figure 11.1). Typically, lightning discharges produce 30,000 A of current at up to 100 million volts and emits light, radio waves, X-rays, and even gamma rays [6]. Plasma temperatures in lightning can approach 28,000 K and electron densities may exceed 10^{24} m^{-3}.

In fact, lightning is a massive electrical discharge in the atmosphere that creates a jagged column of plasma. It is interesting to note that part of a comet's streaming tail is plasma from gas ionized by sunlight and other unknown processes. Our Sun is a 1.5-million-km ball of plasma, heated by nuclear fusion. Figure 11.2 is an illustration of the Solar Max image of the corona, which is nothing but plasma, heated by nuclear fusion. In an effort to harness fusion energy on Earth, physicists are developing devices that can create and confine very hot plasmas in magnetic fields. In space, plasma processes are widely responsible for shielding Earth from cosmic radiation. The majority of the Sun's influence on Earth occurs by energy transfer through the ionized layers of the upper atmosphere.

Heating a gas may ionize its molecules or atoms. This reduces or increases the number of electrons, thus turning it into plasma, which contains charged particles, positive ions, and negative electrons [7]. Ionization can be induced by other methods such as a strong

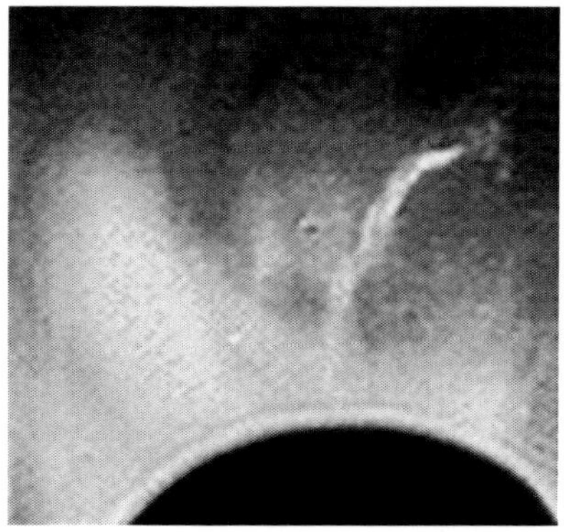

FIGURE 11.2
The photograph of the Solar Max image of the corona, which is plasma, heated by nuclear fusion. Plasma comprises the major state of matter of the Sun.

electromagnetic field applied with a laser or microwave generator and is accompanied by the dissociation of molecular bonds, if present [8]. When air or gas is ionized, plasma produces similar conductive properties to that of metals. Plasma is considered as the most abundant form of matter in the universe because most of the stars are in a plasma state [9,10]. The presence of charge carriers makes the plasma electrically conductive, so that it responds largely to electromagnetic fields. Plasma has thus shown properties unlike solids, liquids, or gases and is taken a completely distinct state of matter. Similar to gas, plasma does not have a definite shape or a definite volume unless kept in a container under the influence of a magnetic field and may form structures such as filaments, beams, and double layers. Some common plasma is observed in stars, and plasma is the most common state of matter. The majority of plasma is in the rarefied intergalactic plasma and in stars. Knowledge about plasma has been enriched from the pursuit of controlled nuclear fusion and fusion power. Figure 11.3 reveals an artist's rendition of the Earth's plasma fountain. The figure appears to be highly interesting showing oxygen, helium, and hydrogen ions,

FIGURE 11.3
An artist's rendition of the Earth's plasma fountain.

which gush into space from regions near the Earth's poles. The faint yellow area exhibits the North Pole representing gas lost from Earth into space. The green area, on the other hand, is the aurora borealis where plasma energy pours back into the atmosphere [11].

11.3 Properties and Parameters

Plasma is loosely referred to as an electrically neutral medium of positive and negative particles. This process will cause the overall charge of plasma to be roughly zero. It is interesting to note that although they are unbound, they are not *free*. When the charges travel, they generate electrical currents with magnetic fields. As a result, they are affected by each other's fields, which govern their collective behavior with many degrees of freedom [8,12]. A definition can have three criteria as pointed out below [13,14]:

1. *The plasma approximation*: Charged particles must be sufficiently close so that each particle influences nearby charged particles, rather than just interacting with the closest particle. These collective effects are a notable feature of plasma. The plasma approximation is valid when the number of charge carriers within the sphere of influence the particle becomes greater than unity and will exhibit a collective behavior of the charged particles.
2. *Bulk interactions*: The Debye screening length is short in comparison to the physical size of the plasma, which means that interactions in the bulk of the plasma are more important than those at its edges, where boundary effects may occur. When this condition is satisfied, the plasma becomes quasineutral.
3. *Plasma frequency*: The electron plasma frequency, measuring the plasma oscillations of the electrons, is large in comparison to the electron-neutral collision frequency. This measures the frequency of collisions between electrons and neutral particles.

11.3.1 Comparison of Gas and Plasma Phases

Plasma is popularly referred to as the fourth state of matter after solid, liquids, and gases [15,16]. Although it is highly related to the gas phase, in that it also has no definite form or volume, it differs in a number of ways [17]. Some important distinctions between the two are mentioned below:

Property	Gas	Plasma
Electrical conductivity	Too low; air is a good insulator until it breaks down into plasma at electric field strengths above 30 kV/cm [16]	Generally very high; for many purposes, the conductivity of plasma may be treated as infinite
Independently acting species	One; all gas particles behave in a similar way, influenced by gravity and by collisions with each other	Two or three; electrons, ions, protons, and neutrons can be distinguished by the sign and value of their charge, so that they can behave independently

(*Continued*)

Property	Gas	Plasma
Velocity distribution	Maxwellian; collisions generally lead to a Maxwellian velocity distribution of all gas particles	Often non-Maxwellian; collisional interactions are weak in hot plasmas and external forcing can drive the plasma far from local equilibrium
Interactions	Binary; two-particle collisions are the rule, three-body collisions extremely rare	Collective; waves or organized motion of plasma are very important, as the particles can interact at long ranges through the electric and magnetic forces

11.3.2 Astrophysical Plasma

Plasmas are the most common phase of ordinary matter in the universe, both by mass and volume [18]. Our Sun and all other stars, as well as intergalactic space, are made of plasma and also most of the interstellar space is filled with a plasma. Our solar system and interplanetary space is filled with the plasma from the solar wind, which extends from the Sun out to the heliopause. Even the black holes, which are not directly visible, are fuelled by accreting ionizing matter, by plasma [19], and they are related to the astrophysical jets of luminous ejected plasma [20], for example, M87's jet [21]. Dust and small grains within plasma also pick up a net negative charge, so that they in turn may act like a very heavy negative ion component of the plasma. The present day view is that about 96% of the total energy density in the universe is not plasma or any other form of ordinary matter but a combination of cold dark matter and dark energy. In solar system, the density of ordinary matter is sufficiently higher than average and also much higher than that of either dark matter or dark energy. The planet Jupiter and its satellites account for most of the non-plasma, only about 0.1% of the mass and 10%–15% of the volume within the orbit of Pluto. Some common forms of plasma are listed below under different categories.

Artificially Produced	Terrestrial Plasmas	Space and Astrophysical Plasmas
1. Inside fluorescent lamps, neon signs [21]	1. Lightning	1. The Sun and other stars wherein plasmas heated by nuclear fusion
2. Rocket exhaust and ion thrusters	2. St. Elmo's fire	2. The solar wind
3. The area in front of a spacecraft's heat shield during re-entry into the atmosphere	3. Upper-atmospheric lightning, e.g., Blue jets, Blue starters, Gigantic jets, ELVES	3. The interplanetary medium, i.e., the space between planets
4. Inside a corona discharge ozone generator	4. Sprites	4. The interstellar medium, i.e., the space between star systems
5. Fusion energy research	5. The ionosphere	5. The intergalactic medium, i.e., the space between galaxies
6. Plasmas used in semiconductor device fabrication	6. The plasmasphere	6. The Io-Jupiter flux tube
7. Laser-produced plasmas found when high-power lasers interact with materials	7. The polar aurorae	7. Accretion disks
8. Inductively coupled plasmas formed typically in argon gas	8. Some flames [22,23]	8. Interstellar nebulae
9. Magnetically induced plasmas using microwaves as a resonant coupling method	9. The polar wind, a plasma fountain	9. Cometary ion tail

11.4 Complex Plasma Phenomena

Plasma behaviors are extraordinarily varied [24]. The spontaneous formation of interesting spatial features on a wide range of length scales is a manifestation of plasma complexity. The features are interesting as they are very sharp, spatially intermittent or have a fractal form. Many of these features were first investigated in the laboratory and have subsequently been identified throughout the universe.

11.4.1 Ranges of Plasmas

Plasma parameters can take on values changing by many orders of magnitude, but the properties of plasmas with apparently disparate parameters are very similar. The chart shown in Figure 11.4 presents only conventional atomic plasmas and not exotic phenomena like quark gluon plasmas.

If we look at the range of plasmas, we find that the density increases upward and temperature increases toward the right. The free electrons in a metal may be considered an electron plasma [25,26]. Typical ranges of plasma parameters are given below, including the orders of magnitude (OOM).

Dominant Features	Terrestrial Plasmas	Cosmic Plasmas
Size in meter	10^{-6} m (lab plasmas) to 10^{2} m (lightning) (~8 OOM)	10^{-6} m (spacecraft sheath) to 10^{25} m (intergalactic nebula) (~31 OOM)
Lifetime in second	10^{-12} s (laser-produced plasma) to 10^{7} s (fluorescent lights) (~19 OOM)	10^{1} s (solar flares) to 10^{17} s (intergalactic plasma) (~16 OOM)
Density in particles/cubic meter	10^{7} m^{-3} to 10^{32} m^{-3} (inertial confinement plasma)	1 m^{-3} (intergalactic medium) to 10^{30} m^{-3} (stellar core)
Temperature in kelvin	~0 K (crystalline nonneutral plasma [25]) to 10^{8} K (magnetic fusion plasma)	10^{2} K (aurora) to 10^{7} K (solar core)
Magnetic fields in tesla	10^{-4} T (lab plasma) to 10^{3} T (pulsed-power plasma)	10^{-12} T (intergalactic medium) to 10^{11} T (near neutron stars)

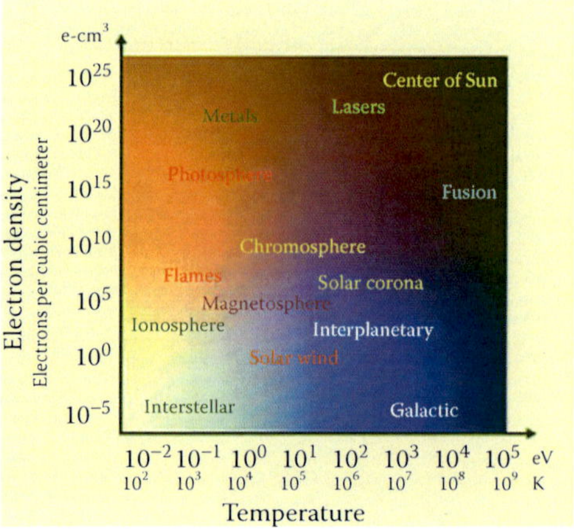

FIGURE 11.4
The chart showing only conventional atomic plasmas and not exotic phenomenon.

11.5 Energy Considerations

The average gravitational potential energy of an atom of mass m within a spherical mass M with a density distribution ~$1/r^2$ and radius R is

$$E_{pot} = -Gm(M/R) \qquad (G = \text{gravitational constant}). \tag{11.1}$$

For a hydrogen atom ($m = 1.67 \times 10^{-24}$ g), the solar mass $M = 1.99 \times 10^{33}$ g, and the solar radius $R = 6.96 \times 10^{10}$ cm, which yields a potential energy of

$$E_{pot} = -3.2 \times 10^{-9} \text{ erg}. \tag{11.2}$$

According to the virial theorem of mechanics, the kinetic energy is related to the potential energy by

$$E_{kin} = -E_{pot}/2 = 1.6 \times 10^{-9} \text{ erg} = 996 \text{ eV}. \tag{11.3}$$

As $E_{kin} = kT$ (where k is the Boltzmann constant), this corresponds to a temperature:

$$T = 1.1 \times 10^7 \text{ K}, \tag{11.4}$$

which is more than enough to account for the temperature of the solar corona.

11.6 Density Consideration

For inelastic collisions to happen, we need individual neutral atoms: The average distance between two nuclei in the plasma has to be greater than twice the Bohr radius r_0 ($r_0 = 5 \times 10^{-9}$ cm), which means the gas density has to be less than

$$n_0 = 1/[(4/3)\pi(2r_0)^3] = 2.4 \times 10^{23} \text{ cm}^{-3}. \tag{11.5}$$

For densities greater than this, radiative processes are not possible at all. Under that condition, one has hot but dark plasma *soup* of protons and electrons. The gas density in the Sun can be obtained in the following way: for a homogeneous sphere, it would be given by

$$n_h = M/[(4/3)\pi R^3 m], \tag{11.6}$$

where:
- M is the total solar mass
- R the solar radius
- m the weight of a hydrogen atom

This gives

$$n_h = 8.7 \times 10^{23} \text{ cm}^{-3}. \tag{11.7}$$

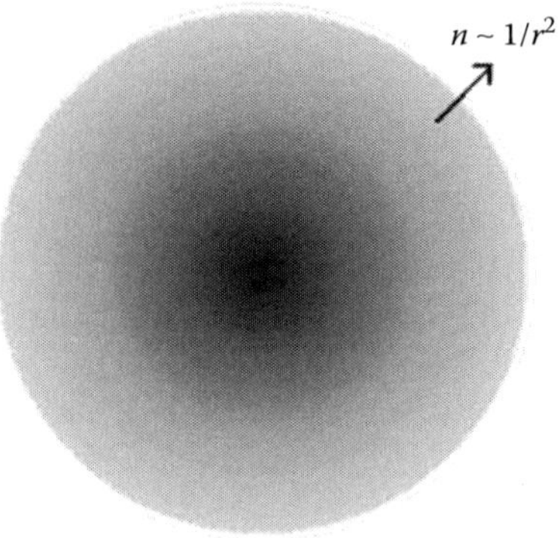

FIGURE 11.5
Gas ball in hydrostatic equilibrium according to classical physics (density decreases slowly outward).

However, the assumption of a homogeneous gas sphere is not adequate because a gas sphere in hydrostatic equilibrium necessarily has a density decreasing with the distance r from its center ~$1/r^2$. The density $n(R)$ near the surface of the Sun (located at a radius R) has therefore to be found by the equation

$$M = n(R)m4\pi \,_0\!\int^R dr'\, r'^2(R/r')^2 = n(R)m4\pi R^3, \tag{11.8}$$

or

$$n(R) = M/(4\pi R^3 m) = 2.9 \times 10^{23}\ \text{cm}^{-3}. \tag{11.9}$$

As this is virtually identical to the maximum density for inelastic collisions, the latter can only occur in a very narrow layer near the surface of the Sun. As the density increases toward the center of the Sun, no inelastic collisions can occur below the photosphere and, consequently, the plasma will not be cooled but have the *gravitational temperature* of 1.1×10^7 K. The situation is illustrated schematically in Figure 11.5, which shows the cross section through a gas sphere in classical hydrostatic equilibrium according to classical physics. It appears from the figure that the density decreases slowly outward. Figure 11.6, on the other hand, reveals the additional effect of quantum mechanical inelastic collisions. It is seen from the figure that there is a strong density decrease in collision region.

11.7 Inelastic Collision Process

The production of striations in glow discharges and artificial ionospheric airglow experiments shows the existence of a resonant collisional excitation process, with a cross section identical to the resonant scattering of radiation. Although in the context of these phenomena we deal with inelastic collisions of electrons, we can consider that the same process generally applies to ions as well, with the necessary energies scaled upward by a factor m_p/m_e, where m_p represents the proton mass and m_e the electron mass. Due to the inelastic collision, the proton will lose all of its kinetic energy but will subsequently be heated again by elastic collisions with the high-energy electrons. Elastic collisions should indeed

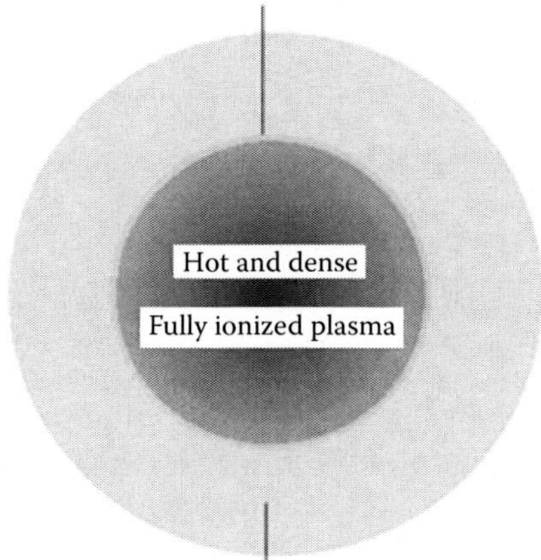

FIGURE 11.6
Gas ball in hydrostatic equilibrium plus cooling by inelastic collisions (strong density decrease in the collision region).

equalize the temperatures between the electrons and protons after a number of collisions, but the loss of kinetic energy of an electron in the course of a close Coulomb collision indicates that the electron has a high probability to recombine with the proton. With this expressed, it loses all its kinetic energy on recombination, apart from the fraction that has been transferred in this collision. Classical mechanics shows the relative collisional energy transfer between two very different masses. The mass ratio between the masses is a fraction 1/1836 of the electron temperature of 1.1×10^7 K, which results in the photospheric temperature in agreement with the observed value:

$$T_{ph} \approx 6000°K. \tag{11.10}$$

It is indeed the recombination of the electrons and protons that produces the radiative output of the photosphere and not the cooling of the inelastic collisions with neutral atoms protons. For the density to diminish from a value of around 10^{23} cm^{-3} at the Sun's surface to about 10^8 to 10^5 cm^{-3} in the solar corona, the inelastic collision process has to reduce the flux of protons by a factor 10^{-15} to 10^{-18}. Following the exponential absorption law, this needs nearly 35–40 mean free paths:

$$l_{ex} = 1/(N_n \sigma_{ex}), \tag{11.11}$$

where:
- N_n is the neutral atom density
- σ_{ex} the inelastic collision cross section

So, l_{ex} should have a value of around 1/40 of the scale height of the photosphere. This result in

$$l_{ex} = 5 \text{ km}. \tag{11.12}$$

A neutral density of 10^8 cm^{-3} and an excitation cross section of 10^{-14} cm^2 would yield this length approximately, and both of these values can be shown to be reasonable estimates for the photosphere. Inelastic collisions for the protons can occur in very narrow energy bands coinciding with the usual spectral lines and most protons will not suffer any energy loss in this respect.

11.8 The Plasma Universe

More than 99% of matter in the universe exists in the plasma state although nature produces plasma on Earth's surface rarely. Plasma processes have important roles in the behavior of stars, interstellar clouds, comets, the aurora, and even our upper atmosphere. To understand astrophysical and geophysical phenomena, they have a clear understanding of how matter behaves in the plasma state. Ground-based radio wave observations showed that a region of plasma exists above Earth's electrically neutral atmosphere. The ionosphere effectively reflects most radio waves back to Earth and it is this process that led to the region's discovery. Figure 11.7 reveals the plasma galaxy including the magnetic field and the galactic jet.

The nature of the ionospheric plasma is very thin and has sharp distinctions, which differ in composition and density. The F1 layer of the ionosphere is situated at about 200 km and the F2 layer around 300–400 km. Of these two layers, in the F2 layer, the plasma is most dense. There are rarely more than 1 million electron–ion pairs in a cubic centimeter of space; in comparison, the neutral gas density for the same region is typically 1 billion particles per cubic centimeter. The plasma in these layers, consisting of electrons and atomic oxygen ions, is sustained by the ionizing action of solar UV radiation on the neutral atmospheric gas. Plasma is highly influenced by both magnetic and electric forces. Plasma particles affect the distribution of magnetic and electric fields. Beyond the magnetopause, energetic plasma from the Sun rushes past Earth with a speed from 300 to 1000 km/s. Most of this solar wind is deflected around Earth, whereas some other penetrates the magnetosphere. The interaction between the solar wind and the magnetosphere plasma behaves like an electric generator creating electric fields deep inside the magnetopause. The fields so formed make a general circulation of the plasma and accelerate some electrons and ions to higher energies. The visible manifestation of the high-energy electrons is found in the auroras, the colorful northern and southern lights, which appear at 90–160 km above

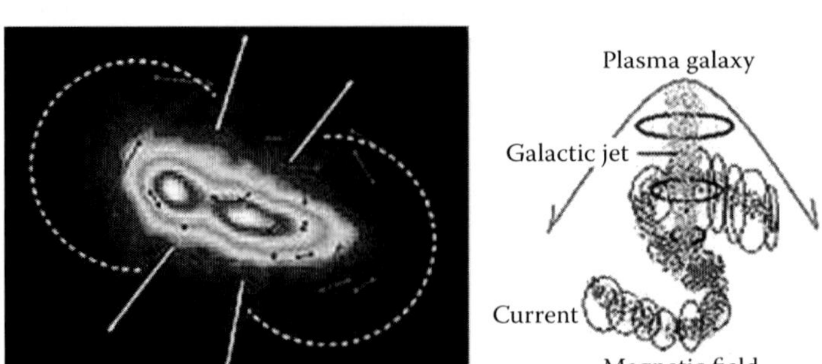

FIGURE 11.7
The plasma Galaxy Plasma dominates Earth's environment in the ionosphere.

FIGURE 11.8
The northern lights.

Earth. The aurora colors are obtained by the nature of the atoms struck by magnetosphere electrons and the energies of the collisions. A typical 3-h aurora, covering a million square kilometers, discharges nearly 100 million kW h of electric energy into Earth's immediate environment. This is considered as sufficient energy and is roughly equivalent to 6 days of energy output by a large nuclear power plant.

Characterizing the magnetic and plasma environments in Earth's neighborhood, scientists have recognized the plasma processes in the rest of the universe. In fact, auroras have been spotted on Jupiter, and the same types of phenomena appear to occur in the magnetospheres of both Saturn and Uranus. Many high-energy X-rays and gamma rays as detected by astronomical observations come from magnetized plasmas near stars, galaxies, and other objects. A visual image of the universe shows only the superficial appearances, but investigations through plasma reveals the invisible structure of space and the processes that may have formed the solar system from dust and plasma. The northern light is shown in Figure 11.8.

11.8.1 Degree of Ionization

The term *plasma density* by itself refers to the *electron density*, which is the number of free electrons per unit volume. The degree of ionization of a plasma is the proportion of atoms that have lost or gained electrons and is controlled largely by the temperature. Even a partially ionized gas where only 1% of the particles are ionized can have the characteristics of plasma.

The degree of ionization, α, is defined as

$$\alpha = n_i/(n_i + n_a) \tag{11.13}$$

where:

n_i is the number density of ions
n_a is the number density of neutral atoms

The electron density is associated with this by the average charge state $\langle Z \rangle$ of the ions through $n_e = \langle Z \rangle n_i$, where n_e is the number density of electrons.

11.9 Nonthermal Plasma

Generally, plasma temperature is measured in kelvin or electronvolt. It is practically a measurement of the thermal kinetic energy per particle. Very high temperatures are required to sustain ionization, which is a defining feature of plasma. The degree of plasma ionization is found by the *electron temperature* relative to the ionization energy in a relationship called the Saha equation [27]. At low temperatures, ions and electrons tend to recombine into bound states and the plasma will eventually become a gas. In most of the cases, the electrons are sufficiently close to thermal equilibrium that their temperature is relatively well defined, even when there is a significant deviation from a Maxwellian energy distribution function. Due to the large difference in mass, the electrons come to thermodynamic equilibrium among themselves much faster than they come into equilibrium with the ions or neutral atoms. For this reason, the *ion temperature* may be very different from the *electron temperature*, which is especially common in weakly ionized technological plasmas, where the ions are frequently noted near the ambient temperature.

11.9.1 Thermal versus Nonthermal Plasmas

Based on the relative temperatures of the electrons, ions, and neutrals, plasmas are categorized as *thermal* and *nonthermal*. Thermal plasmas have electrons and heavy particles at the same temperature, whereas nonthermal plasmas have the ions and neutrals at a much lower temperature. A plasma is sometimes referred to as being *hot* if it is almost ionized or *cold* if only a small fraction of the gas molecules are ionized. In *cold* plasma, the electron temperature is typically several thousand degrees Celsius. Plasmas used in *plasma technology* are generally cold plasmas in the sense that only a small fraction of the gas molecules are ionized.

11.10 Plasma Potential

As plasmas are very good electrical conductors, electric potentials play a vital role. The potential exists on average in the space between charged particles. This is referred to as the *plasma potential* or the *space potential*. If an electrode is inserted into plasma, its potential will lie considerably below the plasma potential owing to the Debye sheath [28]. The good electrical conductivity of plasmas makes their electric fields very low. This provides the concept of *quasi-neutrality*, which suggests that the density of negative charges is nearly equal to the density of positive charges over large volumes of the plasma. However, on the scale of the Debye length, there can be charge imbalance. In the special case, the charge separation can extend some tens of Debye lengths. The magnitude of the potentials and electric fields are found by finding the net charge density. An example is to assume that the electrons satisfy the Boltzmann relation,

$$n_e \propto e^{e\Phi/(k_B T_e)}. \tag{11.14}$$

Differentiating this relation provides a means to find the electric field from the density,

$$\vec{E} = [(k_B T_e)/e][\nabla n_e / n_e]. \quad (11.15)$$

It is possible to create plasma that is not quasineutral. The density of nonneutral plasma must be very low; otherwise it will be dissipated by the repulsive electrostatic force. In astrophysical plasmas, Debye screening prevents electric fields from directly affecting the plasma over large distances, greater than the Debye length. However, the existence of charged particles causes the plasma to be affected by magnetic fields, which contributes to magnetization.

11.10.1 Magnetization

Plasma with a strong magnetic field influences the motion of the charged particles. Such a particle is said to be magnetized. A particle on average completes at least one gyration around the magnetic field before making a collision,

$$\omega_{ce}/\nu_{coll} > 1, \quad (11.16)$$

where:

ω_{ce} is the *electron gyrofrequency*
ν_{coll} is the *electron collision rate*

The electrons are magnetized, whereas the ions are not [29]. Magnetized plasmas are anisotropic, meaning that their properties in the direction parallel to the magnetic field are different from those in the direction perpendicular to it. While electric fields in plasmas are small for high conductivity, the electric field related to plasma moving in a magnetic field is given by

$$E = -v \times B$$

where:

E is the electric field
v is the velocity
B is the magnetic field

11.10.2 Filamentation

Striations or string-like structures [30], also called Birkeland currents, are found in many plasmas, like the plasma ball, the aurora [31], lightning [32], electric arcs, solar flares [33], and supernova remnants [34]. They are also found to be related to larger current densities and the interaction with the magnetic field can form a magnetic rope structure [35]. High-power microwave breakdown at atmospheric pressure leads to the formation of filamentary structures [36]. One interesting aspect of the filamentation-generated plasma is the relatively low ion density due to defocusing effects of the ionized electrons [37].

11.10.3 Electric Fields and Circuits

Quasi-neutrality of plasma requires that plasma currents close on themselves in electric circuits following Kirchhoff's circuit laws and possess a resistance and inductance. These circuits must be treated as a strongly coupled system, with the behavior in each plasma region dependent on the entire circuit. It is this strong coupling between system elements, together with nonlinearity, that may lead to complex behavior. Electrical circuits in plasmas store inductive energy and, should the circuit be disrupted by plasma instability, the inductive energy will be released as plasma heating and acceleration. This gives a common explanation for the heating that takes place in the solar corona. Electric currents, in particular, and magnetic-field-aligned electric currents are also observed in the Earth's aurora and in plasma filaments.

11.11 Ultracold Plasma

Ultracold plasmas are produced in a magneto optical trap. This process is accomplished by trapping and cooling neutral atoms to temperatures of 1 mK or lower to ionize the atoms, thus giving each of the outermost electrons just enough energy to escape the electrical attraction of its parent ion. One advantage of ultracold plasmas is their well-characterized and tunable initial conditions, including their size and electron temperature. Adjusting the wavelength of the ionizing laser, the kinetic energy of the liberated electrons can be tuned as low as 0.1 K, a limit set by the frequency bandwidth of the laser pulse. One of the metastable states of a strongly nonideal plasma is the Rydberg matter [38], which forms on condensation of excited atoms.

11.11.1 Nonneutral Plasma

The strength and range of the electric force and the good conductivity of plasmas ensure that the densities of positive and negative charges in any sizeable region are equal. Plasma with a significant excess of charge density or, in the extreme case, composed of a single species is called nonneutral plasma. In such a plasma, electric fields play a vital role.

11.11.2 Dusty Plasma and Grain Plasma

Dusty plasma contains tiny charged particles of dust, typically found in space. The dust particles acquire high charges and interact with each other. Plasma that contains larger particles is called grain plasma. Under laboratory conditions, dusty plasmas are also termed as complex plasmas [39].

11.11.3 Impermeable Plasma

Impermeable plasma is a type of thermal plasma that acts like an impermeable solid with respect to gas or cold plasma and can be physically pushed [40]. The spectroscopic data on the characteristics of plasma were claimed to be difficult to get, due to the high-pressure and the passive effect of plasma. Different nanostructures also suggested the effective confinement.

11.12 Plasma Modeling

To describe the state of plasma, we need to write down all the particle locations and velocities and describe the electromagnetic field in the plasma region. However, it is generally not necessary to keep track of all the particles in plasma. The complex self-constricting magnetic field lines and current paths in a field-aligned Birkeland current that can develop in a plasma is shown in Figure 11.9 [41].

11.12.1 Fluid Model

Fluid models explain plasmas in terms of smoothed quantities, such as density and averaged velocity around each position. One simple fluid model, magnetohydrodynamics, considers the plasma as a single fluid governed by a combination of Maxwell's equations and the Navier–Stokes equations [42]. A more general description is given in the two-fluid plasma picture where the ions and electrons are considered separately.

11.12.2 Kinetic Model

The kinetic model describes the particle velocity distribution function at each point in the plasma and so does not require or assume a Maxwell–Boltzmann distribution. A kinetic description is essential for collision-less plasmas. There are two common approaches to the kinetic description of plasma. One is based on a *smoothed distribution function* on a grid

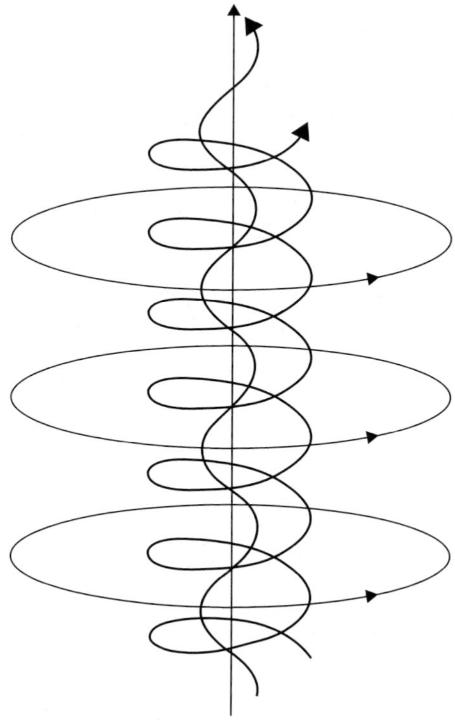

FIGURE 11.9
The complex self-constricting magnetic field lines and current paths.

in velocity and position. The other is the *particle-in-cell technique,* which includes kinetic information by following the trajectories of a large number of individual particles.

11.13 The Sun's Atmosphere

Most of the visible white light comes from the photosphere. Both the chromosphere and the corona also emit white light and can be found when the light from the photosphere is blocked out, as occurs during a solar eclipse. The Sun emits electromagnetic radiation at many other wavelengths also. Different types of radiation, for example, radio, UV, X-rays, and gamma rays originate from different parts of the Sun. Scientists use special type of instruments for detecting this radiation and study different parts of the solar atmosphere. The solar atmosphere is so hot that the gas is mainly noted in a plasma state [43]. In this charged state, the solar atmosphere is largely influenced by the strong solar magnetic fields that thread through it. These magnetic fields and the outer solar atmosphere extend out into interplanetary space as part of the solar wind. The different zones of the Sun with some important events originating from those zones have been indicated in Figure 11.10.

11.13.1 Corona and Chromosphere of the Sun's Atmosphere during Eclipse

As we have mentioned, the lower region of the solar atmosphere is known as the chromosphere. This name comes from the Greek root chroma, which means color. In fact, it appears bright red. The outer atmosphere of the Sun contains the chromosphere and corona, which are visible to the human eye only during a solar eclipse. In these layers,

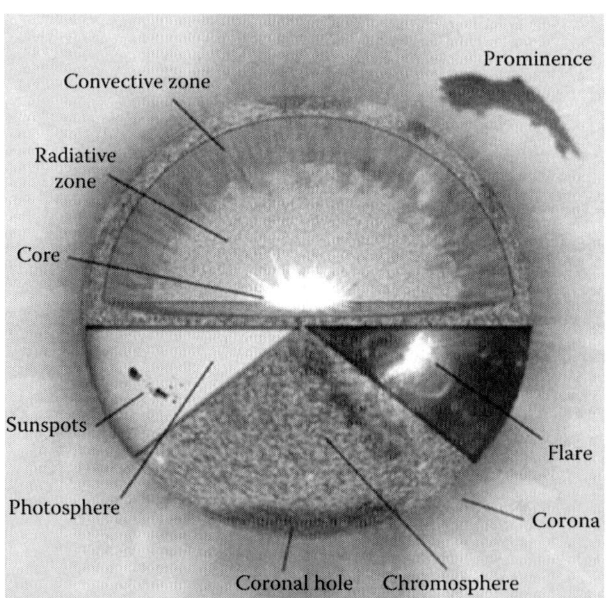

FIGURE 11.10
The different zones of the Sun with some important events.

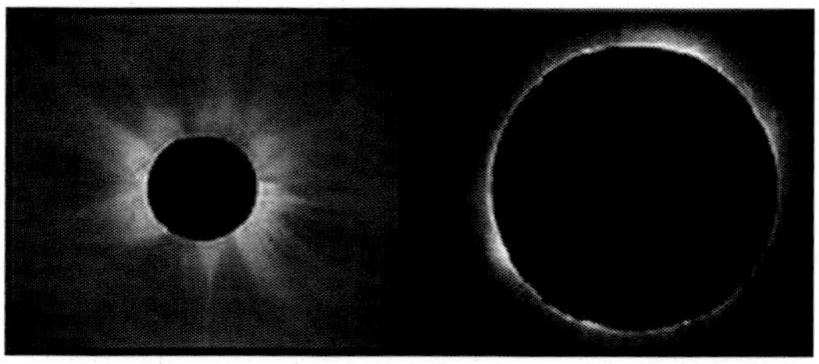

FIGURE 11.11
The two main regions of the Sun's atmosphere as seen during an eclipse: corona (left) and chromosphere (right). (Courtesy of NCAR's High Altitude Observatory, National Center for Atmospheric Research (NCAR), Boulder, CO; Courtesy of Luc Via tour 1999.)

an enormous increase in temperature from a few thousand kelvin up to a million kelvin can be found [44] when viewed during a solar eclipse. A thin transition region where the temperatures are enhanced sharply separates the Chromospheres from the Coronae. The uppermost portion of the Sun's atmosphere, the corona, is much hotter than the Sun's surface. The upper corona gradually turns into the solar wind. This is nothing but a flow of plasma that moves outward through our solar system into interstellar space. The solar wind may be taken as an extension of the Sun's atmosphere that engulfs all of the planets [45]. Figure 11.11 reveals the two main regions of the Sun's atmosphere as seen during an eclipse: corona (left) and chromosphere (right).

11.14 Dynamics of the Sun's Outer Atmosphere

The heating mechanism responsible for the rise in temperature is indeed a focus of research today. To determine the spectrum of the corona, satellites such as SOHO and the Hinode Space Observatory are vitally important, as they are capable of detecting radiation that is predominantly in the extremely short-wave UV range. It may be pointed out that the theoretical magnetohydrodynamic models range from 1D models for chromospheric radiation, to 2D models for funnel-shaped structures, up to 3D models for the active regions of the corona [42]. The 3D magnetohydrodynamic model of the corona showing the Doppler shift of Ne VIII is presented in Figure 11.12.

FIGURE 11.12
The 3D magnetohydrodynamic model of the corona showing the Doppler shift of Ne VIII. (Courtesy of Kiepenheuer Institute for Solar Physics, Freiburg, Germany.)

11.15 The Solar Atmosphere and Plasma

The Sun's atmosphere is composed of several layers. Of those, the main layers are the photosphere, the chromosphere, and the corona. It is in these outer layers that the Sun's energy, which has bubbled up from the interior layers, is detected as sunlight. The lowest layer of the Sun's atmosphere is the photosphere, which is about 500 km thick. This layer is where the Sun's energy is released as light. The next layer is the chromospheres, which emits a reddish glow as superheated hydrogen burns off. However, the red rim can only be found during a total solar eclipse. At other times, light from the chromosphere becomes too weak to be observed against the brighter photosphere. The third layer of the Sun's atmosphere is the corona, which can only be seen during a total solar eclipse. It looks as white streamers or plumes of ionized gas that flow outward into space. Temperatures in the Sun's corona can get as high as 3.5 million degrees Fahrenheit. As the gases cool, they become the solar wind. Due to the distance from the Sun to Earth, light arrives at our planet in about 8 min. The image of the solar corona, taken by the SECCHI outer coronagraph (COR2) on the STEREO, is presented in Figure 11.13.

11.16 River of Plasma, Guarding against the Sun

Earth's magnetosphere stretches from the planet's core out into space, where it meets the solar wind (a stream of charged particles) emitted by the Sun. For the major part, the magnetosphere behaves as a shield for protecting the Earth from this high-energy solar activity. This field comes into contact with the Sun's magnetic field in a process called *magnetic reconnection*. Powerful electrical currents from the Sun can stream into Earth's atmosphere, whipping up geomagnetic storms and space weather phenomena that can affect high-altitude aircraft, as well as astronauts on the International Space Station. The Earth's

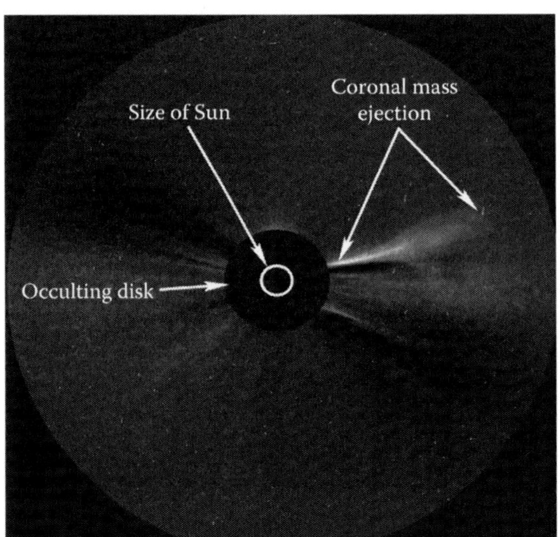

FIGURE 11.13
Image of the solar corona taken by the SECCHI outer coronagraph (COR2) on the STEREO. (Courtesy of NASA/STEREO, Coronagraphs, http://stereo.gsfc.nasa.gov/classroom/coronagraphs.shtml [accessed August 12, 2016].)

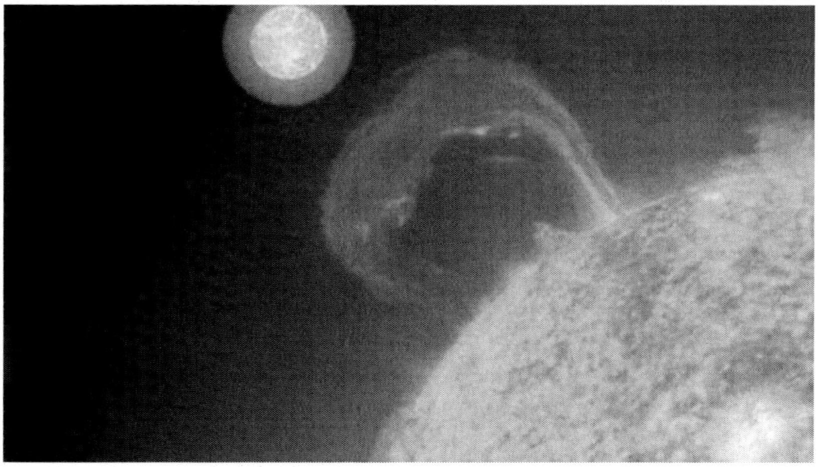

FIGURE 11.14
NASA images used in a photo composite by Christine Daniloff (MIT). (Courtesy of MIT News, http://news.mit.edu/2014/a-river-of-plasma-guarding-against-the-sun-0306 [accessed August 12, 2016].)

magnetic field protects life on the surface from the full impact of these solar outbursts. These plasmas get pulled into space and slow down the reconnection process, causing the impact of the Sun on the Earth to be less violent [46]. A group of scientists at MIT and NASA have identified a process in the Earth's magnetosphere that reinforces its shielding effect, keeping incoming solar energy at bay. By combining observations from the ground and in space, the team found a plume of low-energy plasma particles that hitches a ride along magnetic field lines, where the planet's magnetic field connects with that of the Sun. Scientists call this the *merging point*. The presence of cold, dense plasma slows magnetic reconnection, thereby blunting the Sun's effects on Earth. A NASA image used in a photo composite is presented in Figure 11.14.

11.17 Solar and Space Plasma

By observing the Sun's surface or by studying an active aurora on Earth, we may try to understand the fourth state of matter. By using space-based measurements, we may get information about how energy flows through the system and how it is converted and also what causes quiet plasma to erupt into an explosive energy release. Investigating different plasma conditions we can get idea about space plasmas [47]. The plasma between the Earth and the Sun is so tenuous that we cannot see it. For that reason, we send space probes to measure it directly. In contrast, temperatures of plasmas nearer the Sun are extremely high for instruments to survive. However, because the plasma is dense at the Sun, the plasma can be remotely imaged by cameras [47]. The composite image shown in Figure 11.15 demonstrates an intense coronal mass ejection (CME). The Space Plasma group at LASP observed fundamental plasma processes throughout our solar system such as reconnection, shocks, Alfven waves, and ion heating.

One important object of LASP research is to understand space plasma interactions and processes, including (i) magnetic reconnection and particle energization and (ii) Alfven waves, collision-less shocks, turbulence, parallel (to **B**) electric fields, and instabilities.

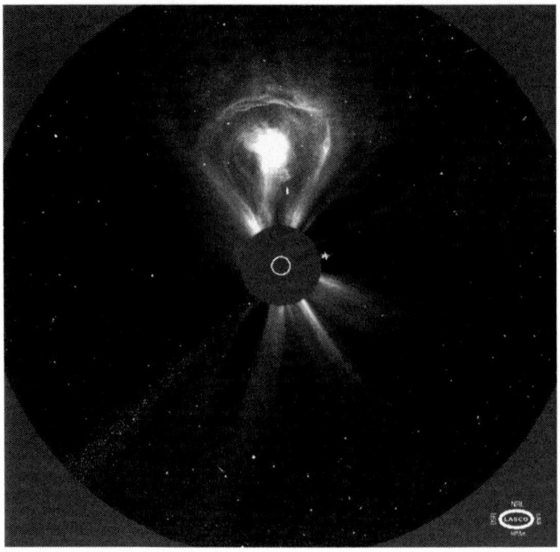

FIGURE 11.15
Sun Storm: a coronal mass ejection. (Courtesy of ESA/NASA/SOHO, http://www.nasa.gov/content/goddard/what-to-see-in-cme-images/#.V72KTfl9671 [accessed August 12, 2016].)

11.17.1 Effects of Solar Plasma on Earth

A giant wave of super-hot solar plasma larger than the Earth erupted from the Sun in a spectacular display captured by a NASA spacecraft (Figure 11.16). The huge solar prominence was recorded in amazing detail by the high-definition cameras aboard NASA's Solar Dynamics Observatory before escaping the Sun. Once it started breaking away, the process took about 10 h only before it was out of sight [48]. The prominence stretched out many times than the size of Earth.

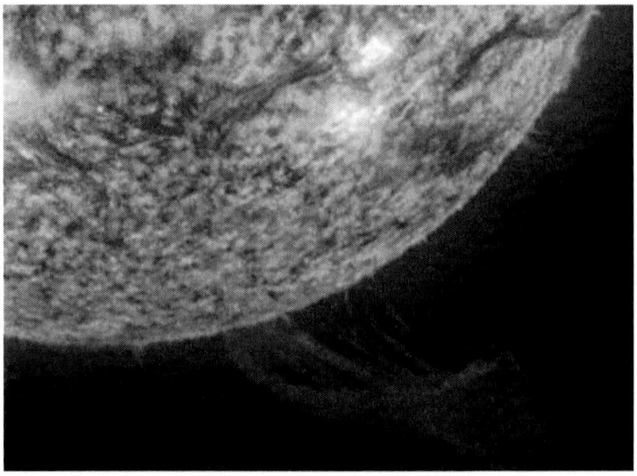

FIGURE 11.16
Solar prominence was recorded in amazing detail by the high-definition cameras aboard NASA's Solar Dynamics Observatory before escaping the Sun. (Courtesy of NBC/Steele Hill, *NBC News*, http://www.nbcnews.com/id/49483655/ns/technology_and_science-space/t/sun-unleashes-solar-plasma-wave-larger-earth/ [accessed August 12, 2016].)

References

1. Eliezer, S. (2001). *The Fourth State of Matter: An Introduction to Plasma Science*, p. 2. Boca Raton, FL: CRC Press.
2. William, C. (1878). On the illumination of lines of molecular pressure, and the trajectory of molecules. *Philosophical Transactions of the Royal Society of London 170*: 135–164. doi:10.1098/rstl.1879.0065.
3. Thomson, J. J. (1897). Cathode rays. *Philosophical Magazine* 44, 293.
4. Langmuir, I. (1928). Oscillations in ionized gases. *Proceedings of the National Academy of Sciences 14* (8): 627.
5. Brown, S. C. (1978). Chapter 1: A short history of gaseous electronics. In Hirsh, M. N. and Oskam, H. J. (Eds.), *Gaseous Electronics*, Vol. 1, pp. 1–18. New York: Academic Press.
6. NASA News (February 17, 2005). Flashes in the sky: Earth's gamma-ray bursts triggered by lightning. http://www.nasa.gov/vision/universe/solarsystem/rhessi_tgf.html#.V5H2hkt961s (accessed on August 06, 2016).
7. Luo, Q.-Z. et al. (1998). Shock formation in a negative ion plasma. *Physics of Plasmas* 5(8): 2868–2870.
8. Sturrock, P. A. (1994). *Plasma Physics: An Introduction to the Theory of Astrophysical, Geophysical & Laboratory Plasmas*. CA: Stanford University.
9. Ionization and plasmas. The University of Tennessee, Knoxville Department of Physics and Astronomy. http://csep10.phys.utk.edu/OJTA2dev/ojta/course1/light/atomframe/ionization_tl.html (accessed August 12, 2016).
10. Zavisa, J. (2002). How Lightning Works. HowStuffWorks. http://www.howstuffworks.com/lightning.html (accessed on November 04, 2002).
11. Plasma fountain Source, press release: Solar Wind Squeezes Some of Earth's Atmosphere into Space. http://pwg.gsfc.nasa.gov/istp/news/9812/solarwind.html (accessed on August 06, 2016).
12. Hazeltine, R. D. and Waelbroeck, F. L. (2004). *The Framework of Plasma Physics*. Boulder, CO: Westview Press.
13. Dendy, R. O. (1990). *Plasma Dynamics*. Oxford, UK: Oxford University Press.
14. Hastings, D. and Garrett, H. (2000). *Spacecraft-Environment Interactions*. New York: Cambridge University Press.
15. Yaffa Eliezer, S. E. (1989). *The Fourth State of Matter: An Introduction to the Physics of Plasma*. 226 pp, p. 5. Philadelphia, PA: Adam Hilger.
16. Bittencourt, J. A. (2004). *Fundamentals of Plasma Physics*. New York: Springer. p. 1.
17. Hong, A. (2000). Dielectric strength of air. *The Physics Factbook*. http://hypertextbook.com/facts/2000/AliceHong.shtml (accessed November 12, 2013).
18. Gurnett, D. A. and Bhattacharjee, A. (2005). *Introduction to Plasma Physics: With Space and Laboratory Applications*. Cambridge, UK: Cambridge University Press. p. 2;
    Scherer, K; Fichtner, H.; Heber, B. (2005). *Space Weather: The Physics Behind a Slogan*. Berlin: Springer. p. 138.
19. Mészáros, P. (2010). *The High Energy Universe: Ultra-High Energy Events in Astrophysics and Cosmology*. Cambridge, UK: Cambridge University Press.
20. Raine, D. J. and Thomas, E. G. (2010). *Black Holes: An Introduction*, p. 160. New York: Springer-Verlag.
21. NASA. NASA Astronomy Picture of the Day. nasa.gov (December 11, 2004). https://apod.nasa.gov/apod/ap141211.html (accessed August 12, 2016).
22. IPPEX. IPPEX Glossary of Fusion Terms. https://ippex.pppl.gov/fusion/glossary.html (accessed August 12, 2016).
23. Rogoff, G. (2008). Plasma and flames—The burning question, *Coalition for Plasma Science*, Rivernberg. P. (ed.). http://www.plasmacoalition.org/plasma_writeups/flame.pdf (accessed on August 06, 2016).

24. von Engel, A. and Cozens, J. R. (1976). Flame plasma, *Advances in Electronics and Electron Physics*, L. L. Marton (ed.), p. 99. New York: Academic Press.
25. Peratt, A. L. (1996). Advances in numerical modeling of astrophysical and space plasmas. *Astrophysics and Space Science* 242 (1–2): 93–163.
26. The Non-neutral Plasma Group at the University of California, San Diego, CA. Nonneutral plasmas. http://sdpha2.ucsd.edu (accessed on August 06, 2016).
27. Dwight, N. R. (1983). *Introduction to Plasma Theory*. Hoboken, NJ: John Wiley & Sons.
28. Grossmann, J. M. et al. (1995). Gap formation processes in a high-density plasma opening switch. *Physics of Plasmas* 2: 299.
29. Fitzpatrick, R. (2014). *Plasma Physics: An Introduction*. Boca Raton, FL: CRC Press.
30. Dickel, J. R. (1990). The filaments in supernova remnants: Sheets, strings, ribbons, or ? *Bulletin of the American Astronomical Society* 22: 832.
31. Grydeland, T. (2003). Interferometric observations of filamentary structures associated with plasma instability in the auroral ionosphere. *Geophysical Research Letters* 30 (6): 71-1–71-4.
32. Moss, G. D. et al. (2006). Monte Carlo model for analysis of thermal runaway electrons in streamer tips in transient luminous events and streamer zones of lightning leaders. *Journal of Geophysical Research* 111: A02307.
33. Doherty, L. R. and Menzel, D. H. (1965). Filamentary structure in solar prominences. *The Astrophysical Journal* 141: 251.
34. Hester et al. (1995). WFPC2 studies of the crab nebula. I. HST and ROSAT imaging of the synchrotron nebula. *The Astrophysical Journal* 448: 240–263.
35. Zhang, Y. A. et al. (2002). A rope-shaped solar filament and aIIIb flare. *Chinese Astronomy and Astrophysics* 26 (4): 442.
36. Boeuf, J. P. et al. (2010). Theory and modeling of self-organization and propagation of filamentary plasma arrays in microwave breakdown at atmospheric pressure.*Physical Review Letters* 104: 015002.
37. Talebpour, A. et al. (2000). Focusing limits of intense ultrafast laser pulses in a high pressure gas: Road to new spectroscopic source. *Optics Communications* 183 (5–6): 479.
38. Wikipedia. Rydberg. matter http://en.wikipedia.org/wiki/Rydberg_matter (last modified September 24, 2016).
39. Morfill, G. E. (2009). Complex plasmas: An interdisciplinary research field. *Review of Modern Physics* 81 (4): 1353–1404.
40. Alfvén, H. and Smårs, E. (1960). Gas-insulation of a hot plasma. *Nature* 188 (4753): 801–802.
41. Alfvén, H. (1976). *Evolution of the Solar System*. Washington, D.C.: Scientific and Technical Information Office, National Aeronautics and Space Administration.
42. Temam, R. (2001). *Navier–Stokes Equations, Theory and Numerical Analysis*, pp. 107–112. Chelsea, London: American Mathematical Society.
43. Charbonneau, P. (2014). Solar dynamo theory. *Annual Review of Astronomy and Astrophysics* 52: 251–290.
44. Zirker, J. B. (2002). *Journey from the Center of the Sun*, pp. 15–34. Princeton, NJ: Princeton University Press.
45. Nicole, M-V. (2007). *Basics of the Solar Winds*. New York: Cambridge University Press.
46. Tony, P. (25 October 2013). The efects of sace wather on aiation. *Science News*. NASA. http://science1.nasa.gov/science-news/science-at-nasa/2013/25oct_aviationswx/(accessed August 10, 2016).
47. Alfvén, H. (1971). Physics, space research and the origin of the solar system. *Science* 172: 991–994.
48. Hansteen, V. H. et al. (1997). The role of helium in the outer solar atmosphere. *The Astrophysical Journal* 482 (1): 498–509.

12

Climate Changes in the Solar System

12.1 Introduction

We dwell on the surface of the land; we sail across the surface of the sea, but we live at the bottom of the atmosphere and are becoming familiar with the varying climatology of our Earth from clear sunshine of blue skies to the dark shadows cast by the clouds. The different seasons of the year bring us a wide range of surface temperatures that are just right for making our beloved Earth habitable. Evidence shows that CO_2 is not the principle driver for warming of the planet. The other planets and moons show warming, despite the fact that they have no anthropogenic emissions of greenhouse gases, pointing to the Sun as the dominating influence in determining climate throughout the solar system. If this is happening throughout the solar system, it must be the Sun causing the rise in temperatures including here on Earth. We may take a wider view and move away from local earth-bound data to what is happening in our solar system. If there is sufficient evidence to support that climate change is occurring to planets that surround us, we have then no alternative but to make us believe that we might experience a serious problem in the future.

12.2 Planetary Climate Change

It is curious that the existing theory of solar system climatology largely depends on sparse information, which only provides scattered knowledge about the climates on other planets. Over the last 50 years, scientists using various instruments, including satellites, have measured the Sun's output and have found that the Sun's output has diminished slightly, radiating less heat. We can also look at the 22-year solar cycle, 11 years of an increase and 11 years of declining activity. Planetary research shows that there is enough evidence to show that it is not just the Earth that is going through climate change. Our entire solar system is in a period of a rapid change due to solar activity, the Sun being a long period variable star. The Sun, acting like the central brain of the solar system, orchestrates this change. This issue is now indeed a very challenging one to both scientists and engineers as well as for the Earth's inhabitants in general. As an interesting example, we may note that on May 19, 2011, both NASA's Cassini spacecraft and a European Southern Observatory ground-based telescope were able to track the growth of a giant early spring storm in Saturn's northern hemisphere (Figure 12.1). It was so powerful that it stretched around the entire planet. The rare storm had been wreaking havoc for months and shooting plumes of gas high into the atmosphere of the planet [1].

FIGURE 12.1
Giant early-spring storm in Saturn's northern hemisphere. (Courtesy of NASA JPL/University of Arizona, http://www.nasa.gov/mission_pages/cassini/multimedia/pia14119.html [accessed September 10, 2016].)

12.3 Changes in Earth's Climate

Figure 12.2 illustrates global land–ocean mean temperature changes from 1880 to 2013 relative to the mean temperatures from 1951 to 1980 [2]. The black dotted line in the figure represents the annual mean and the wavy line running through the center is the 5-year moving average, whereas there was uncertainty estimates in the early 1890's and around 1943 and 1944. The figure clearly exhibits the temperature anomaly.

Figure 12.3 exhibits the 10-year average (2000–2009) global mean temperature anomaly relative to the 1951–1980 mean. The largest temperature increases are in the Arctic and the Antarctic Peninsula [2].

This map shows the 10-year average (2000–2009) temperature anomaly relative to the 1951–1980 mean. Earth has been in radiative imbalance where less energy leaves the atmosphere than enters it. A major part of this extra energy has been absorbed by the oceans [3].

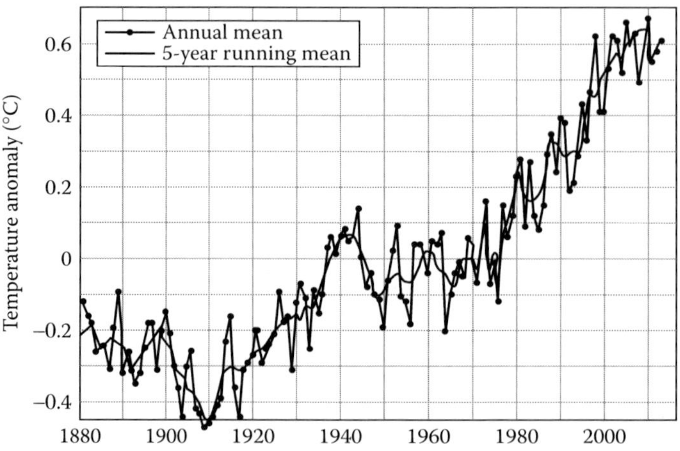

FIGURE 12.2
Global land–ocean temperature indexes. (Courtesy of NASA/GISS, GISS surface temperature analysis, http://data.giss.nasa.gov/gistemp/graphs_v3/ [accessed August 12, 2016].)

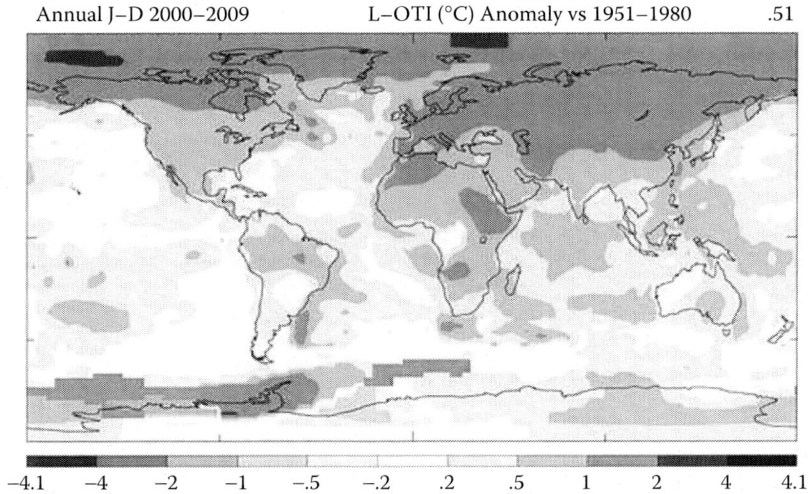

FIGURE 12.3
The largest temperature increases are in the Arctic and the Antarctic Peninsula. (Courtesy of NASA/GISS, GISS surface temperature analysis, http://data.giss.nasa.gov/gistemp/ [accessed August 12, 2016].)

It is likely that human activities significantly contributed to this increase in ocean heat content [4], but it is still not completely proved. As previously mentioned, the solar activity as well as geologic events plays a significant part in the temperature increase. The Earth's average surface temperature enhanced by (0.74 ± 0.18) °C over the period 1906–2005. The rate of warming during the last half of that period was almost double for that period as a whole, (0.13 ± 0.03) °C per decade versus (0.07 ± 0.02) °C per decade. The urban heat island effect is too small and is estimated to account for less than 0.002°C of warming per decade since 1900 [5]. Temperatures in the lower part of the troposphere have increased between 0.13 and 0.22°C per decade since 1979, according to satellite temperature measurements. Climate proxies reveal the temperature to have been relatively stable over the one or two thousand years before 1850, with regionally varying fluctuations such as the Medieval Warm Period and the Little Ice Age [6]. Figure 12.4 shows the energy change inventory (1971–2010).

The climate system can respond to changes in external forcings [7, 8], which can *push* the climate in the direction of warming or cooling [9]. Examples of external forcings include

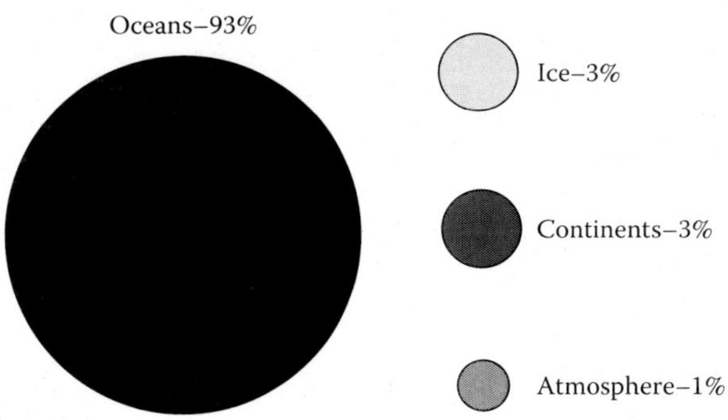

FIGURE 12.4
Energy change inventory (1971–2010) (Courtesy of Wikimedia Commons, https://commons.wikimedia.org/wiki/File:Energy_change_inventory,_1971-2010.svg [last modified May 16, 2016].)

changes in atmospheric composition, solar luminosity, volcanic eruptions, and variations in Earth's orbit around the Sun [10]. Orbital cycles change slowly over tens of thousands of years, and, at present, are in an overall cooling trend, which would be expected to lead toward a glacial period within the current ice age, but the twentieth-century instrumental temperature record reveals a sudden rise in global temperatures [11]. Reduced stratospheric ozone has a slight cooling influence on surface temperatures, whereas increased tropospheric ozone has a somewhat larger warming effect [12]. The influences of particulates including black carbon are prominent in the tropics and subtropics, particularly in Asia, whereas the effects of greenhouse gases have dominant role in the extra tropics and southern hemisphere [13].

12.3.1 Conflicting Claims on Global Warming

Earth is heating up lately, which is similar to Mars, Pluto, and other worlds in our solar system. This fact leads many scientists to speculate that a change in the Sun's activity is the common thread linking all these events. Some others argue that such claims are misleading and produce a false impression that rapid global warming (Figure 12.5), as Earth is experiencing, is a natural phenomenon. Evidence suggests that fluctuations in solar activity can affect climate on Earth and that it has done so in the past. A good number of climate scientists and astrophysicists agree that the Sun is not to blame (but there are many scientists who are in disagreement [14]) for the current and historically sudden uptick in global temperatures on Earth, which appears to be mostly a mess created by our own species.

12.3.2 Additional Thoughts on Earth's Climate

NOTE: The subject of climate change is a highly controversial one and still open to the pros and cons. The following are thoughts for the reader to think about and either accept, decline, or include in his or her own theories. For example, if we look at concrete or the dark asphalt streets, we can see that they retain the daytime heat even after the Sun has set. At the same time, our oceans and other bodies of water retain the warmth of the summer Sun. Let us also look at the increased building cycles of industry, such as apartment buildings, office buildings, factories, and so on, and also the depletion of our forests. All this contributes to

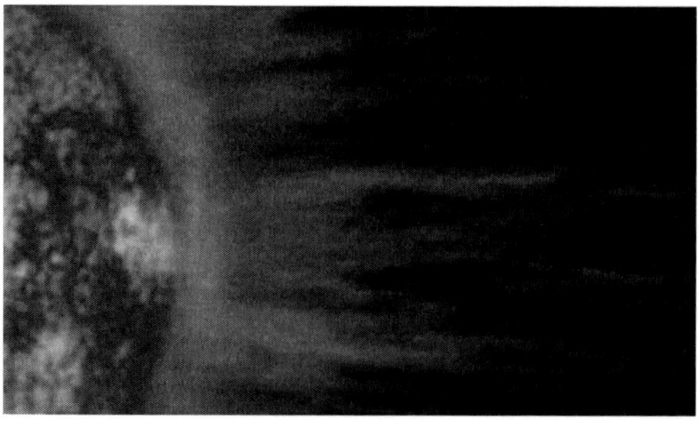

FIGURE 12.5
Conflicting claims on global warming. (Courtesy of NASA/ESA/SOHO, http://www.space.com/11506-space-weather-sunspots-solar-flares-coronal-mass-ejections.html [accessed August 12, 2016].)

the warming of the earth even with the reduced activity of the Sun. If we think about the previous paragraphs, we see the Sun in a declining phase, but the planet is still warm from the earlier increase, thus melting ice, and so on. When the heat dissipates, the earth will cool, thus the increase in colder temperatures and the increase in ice. So, we see a lagging in the response to the varying climate. Let us take it a step further, the possibility of geologic effects from volcanic forces: magma movements above and below the surface. For example, the super volcano theory at Yellowstone Park in Wyoming, the United States. Another great sources do heating and gasses that all contribute to the climate.

12.4 The Planets and Their Climates

As discussed in the previous paragraphs, it is hard to ignore the discrepancy between the facts that the Sun is causing the rise in temperatures. But if the Sun's output has leveled off or even diminished, then what is dominating other planets to warm up. The planets and moons, claimed to be warming, are roughly eight out of dozens of large bodies in the solar system. Some of the planets, like Uranus, may be cooling. Uranus has an 84-year orbit and 98° axial tilt, so its seasons are extreme. The outer planets have vastly longer orbital periods than that of Earth, so any climate change in them may be seasonal [15]. Saturn and its moons take 30 Earth years to orbit the Sun, and, so, three decades of observations equates to only 1 Saturnian year. Neptune has not yet completed a single orbit since its discovery in 1846. A round-up of the planets experiencing climate change is pointed out below [15]:

1. *Mars:* The notion that Mars is warming came from a conflation of weather and climate. Based on two pictures taken 22 years apart, assumptions were made that have not proved to be reliable. Presently, there is no evidence to support claims that Mars is warming at all. The considerable heating and cooling on the Earth and on Mars are always in parallel. On Earth, wobbles, known as Milankovitch cycles, are thought to contribute to the onset and disappearance of ice ages [16]. It is believed that what drives climate change on Mars are orbital variations, and the Earth also goes through orbital variations similar to that of Mars.
2. *Jupiter:* The notion that Jupiter is warming is primarily based on predictions because no warming has actually been observed. Climate models predict temperature increases along the equator and cooling at the poles. It is assumed that these changes will be catalyzed by storms that merge into one superstorm, inhibiting the ability of the planet to mix heat. Skeptical arguments have ignored the fact this is not a phenomenon we have found and that the modeled forcing is storm and dust movements, not changes in solar radiation.
   a. *Io:* Jupiter's moon Io is the most volcanically active body in our solar system. However, Io is a very tiny moon that is enormously influenced by the gravity of the giant planet Jupiter. The gravitational attraction of Jupiter and its other moons exert such strong *pulls* on Io that it deforms continuously from strong internal tides. These tides produce a tremendous amount of internal friction. This friction heats the moon and enables the intense volcanic activity. The heating caused by the tidal forces does change the climate of Io, to what extreme; there are no data that can be seen at this time.

3. *Neptune:* Changes in luminosity on the surface of both Neptune and its largest moon, Triton, have been observed, showing warming caused by increased solar activity. This brightening is due to the very slow seasonal changes of the planet. Neptune's summer in the southern hemisphere brings more sunlight, as it does every 164 years.
4. *Pluto:* The warming of Pluto is not yet understood. Pluto's existence has only been known for a third of its 248 year orbit, and it has never been visited by a space probe. The *evidence* for climate change consists of two observations made in 1988 and 2002, which is equivalent to observing the Earth's weather for just 3 weeks out of the year. Some theories suggest its highly elliptical orbit may play a role, as could the large angle of its rotational axis. It has also suggested that the length of Pluto's orbit is a key factor, as with Neptune. Sunlight at Pluto is 900 times weaker compared with that on Earth. Claims that solar system bodies are heating up due to increased solar activity are not proper. Only Pluto and Neptune are showing increased brightness. Heating attributed to other solar bodies has not been proved.

12.5 Changing Weather on Mars and Pluto

Mars may be going through a period of climate change, as suggested by new findings from NASA's Mars Odyssey orbiter. Odyssey has been mapping the distribution of materials on and near Mars' surface from 2002, nearly a full annual cycle on Mars [17]. Besides tracking seasonal changes, like the advance and retreat of polar dry ice, the orbiter is returning evidence useful for learning about longer-term dynamics. New gullies that did not exist in mid-2002 have appeared on a Martian sand dune. In addition, boulders tumbling down a Martian slope left tracks that were not there 2 years ago. New impact craters formed since the 1970s noticed changes to age-estimating models. South Pole has shrunk from the previous year's size, indicating a climate change. The extended life of NASA's Mars Global Surveyor has contributed to these new findings. Figure 12.6 shows that the Martian ices are shrinking dramatically.

Astronomers suggested that Pluto is undergoing global warming. Evidence points to a three-fold increase in the planet's atmospheric pressure during the past 14 years [18], suggesting a stark temperature rise. The change is similar to a seasonal event, much as seasons on Earth change as the hemispheres alter their inclination to the Sun.

12.5.1 Warming on Triton, Neptune's Largest Moon

Observations obtained by NASA's Hubble Space Telescope and ground-based instruments show that Neptune's largest moon, Triton, seems to have heated up considerably since the Voyager space probe visited it in 1989. The warming trend is causing part of Triton's surface of frozen nitrogen to turn into gas, hence making its thin atmosphere denser [19]. It may be noted that there may not be enough on Neptune's largest moon. According to observations made by NASA's Hubble Space Telescope and several ground-based instruments, temperatures on Neptune's largest moon has increased dramatically as the Voyager space probe swung by in 1989. It is believed that Triton's surface of frozen nitrogen is turning into gas, making its thin atmosphere denser by the day. Global warming on Triton is shown in Figure 12.7.

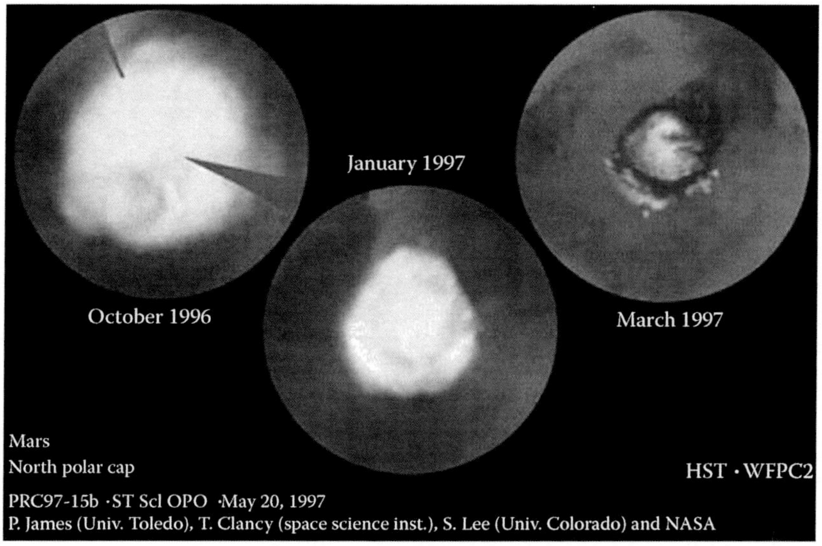

FIGURE 12.6
Martian ices shrinking dramatically. (Courtesy of NASA/ESA, Seasonal changes in Mars' north polar ice cap, http://spacetelescope.org/images/opo9715b/ [accessed August 12, 2016].)

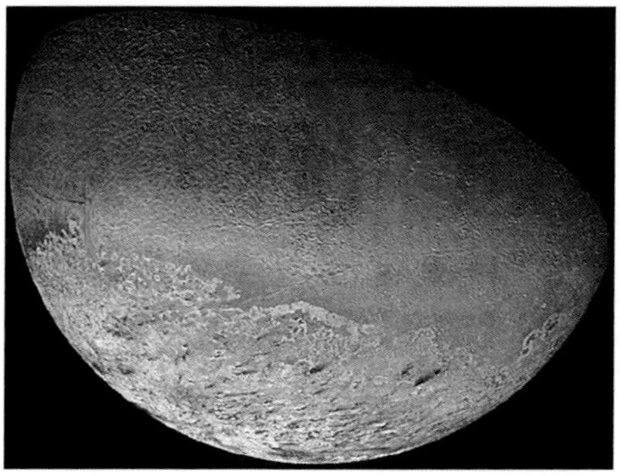

FIGURE 12.7
Possible global warming effects on Triton, Neptune's largest moon. (Courtesy of NASA/JPL/U.S. Geological Survey, PIA00317: Global color mosaic of Triton, http://photojournal.jpl.nasa.gov/catalog/PIA00317 [accessed August 12, 2016].)

12.6 Prediction of a Global Climate Change on Jupiter

In 1979, the Voyager spacecraft photographed Jupiter's atmosphere. The images are characterized by 12 zonal jet streams and about 80 vortices, the largest of which are the Great Red Spot and three White Ovals that were formed in the 1930s [20]. The Great Red Spot has been seen continuously since 1665 as well as the White Ovals [21], which are somewhat similar. The disappearance of the White Ovals (1997–2000) was unexpected but later explained to be part of a recurring climate cycle that will cause most of Jupiter's vortices to disappear within the next decade (Figure 12.8). It has been noted that Jupiter loses cloud belt, signaling a solar system change that may also be related to a climate change on Jupiter (Figure 12.8).

FIGURE 12.8
Global climate change on Jupiter and the loss of the cloud belts.

12.7 Satellites Measurement of Solar Activity

Output from the Sun has been precisely measured by satellites since 1978 [22]. Satellite observations of total solar irradiance from 1979 to 2006 are shown in Figure 12.9. All these measurements show that the Sun's output has not increased and so the warming cannot be attributed to an increase in solar energy reaching the Earth. Climate models have been utilized to find the role of the Sun in recent climate change [23] but are unable to reproduce the rapid warming found in recent decades when they only take into account variations in solar output and volcanic activity.

Models are able to simulate the observed twentieth-century changes in temperature, when they include external forcing, including human influences and natural forcing. Another line of evidence against the Sun having caused recent climate change comes from looking at how temperatures at different levels in the Earth's atmosphere have altered [24]. Models and observations suggest that greenhouse warming causes heating in the lower atmosphere but cooling in the upper atmosphere [24,25]. Depletion of the ozone layer has also resulted in a strong cooling effect in the stratosphere. If the Sun were responsible for observed warming, then warming of both the troposphere and stratosphere would be

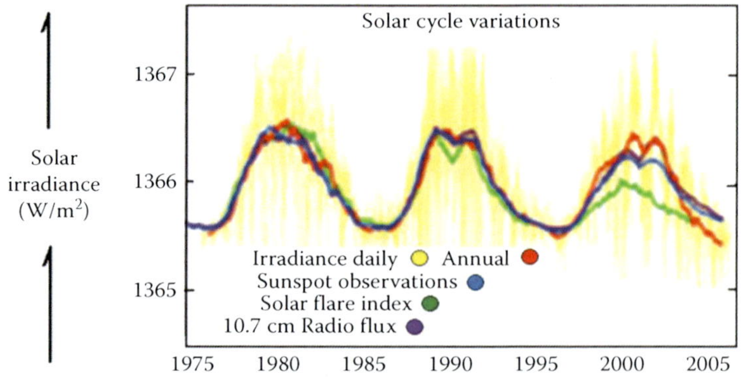

FIGURE 12.9
Satellite observations of total solar irradiance from 1979 to 2006.

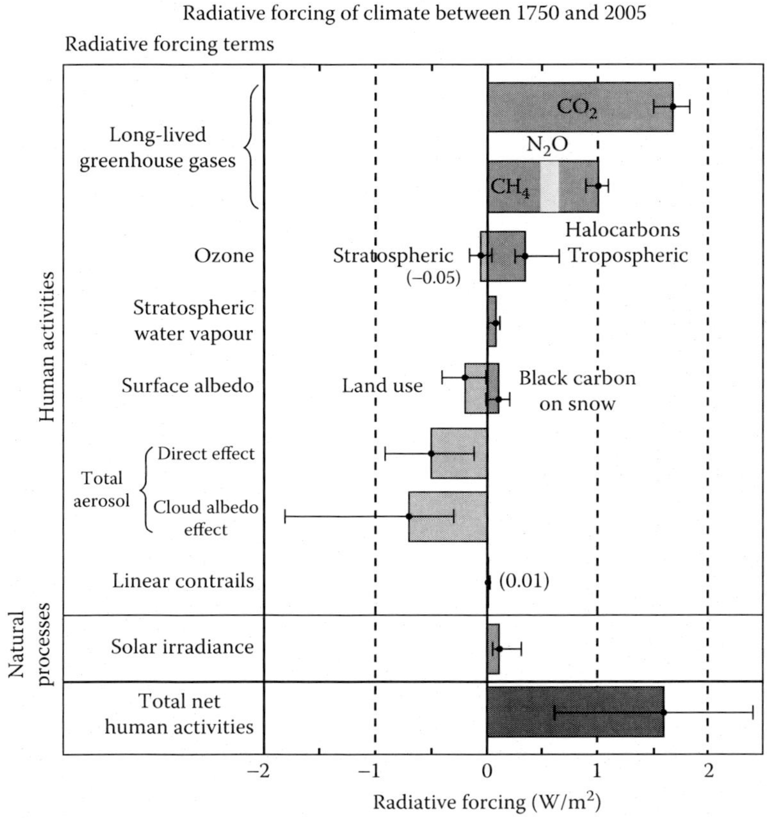

FIGURE 12.10
Contribution of natural factors and human activities to radiative forcing of climate change [25]. (Courtesy of CO2Now.org, https://www.ipcc.ch/publications_and_data/ar4/wg1/en/faq-2-1.html [accessed August 07, 2016].)

expected [26]. Contribution of natural factors and human activities to radiative forcing of climate change is shown in Figure 12.10.

In the figure, the radiative forcing values are for the year 2005, relative to the pre-industrial era (1750) [27]. The contribution of solar irradiance to radiative forcing is 5% the value of the combined radiative forcing owing to increases in the atmospheric concentrations of carbon dioxide, methane, and nitrous oxide [28].

12.8 Solar Irradiance Measurements

Full spectrum cavity radiometers are monitored to get the total electromagnetic radiation output from the Sun, called the solar constant [29,30]. The average value of the solar constant at the mean distance of the Earth from the Sun is about 1370 W/m^2. This has revealed an intrinsic solar variation of just under 0.1% in the last two decades (Figure 12.11). Further work has been attempted to identify the source of these changes [30].

It is believed that most solar irradiance variations originate in magnetic activity on the solar surface. Nearly 80% of the measured variation can be accounted for by the two main visible features on solar disk, such as sunspots and faculae. Sunspots are

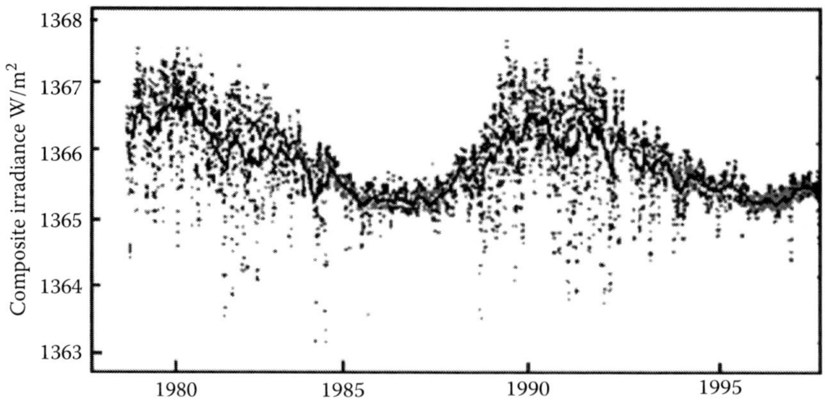

FIGURE 12.11
Solar irradiance variations as recorded by a composite of satellite sensors for the latter part of the twentieth century (the continuous line is an 81-day running average).

small dark areas on the solar disk, whereas faculae are bright areas that form complex networks. Both these phenomena occur around *active regions*, where the magnetic field strengths are very high. Around sunspots the magnetic field strength can rise to thousands of gauss, whereas the average field is of the order of 1 G. The presence of faculae is highly correlated with the presence of sunspots and so with the sunspot number. Over long timescales of months or more, it appears that increased irradiance is due to the bright faculae that exceed the deficit, due to the darker and cooler sunspots as seen in Figure 12.12. Thoughts and consensus are that solar irradiance from one minimum to another [31] (an 11-year period), and when no spots or faculae are visible, a small change such as a nonmagnetic variation is within the noise level (maybe 0.02%). Therefore, it is

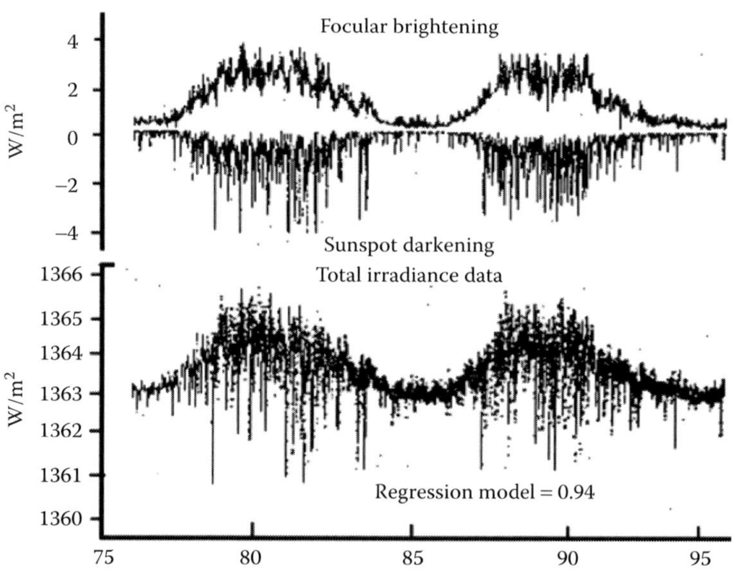

FIGURE 12.12
Solar satellite irradiance data showing modeled contributions due to sunspots and faculae separately.

assumed that the sunspot number is a good proxy for solar irradiance changes and historical variations. This number has led to a belief that the total variation in solar irradiance over timescales of thousands of years is unlikely to exceed 1%.

An alternative way to approach the problem is to observe a sample of Sun-like stars for irradiance fluctuations. It has been reported that the average G-type (Sun-like) star shows a 4% irradiance variation over a timescale of decades [32]. Also, this variation may be due to nonmagnetically related activity and is thus a phenomenon that we do not see at present in our Sun.

12.9 Radiation Fluctuation of the Sun

The radiation output of the Sun does fluctuate over the 11-year solar cycle. However, the change is only about one-tenth of 1%, which is not enough to affect Earth's climate in dramatic ways, and certainly not fully responsible of our planet's current warming trend. Small changes in solar output and variations from one decade to the next are only on the order of a fraction of a percent and can provide a detectable signal in the surface temperature record.

12.10 The Maunder Minimum

This is not to say that solar fluctuations never influence climatology of our earth in substantial ways. During a 75-year period beginning in 1645, astronomers detected practically no sunspot activity on the Sun [33]. This is called the *Maunder Minimum* [34]. This event coincided with the coldest part of the Little Ice Age, a 350-year cold spell that gripped much of Europe and North America. However, subsequent investigations have cast doubt on this relationship. Also, new estimates of the total change in the brightness of the Sun, during the Maunder Minimum, conclude that it was only fractions of a percent and perhaps not enough to create the global cooling commonly attributed to it.

12.11 Similarities of Mars, Venus, and Earth

Mars' climate is very similar to that of Earth, with water ice and carbon dioxide caps at the poles, seasonal snow, and dust storms. This is because Mars has a similar axial tilt as the Earth, which creates similar seasonal temperature variations [35]. However, the colder average temperatures and the thin atmosphere mean liquid water can only exist on the surface around mid-summer and at the lowest elevations. The thin atmosphere means the surface is exposed to intense UV radiation. From that point of view, Mars may not be habitable today. In Figure 12.13, the Hubble image of Mars show the planet engulfed in a global dust storm, with its polar caps peeking through.

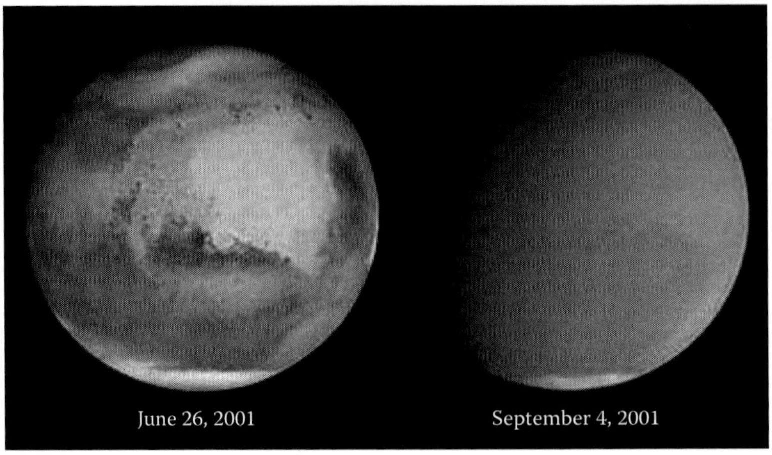

FIGURE 12.13
Hubble image of Mars engulfed in a global dust storm, with its polar caps peeking through. (Courtesy of NASA, J. Bell (Cornell), M. Wolff (Space Science Institute), Hubble Heritage Team (STScI/AURA), Mars dust storm, https://airandspace.si.edu/multimedia-gallery/3941hjpg [accessed August 12, 2016].)

Some evidences point to Mars being wet and warm early in its history. Water-carved channels, minerals formed by interaction with groundwater, river delta deposits, and a shoreline all the way around the northern lowlands, all point to liquid water on the surface sometime in the distant past. So, why was Mars so much warmer and wetter than it is today and why did it change were some of the unanswered fundamental questions about climate change. Early Mars had a thicker atmosphere, made of mostly CO_2 like it is today, which would have warmed the surface through the greenhouse effect. One way to understand the early climate of Mars is to study the oldest rocks and to look at more recent climate changes, which are preserved in the polar ice caps. Just as ice cores on Earth provide a record of annual changes in climate, the thick stacks of polar ice on Mars have internal layering, which indicates that they were built up one layer at a time.

12.11.1 Venus

Venus is almost the same size as Earth and only slightly closer to the Sun. However, its axis does not tilt relative to the Sun, and, so, it does not have seasons like Earth and Mars. Venus is dry and hot, despite its clouds reflecting 80% of the sunlight that it receives, as it very effectively traps the remaining 20%. We know less about ancient Venus because the surface of Venus is relatively young, only ~1 billion years old. With 100 times the atmosphere of Earth, its runaway greenhouse effect boiled all the water off the surface [35]. Some of that water is bound to sulfur that mixes and forms sulfuric acid clouds that circle the planet. These clouds are broken down in the atmosphere and eventually removed by the solar wind. Figure 12.14 shows the swirling cloud formations.

Due to many similarities with Earth, Venus is formed from the some of the same material. The key to their different climates may be for Earth's plate tectonics, which buries carbon-rich sedimentary rocks. To understand when and why it became so hot and dry, Venus keeps all of its CO_2 in the atmosphere. In addition, isotopic ratios of elements like carbon and hydrogen also point the way.

FIGURE 12.14
Clouds swirl around the south pole of Venus. (Courtesy of European Space Agency, https://airandspace.si.edu/stories/editorial/climate-change-solar-system [accessed August 12, 2016].)

12.12 Similarities of Titan and Earth

Titan is interesting, as its atmosphere is mainly made by nitrogen, just like the Earth. The remainder is mostly methane, which breaks down in the atmosphere and has to be replenished every ~50 million years. This implies some ongoing process. Titan gets 100 times less sunlight than the Earth. So, its surface is frigid, and cold enough that water ice is as hard as rock [35]. Titan is not currently habitable for life as we know, but it is the only other place in the solar system with rain. Many scientists believe that in another 5 billion year, the Sun will become a red giant star, and Titan probably will be warm enough to have liquid water on its surface, making it habitable at last. Titan, the only moon in the solar system with a thick atmosphere (outermost layer), is shown in Figure 12.15.

12.13 Solar Variability and a Balance with Climate Change

The Sun has powered everything on Earth since life began, including its climate. The Sun has also delivered an annual and seasonal impact, changing the character of each hemisphere as the Earth's orientation shifts through the year. However, from the Industrial Revolution, new forces have started to exert large-scale influence on Earth's climate. For the last 50 years or more, NASA scientists have investigated the unique relationship between the Sun and Earth. Using space-based tools, like the Solar Radiation and Climate Experiment, they have investigated how much solar energy illuminates Earth and explored what happens to that energy once it penetrates the atmosphere. The amount of energy that reaches the Earth's outer atmosphere is called the total solar irradiance. This

FIGURE 12.15
Titan is the only moon in the solar system with a thick atmosphere, imaged by Cassini. (Courtesy of NASA/JPL/Space Science Institute, Titan, http://solarviews.com/eng/titan.htm [accessed August 12, 2016].)

is highly variable over many different timescales, ranging from seconds to centuries due to changes in solar activity. The Sun goes through roughly an 11-year cycle (full cycle of 22 years) of activity, from stormy to quiet and back again. As mentioned earlier, solar activity frequently occurs near sunspots, dark regions on the Sun caused by concentrated magnetic fields. The solar irradiance measurement is much higher during solar maximum, when sunspot cycle and solar activity is high, versus solar minimum, when the Sun is quiet and there are usually no sunspots [36].

12.13.1 Solar Energy

The Sun's energy received at the top of Earth's atmosphere has been measured by satellites since 1978. It has followed its natural 11-year cycle of small ups and downs, but with no net increase as evident from the bottom as seen in Figure 12.16. However, over the same period, global temperature has raised remarkably as we can find from the top of the figure. Climate is influenced by natural changes that affect how much solar energy reaches Earth. These changes include changes within the Sun and changes in Earth's orbit. Changes occurring in the Sun itself can affect the intensity of the sunlight that reaches the surface of the Earth [37].

Figure 12.17 shows some of the recent effects of solar energy (heat). Changes in the Sun's intensity have influenced Earth's climate in the past, for example, the so-called *Little Ice Age* between the seventeenth and nineteenth centuries may have been partially caused by a low solar activity phase from 1645 to 1715, which coincided with cooler temperatures. The *Little Ice Age* refers to a slight cooling of North America, Europe, and probably other areas around the globe.

However, solar activity has been relatively constant, aside from the 11-year cycle, since the mid-twentieth century and, therefore, does not explain the recent warming of Earth. Similarly, changes in the shape of Earth's orbit as well as the tilt and position of Earth's axis affect temperature on relatively long timescales and hence cannot explain the recent warming. Earth is only one small recipient of the Sun's energy. The Sun's rays extend far out into the solar system, illuminating all the other planets. Using SORCE [38], scientists have learned that about 1361 W/m^2 of solar energy reaches Earth's outermost atmosphere

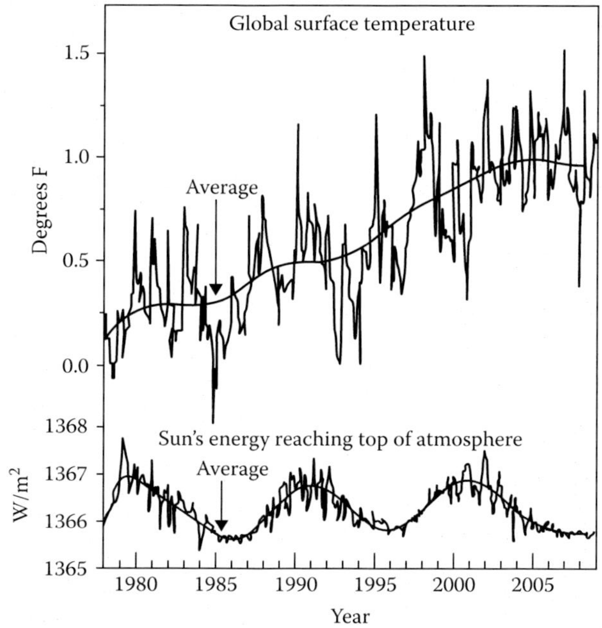

FIGURE 12.16
Changes in the shape of Earth's orbit as well as the tilt and position of Earth's axis can also affect the amount of sunlight reaching Earth's surface. (Courtesy of US Environmental Protection Agency, Causes of climate change, https://www3.epa.gov/climatechange/science/causes.html [accessed August 12, 2016].)

FIGURE 12.17
The role of the Sun's energy changes in solar energy continue to affect climate.

during the period of Sun's quietest period. But when the Sun becomes active, 1.3 W/m^2 (0.1%) more energy reaches Earth. Earth's climate depends on the delicate balance between incoming solar radiation, and outgoing thermal radiation as well as on the composition of Earth's atmosphere. Even small changes in these parameters can affect climate. Around 30% of the solar energy that strikes Earth is reflected back into space. Clouds, atmospheric aerosols, snow, ice, sand, ocean surface, and even rooftops play a role in deflecting the incoming rays. The rest 70% of solar energy is absorbed by land, ocean, and atmosphere. Greenhouse gases block nearly 40% of outgoing thermal radiation that emanates from Earth. The resulting imbalance between incoming solar radiation and outgoing thermal radiation will likely cause Earth to heat up over the next century, accelerating the melting polar ice caps, causing sea levels to rise and increasing the probability significantly of violent and more active global weather patterns.

References

1. G. D. Krebs. Cassini / Huygens. Gunter's Space Page. http://science.nasa.gov/science-news/science-at-nasa/2011/19may_saturnstorm/ (retrieved on June 15, 2016).
2. NASA/GISS. GISS surface temperature analysis. http://data.giss.nasa.gov/gistemp/graphs_v3/ (accessed on August 07, 2016).
3. IPCC. Climate Change 2014: Mitigation of climate change—Summary for policymakers, trends in stocks and flows of greenhouse gases and their drivers, in *Contribution of Working Group III to the Fifth Assessment Report of the Intergovernmental Panel on Climate Change*, O. Edenhofer, R. Pichs-Madruga, Y. Sokona, E. Farahani, S. Kadner, K. Seyboth, A. Adler, I. Baum, S. Brunner, P. Eickemeier, B. Kriemann, J. Savolainen, S. Schlömer, C. von Stechow, T. Zwickel, and J. C. Minx (eds.). Cambridge, UK and New York, NY: Cambridge University Press. p. 6.
4. Rhein, M. et al. (June 7, 2013). Box 3.1, in: Chapter 3: Observations: Ocean (final draft accepted by IPCC Working Group I). pp. 11–12. In IPCC AR5 WG1 2013.
5. IPCC (November 11, 2013). D.3 detection and attribution of climate change, in *Summary for Policymakers* (finalized version). p. 15. In IPCC AR5 WG1 2013.
6. Trenberth, K. E. et al. (2007). Ch. 3: Observations: Atmospheric surface and climate change, Section 3.2.2.2: Urban heat islands and land use effects. p. 244. In Fourth Assessment Report of the Intergovernmental Panel on Climate Change, 2007.
7. Meehl, G. A. et al. (2005). How much more global warming and sea level rise. *Science* 307 (5716): 1769–1772.
8. Hansen, J. et al. (2002). Climate forcings in Goddard Institute for space studies SI2000 simulations. *Journal of Geophysical Research* 107(D18): 4347. doi:10.1029/2001JD001143.
9. Pew Center on Global Climate Change/Center for Climate and Energy Solutions. (September 2006). *Science Brief 1: The Causes of Global Climate Change*. Arlington, VA: Center for Climate and Energy Solutions. p. 2.
10. Division on Earth and Life Studies, US National Research Council (US NRC). (2012). *Climate Change: Evidence, Impacts, and Choices*, 36pp. Washington, D.C: The National Academies Press.
11. Hegerl, G. C. et al. (2007). Chapter 9: Understanding and Attributing climate change, Section 9.4.1.5: The influence of other anthropogenic and natural forcings, in fourth Assessment Report of the Intergovernmental Panel on Climate Change, 2007, pp. 690–691.
12. Npr.org (2010). Talk of the nation: Americans fail the climate quiz. http://www.npr.org/2010/12/03/131785444/Americans-Fail-The-Climate-Quiz(retrieved on December 27, 2011).
13. Ramanathan, V. et al. (2008). Report summary, in *Atmospheric Brown Clouds: Regional Assessment Report with Focus on Asia*. United Nations Environment Programme.
14. Anderegg, W. R. L. et al. (2010). Expert credibility in climate change. *Proceedings of the National Academy Sciences of the United States of America*, 107(27): 12107–12109.
15. Lissauer, J. J. and Stevenson, D. J. (2006). Formation of giant planets. in *Protostars and Planets V*, B. Reipurth, D. Jewitt, and K. Keil (eds.). Tucson: University of Arizona Press.

16. Williams, D.M. and Pollard, P. (2002). Earth-like worlds on eccentric orbits: excursions beyond the habitable zone. *International Journal of Astrobiology* 1: 21–29. doi:10.1017/s1473550402001064.
17. Vidal-Madjar, A. et al. (1978) Is the solar system entering a nearby interstellar cloud? *The Astrophysical Journal* 223: 589-600.
18. Buie, M. W. et al. (2010). Pluto and Charon with the hubble space telescope: I. Monitoring global change and improved surface properties from light curves. *The Astronomical Journal* 139(3): 1117–1127.
19. Cruikshank, D. P. et al. (2006). Ethane on Pluto and Triton. *Bulletin of the American Astronomical Society*, Vol. 38, p.518. American Astronomical Society, DPS meeting #38, #21.03.
20. Marcus, P. S. (2004). Prediction of a global climate change on Jupiter. *Nature* 428: 828–831.
21. Reese, E. J.; Gordon, G. H. (1966). Recent measures of the latitude and longitude of Jupiter's red spot. *Icarus* 5(1-6): 266–273.
22. U.S. Environmental Protection Agency. (2009). 3.2.2 Solar Irradiance, in *Volume 3: Attribution of Observed Climate Change*. Endangerment and Cause or Contribute Findings for Greenhouse Gases under Section 202(a) of the Clean Air Act. EPA's Response to Public Comments. US Environmental Protection Agency. Archived from the original June 16, 2011.
23. US National Research Council (US NRC) (2008). *Understanding and Responding to Climate Change: Highlights of National Academies Reports*, p. 6. Washington, D.C.: The National Academies Press.
24. Hegerl, G. C. et al. (2007). Understanding and attributing climate change, frequently asked question 9.2: can the warming of the 20th century be explained by natural variability? in *Climate Change 2007: The Physical Science Basis. Contribution of Working Group I to the Fourth Assessment Report of the Intergovernmental Panel on Climate Change*, S. Solomon, D. Qin, M. Manning, Z. Chen, M. Marquis, K. B. Averyt, M. Tignor, and H. L. Miller (eds.). Cambridge, UK: Cambridge University Press. p. 702.
25. Simmon, R. and Herring, D. (November 2009). Notes for slide number 7, titled 'Satellite evidence also suggests greenhouse gas warming,' in presentation, 'Human contributions to global climate change'. Presentation library on the U.S. National Oceanic and Atmospheric Administration's Climate Services website. Archived from the original on July 3, 2011.
26. Randel, W. J. et al. (2009). An update of observed stratospheric temperature trends. *Journal of Geophysical Research* 114 (D2): D02107.
27. Ramanathan, V. et al. (2008). Part III: Global and future implications, in *Atmospheric Brown Clouds: Regional Assessment Report with Focus on Asia*. United Nations Environment Programme, Nairobi, Kenya.
28. IPCC, *Summary for Policymakers, Human and Natural Drivers of Climate Change*, Figure SPM.2, in IPCC AR4 WG1 2007.
29. Balasubramaniam, K. S. et al. (eds.). (1998). *Synoptic Solar Physics*. San Francisco, CA: Astronomical Society of the Pacific.
30. Lean, J. and Frohlich, C. (1998). Solar total irradiance variations, *in Synoptic Solar Physics*, ASP Conference Series, K. Balasubramaniam, J. Harvey, and D. Rabin (eds.), Vol. 140, pp. 281–292.
31. Willson, R. C. (1997). Total solar irradiance trend during solar cycles 21 and 22. *Science* 277: 1963.
32. Lockwood, G. W. (1998). Luminosity and chromospheric variations of solar analog stars, in *Synoptic Solar Physics*, ASP Conference Series, K. S. Balasubramaniam, J. W. Harvey, and D. M. Rabin (eds.), Vol. 140, pp. 261–269.
33. Weber, B. and J. A. Eddy (2009). Solar Detective, Dies at 78. *New York Times*, June 17, 2009. http://www.nytimes.com/2009/06/18/us/18eddy.html?_r=0 (accessed on August 07, 2016).
34. Eddy, J. A. (1976). The Maunder Minimum. *Science* 192(4245): 1189–1202.
35. Goodman, J. C. (September 22, 1997). The Past, Present, and Possible Future of Martian Climate. MIT. Archived from the original on November 10, 2010. https://translate.googleusercontent.com/translate_c?depth=1&hl=en&prev=search&rurl=translate.google.co.in&sl=eo&u=https://web.archive.org/web/20101110051940/http://www.mit.edu/people/goodmanj/terraforming/terraforming.html&usg=ALkJrhhVFOhAmudmnuaVLHMw2Trihca9_A (accessed on August 07, 2016).

36. Willson, R. C. (2014). ACRIM3 and the total Solar irradiance database. *Astrophysics and Space Science* 352: 341–352.
37. USGCRP. (2009). *Global Climate Change Impacts in the United States,* Thomas R. Karl, Jerry M. Melillo, and Thomas C. Peterson (eds.). United States Global Change Research Program. New York, NY: Cambridge University Press.
38. Burlaga, L.F. (2001). Magnetic fields and plasmas in the inner heliosphere: Helios results. *Planetary and Space Science* 49(14–15): 1619–1627.

13

Observed Planetary Climate Changes

13.1 Introduction

The general definition of climate change means a variation in the statistical properties of the climate system when considered for a long period of time, regardless of cause [1]. Sometimes, the term *climate change* can also be used to refer to the climate change caused by human activity, as opposed to changes in climate that may have resulted as part of Earth's natural processes [2]. NASA satellites have verified some of the characteristics of the climate change on Earth that have largely supported the theories and data that aid scientists in determining the big picture about our planet and its climate on a global scale. Going through the climate data collected over many years, a clear signal of a changing climate is noticed. The graph shown in Figure 13.1 is based on the comparison of atmospheric samples contained in ice cores and more recent direct measurements. It provides evidence that atmospheric CO_2 has increased since the industrial revolution.

On Earth, climate change has been observed and it has been estimated that it is a lasting change in the statistical distribution of weather patterns, ranging from decades to millions of years. It may be a variation in average weather conditions or in the distribution of weather around the average conditions. Climate change is noted due to many factors such as biotic processes, solar radiation (received by Earth), plate tectonics, and volcanic eruptions. Some typical human activities have also been noted for recent climate change and possibly been responsible for *global warming* [1]. In recent years, several reports have been circulated about global warming occurring on other planets and their satellites in the solar system. This chapter reviews the actual observations and the estimations regarding these planets and moons related to the climate changes with time.

13.2 Earth, Climate, and Causes

The Earth's climate has been altered throughout the history. As of the last 650,000 years, there have been seven cycles of glacial advance and retreat, with the abrupt end of the last ice age around 7000 years ago. This end marked the start of the modern climate era and of human civilization. Most of the climate changes are associated with very small variations in Earth's orbit that change the amount of solar energy our planet receives. The present day warming trend has significance because some of it is very likely human-induced (including natural processes, i.e., Geologic, etc.) and proceeding at a rate that has been unprecedented in the past 1300 years [3]. The rate at which energy is received from the Sun and the

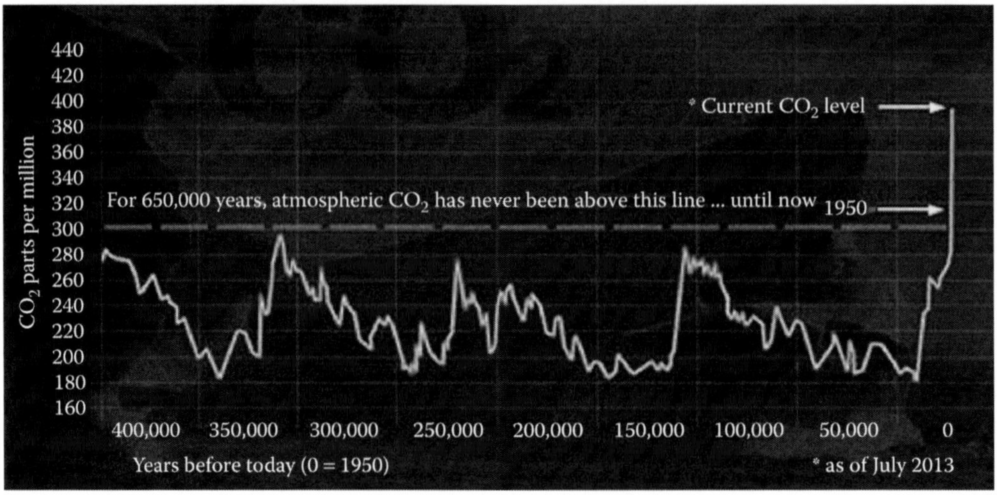

FIGURE 13.1
Evidence that atmospheric CO_2 has increased since the industrial revolution. (Data from J. R. Petit et al., Climate and atmospheric history of the past 420,000 years from the Vostok ice core, Antarctica, *Nature*, 399, 429–436, 1999.)

rate at which it is lost to space find the equilibrium temperature and climate of Earth. This energy is derived from the distribution of winds, ocean currents, and other mechanisms that affect the climates of different regions. Factors that are responsible to shape climate are called *forcing mechanisms* [4]. This mechanism includes processes such as changes in solar radiation, changes in the Earth's orbit, mountain-building and continental drift, and also variations in greenhouse gas concentrations. There are certain climate change feedbacks that can either amplify or diminish the initial forcing. Some parts of the climate system, such as the oceans and ice caps, respond slowly in reaction to climate forcing, while a few others respond rapidly. The total solar irradiance or flux of solar energy arriving at the Earth is well determined and has varied by only about 0.1% in the last few decades. Believers have pointed out that this can directly produce a temperature change of about 0.07°C on Earth [5]. This is considered as the only possible Sun–climate link. Evidences support the hypothesis that solar influences have indirect effects on the Earth's climate, with the potential of contributing significantly to the observed climate change. Some solar output changes by more than 0.1%, such as UV solar radiation and magnetic activity. One proposed mechanism suggests that variations in solar magnetic activity affect the amount of galactic cosmic rays reaching the Earth. This activity indirectly influences climate due to cosmic rays affecting the formation of clouds that reflect more or less sunlight back to space (depending on solar activity), causing a change in the Earth's climate. Such indirect mechanisms could in principle produce most of the current observed change in global temperature. Investigations show correlations supporting these claimed mechanisms but parts of the casual link have yet to be confirmed.

13.2.1 Solar Activity as Related to Global Climate

The Sun is considered as the predominant source of energy input to the Earth. Both long- and short-term changes in solar intensity are known to affect global climate. Changes in solar activity during the last several centuries are found to be based on observations of sunspots and beryllium isotopes [6]. The period of extraordinarily few sunspots in the

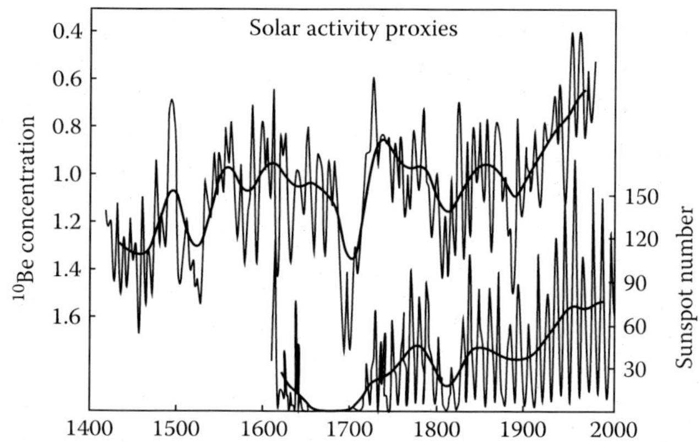

FIGURE 13.2
Variations in solar activity during the last several centuries based on observations of sunspots and beryllium isotopes. (Courtesy of Wikipedia, Maunder minimum, https://en.wikipedia.org/wiki/Maunder_Minimum [last modified July 5, 2016].)

late seventeenth century was the Maunder minimum. Figure 13.2 reveals the variations in solar activity during the last several centuries based on observations of sunspots and beryllium isotopes.

It is generally believed that 3 to 4 billion years ago the Sun emitted only 70% as much power as it does today. If the atmospheric composition had been the same as today, liquid water should not have existed on Earth. Over the following nearly 4 billion years, the energy output of the Sun has increased and the atmospheric composition has changed. The Great Oxygenation Event,* which is the oxygenation of the atmosphere around 2.4 billion years ago, was the most significant alteration. Over the next 5 billion years as the Sun ages and progresses toward a red giant phase and then a white dwarf phase, there will be a large variation in climate. The red giant phase will possibly end any life existing on Earth. Interestingly, an investigation [7] concludes that the effects of solar variability on temperature throughout the atmosphere may be contrary to current expectations. In an August 2011 Press Release [8], The European Organization for Nuclear Research, known as CERN, announced the initial results from its CLOUD experiment in the publication in *Nature*. The findings indicate that ionization from cosmic rays significantly enhances aerosol formation in the presence of sulfuric acid and water. In the lower atmosphere where ammonia is also essential, this presence is insufficient to account for aerosol formation and additional trace vapors being involved. The next step is to find out more about these trace vapors, including whether they are of natural or of human origin.

13.2.2 Other Associated Causes

In general, scientists define the five components of Earth's climate system to include atmosphere, hydrosphere, cryosphere, lithosphere, and biosphere [9]. Natural changes in the climate system, the so-called internal forcing, result in internal *climate variability* [10]. Examples include the type and distribution of species and changes in ocean currents.

* In the 1860s, physicist John Tyndall recognized the Earth's natural greenhouse effect and suggested that slight changes in the atmospheric composition could bring about climatic variations. In 1896, a seminal paper by Swedish scientist Svante Arrhenius first speculated that changes in the levels of carbon dioxide in the atmosphere could substantially alter the surface temperature through the greenhouse effect [59].

FIGURE 13.3
Republic of Maldives: vulnerable to a sea level rise.

Forcing mechanisms may be either *internal* or *external*. Internal forcing mechanisms are natural processes within the climate system itself, such as thermohaline circulation (THC), which is a part of the large-scale ocean circulation that is driven by global density gradients created by surface heat and freshwater fluxes [11]. Whether the initial forcing mechanism is internal or external, the response of the climate system might be changeable causing an abrupt response of the system though the full response to forcing mechanisms might not be developed completely for centuries or more. Changes in climate have happened very rapidly, geologically, in tens of years and not in millions or even thousands [12]. It is important to note that in the last century the global sea level rose about 17 cm. However, the rate in the last decade was approximately double that in the last century [13]. Figure 13.3 shows a very interesting photograph of the Republic of Maldives: vulnerable to sea level rise due to melting of glaciers.

13.2.3 Earths Movements Related to Atmospheric Changes

The Milankovitch theory [14] describes the collective effects of changes in the Earth's movements on its climate. In addition, slight changes in the orbit of the Earth have led to changes in the seasonal distribution of sunlight reaching the Earth's surface and how it is distributed across the globe. There is a very little variation in the (area-averaged) annually averaged sunshine. But there can be strong variations in the geographical and seasonal distribution. The three types of orbital variations are the (1) changes in Earth's eccentricity, (2) changes in the tilt angle of Earth's axis of rotation, and (3) precession of Earth's axis. All these are combined to produce Milankovitch cycles that have a large impact on climate. They are most notable for their correlation with glacial and interglacial periods [15], their correlation with the advance and retreat of the Sahara [16], and for their appearance in the geological record [17]. The Intergovernmental Panel on Climate Change (IPCC) has noted that CO_2 followed temperature change with a lag of hundreds of years and that as a feedback it amplified the temperature change [18]. The depths of the ocean have a lag time in changing temperature. On seawater temperature variation, the solubility of CO_2 in the oceans changes and also other factors impacting air–sea CO_2 exchange [19]. Figure 13.4 reveals the Milankovitch cycles from 800,000 years ago in the past to 800,000 years in the future, while Figure 13.5 shows the changes in CO_2, temperature, and dust from the Vostok ice core over the last 450,000 years.

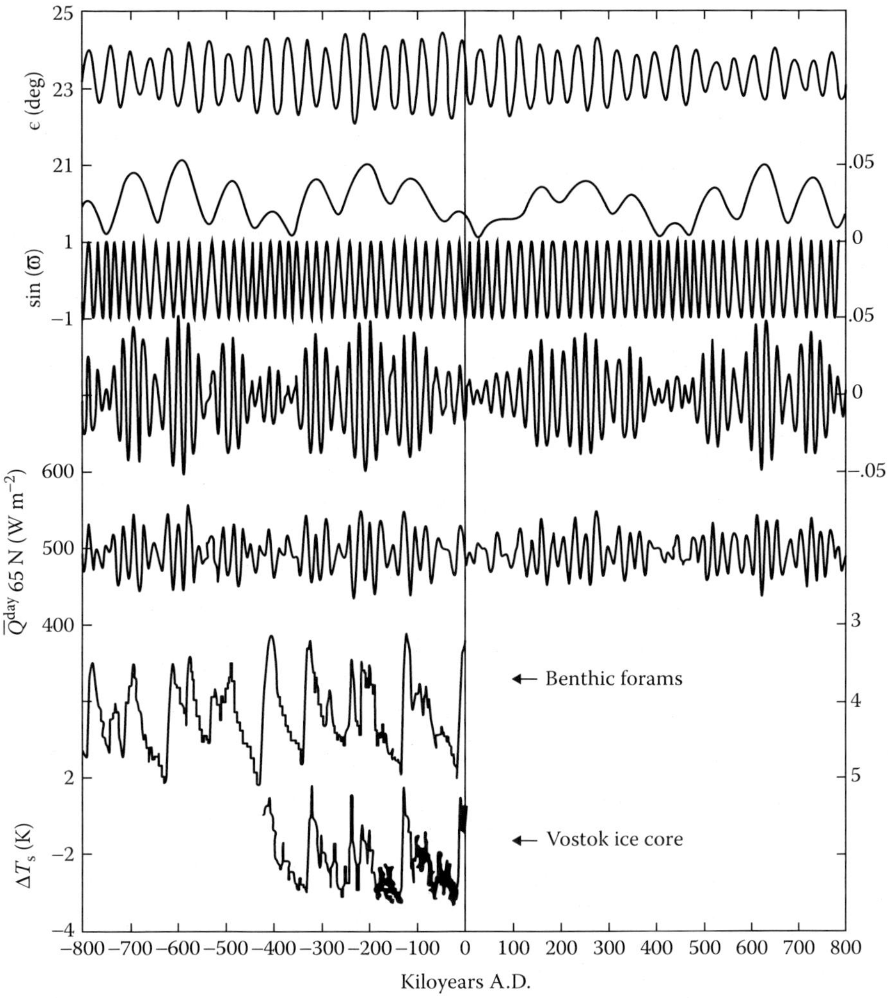

FIGURE 13.4
Milankovitch cycles from 800,000 years ago in the past to 800,000 years in the future.

13.2.4 Ocean Temperature Variability

The ocean is a basic and important part of the climate system. Its various changes with very high thermal inertia occur at longer timescales than those in the atmosphere. Figure 13.6 reveals the Pacific decadal oscillation from 1925 to 2000.

Short-term fluctuations (years to a few decades) such as the El Niño-Southern oscillation, the Pacific decadal oscillation, the North Atlantic oscillation, and the Arctic oscillation represent the climate variability rather than the climate change. On longer timescales, alterations to ocean processes, such as THC, play a vital role in redistributing the heat by carrying out a very slow and extremely deep movement of water and the long-term

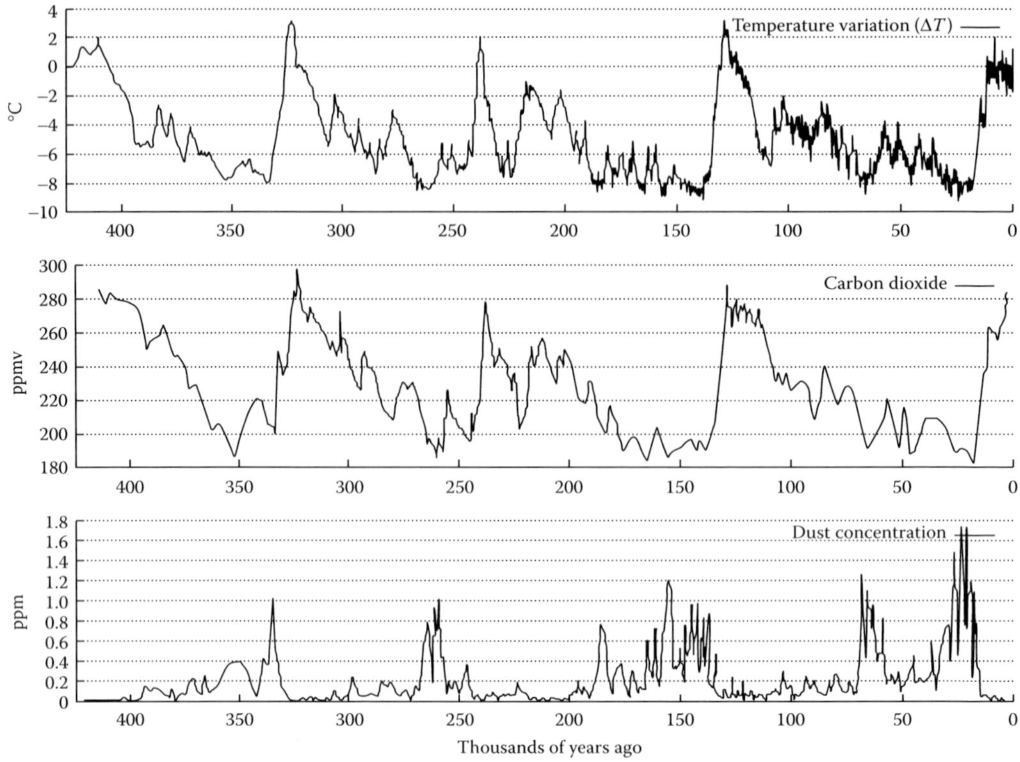

FIGURE 13.5
Variations in CO_2, temperature, and dust from the Vostok ice core over the last 450,000 years.

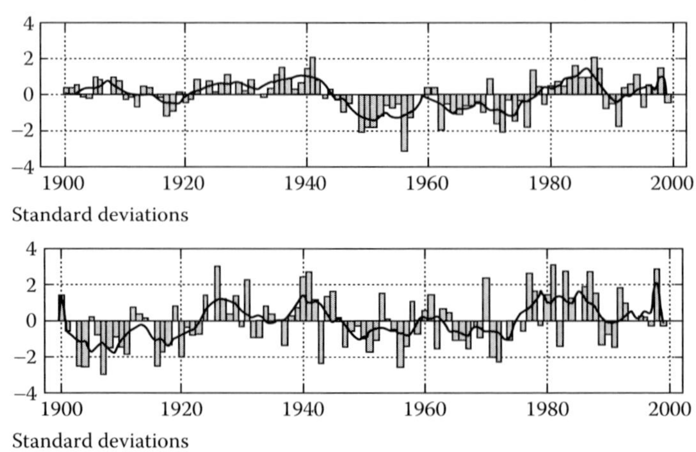

FIGURE 13.6
Pacific Decadal Oscillation 1925 to 2010.

redistribution of heat in the world's oceans. A schematic of modern THC is shown in Figure 13.7. Tens of millions of years ago, continental plate movement formed a land-free gap around Antarctica, permitting formation of the Antarctic Circumpolar Current (ACC) that keeps warm waters away from Antarctica.

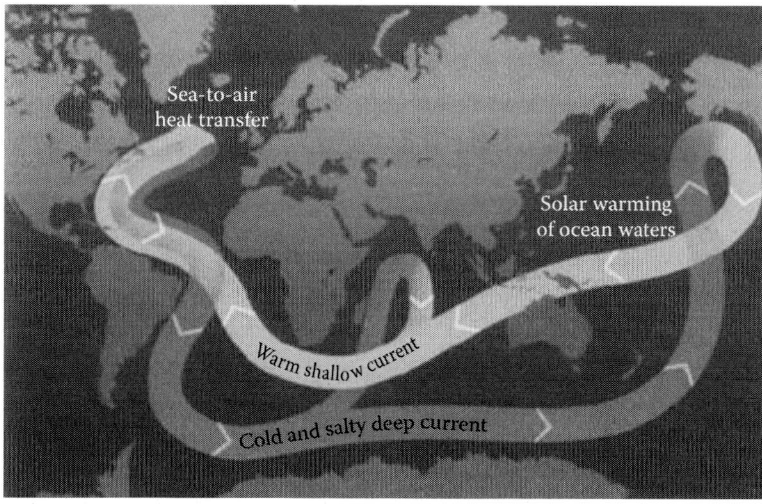

FIGURE 13.7
A schematic of modern THC.

Life affects climate through its role in the carbon and water cycles. Mechanisms such as albedo, evapotranspiration, cloud formation, and weathering are related to these cycles [20,21]. Examples of how life may have affected past climate include glaciations 2.3 billion years ago triggered by the evolution of oxygenic photosynthesis [22,23], glaciations 300 million years ago ushered in by long-term burial of decomposition-resistant detritus of vascular land plants [24,25], termination of the Paleocene-Eocene Thermal Maximum 55 million years ago by flourishing marine phytoplankton [26,27], reversal of global warming 49 million years ago by 800,000 years of arctic azolla blooms [28,29], and global cooling over the past 40 million years driven by the expansion of grass–grazer ecosystems [30,31].

13.3 Earth's Ice Sheets—A Quandry

It has been noted that the Greenland and Antarctic ice sheets have diminished in mass. Data from NASA's Gravity Recovery and Climate Experiment suggest that Greenland has lost 150 to 250 cubic kilometers of ice per year between 2002 and 2006, whereas Antarctica has lost about 152 km^3 of ice between 2002 and 2005. Both the extent and thickness of Arctic sea ice have declined very quickly over the past several decades [32]. The disappearing snowcap of Mount Kilimanjaro from space has been presented in Figure 13.8.*

The use of the word *quandry* expresses what we see here in an article in the *Washington Post* titled "Antarctic sea ice hit 35-year record high Saturday" by Jason Samenow [33]. The sea ice of Antarctica developed a record over the value of October 1978 with an expansion of about 19.51 million km^2 rather than shrinking in spite of world warming. The under ice water gets warmed and not cooled. The Southern Ocean was warming with

* Further pictures of varying ice can also be seen at http://www.bing.com/images/search?q=kilimanjaro+from+space&id=1C1C7073447E069D37C9 48CCB622D663A7A912F2&FORM=IQFRBA.

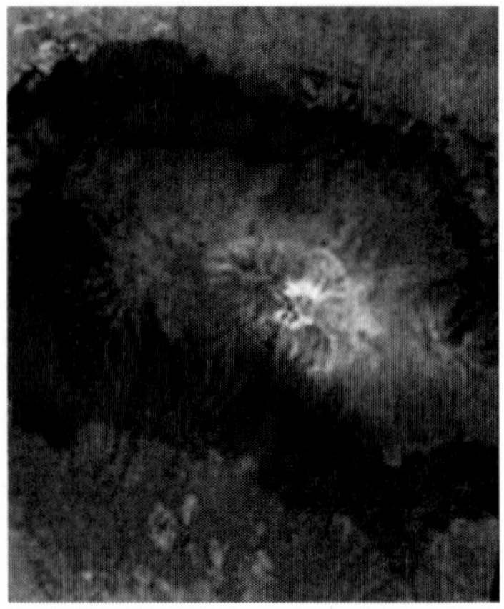

FIGURE 13.8
The disappearing snowcap of Mount Kilimanjaro, from space.

the ice and the rate of increase of ice became small. Strengthening and converging winds around the South Pole can possibly explain largely the increase in ice volume. Water surroundings with thin ice to the intense cold winds appear to be responsible for more ice growth. The computer modeling simulates the declining of ice level with the stratospheric ozone depletion. Following some observations, it was concluded that the recovery of the stratospheric ozone layer that made the Antarctic warm causing ice melt was a complex dependence on the ozone depletion along with climate warming from greenhouse gases and natural variability. However, the Arctic temperature and ice extent are easily related. This contradictory behavior of ice of Antarctic and Arctic is not a paradox that can be realized from the topographical concept that Arctic Ocean surrounded by land with sea ice decreases in the summer, whereas the Antarctic land surrounded by ocean with sea ice increases in the winter [33].

13.4 What We Know

- Three major global surface temperature reconstructions reveal that Earth has warmed since 1880 [34].
- Most of this warming has recorded since the 1970s, with the 20 warmest years having occurred since 1981 and with all 10 of the warmest years occurring in the past 12 years [35].
- The 2000s witnessed a solar output decline resulting in an unusually deep solar minimum in 2007–2009; surface temperatures continue to increase [36].
- The oceans have absorbed a major part of this increased heat, with the top 700 m of ocean showing warming of 0.302°F since 1969 [37].
- The global warming community holds to the anthropogenic global warming hypothesis [38], which involves multiple claims: (a) anomalous global warming is now occurring on Earth; (b) unnatural levels of global warming will

occur during the next century; (c) this warming is the result of anthropogenic greenhouse gas emissions; (d) the consequences of warming will be disastrous; and (e) specific, immediate public policy actions are required to mitigate this coming disaster.

13.5 Climate Observations in the Planets

Our ability is very limited when it comes to making long-term comparative climate observations elsewhere in the solar system. The outer planets have orbital periods of decades, which mean that their seasonal cycles occur over decades, so that changes we observe on shorter time periods may just be seasonal.

13.5.1 Jupiter

Jupiter was observed by several spacecraft like Pioneers 10 and 11 in 1972, Voyagers 1 and 2 in 1979, and New Horizons in 2007. The Galileo orbiter provided long-period observations of Jupiter from 1995 to 2003. Further, the Hubble Space Telescope has been able to investigate Jupiter for years. The atmosphere of Jupiter is highly dynamic, with cloud belts and circular storms, the largest of which is the Great Red Spot, known for 300 years. Some noteworthy changes in the atmosphere of Jupiter have been found in recent years. Three storms merged in 1998–2000, producing what became a new Red Spot in 2005 [39]. Moreover, Baines et al. [40] have described recent changes in Jupiter's clouds. The internal energy and dynamics of Jupiter are responsible for the mechanisms showing a variety of changes. Periodic changes and upheavals have been noted in Jupiter's atmosphere for many decades. In Figure 13.9, we show two photographs of Jupiter as

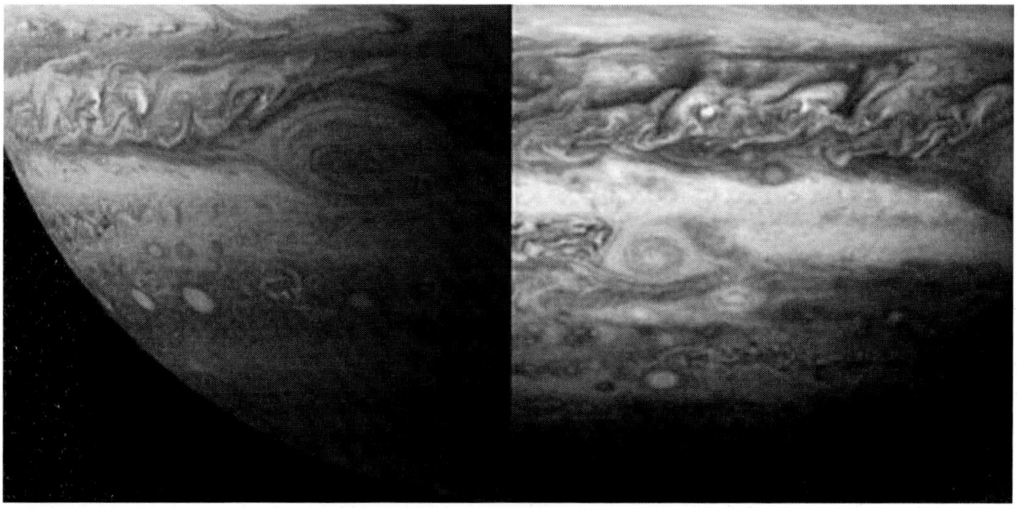

FIGURE 13.9
Jupiter in 2000 and 2006. (Courtesy of NASA/JPL/SSI/ESA/A. SimonMiller, Hubble's planetary portrait captures new changes in Jupiter's great Red Spot, http://www.nasa.gov/press-release/goddard/hubble-s-planetary-portrait-captures-new-changes-in-jupiter-s-great-red-spot [accessed August 12, 2016].)

found in 2000 and 2006. The Great Red Spot is at upper right in both the images. The left image shows several white spots (left center) that merged to form the new Red Spot, as found in the right image (left of center).

13.5.2 Mars

Different spacecraft have been used to observe Mars from orbit since 1971 to the present. In 2001, Malin et al. [41] observed that images of Mars' south polar cap taken one Martian year apart have revealed small retreats (of about 1–3 m) in the cover of frozen carbon dioxide. This frozen CO_2 sublimes directly from ice to gas in the thin atmosphere of Mars. Observations further showed continuing retreat, resulting in expanding pits in the residual polar cap [42–44]. Fenton et al. [45] identified trends in variations in the reflectivity of the surface dust on Mars from 1976 to 2000. From observed albedo variations, they used models for estimating a global annual air temperature increase of 0.65°C. The direct cause of this predicted temperature variation is a change in the distribution of darker dust on the surface of Mars. This change may be the cause of the warming and could be a factor in the retreat of Mars' south polar ice. The main cause of this climate change could be slight shifts in Mars' axial tilt or orbital eccentricity. Such changes have been proposed as key drivers in changing Earth's climate between glacial and interglacial conditions, starting and ending ice ages. A photograph of Mars in the 1970s is shown in Figure 13.10, while changes in Mars' south polar cap from 1999 to 2005 are shown in Figure 13.11.

13.5.3 Titan

Voyager 1 and 2 flybys in 1980 and 1981 and Cassini in 2004 detected that Titan (Saturn's largest moon) has a thick atmosphere consisting of nitrogen and methane. Using data from the Hubble Space Telescope, Lorenz et al. [46] found that the relative brightness of the northern and southern hemispheres had changed from 1994 to 1997. The amount and timing of the changes were found to be different at different wavelengths. These changes are observed to be seasonal. Titan's seasons have a 29-year cycle, corresponding to Saturn's orbital period about the Sun. Cassini has clearly shown that Titan's atmosphere is largely

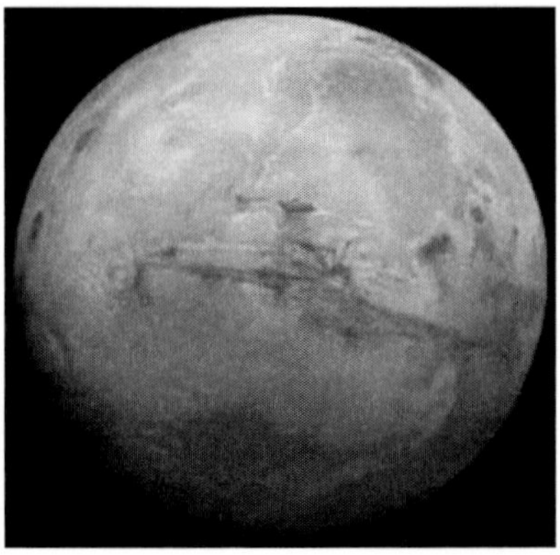

FIGURE 13.10
Mars in the 1970s. (Courtesy of NASA, Viking project, http://www.nasa.gov/multimedia/imagegallery/image_feature_83.html [accessed on August 07, 2016].)

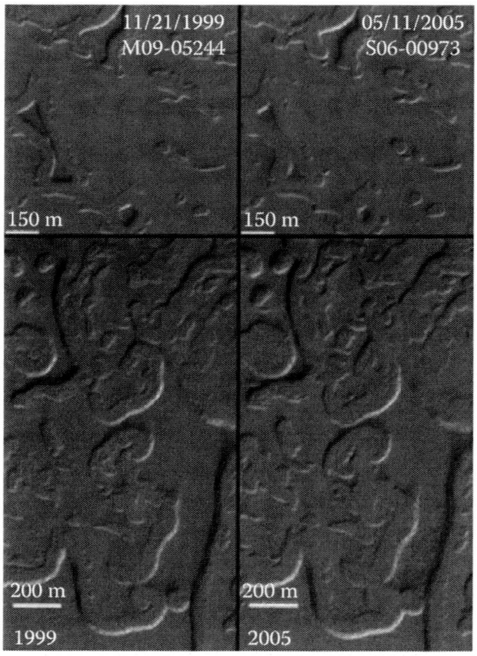

FIGURE 13.11
Changes in Mars' south polar cap from 1999 to 2005. (Courtesy of NASA, Four Mars years of south polar changes, https://www.nasa.gov/vision/universe/solarsystem/mgs-092005-imagesc.html#.V7dQC6LINOY [accessed on August 12, 2016].)

dynamic and has given evidence of methane-driven weather, similar to the water-driven weather on Earth. The changes in Titan's brightness represent changes in haze particles, probably from transport between hemispheres as the seasons change. Figure 13.12 reveals Titan in 1981 and 2005. Color differences in the two images mainly result from the different cameras used and differences in processing.

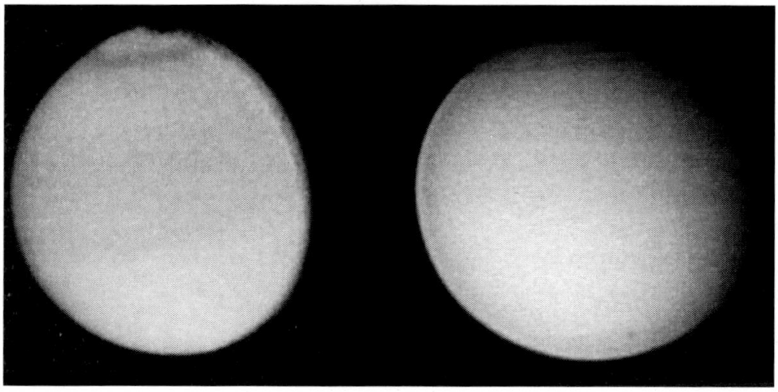

FIGURE 13.12
Titan in 1981 and 2005. (Courtesy of NASA/JPL, http://www.universetoday.com/15429/saturns-moon-titan/# [accessed August 12, 2016].)

13.5.4 Uranus

Uranus is a gas giant with a very thick atmosphere. The only close-up visit to Uranus was a flyby by Voyager 2 in 1986. However, earth-bound telescopes have shown slight changes in the brightness of Uranus over the last 50 years [47–49]. These variations are explainable as seasonal changes. The seasonal cycle of Uranus is 84 years, matching its orbital period. Uranus has a high axial tilt, such that by the 1980s, the south pole of Uranus is presented almost face on toward the Sun and at the same time to the Earth. The extreme changes are assumed to contribute to changes in the cloud cover on Uranus. Figure 13.13 reveals Uranus in 1986 and 2005. Color differences in the two images mainly result from the different cameras used. In the photographs the differences in bright cloud cover are real.

13.5.5 Neptune

Similar to Uranus, Neptune is a gas giant with a thick atmosphere of hydrogen, helium, and methane. Voyager 2 flyby in 1989 is the only spacecraft to view a close-up of Neptune. In addition, Neptune has been long observed by earth-bound telescopes. Considering these decade-long observations, some investigations have reported on the variations in Neptune's brightness over the last 50 years [50–52]. Given that Neptune's seasons change with a 164-year period, matching its orbital period about the Sun, seasonal changes are assumed to contribute to the brightness changes. These changes may be either through changes in the clouds in Neptune's atmosphere, changes in the portion of Neptune visible from Earth, or both. However, Hammel and Lockwood [49] proposed that the details of the variations cannot be interpreted as seasonal changes alone. They noted correlation between the changes in solar activity along with correlations between Neptune's brightness and the average global temperature on Earth. If there is a close relation between the Sun and the climate of Neptune, it would be indirect. The available data indicate about 12% brightening of Neptune from 1950 to 2005, which would necessarily involve variations in Neptune's clouds. Figure 13.14 shows Neptune in 1989 and 2005. Differences in the two images mostly represent the different cameras used for photograph.

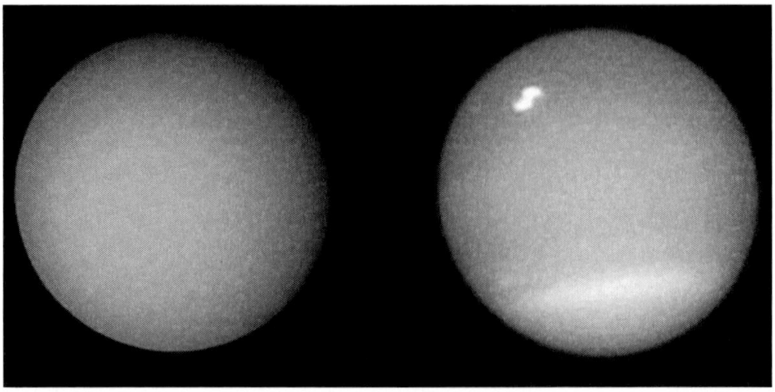

FIGURE 13.13
Uranus in 1986 and 2005. (Courtesy of NASA/ESA/L. Lamy [Observatory of Paris, CNRS, CNES], https://www.nasa.gov/mission_pages/hubble/science/uranus-aurora.html [accessed August 12, 2016].)

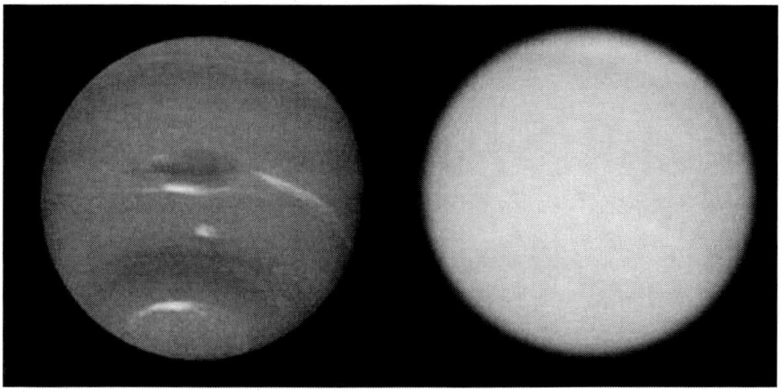

FIGURE 13.14
Neptune in 1989 and 2005. (Courtesy of NASA, http://www.nasa.gov/multimedia/imagegallery/image_feature_552.html [accessed on August 07, 2016]; http://www.nasa.gov/jpl/voyager/triton-20140821/ [accessed on August 07, 2016].)

13.5.6 Triton

Triton is the largest moon of Neptune with a very thin atmosphere consisting of nitrogen and methane. This was found during Voyager 2's brief flyby in 1989, using ground-based telescopes and the Hubble Space Telescope. In addition, stellar occultation data have permitted astronomers to analyze the properties of its atmosphere. Elliot et al. [32] reported that Triton's atmosphere had thickened between 1989 and 1997. They observed that at an altitude of 48 km above Triton's surface, the atmospheric pressure increased about 65% from 1995 to 1997, with the temperature increase of 2°C in that period. This implied a doubling of the pressure at Triton's surface from 1989 to 1997, with a corresponding surface temperature increase of about 2°C. These variations are assumed to be related to seasonal changes. Triton, like Neptune, has seasons with a 164-year cycle. Presently, (at the time of this writing) the south pole of Triton is in summer that is responsible for the warming of the surface resulting in vaporizing nitrogen frosts on the surface. Also, darker ices below absorb more sunlight and further warm Triton. The nitrogen contributes to a thicker atmosphere and freezes as frost in the northern hemisphere where it is winter. Figure 13.15 reveals Triton in 1989. The bright bluish areas toward the top in the figure are frost deposits.

13.5.7 Pluto

Pluto has a thin atmosphere consisting of nitrogen and methane. The surface is covered with various ices consisting of methane, nitrogen, and carbon monoxide. The orbital period of Pluto is 248 years, and its eccentric orbit results in much more solar radiation near perihelion. Like Triton, astronomers have used stellar occultation data to find Pluto's atmosphere. From 1988 to 2002, such observations showed that the pressure in the atmosphere of Pluto is at a fixed height and has doubled [52,53]. This fixed height is approximately 10–90 km above Pluto's surface, depending on where its surface is. Seasonal changes may be important here, given that the peak of Southern Hemisphere summer on Pluto is yet approaching. Changes on Pluto involve the evaporation of surface ices from a warm zone that adds to atmospheric gases. These gases can migrate and refreeze in cooler locations. This causes variations in the bright or dark regions on Pluto's surface and changes in its overall reflectivity [54–56]. The atmosphere of Pluto

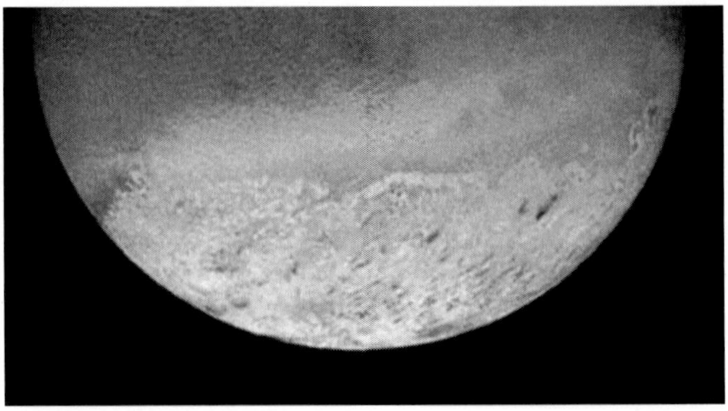

FIGURE 13.15
Triton in 1989. (Courtesy of NASA/JPL/U.S. Geological Survey, PIA00317: Global color mosaic of Triton, http://photojournal.jpl.nasa.gov/catalog/PIA00317 [accessed August 12, 2016].)

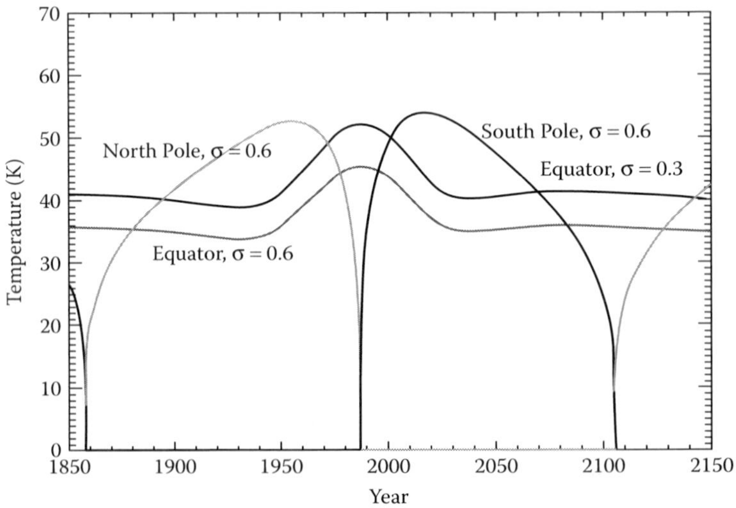

FIGURE 13.16
Estimated seasonal changes in surface temperature on Pluto at various locations.

was found during another stellar occultation in 2006, exhibiting that the atmosphere has become more transparent since 1988 but apparently no additional warming. The warming beyond Pluto's perihelion may be interpreted by high thermal inertia or darkening of surface materials [57]. Estimated seasonal changes in surface temperature on Pluto at different locations are shown in Figure 13.16.

13.6 The Facts

The changes observed in recent years on Jupiter, Titan, Uranus, and Triton can be explained as normal and seasonal changes. However, changes on Mars, Neptune, and Pluto are not yet fully understood: (a) changes in Mars' south polar cap may represent climate change, but this may be due to phenomena particular to Mars; (b) changes in Neptune's brightness appear to correlate with changes in the Sun but the linking mechanism is not known; (c) changes in Pluto's atmosphere seem to be seasonal but further study is required. For Mars, Neptune, and Pluto, our understanding is sufficient. They may exhibit little or no connection to mechanisms for natural climate change on Earth.

References

1. Intergovernmental Panel on Climate Change (2001). Appendix: Glossary, in *Climate Change 2001: The Scientific Basis,* A. P. M. Baede (ed.). Cambridge, UK: Cambridge University Press. http://www.grida.no/climate/ipcc_tar/wg1/518.htm (accessed on August 07, 2016).
2. D. King et al. (2011). *International Climate Change Negotiations: Key Lessons and Next Steps,* 50 pp. Oxford, UK: Smith School of Enterprise and the Environment, University of Oxford.
3. B. D. Santer et al. (1996). A search for human influences on the thermal structure of the atmosphere. *Nature* 382: 39–46.
4. United States Environmental Protection Agency (US EPA). Glossary of climate change terms. https://www3.epa.gov/climatechange/glossary.html (accessed on August 02, 2016).
5. J. Knight et al. (2009). Do global temperature trends over the last decade falsify climate predictions? *Bulletin of American Meteorological Society* 90(8): S75–S79.
6. A. Aldahan et al. (2008). Atmospheric impact on beryllium isotopes as solar activity proxy. *Geophysics Research Letters* 35: L21812.
7. J. D. Haigh et al. (2010). An influence of solar spectral variations on radiative forcing of climate. *Nature* 467: 696–699.
8. J. Kirkby et al. (2011). Role of sulphuric acid, ammonia and galactic cosmic rays in atmospheric aerosol nucleation. *Nature* 476: 429–433. doi:10.1038/news.2011.504.
9. NASA Earth Observatory. 2011. Glossary (Climate System: The five physical components [atmosphere, hydrosphere, cryosphere, lithosphere, and biosphere] that are responsible for the climate and its variations). http://earthobservatory.nasa.gov/Glossary/?mode=all (accessed on August 07, 2016).
10. S. Solomon et al. (eds.) (2007). What are climate change and climate variability?, Section 9.1.1, in *Contribution of Working Group I to the Fourth Assessment Report of the Intergovernmental Panel on Climate Change, 2007.* New York: Cambridge University Press.
11. S. Rahmstorf (2003). The concept of the thermohaline circulation. *Nature* 421(6924): 699.
12. National Research Council (NRC). (2006). *Surface Temperature Reconstructions for the Last 2,000 Years.* Washington, DC: National Academy Press.
13. J. A. Church and N. J. White (2006). A 20th century acceleration in global sea level rise. *Geophysical Research Letters* 33: L01602, doi:10.1029/2005GL024826.
14. J. D. Hays et al. (1976). Variations in the Earth's orbit: Pacemaker of the ice ages. *Science* 194(4270): 1121–1132.
15. J. A. Rial (2003). Earth's orbital eccentricity and the rhythm of the Pleistocene ice ages: The concealed pacemaker. *Global and Planetary Change* 41(2): 81–93.
16. C. J. Hansen and D. A. Paige (1996). Seasonal nitrogen cycles on Pluto. *Icarus* 120, 247–265.
17. A. S. Gale (1989). A Milankovitch scale for Cenomanian time. *Terra Nova* 1(5): 420.
18. S. Solomon et al. (eds.) (2007). What caused the ice ages and other important climate changes before the industrial era?, FAQ 6.1, in *Contribution of Working Group I to the Fourth Assessment Report of the Intergovernmental Panel on Climate Change, 2007.* New York: Cambridge University Press.
19. S. Solomon et al. (eds.) (2007). What caused the low atmospheric carbon dioxide concentrations during glacial times? Box 6.2, in *Contribution of Working Group I to the Fourth Assessment Report of the Intergovernmental Panel on Climate Change, 2007.* New York: Cambridge University Press.

20. D. V. Spracklen et al. (2008). Boreal forests, aerosols and the impacts on clouds and climate. *Philosophical Transactions of the Royal Society A: Mathematical, Physical and Engineering Sciences* 366(1885): 4613–4626.
21. D. W. Schwartzman and T. Volk (1989). Biotic enhancement of weathering and the habitability of Earth. *Nature* 340(6233): 457.
22. R. E. Kopp et al. (2005). The Paleoproterozoic snowball Earth: A climate disaster triggered by the evolution of oxygenic photosynthesis. *Proceedings of the National Academy of Sciences* 102(32): 11131–11136.
23. J. F. Kasting and J. L. Siefert (2002). Life and the Evolution of Earth's Atmosphere. *Science* 296(5570): 1066–1068.
24. C. I. Mora et al. (1996). Middle to late Paleozoic atmospheric CO_2 levels from soil carbonate and organic matter. *Science* 271(5252): 1105.
25. R. A. Berner (1999). Atmospheric oxygen over Phanerozoic time. *Proceedings of the National Academy of Sciences* 96(20): 10955–10957.
26. S. Bains et al. (2000). Termination of global warmth at the Palaeocene/Eocene boundary through productivity feedback. *Nature* 407(6801): 171–174.
27. J. C. Zachos and G. R. Dickens, (2000). An assessment of the biogeochemical feedback response to the climatic and chemical perturbations of the LPTM. *GFF* 122: 188.
28. E. N. Speelman et al. (2009). The Eocene Arctic Azollabloom: Environmental conditions, productivity and carbon drawdown. *Geobiology* 7(2): 155–170.
29. H. Brinkhuis et al. (2006). Episodic fresh surface waters in the Eocene Arctic Ocean. *Nature* 441(7093): 606–609.
30. G. J. Retallack (2001). Cenozoic expansion of grasslands and climatic cooling. *The Journal of Geology* 109(4): 407.
31. J. F. Dutton and E. J. Barron (1997). Miocene to present vegetation changes: A possible piece of the Cenozoic cooling puzzle. *Geology* 25: 39.
32. J. L. Elliot et al. (1998). Global warming on Triton. *Nature* 393: 765–767.
33. J. Samenow (September 23, 2013). Antarctic sea ice hit 35-year record high Saturday. *The Washington Post.* https://www.washingtonpost.com/news/capital-weather-gang/wp/2013/09/23/antarctic-sea-ice-hit-35-year-record-high-saturday/ (accessed September 28, 2016).
34. NASA. Global climate change: Vital signs of the planet. http://climate.nasa.gov/evidence/ (accessed August 12, 2016).
35. T. C. Peterson et al. (2009). State of the Climate in 2008. *Bulletin of the American Meteorological Society* (Special Supplement) 90(8): S17–S18.
36. I. Allison et al. (2009). *The Copenhagen Diagnosis: Updating the World on the Latest Climate Science.* Sydney, Australia: UNSW Climate Change Research Center. p. 11.
37. Levitus et al. (2009). Global ocean heat content 1955–2008 in light of recently revealed instrumentation problems. *Geophysical Research Letters* 36: L07608.
38. P. Wood and R. Peterson (2015). The global warming debate. *Sustainability: Higher Education's New Fundamentalism.* New York: National Association of Scholars, pp. 98–127.
39. P. Marcus et al. (2006). Velocities and temperatures of Jupiter's Great Red Spot and the New Red Oval and their implications for global climate change. *Bulletin of the American Astronomical Society* 38: 554.
40. K. H. Baines et al. (2007). Polar lightning and decadal-scale cloud variability on Jupiter. *Science* 318: 226–229.
41. M. C. Malin et al. (2001). Observational evidence for an active surface reservoir of solid carbon dioxide on Mars. *Science* 294: 2146–2149.
42. J. L. Benson and P. B. James (2005). Yearly comparisons of the Martian polar caps: 1999–2003 Mars Orbiter Camera observations. *Icarus* 174: 513–523.
43. P. C. Thomas et al. (2005). South polar residual cap of Mars: Features, stratigraphy, and changes. *Icarus* 174: 535–559.
44. P. B. James et al. (2007). MOC observations of four Mars year variations in the south polar residual cap of Mars. *Icarus* 192: 318–326.

45. L. K. Fenton et al. (2007). Global warming and climate forcing by recent albedo changes on Mars. *Nature* 446: 646–649.
46. R. D. Lorenz et al. (1999). Seasonal change on Titan observed with the Hubble Space Telescope WFPC-2. *Icarus* 142: 391–401.
47. G. W. Lockwood and D. T. Thompson (2002). Photometric variability of Neptune, 1972–2000. *Icarus* 156: 37–51.
48. G. W. Lockwood and M. Jerzykiewicz (2006). Photometric variability of Uranus and Neptune, 1950–2004. *Icarus* 180: 442–452.
49. H. B. Hammel and G. W. Lockwood (2007). Long-term atmospheric variability on Uranus and Neptune. *Icarus* 186: 291–301.
50. G. W. Lockwood and M. Jerzykiewicz (2006). Photometric variability of Uranus and Neptune, 1950–2004. *Icarus* 180: 442–452.
51. G. W. Lockwood and D. T. Thompson (1999). Photometric variability of Uranus, 1972–1996. *Icarus* 137: 2–12.
52. H. B. Hammel and G. W. Lockwood (2007). Suggestive correlations between the brightness of Neptune, solar variability, and Earth's temperature. *Geophysical Research Letters* 34: L08203.
53. J. L. Elliot et al. (2003). The recent expansion of Pluto's atmosphere. *Nature* 424: 165–168.
54. B. Sicardy et al. (2003). Large changes in Pluto's atmosphere as revealed by recent stellar occultations. *Nature* 424: 168–170.
55. M. E. Brown (2002). Pluto and Charon: formation, seasons, composition. *Annual Review of Earth and Planetary Science* 30: 307–345.
56. J. M. Pasachoff et al. (2005). The structure of Pluto's atmosphere from the 2002 August 21 stellar occultation. *The Astrophysical Journal* 129: 1718–1723.
57. J. L. Elliot et al. (2007). Changes in Pluto's atmosphere: 1988–2006. *The Astronomical Journal* 134: 1–13.
58. J. R. Petit et al. (1999). Climate and atmospheric history of the past 420,000 years from the Vostok ice core, Antarctica. *Nature* 399: 429–436.
59. NASA. Climate change: How do we know? http://climate.nasa.gov/evidence/ (accessed on August 25, 2016).

14

Organic Life Limitations in Planetary Systems

14.1 Introduction

A clear understanding about the origin of the Earth and its ability to support life is the main objective for scientists engaged in the field of astronomy. Those studying this subject gain knowledge about orbits, chemical properties, and physical history that help in the process of understanding how it is all possible. The cooler objects (thermally) and dust in the outer solar system provide evidence of conditions in the early solar system and can be directly compared to cool objects and dust observed around other stars. An absolute requirement for life is an energy source. For planetary habitability, there are many geophysical, geochemical, and astrophysical criteria that must be met before an astronomical body can support life. In the astrobiology roadmap, NASA has pointed out the principal habitability criteria as "extended regions of liquid water, conditions favorable for the assembly of complex organic molecules and energy sources to sustain metabolism" [1].

14.2 Planets and the Origins of Life

As we have already pointed out, the first planet outside of our solar system was discovered in 1992. Since then, scientists have discovered a multitude of planets around other stars and have come to the realization that planets are quite common. Most of the known planets are large gas giants such as Jupiter. Recent discoveries have included many smaller planets; some of those are few times more massive than Earth. The ultimate target of the search is to identify planets orbiting in the habitable zone (HZ) of their star, where it is possible for liquid water and perhaps even life to exist. Figure 14.1 shows HR8799 and three of its planets (b, c, and d), which are thought to be gas giants such as Jupiter but more massive. They orbit their host star (marked with X) at roughly 24, 38, and 68 times the distance between our Earth and Sun, respectively. A fourth gas giant (e) has since been discovered with the closest orbit in the system. The location of HR8799 is about 129 light-years away from the Earth.

To trace the origins of the Earth and life in the universe, astronomers dedicated their study of planet formation and evolution, including the material around stars where planets form. Out of the many problems they attempted, a key issue is to understand how the building blocks of planets are assembled. They have no clear idea if all planets in a planetary system form in place or travel inward after formation. It is also unknown how planets reach their ultimate orbits or how large planets affect the smaller ones in solar

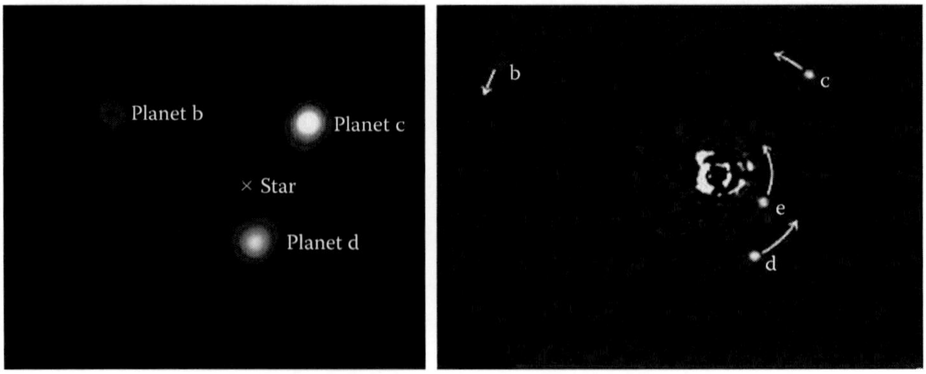

FIGURE 14.1
HR8799 and three of its planets (b, c, and d); a fourth gas giant (e), as shown on the right-hand side, has since been discovered with the closest orbit in the system. (Courtesy of NASA/JPL-Caltech/Palomar Observatory, http://www.leviathanastronomy.com/extrasolar-planets.html [accessed September 10, 2016].)

systems like our own. However, careful investigations have provided much valuable information in recent years. The icy bodies and dust when our solar system was very young can be directly compared to the objects and dust observed around other stars. Highly sensitive instruments on the James Webb Space Telescope (JWST) were used to get infrared images of giant planets and planetary systems. These images were used to characterize age and mass. In addition, to investigate the characteristics of planets outside our solar system, scientists attempted to learn more about our own home. Studying the chemical and physical properties of the small and large bodies that came together to form the Earth may help us to gather information about how life developed on Earth. The Webb Space Telescope is, in fact, powerful enough to identify and characterize comets and other icy bodies in the outermost reaches of our solar system and lead us to our origins on Earth. The goal of the JWST program is being used for research and the study of planetary systems and origins of life. In addition, for determining the physical and chemical properties of planets and to study the potential for the origins of life in those systems. The systems must provide near and mid-IR imaging and spectroscopy to find these objects including imaging and spectroscopic characterization of Mars and the outer planets, Kuiper belt objects, comets, dwarf planets, and icy moons.

14.2.1 Origins of Earth and Life in the Universe

Research has gained us knowledge pertaining to planet formation, evolution, and planet formation. Most estimates are inferred on as to the effect of a repositioned orbit scenario would have on the habitability of Earth or Venus. However, following the extended HZ theory, a planet with a denser atmosphere than Earth orbiting in the extended HZ might possess liquid water theoretically. Gliese 667 Cd [2] and Gliese 581 d are examples of highly massive planets considered to have the potential for possessing terrestrial surfaces and atmospheres that might induce global warming sufficient for habitability [3,4]. Table 14.1 shows the estimated values of the circumstellar-HZ boundaries of the solar system including the techniques implemented for the purpose.

It has been noted that 50% of stars have a planet of Earth size or larger in a close orbit. By adding larger planets, which have been detected in wider orbits and up to the orbital distance of the Earth, this number increases to 70%. Extrapolating from Kepler's observations

TABLE 14.1

Estimates of the Circumstellar-Habitable-Zone Boundaries of the Solar System

Inner Edge (AU)	Outer Edge (AU)	Year	Technique Implemented and Related Information
0.725	1.24	Dole [5] 1964	Used optically thin atmospheres and fixed albedo. Places the aphelion of Venus just inside the zone.
	1.385–1.398	Budyko [6] 1969	Based on investigations of ice albedo feedback models to find the point at which Earth would experience global glaciations. This estimate was supported in the studies by Sellers [7] and North [8].
0.88–0.912		Rasool and De Bergh [9] 1970	Based on studies of the atmosphere of Venus, it was concluded that this is the minimum distance at which Earth would have formed stable oceans.
0.95	1.01	Hart [10] 1979	Based on computer modeling and simulations of the evolution of Earth's atmospheric composition and surface temperature.
	3.0	Fogg [11] 1992	Used the carbon cycle for estimating the outer edge of the circumstellar habitable zone.
	1.37	Kasting et al. [12] 1993	Noted the cooling effect of cloud albedo.
	2.0	Spiegel et al. [13] 2010	Proposed that seasonal liquid water is possible to this limit when combining high obliquity and orbital eccentricity.
0.75		Abe et al. [14] 2011	Found that land-dominated *desert planets* with water at the poles could exist closer to the Sun than watery planets like Earth.
0.77–0.87	1.02–1.18	Vladilo et al. [15] 2013	Inner edge of circumstellar habitable zone is closer and outer edge is farther for higher atmospheric pressures; determined minimum atmospheric pressure needed to be 15 millibar.
0.99	1.688	Kopparapu [16] 2013	Revised estimates utilizing updated runaway greenhouse and water loss algorithms. This applies to a planet with Earth-like atmospheric composition and pressure.
0.5		Zsom et al. [17] 2013	Estimate based on different possible combinations of atmospheric composition, pressure, and relative humidity of the planet's atmosphere.

and the results from other detection techniques, it appears that all Sun-like stars have planets. By doing a simulation of the Kepler survey and focusing on the false positives, they can only account for 9.5% of the huge number of Kepler candidates while the rest are bona-fide planets. Artist's illustration representing the variety of planets, being detected by NASA's Kepler spacecraft, is given in Figure 14.2.

Planets are grouped into five different sizes. It has been reported that about 17% of stars have a planet 0.8 to 1.25 times the size of Earth in an orbit of 85 days or less. About one-fourth of stars have a super-Earth (1.25 to 2 times the size of Earth) in an orbit of 150 days or less. Larger planets can be detected at greater distances more conveniently. The same fraction of stars has a mini-Neptune (2 to 4 times Earth) in orbits up to 250 days long. Larger planets are much less common. Only about 3% of stars have a large Neptune (4 to 6 times Earth) and only 5% of stars have a gas giant (6 to 22 times Earth) in an orbit of 400 days or less.

FIGURE 14.2
Artist's illustration representing the variety of planets, being detected by NASA's Kepler spacecraft. (Courtesy of NASA, https://www.nasa.gov/sites/default/files/images/718518main_loresworm.jpg [accessed September 10, 2016].)

14.3 Planetary Habitability of Natural Satellites

Planetary habitability means the measurement of a planet's or a natural satellite's potential to develop and sustain life. Life may be developed directly on a planet or satellite or be transferred to it from another body, a theoretical process known as panspermia. As the existence of life beyond Earth is currently unknown, planetary habitability is largely an extrapolation of conditions on Earth and the characteristics of the Sun and solar system that appear favorable to flourishing life—in particular those factors that have sustained complex, multicellular organisms, and not just simpler, unicellular creatures. In this regard, research and theory is a component of planetary science and the emerging discipline of astrobiology. Understanding planetary habitability is essentially an extrapolation of the Earth's conditions (Figure 14.3).

To find the habitability potential of a body, observations concentrate on its bulk composition, atmosphere, orbital properties, and potential chemical interactions. Stellar characteristics of importance include mass and luminosity, stable variability, and high metallicity. Rocky, terrestrial-type planets, and moons with the potential for Earth-like chemistry are a fundamental focus of astrobiological research; however, more speculative habitability theories occasionally examine alternative biochemistries and other types of astronomical bodies. The idea that planets beyond Earth might host life is an ancient one, though historically it was framed by philosophy as much as physical science. The late twentieth century saw two breakthroughs in the area. The observation and robotic spacecraft exploration of other planets and moons within the solar system has provided critical information on habitability allowing substantial geophysical comparisons between the Earth and other bodies.

FIGURE 14.3
Understanding planetary habitability, partly an extrapolation of the Earth's conditions.

The discovery of extrasolar planets has provided further information for the study of possible extraterrestrial life. These findings confirm that the Sun is not unique among stars in hosting planets and expands the habitability research beyond the solar system. On the basis of Kepler space mission data, astronomers pointed out that there could be as many as 40 billion Earth-sized planets orbiting in the HZs of Sun-like stars and red dwarf stars within the Milky Way galaxy [18,19]. Eleven billions of these estimated planets may be orbiting Sun-like stars [20]. The nearest planet of this category may be 12 light-years away.

14.4 Suitable Star Systems

The concept of planetary habitability begins with stars. Bodies that are generally Earth-like may be plentiful. It is just as important that their larger system be agreeable to life. Under the auspices of SETI's Project Phoenix, scientists Margaret Turnbull and Jill Tarter developed the *HabCat* (or Catalog of Habitable Stellar Systems) in 2002 [21]. The catalog was formed by winnowing the nearly 120,000 stars of the larger Hipparcos Catalog into a core group of 17,000 *HabStars*, and the selection criteria that were used provide a good starting point for understanding which astrophysical factors are necessary to habitable planets [22].

14.4.1 Spectral Class

The spectral class of a star reveals its photospheric temperature, which correlates to overall mass. The appropriate spectral range for HabStars is considered to be *early F* or *G*, to *mid-K*, which corresponds to temperatures of a little more than 7000 K down to a little more than 4000 K. The Sun, being a G2 star, is situated within these bounds. *Middle-class* stars of this category have some characteristics considered interesting to planetary habitability: (a) They live at least a few billion years, permitting life a chance to evolve. More luminous main-sequence stars of the O, B, and A classes usually live less than a billion years and in exceptional cases less than 10 million [23]. (b) Liquid water may exist on the surface of planets that may be able to support life for long periods [24]. (c) They emit high-frequency ultraviolet radiation to trigger important atmospheric dynamics such as ozone formation, but not so much that ionization destroys incipient life [25]. This spectral range accounts for between 5% and 10% of stars in the local Milky Way galaxy. Whether fainter late K and M class red dwarf stars are suitable hosts for habitable planets is the most important question in the whole field of planetary habitability given their prevalence. Gliese 581 c, a *super-Earth*, has been observed orbiting in the *HZ* of a red dwarf and may possess liquid water. However, it is also possible that a greenhouse effect may render it too hot to support life while its neighbor, Gliese 581 d, may be a more likely candidate for habitability [26]. One of the planets, Gliese 163 c, about 6.9 times the mass of Earth and somewhat hotter, was assumed to be within the HZ [27, 28]. Investigations suggest that cooler stars that emit more light in the infrared and near infrared may actually host warmer planets with less ice and incidence of snowball states.

14.5 Stable HZ

The HZ is a shell-shaped region of space surrounding a star wherein a planet could maintain liquid water on its surface. After an energy source, liquid water is assumed as the most important ingredient for life, considering how integral it is to all life systems on Earth. This may reflect the known dependence of life on water. A *stable* HZ implies two important factors. (i) First, the range of an HZ should not vary greatly over time. All stars increase in luminosity with their age and a given HZ migrates outward. If this happens too fast, planets may only have a brief window inside the HZ and so a smaller chance of developing life. Atmospheric conditions including geology also have great impact on a putative HZ range. The proposed parameters of the Sun's HZ have found to fluctuate largely [29]. Second, no large-mass body similar to a gas giant should be present in or relatively close to the HZ, hence disrupting the formation of Earth-like bodies. The matter in the asteroid belt appears to have been unable to accrete into a planet owing to orbital resonances with Jupiter. If the giant Jovian planet had appeared in the region that is now between the orbits of Venus and Mars, Earth would not have developed in its present form. But a gas giant inside the HZ might have habitable moons under proper conditions [30]. In the solar system, the inner planets are terrestrial and the outer ones are gas giants. However, discoveries of extrasolar planets reveal that this arrangement may not be at all common. In fact, many Jupiter-sized bodies have been observed in close orbit about their primary star that is responsible for disrupting potential HZs. Present data for extrasolar planets are likely to be skewed toward that type because they are simpler to identify. Variations in luminosity are common to all stars but the severity of these fluctuations covers a broad range. Most of the stars are relatively stable but a significant minority of variable stars frequently undergoes intense increases in luminosity and hence in the amount of energy radiated toward bodies in orbit. There is strong evidence that even minor changes in the luminosity of the Sun will have significant effects on the climate of the Earth well within the historical era. For example, the Little Ice Age of the mid-second millennium may have been caused by a relatively long-term decline in the luminosity of the Sun [31]. Thus, a star does not have to be a true variable for differences in luminosity to affect habitability.

14.6 High Metallicity

The bulk of material in any star is hydrogen and helium. However, there is a large variation in the amount of heavier elements that stars contain. Spectroscopic investigations of systems where exoplanets have been found confirm the relation between high metal content and planet formation [32]. This relation between high metallicity and planet formation suggests that habitable systems are likely to be found around younger stars, as stars that formed early in the universe have low metal content.

14.7 Planetary Characteristics

The main assumption about habitable planets is that they are terrestrial. These planets within one order of magnitude of Earth's mass are approximately composed of silicate rocks. In addition, these planets have not accreted the gaseous outer layers of hydrogen and helium as observed on gas giants. The idea that life could evolve in the top cloud layers of giant planets has not been decisively ruled out, though it is assumed very unlikely, as they have no surface and also their gravity is sufficiently large [33]. The natural satellites of giant planets remain valid candidates for hosting life [34]. The Kepler Space Observatory Mission team, in February 2011, released a list of 1235 extrasolar planet candidates, 54 of which may be in the HZ [35,36]. Six of them are smaller than Earth. A subsequent investigation noted that one of these candidates (KOI 326.01) is much larger and hotter than that reported initially [37]. On the basis of the observations, the Kepler team estimated that there are at least 50 billion planets in the Milky Way, of which at least 500 million are in the HZ [38]. Moons of some gas giants could potentially be habitable, as shown in Figure 14.4.

For the purpose of analyzing the fact of the appropriate environments that can support life, a distinction is usually made between simple, unicellular organisms (such as bacteria)

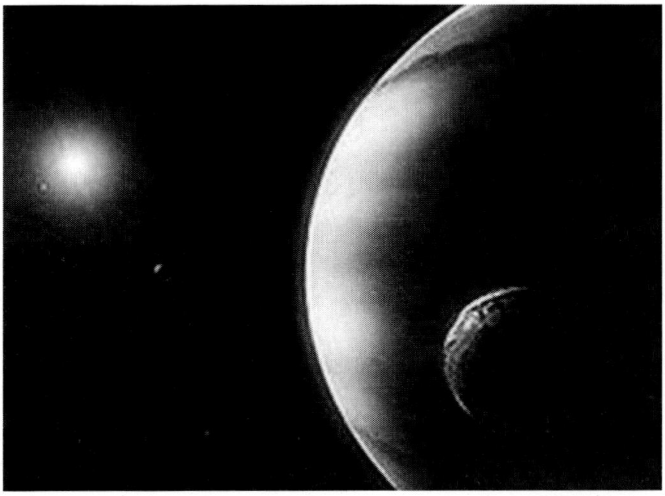

FIGURE 14.4
Moons of some gas giants could potentially be habitable. (From Cowen, R., A shifty moon, *Science News*, 2008. https://www.sciencenews.org/article/shifty-moon?mode=magazine&context=657 [accessed August 12, 2016].)

and complex metazoans (animals). The planetary characteristics are assumed very crucial for life but in every case multicellular organisms are pickier than unicellular life.

14.7.1 Mass

Generally, low-mass planets are poor candidates for life mainly for two reasons. First, their gravity is less and that makes atmosphere retention difficult. Constituent molecules are likely to reach escape velocity and be lost to space when buffeted by solar wind or stirred by collision. Planets without a thick atmosphere have little insulation and poor heat transfer across their surfaces. For example, Mars with its thin atmosphere is colder than the Earth than if it was at a similar distance from the Sun and also it would give less protection against meteoroids and high-frequency radiation. Also where an atmosphere is less dense than 0.006 Earth atmospheres, water cannot exist in liquid form as the required atmospheric pressure, 4.56 mm Hg (608 Pa) (0.18 inch Hg), does not occur. Usually, the temperature range at which water is liquid is smaller at low pressures. Figure 14.5 reveals a photograph of Mars with its rarefied atmosphere, colder than the Earth.

These bodies tend to lose the energy left over from their formation at a very fast rate and end up geologically dead, lacking volcanoes, earthquakes, and tectonic activity that supply the surface with life-sustaining material and the atmosphere with temperature moderators such as carbon dioxide. Plate tectonics appears crucial, at least on Earth. It also fosters biodiversity through continent creation and increased environmental complexity and assists in creating the convective cells required for generating the magnetic field of the earth [39]. *Low mass* is a relative term. The Earth has low mass when compared to the gas giants of the solar system, but it has the largest diameter and is densest of all terrestrial bodies. It is sufficiently large to maintain an atmosphere through gravity alone and also large enough that its molten core remains a heat engine, driving the diverse geology of the surface. Mars, by contrast, is almost geologically dead and has lost much of its atmosphere [40]. It may be concluded that the lower mass limit for habitability lies somewhere between Mars and Earth or Venus. In general, 0.3 Earth masses has been offered as a rough dividing line for habitable planets [41]. Earth lies on the lower boundary of habitability.

FIGURE 14.5
Mars with its rarefied atmosphere, colder than the Earth. Second, smaller planets have lower diameters and so higher surface-to-volume ratios than the larger one.

If it were any smaller, plate tectonics would be impossible. Venus, which has 85% of the mass of Earth, exhibits no signs of tectonic activity. On the other hand, *super-Earths*, terrestrial planets with higher masses than Earth, would have greater levels of plate tectonics and hence be firmly kept in the habitable range. Exceptional situations offer exceptional cases. Jupiter's moon Io, which is smaller than any of the terrestrial planets, is volcanically dynamic due to the gravitational stresses induced by its orbit and its neighbor Europa. Europa may have a liquid ocean or icy slush underneath a frozen shell owing to power generated from orbiting a gas giant. Saturn's Titan has an outside chance of harboring life, as it has retained a thick atmosphere and has seas of liquid methane on its surface. Organic-chemical reactions that only need minimum energy are possible in these seas but whether any living system can be based on such minimal reactions is not clear and would seem unlikely. These satellites are exceptions but they prove that mass, as a criterion for habitability, cannot be considered definitive at this stage of our knowledge. A larger planet is likely to get a more massive atmosphere. A combination of greater escape velocity to retain lighter atoms and enhanced plate tectonics may largely increase the atmospheric pressure and temperature at the surface in comparison to Earth. The enhanced greenhouse effect of this heavy atmosphere would indicate that the HZ should be out from the central star for such massive planets. Again, a larger planet is likely to have a high iron core that permits for a magnetic field. This field would protect the planet from stellar wind and cosmic radiation. It may be noted that the mass is not the only criterion for producing a magnetic field, as the planet must also rotate too fast for producing a dynamo effect within its core [42].

14.7.2 Orbit and Rotation

Stability is an important consideration in evaluating the effect of orbital and rotational characteristics on planetary habitability. Orbital eccentricity is the difference between a farthest and closest approach of a planet to its parent star divided by the sum of said distances. It is a ratio of the shape of the elliptical orbit. The greater the eccentricity, the higher the temperature fluctuation on a planet's surface. Living organisms can only stand so much variation, particularly if the fluctuations overlap both the freezing and boiling point of the planet's main biotic solvent (e.g., water on Earth). If the Earth's oceans were alternately boiling and freezing solid, it is hard to imagine life as we know it having evolved [9]. The orbit of the Earth is almost circular, with an eccentricity of less than 0.02; other planets in the solar system (except Mercury) have eccentricities that are similarly benign. Data gathered on the orbital eccentricities of extrasolar planets show that 90% have an orbital eccentricity larger than that obtained within the solar system and the average is fully 0.25 [10]. It means that the majority of planets have highly eccentric orbits and the average distance from their star is deemed to be within the HZ. The movement of a planet around its rotational axis must reveal certain criteria if life is to have the opportunity to evolve. The significant assumption is that the planet should have moderate seasons. If there is little or no axial tilt relative to the perpendicular of the ecliptic, seasons will not occur. If a planet is radically tilted, seasons will be extreme and make it more difficult for a biosphere to achieve homeostasis. Presently, the exact effects of these changes can only be computer modeling and other investigations that have supported that even extreme tilts of up to 85° do not absolutely preclude life [13]. A drastic variation or a much shorter periodicity would induce climatic effects such as variations in seasonal severity. Other orbital considerations include the following. (a) The planet should rotate quickly so that the day–night cycle is not excessive. If a day takes years, the temperature differential between the day and night side will be pronounced and problems similar to those noted with extreme orbital eccentricity will appear. (b) Change in the direction of the axis rotation should not be pronounced. Precession need not affect habitability as it changes the direction of the tilt, not its degree. But precession tends to accentuate variations owing to other orbital deviations. (c) The planet's rotation should be sufficiently fast, so that a magnetic dynamo may be started in its iron core thus producing a magnetic field. The Earth's moon appears to play a vital role in moderating the climate by stabilizing the axial tilt. It is suggested that a chaotic tilt may be a *deal-breaker* in terms of habitability: A satellite the size of the Moon is not only helpful but also produces stability [42].

14.7.3 Geochemistry

Extraterrestrial life that might exist will be based on the same fundamental biochemistry as found on Earth, as the four elements most vital for life, carbon, hydrogen, oxygen, and nitrogen, are the most common chemically reactive elements in the universe. Simple biogenic compounds, such as simple amino acids (e.g., glycine), have been found in meteorites and in the interstellar medium. These four elements together comprise more than 96% of Earth's collective biomass. Carbon has an ability to bond with itself and to form a massive array of intricate and varied structures, making it an ideal material for the complex mechanisms that form living cells. Hydrogen and oxygen in the form of water compose the solvent in which biological processes take place and are the first reactions that occurred and led to life's emergence. The energy released in the formation of powerful covalent bonds between carbon and oxygen, available by oxidizing organic compounds, is the fuel of all complex lifeforms. These four elements together make up amino acids, which in turn are the building blocks of proteins, the substance of living tissue. In addition, neither sulfur is essential for the building of proteins nor phosphorus is required for the formation of DNA, RNA, and the adenosine phosphates. The relative abundance of the elements in space does not differentiate abundance within planets; of the four life elements, for instance, only oxygen is present in any abundance in the Earth's crust [43]. This can be partly explained by considering that many of these elements, such as hydrogen and nitrogen, along with common compounds, for example, carbon dioxide, carbon monoxide, methane, ammonia, and water are gaseous at warm temperatures. In the hot region near to the Sun, these volatile compounds could not have shown a significant role in the geological formation. They were trapped as gases underneath the newly formed crusts, which were largely made of rocky, involatile compounds such as silica. Volatile compounds through the first volcanoes would have contributed to the formation of the atmospheres of planets. The Miller–Urey experiment revealed that, with the application of energy, amino acids can be formed from the synthesis of the simple compounds within a primordial atmosphere [44]. The vast majority of the water required for life must have come from the outer solar system, away from the Sun's heat, where it could remain solid. It is speculated that comets impacting with the Earth in the early years of Earth's formation would have deposited huge amounts of water along with the other volatile compounds needed for life including amino acids onto the early Earth, providing a start to the origin of life. Hence, while there is strong reason to suspect that the four *life elements* ought to be readily available elsewhere, a habitable system probably also needs a supply of long-term orbiting bodies to seed inner planets. It is believed that without comets there is a possibility that life as we know it would not exist on Earth.

14.7.4 Microenvironments and Extremophiles

One important criterion to habitability is that only a tiny portion of a planet is needed to support life. Astrobiologists concern themselves with *microenvironments*, pointing out that we require a fundamental understanding of how evolutionary forces, such as mutation, selection, and genetic drift operate in microorganisms that act on and respond to changing microenvironments [44]. Extremophiles are Earth organisms that live in niche environments under severe conditions usually assumed inimical to life. In general, unicellular, extremophiles include acutely alkaliphilic and acidophilic organisms and others that can survive water temperatures above 100°C in hydrothermal vents. The discovery of life in extreme conditions has complicated definitions of habitability but generated sufficient excitement among researchers in greatly broadening the known range of conditions under which life can persist. For example, a planet that might otherwise be unable to support an atmosphere given the solar conditions in its vicinity might be able to do so within a deep shadowed rift or volcanic cave [45]. In a similar way, craterous terrain might offer a refuge for primitive life. The Lawn Hill crater [46] has been studied as an astrobiological analog, with researchers suggesting rapid sediment infill created a protected microenvironment for microbial organisms. Similar conditions may have occurred over the geological history of Mars [47]. Figure 14.6 shows the Atacama Desert presenting an analog to Mars and an ideal environment to study the boundary between sterility and habitability.

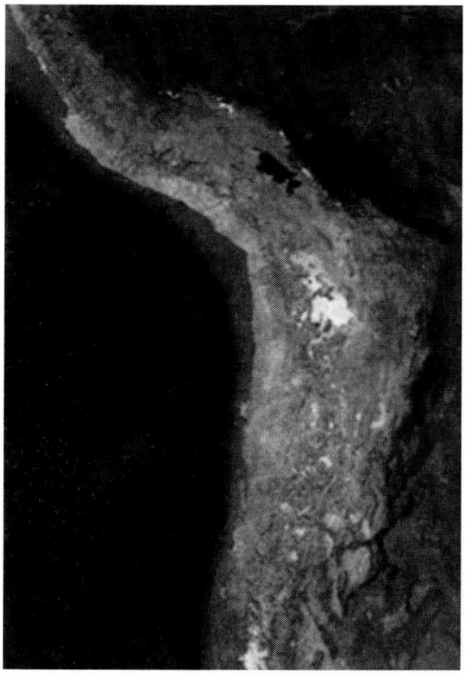

FIGURE 14.6
The Atacama Desert showing an analog to Mars and an ideal environment to study the boundary between sterility and habitability.

Earth environments that cannot support life are still instructive to astrobiologists in defining the limits of what organisms can endure. The Atacama was the subject of study in 2003 that partly replicated experiments from the Viking landings on Mars in the 1970s. No DNA could be recovered from two soil samples, and incubation experiments were also negative for biosignatures [48]. NASA launched the Mars Science Laboratory rover on November 26, 2011, which attempted to find for past or present life on Mars using a variety of scientific instruments.

14.8 Ecological Approaches for Potential Habitability

The two important and remarkable ecological approaches for predicting the potential habitability use 19 or 20 environmental factors, with emphasis on water availability, temperature, presence of nutrients, and protection from solar ultraviolet and galactic cosmic radiation [49,50]. The dominant habitability factors are shown in Table 14.2.

TABLE 14.2

Dominant Habitability Factors

Water	• Activity of liquid water • Past/future liquid (ice) inventories • Salinity, pH, and Eh of available water
Chemical environment	Nutrients: • C, H, N, O, P, S, essential metals, essential micronutrients • Fixed nitrogen • Availability/mineralogy Toxin abundances and lethality: • Heavy metals (e.g., Zn, Ni, Cu, Cr, As, Cd, etc., some essential, but toxic at high levels) • Globally distributed oxidizing soils
Energy for metabolism	Solar (surface and near-surface only) Geochemical (subsurface) • Oxidants • Reductants • Redox gradients
Conducive physical conditions	• Temperature • Extreme diurnal temperature fluctuations • Low pressure (Is there a low-pressure threshold for terrestrial anaerobes?) • Strong ultraviolet germicidal irradiation • Galactic cosmic radiation and solar particle events (long-term accumulated effects) • Solar UV-induced volatile oxidants, for example, O_2^-, O^-, H_2O_2, O_3 • Climate/variability (geography, seasons, diurnal, and eventually, obliquity variations) • Substrate (soil processes, rock microenvironments, dust composition, shielding) • High CO_2 concentrations in the global atmosphere • Transport (aeolian, groundwater flow, surface water, glacial)

Source: Kasting, J. F. and Ackerman, T. P., Climatic consequences of very high carbon dioxide levels in the Earth's early atmosphere, *Science*, 234, 1383–1385, 1986.

14.9 Uninhabited Habitats

An important distinction in habitability is between habitats that contain active life: inhabited habitats and habitats that are habitable for life, but uninhabited [51]. Uninhabited (or vacant) habitats could arise on a planet where there was no origin of life and also no transfer of life to the planet from another inhabited planet but where habitable environments exist. They might also occur on a planet that is inhabited, but the lack of connectivity between habitats might mean that many habitats remain uninhabited. Uninhabited habitats underline the importance of decoupling habitability and the presence of life, which can be stated as the general hypothesis, *where there are habitats, there is life*. The hypothesis is falsifiable by seeing uninhabited habitats and it is experimentally testable. Mars is considered as one plausible world that might harbor uninhabited habitats. Other stellar systems might host planets that are habitable, but devoid of life. In examining the possibility of extraterrestrial life, astronomers had long focused their attention on stars like the Sun. But as planetary systems that resemble the solar system are proving to be rare, they have started to explore the possibility that life might form in systems very unlike our own.

14.10 Habitability of Binary Star Systems

Typical estimates conclude that 50% or more of all stellar systems are binary systems. This may be partly a sample bias, as massive and bright stars tend to be in binaries and are observed and cataloged conveniently. A more precise analysis has focused that the more common fainter stars are generally singular and that up to two thirds of all stellar systems are solitary [52]. The separation between stars in a binary may range from less than one astronomical unit (AU) to several hundred. The gravitational effects will be negligible on a planet orbiting an otherwise suitable star and habitability potential will not be disrupted unless the orbit is highly eccentric. But if the separation is significantly less, a stable orbit may be impossible. If the distance of a planet to its primary exceeds about one-fifth of the closest approach of the other star, orbital stability becomes uncertain [50]. Whether planets might form in binaries had long been unclear, given that gravitational forces might interfere with planet formation. Theoretical work has shown that gas giants can form around stars in binary systems much as they do around solitary stars [53]. One study of Alpha Centauri, the nearest star system to the Sun, indicated that binaries need not be discounted in the search for habitable planets. Centauri A and B have an 11 AU distance at closest approach and both should have stable HZs. A study of long-term orbital stability for simulated planets within the system exhibits that planets within approximately 3 AU of either star may remain stable. The HZ for Centauri A is conservatively estimated at 1.2 to 1.3 AU and Centauri B at 0.73 to 0.74 and so they are well within the stable region [54].

14.11 Habitability of Red Dwarf Systems

If we can determine the habitability of red dwarf stars, it could help to find how common life in the universe might be, as red dwarfs make up between 70% and 90% of all the stars in the galaxy. Relative star sizes and photospheric temperatures are shown in Figure 14.7.

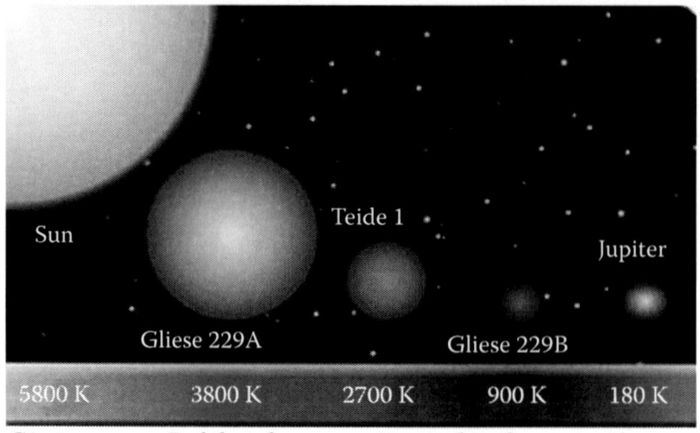

FIGURE 14.7
Relative star sizes and photospheric temperatures.

Any planets around a red dwarf such as the one presented here would have to huddle close to achieving Earth-like temperatures, perhaps inducing tidal locking.

For many years, astronomers ruled out red dwarfs as potential abodes for life. Their small size means that their nuclear reactions proceed very slowly and they emit very little light. Any planet in orbit around a red dwarf would have to huddle very near to its parent star for attaining Earth-like surface temperatures. This is from 0.3 AU for a star like Lacaille 8760 to as little as 0.032 AU for a star like Proxima Centauri [50]. At those distances, the gravity of the star would cause tidal locking. One side of the planet would eternally face the star, while the other would always face away from it. The only alternative in which potential life could avoid either an inferno or a deep freeze would be if the planet had an atmosphere sufficiently thick to transfer the star's heat from the day side to the night side or if there was a gas giant in the HZ, with a habitable moon, that would be locked to the planet instead of the star, permitting a more even distribution of radiation over the planet. It was assumed long past that such a thick atmosphere would prevent sunlight from reaching the surface in the first place preventing photosynthesis. An artist's impression of GJ 667 Cc, a potentially habitable planet orbiting a red dwarf constituent in a trinary star system, is shown in Figure 14.8.

Investigations have shown that the atmosphere of a planet needs only 10% of Earth's atmosphere, for the star's heat to be effectively carried to the night side [55]. This is well within the levels needed for photosynthesis, though water would still remain frozen. Seawater, too, could be effectively circulated without freezing solid if the ocean basins were deep enough for allowing free flow beneath the night side's ice cap. Further research concluded that tidally locked planets in red dwarf systems might at least be habitable for higher plants [56].

14.12 Different Factors Limiting Habitability

It is important to note that the size is not the only factor in making red dwarfs potentially unsuitable for life. On a red dwarf planet, photosynthesis on the night side would be impossible, as it would never find the Sun. On the day side, because the Sun does not rise or

FIGURE 14.8
Artist's impression of GJ 667 Cc, a potentially habitable planet orbiting a red dwarf constituent in a trinary star system.

set, areas in the shadows of mountains would remain so forever. Photosynthesis would be complicated by the fact that a red dwarf produces most of its radiation in the infrared, and on the Earth the process depends on visible light. Numerous terrestrial ecosystems rely on chemosynthesis rather than photosynthesis that would be possible in a red dwarf system. A static primary star position removes the need for plants to steer leaves toward the sun, deal with changing shade/sun patterns, or change from photosynthesis to stored energy during night. Owing to the lack of a day–night cycle, including the weak light of morning and evening, far more energy would be available at a particular radiation level. Red dwarfs are variable and violent. They are frequently covered in starspots that can dim their emitted light by up to 40% for months at a time. Red dwarfs also exhibit gigantic flares that can double their brightness in a matter of minutes [57]. This variation would be highly damaging for life, as it would not only destroy any complex organic molecules that form biological precursors and blow off sizeable portions of the planet's atmosphere. For a planet around a red dwarf star to support life, it would need a rapidly rotating magnetic field for protecting it from the flares. However, a tidally locked planet rotates only very slowly and so cannot produce a geodynamic core. The violent flaring period of a red dwarf's life cycle is estimated to be the first 1.2 billion years of its existence. If a planet forms far away from a red dwarf avoiding tidal locking and then migrates into the star's HZ, and after this turbulent initial period, it is possible that life may have a chance to develop [58].

14.13 Longevity and Ubiquity

There is one major advantage that red dwarfs have over other stars as abodes for life: they live a long time. It took about 4.5 billion years before humanity appeared on Earth and life as we know it. In that period of time (1 to 2.3 billion years), we can see suitable conditions for life [59,60]. Red dwarfs could live for trillions of years because their nuclear reactions are far slower than those of larger stars, meaning that life would have longer to evolve and survive. While the odds of finding a planet in the HZ around any specific red dwarf are slim, the total amount of a HZ around all red dwarfs combined is equal to the total amount around Sun-like stars given their ubiquity [61]. Also, this total amount of HZs will last longer, because red dwarf stars live for hundreds of billions of years or even longer on the main sequence [62].

14.13.1 Massive Stars

Very large stars, greater than ~100 solar masses, could have planetary systems consisting of hundreds of Mercury-sized planets within the HZ. Such systems could also contain brown dwarfs and low-mass stars [63]. However, the very short lifespan of stars of more than a few solar masses would scarcely permit time for a planet to cool. Massive stars are, therefore, eliminated as possible abodes for life [64]. But a massive-star system could be a progenitor of life in another way: the supernova explosion of the massive star in the central part of the system. A supernova disperses heavier elements formed during the phase throughout its vicinity. This could happen when the massive star moves off of the main sequence. Potentially, low-mass stars within the former massive-star system may be enriched with the relatively large supply of the heavy elements due to the supernova explosion. But this describes nothing about what types of planets would form as a result of the supernova material or what their habitability potential would be.

14.14 Galactic Neighborhood

Along with the characteristics of planets as well as their star systems, the wider galactic environment may impact habitability. Scientists considered the possibility that galactic HZs are better suited to life than others; the solar system in which we live, in the Orion Spur, on the Milky Way galaxy's edge is assumed to be in a life-favorable spot [65]. (a) It is not in a globular cluster where immense star densities are inimical to life, given excessive radiation and gravitational disturbance. Globular clusters are primarily composed of older stars. Also, in globular clusters, the great ages of the stars would mean a huge amount of stellar evolution by the host or other nearby stars, which due to their proximity may cause extreme harm to life on any planets, provided that they can form. (b) It is not near the galactic center where star densities increase the likelihood of ionizing radiation (e.g., from magnetars and supernovae). A supermassive black hole is believed to lie at the middle of the galaxy that might prove a danger to any nearby bodies. (c) The circular orbit of the Sun around the galactic center keeps it out of the way of the galaxy's spiral arms where intense radiation and gravitation may again lead to disruption [66]. (d) It is not near an active gamma ray source. As a matter of fact, relative loneliness is ultimately what a life-bearing system demands. If the Sun were crowded among other systems, the chance of being fatally close to dangerous radiation sources would increase significantly. Again, close neighbors might disrupt the stability of different orbiting bodies such as Oort cloud and Kuiper belt objects, which can bring catastrophe if knocked into the inner solar system. A star as metal rich as the Sun would perhaps not have formed in the very outermost regions of the Milky Way given a decline in the relative abundance of metals and a general lack of star formation. Therefore, a *suburban* location such as that the solar system enjoys is preferable to a galaxy's center or farthest reaches [67].

14.15 Hypothetical Types of Biochemistry

Most of the investigations of extraterrestrial life begin with the assumption that advanced lifeforms must have similar needs for life as on Earth. The hypothesis of other types of biochemistry suggests the possibility of lifeforms evolving around a different metabolic mechanism. Earth-like planets may be very rare but non-carbon-based complex life could possibly emerge in other environments. The most frequently mentioned alternative to carbon is silicon-based life, while ammonia is sometimes considered as an alternative solvent to water. Subsequently, more speculative ideas have focused on the bodies altogether different from Earth-like planets.

14.15.1 *Good Jupiters*

Good Jupiters are gas giants, similar to our solar system's Jupiter. They orbit their stars in circular orbits, far away from their HZ but, not to disturb but, to *protect* terrestrial planets in closer orbit in two critical ways. First, they assist to stabilize the orbits and thereby the climates of the inner planets. Second, they maintain the inner solar system relatively free of comets and asteroids that could cause devastating impacts [68]. Jupiter orbits the Sun nearly five times the distance between the Earth and the Sun. This is the rough distance we should expect to find good Jupiter's elsewhere. Research activities [69–72] have shown that Jupiter's role in finding the rate at which objects hit the Earth is significantly complicated than thought. For the long-period comets, Jupiter acts as a shield. It seems to increase the rate at which asteroids and short-period comets are flung toward our planet. Were Jupiter absent, it seems likely that the Earth would experience fewer impacts from potentially hazardous objects. By extension, it is becoming clear that the presence of Jupiter-like planets is no longer necessary as a prerequisite for planetary habitability. The role of Jupiter in the early history of the solar system is accepted as having played an important part in the hydration of our planet. It increased the eccentricity of asteroid belt orbits and enabled many to cross Earth's orbit and supply the planet with important volatiles. Hence, while the gas giants are now helpful protectors, they were once suppliers of critical habitability material. In contrast, Jupiter-sized bodies that orbit too close to the HZ (but not in it) or have a high elliptical orbit that crosses the HZ make it very difficult for an independent Earth-like planet to exist in the system [73].

14.16 Life's Impact on Habitability

A supplement to the factors that support life's emergence is the notion that life itself, once formed, becomes a habitability factor in its own right. An important Earth example was the production of oxygen by ancient cyanobacteria and eventually photosynthesizing plants leading to a radical change in the composition of Earth's atmosphere. This oxygen would prove fundamental to the respiration of animal species. Habitability cannot be separated from life already in existence on a planet. Planets that are geologically and meteorologically alive are much more likely to be biologically alive [74] as well and *a planet and its life will co-evolve.*

14.16.1 Location in Time and Extinction Events

Similar to living organisms, habitable planets and their stars have finite lifespans and their average life expectancy is generally based on the life of the latter: as a Sun-like star grows older, it gradually gets bigger, hotter, and brighter. Presently, the Sun is 70% brighter than when it was first formed. Consequently, the time when life evolves on a habitable planet and how long its star will remain in the main sequence, and before it gets too hot, must be taken into consideration when searching for Earth-like planets. Complex lifeforms such as plants and animals take longer time to evolve than microbes, with a minimum of at least four billion years after the planet forms. If the history of Earth was compressed into one year, all existing plants and animals would not appear until less than a second. So, habitable planets are best for plants and animals during middle age. If the planet is too young, it would be very hot and molten from the impacts occurring while the planet was forming and so complex life cannot survive. On the other hand, if life evolved too early, it would have been wiped out by violent collisions between the planet and space debris during formation. If the planet is too old, then its star's evolution would have caused it to begin its transformation into a red giant and so it would be experiencing a runaway greenhouse effect that would now make it too hot to support life. If life evolved too late, just right before the runaway greenhouse warming starts, there will not be sufficient time for plants and animals to evolve at all as unlike microbes cannot survive in extremely hot temperatures. But on smaller timescales, habitable planets may be vulnerable to catastrophic events that would potentially trigger a mass extinction event, such as an asteroid collision or a sudden change in climate. Due to the background extinction rate, the maximum life expectancy for any given existing plant or animal on Earth is nearly 10 million years. On the other hand, microbes can survive for hundreds of millions years. Mass extinctions can potentially hamper a habitable planet's ability for supporting complex life and planets that suffer fewer extinction events can support plant and animal life than those that suffered more. For planetary systems that are now red giants, any planet that was originally in the same distance as the HZ will now be too hot to support life and eventually be destroyed by the red giant expansion. As it continues to expand, planets that were originally too cold to support life will now be warmed up by the same red giant making life there possible.

14.17 Multidisciplinary Research

To understand the life of the stars and the properties of the numerous planetary systems, multidisciplinary research is essential as many of the parameters related to this challenging problem are closely connected. Conditions for life on a planet largely depend on the distance of the planet from its star, and how big and how luminous the star is. The age of the star, the age of the planet, and whether there are violent solar storm eruptions that can disturb the atmosphere of the planet are significant. Never before has mankind had these fantastic opportunities for investigating the properties of the stars and discovering, observing, and collecting data about solar systems. The joint works for analyzing data from various new angles are important. Instead of looking only for new planets, the researchers will set to work systematically and map the systems thoroughly. Observations of planetary systems and starquakes combined with extensive model calculations will provide completely new information about the properties of the stars. Over a good number of years, international groups of researchers have discovered many solar systems containing possible prerequisites for life as they resemble our own solar system. Technological progress has also boosted opportunities for finding new, interesting solar systems far out in space. The Kepler Mission particularly has led to a number of major discoveries where several Earth-like planets have been observed. However, it is one thing that the planets resemble Earth but another thing altogether that they have a biosphere, or that they are inhabited by living organisms. The fact is that nobody has yet considered the same multidisciplinary approach to space research as the Stellar Astrophysics Center is now doing by combining the study of stars and planets with the study of biology under extreme conditions.

References

1. NASA. Goal 1: Understand the nature and distribution of habitable environments in the Universe. *Astrobiology: Roadmap.* NASA. http://nai.nasa.gov/media/roadmap/2003/g1.html (retrieved August 11, 2007).
2. Ward, P. D. and Brownlee, D. (2000). *Rare Earth: Why Complex Life Is Uncommon in the Universe.* Springer, New York. pp. 191–220.
3. The Heat History of the Earth. *Geolab.* James Madison University. http://csmres.jmu.edu/geollab/Fichter/PlateTect/heathistory.html (retrieved May 11, 2007).
4. Raymond, S. N. et al. (2007). High resolution simulations of the final assembly of Earth-like planets 2: water delivery and planetary habitability. *Astrobiology* 7 (1): 66–84.
5. Dole, S. H. (1964). *Habitable Planets for Man.* Blaisdell Pub. Co., New York. p. 103.
6. Budyko, M. I. (1969). The effect of solar radiation variations on the climate of the Earth. *Tellus* 21 (5): 611.
7. Sellers, W. D. (1969). A global climatic model based on the energy balance of the earth-atmosphere system. *Journal of Applied Meteorology* 8 (3): 392–400.
8. North, G. R. (1975). Theory of energy-balance climate models. *Journal of the Atmospheric Sciences* 32 (11): 2033–2043. http://journals.ametsoc.org/doi/pdf/10.1175/1520-0469(1975)032%3C2033:TOEBCM%3E2.0.CO%3B2 (retrieved August 12, 2016).
9. Rasool, I. and De Bergh, C. (1970). The runaway greenhouse and the accumulation of CO_2 in the Venus atmosphere. *Nature* 226 (5250): 1037–1039.
10. Hart, M. H. (1979). Habitable zones about main sequence stars. *Icarus* 37: 351–357.
11. Fogg, M. J. (1992). An estimate of the prevalence of biocompatible and habitable planets. *Journal of the British Interplanetary Society* 45 (1): 3–12.
12. Kasting, J. F. et al. (1993). Habitable zones around main sequence stars. *Icarus* 101 (1): 108–118.
13. Spiegel, D. S. et al. (2010). Generalized Milankovitch cycles and long-term climatic habitability. *The Astrophysical Journal* 721 (2): 1308.
14. Abe, Y. et al. (2011). Habitable zone limits for dry planets. *Astrobiology* 11 (5): 443–460.
15. Vladilo, G. et al. (2013). The habitable zone of Earth-like planets with different levels of atmospheric pressure. *The Astrophysical Journal* 767 (1): 65.
16. Kopparapu, R. K. (2013). A revised estimate of the occurrence rate of terrestrial planets in the habitable zones around Kepler m-dwarfs. *The Astrophysical Journal Letters* 767 (1): L8.

17. Zsom, A. et al. (2013). Towards the minimum inner edge distance of the habitable zone. *The Astrophysical Journal* 778 (2): 109.
18. Overbye, D. (November 4, 2013). Far-off planets like the Earth dot the galaxy. *The New York Times*. http://www.nytimes.com/2013/11/05/science/cosmic-census-finds-billions-of-planets-that-could-be-like-earth.html?_r=1 (retrieved November 5, 2013).
19. Petigura, E. A. et al. (2013). Prevalence of Earth-size planets orbiting Sun-like stars. *Proceedings of the National Academy of Sciences of the United States of America* 110 (48): 19273–19278.
20. Khan, A. (November 4, 2013). Milky Way may host billions of Earth-size planets. *Los Angeles Times*. http://www.latimes.com/science/la-sci-earth-like-planets-20131105-story.html (retrieved November 5, 2013).
21. Turnbull, M. C. and Tarter, J. C. (2003). Target selection for SETI: 1. A catalog of nearby habitable stellar systems. *The Astrophysical Journal Supplement Series* 145: 181–198.
22. Shriber, M. (Oct. 26, 2009). Detecting life-friendly moons. *Astrobiology Magazine*. NASA. http://www.astrobio.net/news-exclusive/detecting-life-friendly-moons/ (retrieved May 9, 2013).
23. Cowen, R. (June 7, 2008). A shifty moon. *Science News*. https://www.sciencenews.org/article/shifty-moon?mode=magazine&context=657 (accessed August 12, 2016).
24. Rogue planets could harbor life in interstellar space, say astrobiologists. *MIT Technology Review*. February 9, 2011. https://www.technologyreview.com/s/422659/rogue-planets-could-harbor-life-in-interstellar-space-say-astrobiologists/ (retrieved June 24, 2013).
25. Bryner, J. (June 24, 2009). Ocean hidden inside Saturn's moon. Space.com. http://www.space.com/6883-ocean-hidden-saturn-moon.html (retrieved April 22, 2013).
26. Villard, R. (November 18, 2011). Alien life may live in various habitable zones: discovery news. www.news.discovery.com. Discovery Communications LLC. Retrieved April 22, 2013.
27. Kasting, J. (2010). *How to Find a Habitable Planet*. Princeton University Press, New Jersey. p. 127. http://press.princeton.edu/titles/9862.html (retrieved May 4, 2013).
28. Strughold, H. (1953). *The Green and Red Planet: A Physiological Study of the Possibility of Life on Mars*. University of New Mexico Press, New Mexico.
29. Huang, S.-S. (April 1960). Life-supporting regions in the vicinity of binary systems. *Publications of the Astronomical Society of the Pacific* 72 (425): 106–114.
30. Gilster, P. (2004). *Centauri Dreams: Imagining and Planning Interstellar Exploration*. Springer, New York. p. 40.
31. The Goldilocks Zone (Press release). NASA. October 2, 2003. http://science.nasa.gov/science-news/science-at-nasa/2003/02oct_goldilocks/ (retrieved April 22, 2013).
32. Gonzalez, G. et al. (July 2001). The galactic habitable zone I: Galactic chemical evolution. *Icarus* 152 (1): 185–200.
33. Venus. Case Western Reserve University. September 13, 2006. http://burro.cwru.edu/stu/advanced/venus.html (retrieved December 21, 2011).
34. Hadhazy, A. (April 3, 2013). The 'habitable edge' of exomoons. *Astrobiology Magazine*. NASA. http://www.astrobio.net/news-exclusive/the-habitable-edge-of-exomoons/ (retrieved April 22, 2013).
35. Sharp, T. (2012). Atmosphere of the moon. Space.com. http://www.space.com/18067-moon-atmosphere.html (retrieved April 23, 2013).
36. Bolonkin, A. A. (2009). *Artificial Environments on Mars*. Berlin, Heidelberg: Springer. pp. 599–625.
37. A'Hearn, M. F. and Feldman, P. D. (1992). Water vaporization on Ceres. *Icarus* 98 (1): 54–60.
38. Chow, D. (February 2, 2012). Newfound alien planet is best candidate yet to support life, scientists say. Space.com. http://www.space.com/14444-alien-planet-super-earth-habitable-zone.html (retrieved February 3, 2012).
39. von Bloh, W. et al. (2008). Habitability of super-Earths: Gliese 581c and 581d. *Proceedings of the International Astronomical Union* 3 (S249): 503–506.

40. Gilster, P. (2007). Gliese 581d: A habitable world after all? Centauri Dreams.org. http://www.centauri-dreams.org/?p=1625 (retrieved January 18, 2009).
41. Cuntz, M. (2013). S-type and P-type habitability in stellar binary systems: a comprehensive approach: I. Method and applications. *The Astrophysical Journal* 780 (1): 14.
42. Forget, F. and Pierrehumbert, R. T. (1997). Warming early Mars with carbon dioxide clouds that scatter infrared radiation. *Science* 278 (5341): 1273–1276.
43. Mischna, M. et al. (2000). Influence of carbon dioxide clouds on early Martian climate. *Icarus* 145 (2): 546–554.
44. Vu, L. Planets prefer safe neighborhoods (Press release). www.spitzer.caltech.edu. NASA/Caltech. Retrieved April 22, 2013.
45. Buccino, A. P. et al. (2006). Ultraviolet radiation constraints around the circumstellar habitable zones. *Icarus* 183 (2): 491–503.
46. Lindsay, J. and Brasier, M. (2006). Impact Craters as biospheric microenvironments, Lawn Hill Structure, Northern Australia. *Astrobiology* 6 (2): 348–363.
47. Barnes, R. and Heller, R. (2013). Habitable planets around white and brown dwarfs: the perils of a cooling primary. *Astrobiology* 13 (3): 279–291.
48. Yang, J. et al. (2013). Stabilizing cloud feedback dramatically expands the habitable zone of tidally locked planets. *The Astrophysical Journal* 771 (2): L45.
49. Guo, J. et al. (2009). Probability distribution of terrestrial planets in habitable zones around host stars. *Astrophysics and Space Science* 323 (4): 367.
50. Kasting, J. F. and Ackerman, T. P. (1986). Climatic consequences of very high carbon dioxide levels in the Earth's early atmosphere. *Science* 234 (4782): 1383–1385.
51. Cain, F. and Gay, P. (2007). AstronomyCast episode 40: American Astronomical Society Meeting, May 2007. *Universe Today*. http://www.astronomycast.com/2007/06/episode-40-american-astronomical-society-meeting-may-2007/ (retrieved June 17, 2007).
52. Croswell, K. (January 27, 2001). Red, willing and able (Full reprint). *New Scientist*. http://www.kencroswell.com/reddwarflife.html (retrieved August 5, 2007).
53. Alpert, M. (November 7, 2005). Red star rising. *Scientific American*. http://www.scientificamerican.com/article/red-star-rising/ (retrieved January 19, 2013).
54. Research Corporation. (December 19, 2006). Andrew West: 'Fewer flares, starspots for older dwarf stars.' EarthSky. http://earthsky.org/space/fewer-flares-starspots-for-older-dwarf-stars (retrieved April 27, 2013).
55. Christensen, B. (April 1, 2005). Red giants and planets to live on. Space.com. http://www.space.com/920-red-giants-planets-live.html (retrieved April 27, 2013).
56. Lopez, B. et al. (2005). Can life develop in the expanded habitable zones around red giant stars? *The Astrophysical Journal* 627 (2): 974.
57. Lorenz, R. D. et al. (1997). Titan under a red giant sun: A new kind of 'habitable' moon. *Geophysical Research Letters* 24 (22): 2905–2908.
58. Voisey, J. (February 23, 2011). Plausibility check—habitable planets around red giants. *Universe Today*. http://www.universetoday.com/83248/plausibility-check-habitable-planet-around-red-giants/ (retrieved April 27, 2013).
59. Choi, C. Q. (January 9, 2011). Alien life more likely on 'Dune' planets. *Astrobiology Magazine*. http://www.astrobio.net/news-exclusive/alien-life-more-likely-on-dune-planets/ (retrieved August 12, 2016).
60. Abe, Y. et al. (2011). Habitable zone limits for dry planets. *Astrobiology* 11 (5): 443–460.
61. Drake, M. J. (2005). Origin of water in the terrestrial planets. *Meteoritics and Planetary Science* 40 (4): 519–527.
62. Drake, M. J. and Campins, H. (August 2005). Origin of water in the terrestrial planets. in 229th Symposium of the International Astronomical Union. *Asteroids, Comets, and Meteors (IAU S229)* 1 (4). Búzios, Rio de Janeiro, Brazil: Cambridge University Press. pp. 381–394.
63. Kuchner, M. (2003). Volatile-rich Earth-mass planets in the habitable zone. *The Astrophysical Journal* 596 (1): L105–L108.

64. Charbonneau, D. et al. (2009). A super-Earth transiting a nearby low-mass star. *Nature* 462 (7275): 891–894.
65. Kuchner, M. et al. (2007). Mass–radius relationships for solid exoplanets. *The Astrophysical Journal* 669 (2): 1279–1297.
66. Vastag, B. (December 5, 2011). Newest alien planet is just the right temperature for life. *The Washington Post*. https://www.washingtonpost.com/national/health-science/newest-alien-planet-is-just-the-right-temperature-for-life/2011/12/05/gIQAPk1vWO_story.html (retrieved August 12, 2016).
67. Robinson, T. D. and Catling, D. C. (2012). An analytic radiative-convective model for planetary atmospheres. *The Astrophysical Journal* 757 (1): 104.
68. Hamilton, D. P. and Burns, J. A. (1992). Orbital stability zones about asteroids. II—the destabilizing effects of eccentric orbits and of solar radiation. *Icarus* 96 (1): 43.
69. Becquerel, P. (1950). La suspension de la vie au dessous de 1/20 K absolu par demagnetization adiabatique de l'alun de fer dans le vide les plus eléve. *Comptes Rendus de l'Académie des Séances* (in French) 231: 261–263.
70. Horikawa, D. D. (2012). *Anoxia Evidence for Eukaryote Survival and Paleontological Strategies,* 21st edn. A. V. Altenbach, J. M. Bernhard & J. Seckbach (eds.), The Netherlands: Springer. pp. 205–217. http://www.springer.com/us/book/9789400718951 (retrieved January 21, 2012).
71. Kane, S. R. and Gelino, D. M. (2012). The habitable zone and extreme planetary orbits. *Astrobiology* 12 (10): 940–945.
72. Wethington, N. (September 16, 2008). How many stars are in the Milky Way? *Universe Today*. http://www.universetoday.com/123225/how-many-stars-are-in-the-milky-way-2/ (retrieved April 21, 2013).
73. Borenstein, S. (February 19, 2011). *Cosmic Census Finds Crowd of Planets in Our Galaxy*. New York, NY: Associated Press.
74. Choi, C. Q. (March 21, 2011). New estimate for alien Earths: 2 billion in our galaxy alone. Space.com. http://www.space.com/11188-alien-earths-planets-sun-stars.html (retrieved August 12, 2016).

15

Solar Satellites

15.1 Introduction

The Sun is one of the most studied body in our solar system, the reason, it is the closest star and its proximity to Earth makes it a prime candidate compared to all other astrophysical objects outside the solar system, which means that it can be investigated in far greater detail for space research. In addition, at most electromagnetic wavelengths it is the brightest source in the sky. Considering the dominance of the Sun and its overwhelming necessity for life on Earth, solar observations have always got highest priority. This is evident from the large number of spacecraft that have been dedicated to solar observations. The Sun is receiving more and more attention due to the inherent vulnerabilities to the solar-controlled space weather, that is, coronal mass ejection (CME). In 1946, captured World War II German V-2 rockets played a big role in the far ultraviolet (UV) research. Years later, NASA initiated the orbiting solar observatory (OSO) series of unmanned solar satellites as well as the first major manned solar mission from space called Skylab. Early solar programs provided many outstanding observations in the UV, EUV, X-rays, and gamma rays, which provided the foundation of solar space science. Some other nations followed with their own solar missions, either alone as in the case of the Soviet Union or in collaboration with the United States.

15.2 Timeline

Plot of Figure 15.1 reveals the periods of operation of the major past, present, and future solar space missions. The bars that fade out in 2009 exhibit missions those are still operational and are expected to continue as long as the instruments function and funding is available. Future missions have also shown starting at their expected launch dates and extending their planned lifetimes.

15.3 Past Missions

The following subsections cover all past solar-dedicated spacecraft including associated solar instruments which made significant contributions to solar research. We have described briefly the major solar missions with their capabilities and the main scientific results achieved.

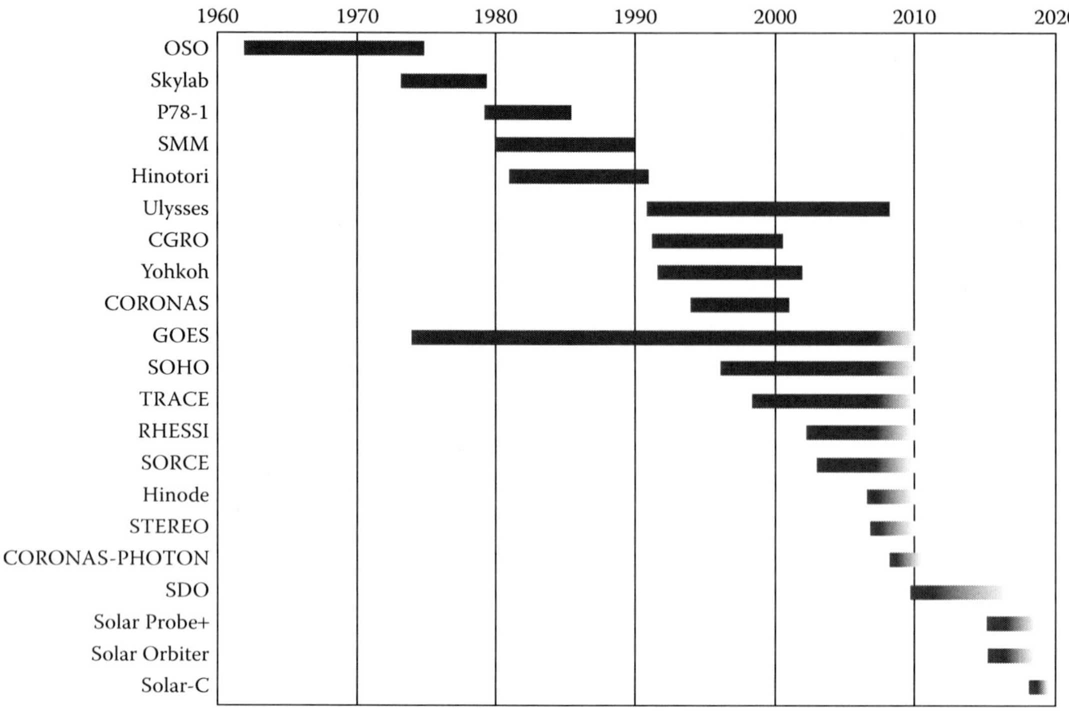

FIGURE 15.1
Plot shows the periods of operation of the major past, present, and future solar space missions.

15.3.1 Orbiting Solar Observatories

In the early space program, NASA initiated the OSO series of dedicated unmanned solar satellites. The solar instruments carried on these spacecraft provided UV and EUV imaging and spectroscopic observations including X-ray and gamma ray spectroscopy. The results of these pioneering observations provide the information related to basic solar space science. Figure 15.2 shows the photograph of an OSO.

The OSO series involved a total of nine spacecraft and were given sequential letter designations (OSO-A through OSO-I) until launch when they were renamed with the next number in the sequence. OSO-C was not launched owing to failure during testing on April 14, 1964, that damaged the spacecraft resulting in the death of three engineers. The other eight successful spacecraft were named OSO-1, that was launched in 1962, through OSO-8, which was launched in 1975. The first seven OSO's were built by Ball Brothers Research Corporation (later renamed as Ball Aerospace). Hughes Aircraft won the follow-on contract for OSO-I, J, and K. OSO-I became OSO-8 at launch on June 21, 1975, but OSO-J and -K were combined due to budgetary constraints causing it to become the solar maximum mission (SMM). All the OSO spacecraft incorporated a "wheel" component spinning at up to 15 rpm for pointing stability and a de-spun platform to carry the imaging instruments. This arrangement used exquisite sub-arc second pointing accuracy and stability before the era of three-axis stabilization techniques used presently. Some other instruments that did not require precise arrangement were mounted in the spinning "wheel" section. Those instruments were X-ray and gamma ray spectrometers pointed perpendicular to the spin axis so that they scanned across the Sun. Similar instruments pointed parallel or

FIGURE 15.2
Orbiting solar observatory.

near-parallel to the spin axis to observe nonsolar sources of astrophysical interest. Solar panels mounted on the de-spun platform gave the power and slip rings between the spinning and de-spun components were used for power and signal connections.

15.3.2 Hinotori

On February 21, 1981, the Japanese launched a dedicated solar spacecraft (Figure 15.3) named Hinotori, which means "phoenix." The satellite carried hard X-ray imaging that was to rival SMM. It was the first mission for using modulation X-ray collimators on a spinning spacecraft to make X-ray images of solar flares utilizing a Fourier transform technique. The same technique was repeated almost 20 years later on NASA's high energy solar spectroscopic imager (HESSI) later called the Reuven Ramaty HESSI (RHESSI). The X-ray

FIGURE 15.3
Hinotori.

images made with Hinotori data at energies above ~25 keV were the first to reveal clear evidence for footpoint brightening during impulsive flares.

15.3.3 Skylab

The biggest early manned program for observing the Sun was Skylab with the Apollo Telescope Mount (ATM) as the main solar observatory (Figure 15.4). A solar panel and part of its external shielding were lost on launch and the astronauts had to rig a "golden umbrella" for maintaining the temperature under control. Skylab operated for 6 years in orbit from launch on May 14, 1973, through re-entry on July 11, 1979.

Skylab carried a large number of astrophysical instruments with eight independent solar instruments on the ATM. These are listed in Table 15.1. Skylab data was credited for many outstanding observations. In 1973, astronauts maintained the instruments and returned the film utilized for recording the images. The major results were presented in reports to promote the study of observations.

FIGURE 15.4
Skylab.

TABLE 15.1
Solar Instruments on Skylab's Apollo Mount

Name	Instrument	Institution	Science Objective
H-alpha 1 and 2	Telescope	HCO & MFSC	Solar emission, solar flares
S020	EUV and X-ray camera	NRA	Solar photography for ionized atoms
S052	White light chronograph	HAO	Solar corona
S054	X-ray telescope	AS&E	Spectrography of solar flares and active regions
S055	UV spectroheliometer	HCO	Chromospheric and coronal EUV photography
S056	X-ray telescope	MSFC	Emission of lower corona
S082a	EUV spectroheliograph	NRL	Structure of inner corona
S082b	UV spectroheliograph	NRL	Corona, chromosphere, and transition region

FIGURE 15.5
Solar Maximum Mission.

15.3.4 Solar Maximum Mission

The SMM (Figure 15.5) was a three-axis stabilized instrument so that all of the instruments pointed to the Sun for arc second accuracy and stability. The payload was selected to concentrate on various observations including understanding of solar flares (energetic explosions), which were revealed by earlier observations. The physical processes involved in the impulsive energy release and subsequent dissipation were widely unknown. With a view to provide information related to this field, imaging and spectroscopy was carried out at UV, EUV, and X-ray wavelengths. All these experiments and observations were needed for examining the emission from plasma heated to temperatures as high as several tens of millions Kelvin. The electrons accelerated during the solar flares to super thermal energies were detected through the bremsstrahlung X-rays which they produced. SMM included the first instrument capable of imaging these X-rays up to energies of 30 keV, well above the energies produced by the hottest plasma. The protons and heavier ions accelerated during the process of flares were identified through the nuclear gamma rays. SMM carried a sensitive gamma ray spectrometer, which made the first pioneering measurements of this high-energy emission on OSO-7 back in 1972.

SMM was launched by a Delta rocket on February 14, 1980, on the rising phase of the 11-year cycle of solar activity. It successfully worked until November 1980. The nonimaging instruments continued to operate as the spacecraft slowly spun about an axis pointed within a few degrees of the Sun's center. In April 1984, the first of many spacecraft rescue missions using the space shuttle, the aspect system was fixed by the astronauts and the SMM was able to give six more years of solar observations, until it re-entered the Earth's atmosphere in December 1989.

15.3.5 P78-1

This spacecraft was of the OSO type with a solar-oriented sail and a rotating wheel section. It was built using spare components from OSO-7. The mission was launched on February 24, 1979. On September 13, 1985, it was destroyed in orbit during a US Air Force test of an anti-satellite weapon (ASAT) launched from an F-15 fighter plane at 80,000 ft (Figure 15.6).

The solar instruments included Bragg crystal spectrometers of high-resolution covering selected wavelengths, a spectrometer, X-ray proportional-counter spectrometers, and a white light coronagraph. This mission was able to successfully record the first observation

FIGURE 15.6
P78-1.

of a halo CME and the discovery of the association of mass ejections with interplanetary shocks. High spectral resolutions of X-ray spectra (solar flares) at different wavelengths were obtained as well as chromospheric signatures and evaporation on X-ray line profiles.

15.3.6 Ulysses

This is the only mission capable of overcoming the limitations of all other measurements restricted to the vicinity of the ecliptic plane. The mission gave unprecedented details about the nature of the solar wind. With a lifetime of 5 years, Ulysses (Figure 15.7) continued to supply new scientific results up to early 2008, when a power malfunction led to the failure of the X-band transmitter and heating systems.

The following instruments were carried by the spacecraft:

- Magnetometer (VHM/FGM)
- Solar Wind Plasma Experiment (SWOOPS)
- Solar Wind Ion Composition Instrument (SWICS)
- Coronal-Sounding Experiment (SCE)
- Gravitational Wave Experiment (GWE)
- Unified Radio and Plasma Wave Instrument (URAP)
- Energetic Particle Instrument (EPAC)
- Interstellar Neutral-Gas Experiment (GAS)
- Low-Energy Ion and Electron Experiment (HISCALE)
- Cosmic Ray and Solar Particle Instrument (COSPIN)
- Solar X-Ray and Cosmic Gamma-Ray Burst Instrument (GRB)
- Dust Experiment (DUST)

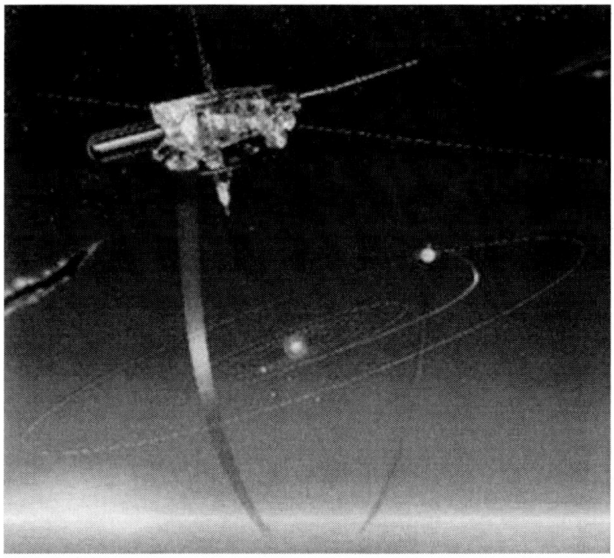

FIGURE 15.7
Ulysses in 1990, the joint venture of ESA–NASA Ulysses mission into a unique out-of-the-ecliptic orbit (Figure 15.8) which goes over the Sun's poles every 6.2 years. (Courtesy of ESA, http://www.esa.int/Our_Activities/Space_Science/Ulysses_overview [accessed August 12, 2016].)

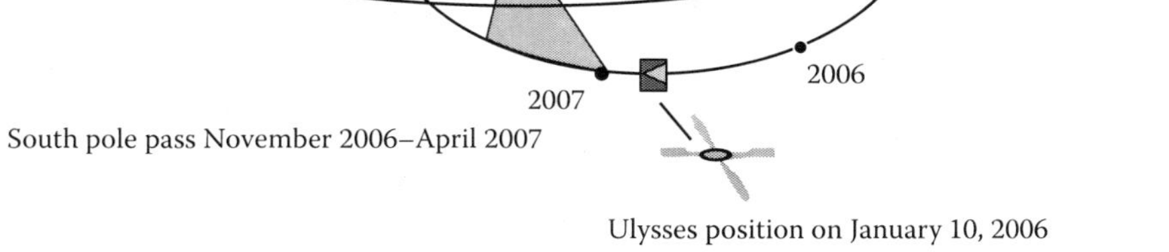

FIGURE 15.8
Ulysses orbit. (Courtesy of NASA, http://ulysses.jpl.nasa.gov/mission/trajectory.html [accessed August 12, 2016].)

15.3.7 Compton Gamma Ray Observatory

Strictly speaking, Compton Gamma Ray Observatory (CGRO) is not a solar satellite. It is the second of NASA's Great Observatories that made many valuable observations of solar flare X-rays and gamma rays (Figure 15.9). Launched in April 1991 and re-entering the Earth's atmosphere in June 4, 2000, CGRO covered nearly a solar cycle with observations spanning an unprecedented six decades of the electromagnetic spectrum, from X-rays as low as 30 keV to gamma rays at 30 GeV.

FIGURE 15.9
CGRO.

TABLE 15.2
CGRO Instruments

Name	Acronym	Institution	Energy Range
Burst and transient source experiment	BATSE	MSFC	<20 keV to 1 MeV
Oriented scintillation spectrometer experiment	OSSE	NRL	50 keV to 10 MeV
Compton telescope	COMPTEL	UNH	1–30 MeV
Energetic gamma experiment telescope	EGRET	GSFC	20 MeV to 30 GeV

The four instruments of CGRO are listed in Table 15.2.

BATSE solar flare data can be obtained at the Solar Data Analysis Center. Recent gamma ray observations continue to be made with the sensitive ESA International Gamma-Ray Astrophysics Laboratory (INTEGRAL) in cooperation with Russia and the United States. The four instruments cover the energy range from 3 keV to 10 MeV. However, it cannot be pointed to the Sun and hence gamma ray flares can only be identified through the thick anticoincidence shield around the high-resolution germanium detectors.

15.3.8 Complex Orbital Observations Near-Earth of Activity of the Sun

Russian CORONAS project (Complex Orbital Observations Near-Earth of Activity of the Sun) was envisioned to include three missions for making observations during various phases of the 11-year solar cycle (Figure 15.10). During the early period of solar space observations, the Russians launched some spacecraft with instrumentation designed especially for investigating the properties of solar flares. These were less technically sophisticated than the NASA instrumentation but, still managed to make those useful for contributions to solar science.

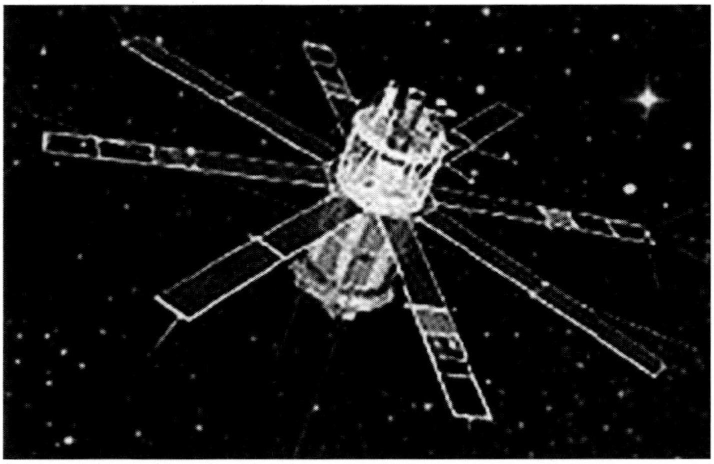

FIGURE 15.10
CORONAS.

15.3.8.1 CORONAS-I

CORONAS-I, the first mission in the series, was launched on March 2, 1994, during solar minimum and re-entered on March 4, 2001. It carried altogether 12 scientific instruments including the RES-K solar X-ray spectrograph, the Helikon solar gamma ray detector, Terek spectroheliometer, the SUVR-SP-C UV radiometer, the DIFOS optical photometer, and other instruments.

15.3.8.2 CORONAS-F

CORONAS-F was launched on July 31, 2001, near solar maximum and re-entered on December 6, 2005. It carried the following instruments including UV, EUV, X-ray and gamma ray spectrometers, radio receivers, and particle counters:

- DIFOS Multichannel Solar Photometer (IZMIRAN, GAO NANU, AIP)
- RES X-Ray Spectroheliograph (FIAN, SAI)
- SPIRIT Full Sun XUV spectroscopy imaging (FIAN, IAS, IAS)
- SRT Solar X-Ray Telescope (FIAN, IAS)
- DIOGENESS X-Ray Spectrometer and Fotometer (CBK PAN, IZMIRAN)
- RESIK X-Ray Spectrometer (CBK PAN, IZMIRAN, MSSL, RAL, NRL)
- SKL Solar Cosmic Rays Complex (NIIYaF, IEP)
- MKL Cosmic Ray Monitor
- SKI Spectrometer of Energy and Ion Chemical Composition
- SONG Solar Neutron and Gamma Ray Spectrometer
- PR-N: X-ray polarimeter
- SPR Solar Spectropolarimeter (FIAN, NIIYaF)
- IRIS Flare Spectrometer (FTI)

- HELICON Gamma Spectrometer (FTI)
- RPS X-Ray Spectrometer (IKI, MIFI)
- AVS Time-Amplitude Spectrometer (MIFI, NIIYaF)
- SUFR Solar UV Radiometer (IPG)
- VUSS Solar UV Spectrophotometer (IPG)

15.3.8.3 CORONAS-PHOTON

CORONAS-PHOTON was the third satellite in the CORONAS series. It was launched on January 29, 2009, into a circular, near-polar orbit (Figure 15.11).

All the scientific instruments on the satellite were turned off due to power supply problems on December 1, 2009, and the mission was declared lost on April 18, 2010. The scientific goals of the project were (1) to study the energy accumulation and the transformation of accelerated energetic particles during solar flares; (2) to investigate the solar activity correlation with physical–chemical processes in the Earth's upper atmosphere; and (3) to examine the acceleration mechanisms, propagation, and interaction of fast particles in the solar atmosphere. The CORONAS-PHOTON payload complex includes the following UV, EUV, X-ray, gamma ray, and cosmic ray instruments:

- Natalya-2M spectrometer MIFI, Moscow, Russia
- RT-2 gamma-telescope TIFR/ICSP/VSSC [14] India
- STEP-F Electron and proton detector Kharkov National University, Ukraine
- SM-8M magnetometer NPP Geologorazvedka/MIFI, Russia
- Pingvin-M (Penguin) polarimeter MIFI, Moscow, Russia
- Konus-RF X-ray and gamma spectrometer Ioffe Institute, Russia
- BRM X-ray detector MIFI, Russia
- FOKA UV-detector MIFI, Russia
- TESIS telescope/spectrometer FIAN, Russia, with SphinX soft X-ray spectrophotometer, SRC PAS, Poland
- Electron-M-Peska charged particles analyser NIIYaF MGU, Russia

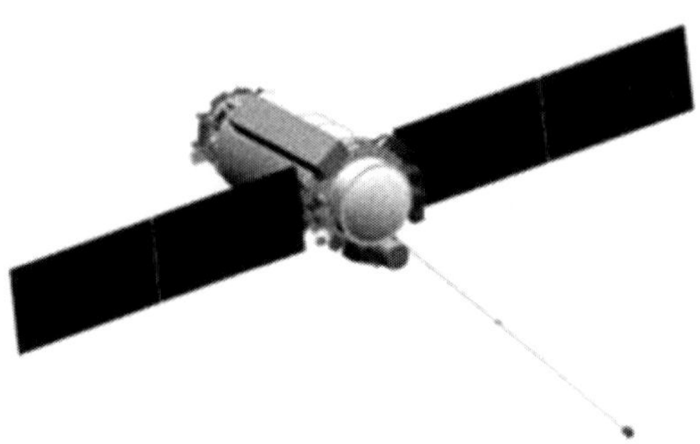

FIGURE 15.11
CORONAS-PHOTON.

Service systems

- Radio transmission system and antennas RNII KP, Russia
- SSRNI science data collection and registration system IKI, Russia

15.3.9 Yohkoh

Yohkoh was the second Japanese solar satellite with US and UK collaboration (Figure 15.12). It was launched on August 30, 1991. All the instruments successfully operated for over 10 years until December 14, 2001, when the spacecraft failed to recover from a "safe-hold" resulting from an annular solar eclipse. The four instruments of Yohkoh solar satellite are given in Table 15.3.

Over 10 years of observations by Yohkoh solar satellite, more that 6 million SXT snapshot images were accumulated and more than 2800 flares were detected with HXT and their spectra recorded with BCS and WBS. The most important scientific result from Yohkoh

FIGURE 15.12
Yohkoh.

TABLE 15.3
Yohkoh Instruments

Name	Acronym	Institution	Energy Range
Soft X-ray Telescope	SXT	United States	~1 keV
Hard X-ray telescope	HXT	Japan	14–93 keV
Bragg Crystal Spectrometer	BCS	United Kingdom	1–10 keV
Wide spectrometer	WBS	Japan	20 MeV to 30 GeV

was the detection of a coronal hard X-ray source, which has become known as the "Masuda Flare." Other remarkable achievements include the establishment of double hard X-ray sources, first observed with the Hard X-ray Imaging Spectrometer on the SMM, as bremsstrahlung emission. This was from nonthermal electrons accelerated in the corona during a flare and propagating to the footpoints of magnetic loops. Thermal sources at the top of magnetic loops were found on a regular basis within soft X-rays during flares. A super-hot (>30 MK) thermal component was often observed in hard X-rays during larger flares [1].

15.4 Current Missions

The following subsections cover present solar-dedicated missions and related solar instruments. These missions are significantly contributing to solar research. We provide the names and capabilities of the instruments and links to the main scientific results.

15.4.1 Solar and Heliospheric Observatory

The Solar and Heliospheric Observatory (SOHO) is an international collaboration between ESA and NASA. Launched on December 2, 1995, SOHO was used for investigating the Sun's deep core to the outer corona and the solar wind (Figure 15.13). The combined gravity of the Earth and Sun keep SOHO locked near the Earth–Sun line, nearly 1.5 million km from Earth [2–6].

SOHO was initially designed for a nominal mission lifetime of two years. SOHO has now been in operation for well over a complete 11-year solar cycle. Control of this spacecraft was lost in June 1998 but restored only after 3 months. All 12 instruments were still usable and interestingly most with no ill effects. However, two of the three on-board gyroscopes failed immediately and a third in December 1998. This demanded new on-board software which was installed in February 1999, which no longer relies on gyroscopes [7–10]. The spacecraft was designed for investigating the internal structure

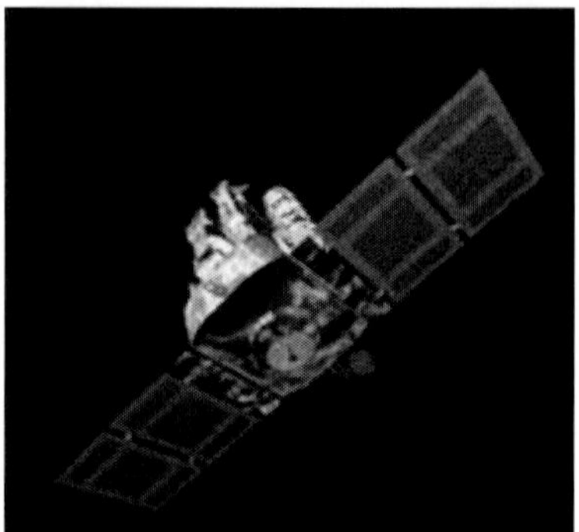

FIGURE 15.13
The SOHO spacecraft.

of the Sun, its extensive outer atmosphere, and the origin of the solar wind, which provided information about the complex interactions between the Sun and the Earth's environment better than ever before. Among its many contributions, SOHO has imaged our star's convective layer and the structure of sunspots. It provided profiles of the temperature, rotation, and gas flows in the interior and also measured the acceleration of the fast and slow solar wind. It was useful in discovering new dynamics of solar phenomena like coronal waves and solar tornadoes. SOHO has revolutionized our ability to forecast space weather and monitoring the total solar irradiance which is useful for understanding the impact of solar variability on Earth's climate. It has become the most prolific discoverer of comets in the history of astronomy, identifying more than 1500 to date [11–14].

SOHO Instruments

1. *Coronal Diagnostic Spectrometer (CDS)*: It detects emission lines from ions and atoms in the solar corona and transition region to provide diagnostic information on the solar atmosphere, particularly of the plasma in the temperature range from 10,000 to more than 1,000,000°C.
2. *Charge, Element, and Isotope Analysis System (CELIAS)*: It samples the solar wind and energetic ions of solar, interplanetary, and interstellar origin. It also analyses the density and composition of particles present in the solar wind and warns of incoming solar storms which could damage satellites in Earth orbit.
3. *Global Oscillations at Low Frequencies (GOLF)*: It can be used to study the internal structure of the Sun by measuring velocity oscillations over the entire solar disc.
4. *Large Angle and Spectrometric Coronagraph (LASCO)*: It observes the outer solar atmosphere from near the solar limb to a distance of 21 million km. LASCO blocks direct light from the surface of the Sun with occulter, creating an artificial eclipse, 24 hours a day, 7 days a week. Further, LASCO has become SOHO's principal comet finder.
5. *Michelson Doppler Imager/Solar Oscillations Investigation (MDI/SOI)*: It records the vertical motion of the Sun's surface at a million different points for each minute. It determines the acoustic waves inside the Sun as they perturb the photosphere. Scientists can examine the structure and dynamics of the Sun's interior with its help. MDI also determines the longitudinal component of the Sun's magnetic field.
6. *Solar UV Measurements of Emitted Radiation (SUMER)*: The instrument is used to get detailed spectroscopic plasma diagnostics like flows, temperature, density, and dynamics of the solar atmosphere, from the chromosphere through the transition region to the inner corona.
7. *Solar Wind Anisotropies (SWAN)*: The instrument is the only remote sensing instrument on SOHO which does not look at the Sun but watches the rest of the sky, measuring hydrogen "blowing" into the solar system from interstellar space. By observing the interaction between the solar wind and this hydrogen gas, SWAN finds how the solar wind is distributed. It can be qualified as SOHO's solar wind "mapper."
8. *UV Coronagraph Spectrometer (UVCS)*: It measures in UV light of the solar corona by producing an artificial solar eclipse. It blocks the bright light from the solar disc and permits observation of the less intense emission from the extended corona. UVCS gives information about the microscopic and macroscopic behavior of the ionized coronal plasma.

9. *Variability of Solar Irradiance and Gravity Oscillations (VIRGO)*: The instrument characterizes solar intensity oscillations and determines the total solar irradiance (termed as the "solar constant") for quantifying its variability over periods of days to the duration of the mission.
10. *Comprehensive Super Thermal and Energetic Particle Analyzer (COSTEP)*: It detects and classifies energetic particle populations of solar, interplanetary, and galactic origin.
11. *Extreme UV Imaging Telescope (EIT)*: It provides full disc images of the Sun at four selected wavelengths in the extreme UV.
12. *Energetic and Relativistic Nuclei and Electron Experiment (ERNE)*: It measures high-energy particles originating from the Sun and the Milky Way. It is a complementary instrument to COSTEP.

15.4.2 Geosynchronous Operational Environmental Satellite (GOES) Series

The National Oceanic and Atmospheric Administration (NOAA) are operating a series of meteorology observing satellites. This is, in fact, a continuation of the synchronous meteorological satellite (SMS) series. The GOES program started in 1975 with the launch of the first operational spacecraft, GOES-A, renamed GOES-1 when it reached the orbit [15,16]. The GOES spacecraft circle the Earth in a geosynchronous orbit means they orbit close to the equatorial plane of the Earth with a speed matching the Earth's rotation (Figure 15.14). This permits them to appear continuously over one position on the surface of the Earth at an altitude of 35,800 km.

With the systematic and continuous observations of terrestrial weather, GOES gives data of space weather through its space environment monitor (SEM) system. The three main components of space weather monitored by GOES are (1) X-rays, (2) energetic particles, and (3) the magnetic field. With the advent of GOES-8, launched in 1995, the basic platform design became "three-axis stabilized." In 2001 GOES-12 was launched with a new X-ray instrument onboard, called the Solar X-ray Imager (SXI) which creates the images of the Sun while the original XRS instrument only generated whole disk flux measurements. Some particulars of the instruments carried are as follows:

1. *X-ray Sensor (XRS)*: It is an ion chamber which provides whole-Sun X-ray fluxes in the 0.5-to-4 and 1-to-8 Angstrom (Å) wavelength bands. The observations give a sensitive means for detecting the start and evolution of solar flares. Two bands are

FIGURE 15.14
GOES 8.

measured to permit the hardness of the solar spectrum to be estimated and so the temperature and emission measure of the emitting hot flare plasma.

2. *Soft X-ray Imager (SXI)*: This is associated with only on GOES-12 and 13. It provides a continuous sequence of full-Sun X-ray images at a 1-min cadence with a 512 × 512 intensified CCD. Broadband filters are employed to get images in some wavelength bands between about 6 and 60 Å.
3. *Energetic Particle Sensor (EPS)*: It is a solid-state detector with pulse-height discrimination for measuring proton, alpha particle, and electron fluxes.
4. *Magnetometer*: It is a twin-fluxgate spinning sensor to measure three mutually perpendicular components of the Earth's magnetic field. A list of GOES spacecraft is given in Table 15.4.

15.4.3 Transition Region and Coronal Explorer

Transition Region and Coronal Explorer (TRACE) is designed for making quantitative observations. It shows the connections between fine-scale magnetic fields and the associated plasma structures in the solar atmosphere (Table 15.5). This is obtained by making coaligned and near simultaneous images at the UV and extreme UV wavelengths.

TABLE 15.4

List of GOES Spacecraft

Program	Launch Date	Status
GOES-1 (A)	10/16/75	Decommissioned
GOES-2 (B)	6/16/77	Decommissioned
GOES-3 (C)	6/16/78	Decommissioned
GOES-4 (D)	9/9/80	Decommissioned
GOES-5 (E)	5/22/81	Decommissioned
GOES-6 (F)	4/28/83	Decommissioned
GOES-G	5/3/86	Failed[a]
GOES-7 (H)	2/26/87	Decommissioned
GOES-8 (I)	4/13/94	Decommissioned
GOES-9 (J)	5/23/95	Decommissioned
GOES-10 (K)	5/25/97	In-operation
GOES-11 (L)	5/23/2000	In-operation
GOES-12 (M)	7/23/2001	In-operation[b]
GOES-13 (N)	5/24/2006	In-orbit (storage)[c]
GOES-O	11/5/2008	
GOES-P	2010	
GOES-Q	Cancelled	
GOES-R	2014	

[a] GOES-G failed for making it into orbit when its Delta rocket lost control after being struck by lightning very soon after lift-off.

[b] The SXI instrument is out of commission indefinitely owing to a GOES 12 X-ray sensor (XRS) anomaly on April 12, 2007.

[c] No GOES 13 SXI images are available after December 2006 owing to a detector anomaly that occurred in conjunction with an X-*class flare*.

TABLE 15.5
TRACE Passbands

Wavelength (Angstrom)	Width (Angstrom)	Observation	Temperature (1000 K)
5000	Broad	Continuum	4–6.4
1700	200	Continuum	4–10
1600	275	C, I, FE, II, continuum	4–10
1550	20	C, IV	60–250
1216	84	H, I, Ly-a	10–30
173	6.4	Fe, IX	160–2000
195	6.5	Fe, XII	500–2000
284	10.7	Fe, XV	1250–1400

The emission in each wavelength band arises from plasma at different temperatures ranging from 4000 K to 4 million K which correspond to various layers of the solar atmosphere from the photosphere, up through the chromosphere and transition region and into the corona. The images are recorded on a 1024 × 1024 CCD with a field of view of 8.5 × 8.5 arc minutes and an angular resolution is one arc second, corresponding to about 725 km spatial resolution on the Sun [17].

15.4.4 Hinode

Hinode was launched from Japan's Uchinoura Space Center on September 22, 2006, into a Sun synchronous orbit at an altitude of nearly 600 km. Hinode (Figure 15.15) is a joint Japan/US/UK mission. Previously, it was known as Solar-B and led by the Japanese Aerospace Exploration Agency's (JAXA) Space Science Research Division. Following the JAXA tradition, the mission was renamed after the spacecraft's first successful orbit from Solar-B to Hinode, which is a Japanese word meaning sunrise [18,19].

Hinode observes the Sun continuously for 9 months at a time. During summer at northern hemisphere each year, Hinode experiences an "eclipse season" when the Sun is eclipsed by Earth for a maximum of 10 min in each 98 min orbit [20,21]. It was designed for

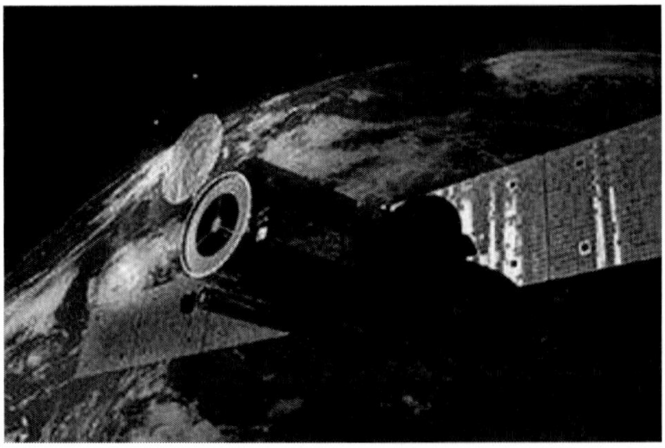

FIGURE 15.15
Hinode.

exploring the magnetic fields in the solar corona and to clarify our understanding of the mechanisms that power the solar atmosphere and drive solar eruptions.

Hinode has a coordinated set of optical, EUV, and X-ray instruments to study the interaction between the Sun's magnetic field and its corona. The three complementary instruments operating together as a highly innovative solar observatory are as follows:

1. *The Solar Optical Telescope/Focal Plane Package (SOT/FPP)*: It is the large optical telescope flown in space with aperture 50 cm in diameter, the angular resolution 0.25″ and the wavelengths covered extend from 480 to 650 nm. SOT includes the focal plane package (FPP) that consists of a vector magnetograph and a spectrograph. The vector magnetograph gives time series of photospheric vector magnetograms, Doppler velocity, and photospheric intensity.
2. *The Extreme-UV Imaging Spectrometer (EIS)*: It is a two-channel, normal-incidence EUV spectrometer which cover the wavelength ranges 170–210 Å and 250–290 Å, selected to cover solar coronal emission lines. It has a mirror used to build up rastered spectral images of the Sun in up to 25 spectral ranges.
3. *The X-ray Telescope (XRT)*: It is a high-resolution grazing-incidence Wolter telescope. High-resolution soft X-ray images covering the energy range from 0.2 to 2 keV show magnetic field configurations and their evolution, permitting the observation of energy build up, storage, and release process in the corona for any transient event. XRT covers a wide temperature range from 0.5 to 10 million K permitting it to find all the coronal features that are not all visible with any normal incidence telescope.

15.4.5 Reuven Ramaty HESSI

RHESSI (Figure 15.16) is a NASA Small Explorer (SMEX) launched on February 5, 2002, for investigating solar flares through X-ray and gamma ray imaging spectroscopy observations. RHESSI's primary mission is to find the basic physics of particle acceleration and explosive energy release in solar flares. It was believed that the majority of the energy released during a flare is used to accelerate electrons and protons and other ions. The objective of the RHESSI mission is to combine high-resolution imaging in hard X-rays and

FIGURE 15.16
RHESSI.

gamma rays with high-resolution spectroscopy. The end result will be a detailed energy spectrum of the radiation at each point of the image [22].

The RHESSI mission spacecraft is spin-stabilized and in a low-altitude orbit inclined at 38° to the Earth's equator. The only instrument on board is an imaging spectrometer which permits high fidelity color movies to be made of solar flares in X-rays and gamma rays between 3 keV and 17 MeV. This advanced capability is obtained by combining two complementary technologies: (1) Fourier-transform imaging utilizing the spacecraft rotation at ~15 rpm and pairs of fine tungsten grids to modulate the solar X-ray and gamma ray flux, and (2) high-resolution spectroscopy using cooled germanium detectors to very precisely measure the energy of each photon that passes through the grids. RHESSI has been operating properly since launch [23,24].

15.4.6 Solar Radiation and Climate Experiment

The Solar Radiation and Climate Experiment (SORCE) is a NASA-sponsored satellite. It is operated by the Laboratory for Atmospheric and Space Physics (LASP) at Colorado.

SORCE was launched on January 25, 2003, by a Pegasus XL launch vehicle into a near-circular orbit with an inclination of 40° and an altitude of 645 km. The mission which provides measurements of incoming X-ray, UV, visible, near-infrared, and total solar radiations (Figure 15.17). Measurements given by SORCE address long-term climate change, natural variability, and enhanced climate prediction as well as atmospheric ozone and UV-B radiation [25,26]. These measurements are critical for studying the Sun and its effect on our Earth system.

SORCE carries four instruments as described below:

FIGURE 15.17
SORCE.

1. *Spectral Irradiance Monitor (SIM)*: It is a newly designed spectrometer which provides the first long-duration solar spectral irradiance measurements in the visible and near-infrared. The wavelength is from 300 to 2400 nm with an additional channel to cover the 200–300 nm.
2. *XUV Photometer System (XPS)*: This is evolved from earlier versions flown on SNOE and TIMED, continue on these solar XUV irradiance measurements with improvements to accuracy, spectral image, and temporal change.
3. *Solar Stellar Irradiance Comparison Experiment (SOLSTICE)*: This makes daily solar UV (115–320 nm) irradiance measurements and compares to the irradiance from early type stars. The approach gives an accurate monitor of instrument in-flight performance.
4. *Total Irradiance Monitor (TIM)*: It measures the total solar irradiance (TSI) at the top of the Earth's atmosphere. TIM continues the climate record since 1978 and is used to find the sensitivity of the Earth's climate to the natural effects of solar forcing.

15.4.7 Solar Terrestrial Relations Observatory

Solar Terrestrial Relations Observatory (STEREO) is the third mission in NASA's Solar Terrestrial Probes (STP) program. STEREO A and B were launched together on a Delta II rocket on October 25, 2006. Their objective and design was for detecting the three-dimensional (3D) varying heliosphere with the help of an unprecedented combination of imaging mounted on virtually identical spacecraft flanking the Earth, in its orbit about the Sun. This mission employs two approximately identical observatories—one ahead of Earth aptly called STEREO-A, the other trailing behind called STEREO-B (Figure 15.18). Together, they give the first-ever stereoscopic measurements to study the Sun and the nature of its CMEs. The STEREO mission helps our understanding of the 3D structure of the Sun's corona, especially regarding the origin of CMEs, their evolution in the interplanetary medium and the dynamic coupling between CMEs and the Earth environment [27,28].

FIGURE 15.18
STEREO A and B.

The two STEREO spacecraft were launched together on a Delta II rocket on October 25, 2006, and used a gravity assist from the Moon to slingshot the spacecraft into a heliocentric orbit [29,30]. The spacecraft bus consists of six operational subsystems supporting two instruments and a total of 16 instruments per observatory as pointed out below:

1. *Sun Earth Connection and Coronal and Heliospheric Investigation (SECCHI)*: It is a suite of five scientific telescopes (EUVI, COR1, COR2, HI1, and HI2) which observes the solar corona and inner heliosphere from the surface of the Sun to the orbit of Earth. Coronagraphs COR1 and COR2 observe the inner and outer corona with greater frequency and polarization. The Extreme UV Imager (EUVI) gives full-Sun coverage. The Heliospheric Imager (HI) is the most novel instrument which extends the concept of traditional externally occulted coronagraphs to a new regime, the heliosphere from the Sun to the Earth.
2. *STEREO WAVE (SWAVES)*: It is an interplanetary radio burst tracker which traces the generation and evolution of travelling radio disturbances from the Sun to the orbit of Earth.
3. *The Plasma and SupraThermal Ion and Composition (PLASTIC)*: It provides plasma characteristics of protons, alpha particles, and heavy ions.

15.4.8 Solar Dynamics Observatory

Solar Dynamics Observatory (SDO), launched on February 11, 2010, is the first mission in NASA's living with a star (LWS) program. This program was designed to realize the causes of solar variability and its impacts on Earth. SDO is designed for the Sun's influence on Earth and near-Earth space by observing the solar atmosphere on small scales of space and time and in many wavelengths simultaneously (Figure 15.19).

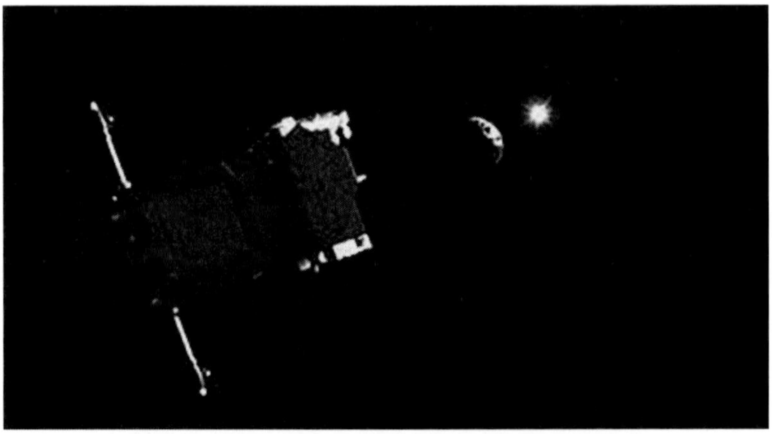

FIGURE 15.19
SDO is a sun-pointing semi-autonomous spacecraft, 4.5 m high and more than 2 m on each side, weighing a total of 3100 kg. The inclined geosynchronous orbit was chosen to permit both continuous observations of the Sun and a continuous and exceptionally high data downlink rate of 130 Megabits per second by using a single dedicated ground station.

It has the following three scientific instruments:

1. *Helioseismic and Magnetic Imager (HMI)*: It extends the capabilities of the SOHO/MDI instrument with continuous full-disk coverage at higher spatial resolution.
2. *Atmospheric Imaging Assembly (AIA)*: It images the solar atmosphere in multiple wavelengths to link variations in the surface to interior variations. Data reveal images of the Sun in 10 wavelengths every 12 s.
3. *EUV Variability Experiment (EVE)*: This measures the solar EUV irradiance with unprecedented spectral resolution, temporal cadence, and precision.

15.5 Future Missions

This section briefly covers future solar-dedicated spacecraft and associated solar instruments which will make significant contributions to solar research. We have outlined the names of the instruments as well as their links to the expected scientific results.

15.5.1 Solar Probe Plus

Solar Probe+ will be a unique mission to explore the last region of the solar system to be visited by a spacecraft, the Sun's outer atmosphere (Figure 15.20). Solar Probe+ was launched in 2015 [31].

Solar Probe+ will sample the near-Sun environment to increase our knowledge of coronal heating and of the origin and evolution of the solar wind. Major science objectives of the Solar Probe+ remain the same as those established for Solar Probe 2005. The four major science objectives are (1) to find the structure and dynamics of the magnetic fields

FIGURE 15.20
Solar Probe Plus.

at the sources of the fast and slow solar wind, (2) to explore dusty plasma phenomena in the near-Sun environment and their influence on the solar wind, (3) to trace the flow of energy which heats the corona and accelerates the solar wind, and (4) to obtain the mechanisms that help to accelerate and transport energetic particles at the Sun and in the inner heliosphere.

15.5.2 Solar Sentinels

It is the NASA component of the joint NASA–ESA mission which includes ESA's Solar Orbiter. The Sentinels mission, originally intended to be launched in 2016, will consist of four similarly instrumented, spinning spacecraft to observe the energetic particle environment and solar wind structures in the inner heliosphere, near the ecliptic plane (Figure 15.21). The spacecraft will operate at near distances to the Sun, inside the orbit of Mercury. It will investigate and characterize how the Sun determines the environment of the inner solar system and how the heliosphere was formed [32,33].

The Sentinels are designed with the following major science objectives [34]: (1) to perform imaging and spectroscopic observations of the source regions on the Sun simultaneously with the in situ measurements, (2) to make simultaneous in situ measurements at multiple locations for capturing the spatial structure and temporal evolution of the phenomena being observed, and (3) to trace the magnetic field from the locations where the in situ measurements are made back to the source regions on the Sun.

15.5.3 Solar Orbiter

Solar Orbiter is a specially designed three-axis stabilized spacecraft. The probe will be approaching as close as 48 solar radii. The Solar Orbiter will see the solar atmosphere with spatial resolution. Over extended periods, the Solar Orbiter will give images and data of the Polar Regions and the side of the Sun not visible from Earth. Following launch in 2017, Solar Orbiter (Figure 15.22) will start its journey to the Sun which will need a cruise phase

FIGURE 15.21
Solar Sentinels.

FIGURE 15.22
Solar Orbiter.

lasting about 3.4 years. During this time, the spacecraft will use gravity assists from Venus and the Earth to get an orbit with a period of 150 days from which it can start its scientific mission.

The smallest face points to the Sun and hence the instruments are protected by a sunshield. The spacecraft will be kept cool by the positioning special "radiators" for dissipating excess heat into space. Presently, all Solar Orbiter's science instruments are in the concept phase. They will be divided into three packages: (1) *Field Package*—Radio and Plasma Wave Analyzer and Magnetometer; (2) *Particle Package*—Energetic Particle Detector, Dust Detector and Solar Wind Plasma Analyzer; and (3) *Solar Remote Sensing Instrumentation*—Visible-light Imager and Magnetograph, EUV Imager and Spectrometer, EUV Imager, Coronagraph, and Spectrometer/Telescope for Imaging X-rays, and also either a Generic Heliospheric Imager or a Wide Angle Coronagraph. The payload will be commissioned during the cruise phase and will be able to give good science data.

References

1. Martens, P. C. H. and Cauffman, D. P. 2002. Multi-wavelength observations of coronal structure and dynamics—Yohkoh 10th anniversary meeting. In *Proceedings of the COSPAR Colloquium*, vol. 13, January 20–24, Kona, HI, Elsevier Science/COSPAR Colloquia Series.
2. Domingo, V. et al. 1995. The SOHO mission: An overview, *Sol. Phys.*, 162:1–37.
3. Gabriel, A. H. et al. 1995. Global oscillations at low frequency from the SOHO mission (GOLF), *Sol. Phys.*, 162:61–99.
4. Fröhlich, C. et al. 1995. VIRGO: Experiment for helioseismology and solar irradiance monitoring, *Sol. Phys.*, 162:101–128.
5. Scherrer, P. H. et al. 1995. The solar oscillations investigation—Michelson Doppler imager, *Sol. Phys.*, 162:129–188.
6. Wilhelm, K. et al. 1995. SUMER—Solar ultraviolet measurements of emitted radiation, *Sol. Phys.*, 162:189–231.

7. Harrison, R. A. et al. 1995. The coronal diagnostic spectrometer for the solar and heliospheric observatory, *Sol. Phys.*, 162:230–299.
8. Delaboudinière, J.-P. et al. 1995. EIT: Extreme-ultraviolet imaging telescope for the SOHO mission, *Sol. Phys.*, 162:291–312.
9. Kohl, J. L. et al. 1995. The ultraviolet coronagraph spectrometer for the solar and heliospheric observatory, *Sol. Phys.*, 162:313–356.
10. Brueckner, G. E. et al. 1995. The large angle spectroscopic coronagraph (LASCO), *Sol. Phys.*, 162:357–402.
11. Bertaux, J. L. et al. 1995. SWAN: A study of solar wind anisotropies on SOHO with Lyman Alpha sky mapping, *Sol. Phys.*, 162:403–439.
12. Hovestadt, D. et al. 1995. CELIAS—Charge, element and isotope analysis system for SOHO, *Sol. Phys.*, 162:441–481.
13. Müller-Mellin, R. et al. 1995. COSTEP—Comprehensive suprathermal and energetic particle analyser, *Sol. Phys.*, 162:483–504.
14. Torsti, J. et al. 1995. Energetic particle experiment ERNE, *Sol. Phys.*, 162:505–531.
15. Donnelly, R. F. et al. 1977. Solar X-ray measurements from SMS-1, SMS-2, and GOES-1, information for data users, *NASA Scientific and Technical Information* (STIN), 7813992.
16. Wagner, W. J. et al. 1987. The solar X-ray imagers (SXI) on NOAA's GOES, *BAAS*, 19:923
17. Handy, B. N. et al. 1999. The transition region and coronal explorer, *Sol. Phys.*, 187:229–260.
18. Kosugi, T. et al. 2007. The Hinode (Solar-B) mission: An overview, *Sol. Phys.*, 243:3–17.
19. Culhane, J. L. et al. 2007. The EUV imaging spectrometer for Hinode, *Sol. Phys.*, 243:19–61.
20. Golub, L. et al. 2007. The X-ray telescope (XRT) for the Hinode mission, *Sol. Phys.*, 243:63–86.
21. Tsuneta, S. et al. 2008. The solar optical telescope for the Hinode mission: An overview, *Sol. Phys.*, 249:167–196.
22. Lin, R. P. et al. 2002. The Reuven Ramaty high-energy solar spectroscopic imager (RHESSI), *Sol. Phys.*, 210:3–32.
23. Smith, D. M. et al. 2002. The RHESSI spectrometer, *Sol. Phys.*, 210:33–60.
24. Hurford, G. J. et al. 2002. The RHESSI imaging concept, *Sol. Phys.*, 210:61–86.
25. Rottman, G. 2005. The SORCE mission, *Sol. Phys.*, 230:7–25.
26. Woods, T. N. and Rottman, G. 2005. XUV photometer system (XPS): Solar variations during the SORCE mission, *Sol. Phys.*, 230:375–387.
27. Kaiser, M. L. et al. 2008. The STEREO mission: An introduction, *SSRv*, 136:5–16.
28. Howard, R. A. et al. 2008. Sun Earth connection coronal and heliospheric investigation (SECCHI), *SSRv*, 136:67–115.
29. Luhmann, J. G. et al. 2008. STEREO IMPACT investigation goals, measurements, and data products overview, *SSRv*, 136:117–184.
30. Bougeret, J. L. et al. 2008. S/WAVES: The radio and plasma wave investigation on the STEREO mission, *SSRv*, 136:487–528.
31. McComas, D. J. et al. 2008. Solar probe plus: Report of the science and technology definition team, http://solarprobe.gsfc.nasa.gov/SolarProbe+Web.pdf.
32. Giacconi, R. and Piero, R. 2008. Cosmic X-ray sources, Scholarpedia, 3(4):4391.
33. Hudson, H. 2008. Solar activity, Scholarpedia, 3(3):3967.
34. Schmidt, W. 2008. Solar telescopes, Scholarpedia, 3(4):4333.

16

Spacecraft Facts and Missions

16.1 Introduction

A spacecraft is defined as a vehicle or device designed for travel or operation outside the earth's atmosphere. It can be utilized for many purposes like meteorology, earth observation, communications, navigation, planetary exploration, space colonization, and even for transportation of humans and cargo. Orbital spacecraft may be recoverable or not. Based on method of re-entry to Earth they may be divided into nonwinged space capsules and winged spaceplanes. Currently, humanity has achieved space flight but only 24 nations actually have spacefaring technology: Russia (Roscosmos, the Russian Space Forces), the United States (NASA, the US Air Force, and several commercial aerospace companies), the member states of the European Space Agency, the People's Republic of China (China National Space Administration), Japan (Japan Aerospace Exploration Agency), and India (Indian Space Research Organization) [1].

16.2 History of Spaceflight

Sputnik (Figure 16.1) was the first artificial satellite [2,3] launched into an elliptical low Earth orbit by the Soviet Union in 1957. This artificial satellite mission was to gain knowledge to assist in identifying the upper atmospheric density layers. In addition, data was gathered about radio-signal distribution in the ionosphere. It also provided the first opportunity for meteoroid detection by the temperature data sent back to Earth.

While Sputnik 1 was the first successful spacecraft to orbit the Earth, other man-made objects had already reached an altitude of 100 km (the height required to count as a spaceflight).

16.3 Past and Present Spacecraft

Up to 2011, only three nations have flown manned spacecraft: USSR, the United States, and China. Subsequently, other countries like India, Japan, Europe, Iran, North and South Korea, Denmark, and Romania have successfully attempted for manned spacecraft. The first manned spacecraft was Vostok 1 (Figure 16.2), which carried Soviet cosmonaut Yuri

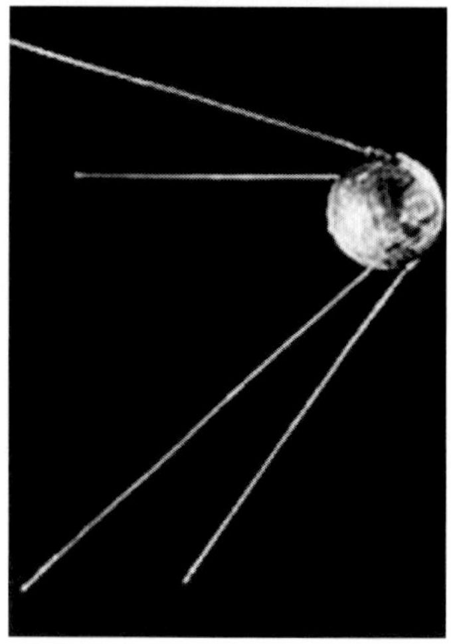

FIGURE 16.1
Sputnik, the first spacecraft.

FIGURE 16.2
Vostok 1, the first manned spacecraft.

Gagarin into space in 1961. There were five other manned missions related to Vostok spacecraft [4]. The second manned spacecraft was Freedom 7, which did a suborbital spaceflight in 1961 carrying astronaut Alan Shepard to an altitude of over 187 km. There were other manned missions using Mercury spacecraft. American manned spacecraft include the Gemini spacecraft, Apollo spacecraft, the Skylab space station, and the space shuttle.

Except for the space shuttle, all the recoverable manned spacecraft were space capsules. As of June 2011, there are more than 2000 spacecraft in orbit.

(a) (b) (c)

FIGURE 16.3
(a) X-15, (b) Buran, and (c) Columbia orbiter. (Courtesy of NASA photos.)

16.4 Spaceplanes

Some reusable vehicles have been designed and constructed for the purpose of manned spaceflight and these are usually named as Spaceplanes. The first example is the North American X-15 Spaceplane that conducted two manned flights and to reach an altitude of over 100 km in the 1960s. Figure 16.3a is a photo of the NASA X-15. The first automatic partially reusable spacecraft was named Buran which was launched by the USSR in 1988. Figure 16.3b is a photo of the USSR Buran. The X-15 was air launched on a suborbital trajectory on July 19, 1963. The first space shuttle to fly into space was named Columbia, followed by Challenger, Discovery, Atlantis, and Endeavour. Endeavour was built for replacing Challenger when it was lost in January 1986. However, Columbia broke up during re-entry in February 2003. Figure 16.3c is a photo of the NASA Columbia. Per the vision for space exploration, the space shuttle was retired in 2011 owing to its old age and high cost of the program.

16.5 Unmanned Spacecraft

We have listed in the following subsection the names and some particulars of unmanned spacecraft, spaceflight, artificial satellites, and space probes including solar system probes, robotic spacecraft, and so on. A photograph of the Hubble Space Telescope is shown in Figure 16.4a while Figure 16.4b and c reveal, respectively, the artist's conception of Cassini–Huygens as it enters Saturn's orbit and the artist's conception of the Phoenix spacecraft as it lands on Mars.

16.5.1 Designed as Manned but Flown as Unmanned-Only Spacecrafts

- Zond/L1—lunar flyby capsule
- TKS—capsule
- Buran Soviet shuttle
- L3—Capsule and lunar lander

(a) (b) (c)

FIGURE 16.4
(a) Hubble Space Telescope, (b) artist's conception of Cassini–Huygens, and (c) artist's conception of the Phoenix on Mars. (Courtesy of NASA photos.)

16.5.2 Semi-Manned—Manned as Space Stations or Part of Space Stations

- Progress—unmanned USSR/Russia cargo spacecraft
- Automated Transfer Vehicle (ATV)—unmanned European cargo spacecraft
- H-II Transfer Vehicle (HTV)—unmanned Japanese cargo spacecraft
- Dragon (spacecraft)—unmanned private spacecraft (Space X)
- TKS—unmanned USSR/Russia cargo spacecraft and space station module

16.5.3 Earth Orbit Satellites

- Explorer 1—first US satellite
- SOHO—orbits the Sun near L1
- Sputnik 1—world's first artificial satellite
- Sputnik 2—first animal in orbit (Laika)
- Syncom—first geosynchronous communications satellite
- Hubble Space Telescope—largest orbital observatory
- X-37—Spaceplane
- Project SCORE—first communications satellite
- Sputnik 5—first capsule recovered from orbit (Vostok precursor) animals survived

16.5.4 Lunar Probes

- Clementine—US Navy mission, orbited Moon, detected hydrogen at the poles
- Kaguya JPN—Lunar orbiter
- Luna 1—first lunar flyby
- Luna 2—first lunar impact
- Luna 3—first images of lunar far side
- Luna 9—first soft landing on the Moon
- Luna 10—first lunar orbiter

- Luna 16—first unmanned lunar sample retrieval
- Lunar Orbiter—very successful series of lunar mapping spacecraft
- Lunar Prospector—confirmed detection of hydrogen at the lunar poles
- Lunar Reconnaissance Orbiter—identifies safe landing sites and located moon resources
- Lunokhod—Soviet lunar rovers
- SMART-1 ESA—lunar impact
- Surveyor—first US soft lander
- Chandrayaan 1—first Indian lunar mission

16.5.5 Planetary Probes

- Akatsuki JPN—a Venus orbiter
- Cassini–Huygens—first Saturn orbiter + Titan lander
- Curiosity rover—Rover sent to mars by NASA in 2012
- Galileo—first Jupiter orbiter + descent probe
- IKAROS JPN—first solar-sail spacecraft
- Mariner 4—first Mars flyby, first close and high resolution images of Mars
- Mariner 9—first Mars orbiter
- Mariner 10—first Mercury flyby, first close up images
- Mars Exploration Rover—a Mars rover
- Mars Express—a Mars orbiter
- Mars Global Surveyor—a Mars orbiter
- Mars Orbiter Mission—India's first Interplanetary probe
- Mars Reconnaissance Orbiter—an advanced climate, imaging, subsurface radar, and telecommunications Mars orbiter
- MESSENGER—first Mercury orbiter (arrival 2011)
- Mars Pathfinder—a Mars lander + rover
- New Horizons—first Pluto flyby (arrival 2015)
- Pioneer 10—first Jupiter flyby, first close up images
- Pioneer 11—second Jupiter flyby + first Saturn flyby (first close up images of Saturn)
- Pioneer Venus—first Venus orbiter + landers
- Vega 1—Balloon release into Venus atmosphere and lander (joint mission with Vega 2), mother ship continued on to flyby Halley's Comet [5]
- Venera 4—first soft landing on another planet (Venus)
- Viking 1—first soft landing on Mars
- Voyager 2—Jupiter flyby + Saturn flyby + first flybys/images of Neptune and Uranus

16.5.6 Other Deep Space Probes

- Cluster
- Deep Space 1
- Deep Impact
- Genesis
- Hayabusa
- Near Earth Asteroid Rendezvous
- Stardust
- STEREO—heliospheric and solar sensing; first images of the entire Sun
- WMAP

16.5.7 Fastest Spacecraft

- Helios I and II Solar Probes (252,792 km/h or 157,078 mph)

16.5.8 Furthest Spacecraft from the Sun

- Pioneer 10 at 116.473 AU as of 2005, traveling outward at about 2.527 AU/year
- Pioneer 11 at 95.559 AU as of 2005, traveling outward at about 2.381 AU/year
- Voyager 1 at 135.680 AU as of July 2008, traveling outward at about 3.586 AU/year
- Voyager 2 at 111.782 AU as of July 2008, traveling outward at about 3.243 AU/year
- New Horizons at 36.046 AU as of 2006, traveling outward at about 3.030 AU/year

16.6 Manned Spacecraft

- Shuguang—Chinese capsule
- Soyuz Kontakt—USSR capsule
- Almaz—USSR space station
- Manned Orbiting Laboratory—US space station
- Altair—US lunar lander of Orion spacecraft

16.6.1 Multistage Spaceplanes

- X-20—US shuttle
- Soviet Spiral Shuttle
- Soviet Buran Shuttle
- ESA Hermes shuttle
- Kliper Russian semi-shuttle/capsule
- Japanese HOPE-X shuttle
- Chinese Shuguang Project 921–3 shuttle

16.6.2 Single Stage to Orbit Spaceplanes

- RR/British Aerospace HOTOL
- ESA Hopper Orbiter
- McDonnell Douglas DC-X (Delta Clipper)
- Roton rotored hybrid
- Lockheed-Martin VentureStar

16.7 Spacecraft under Development

Figure 16.5 exhibits the photograph of the Orion spacecraft which is now under development.

16.7.1 Manned

- (US–NASA) Orion Multi-Purpose Crew Vehicle—capsule
- (US–SpaceX) Dragon—capsule
- (US–Boeing) CST-100—capsule
- (US–Sierra Nevada Corporation) Dream Chaser—orbital spaceplane
- (US–The SpaceShip Company) SpaceshipTwo—suborbital spaceplane
- (US–Blue Origin) New Shepard—VTVL capsule
- (US–XCOR) Lynx rocketplane—suborbital spaceplane
- (Russia–RKA) Prospective Piloted Transport System (PPTS)—capsule
- (Europe–ESA) Advanced Crew Transportation System—capsule
- (Iranian Space Agency) Orbital Vehicle—capsule
- (India–ISRO) Orbital Vehicle—capsule

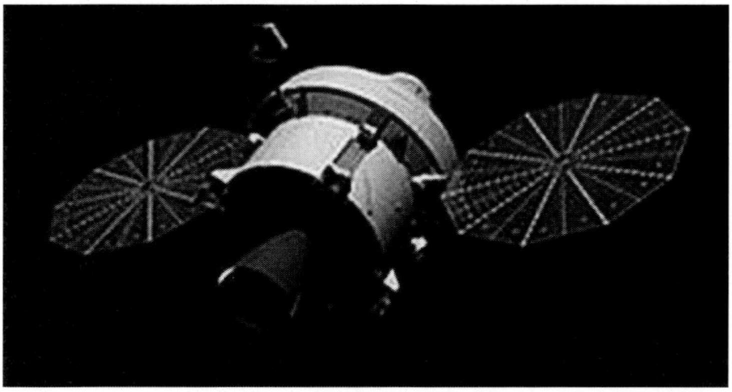

FIGURE 16.5
The Orion spacecraft.

16.7.2 Unmanned

- SpaceX Dragon—cargo delivery to the ISS
- Orbital Sciences Cygnus—cargo delivery to the ISS
- CNES Mars Netlander
- James Webb Space Telescope (delayed)
- ESA Darwin14 probe
- Mars Science Laboratory rover
- Shenzhou spacecraft Cargo
- Terrestrial Planet Finder cancelled probe
- System F6—a DARPA Fractionated Spacecraft demonstrator
- Reaction Engines Limited Skylon (spacecraft)

16.8 Spacecraft Subsystems

A spacecraft system comprises of a number of subsystems, depending on the type of mission profile. The subsystems comprise the spacecraft's "bus" (essential AC and DC) and may include attitude determination and control called ADAC, ADC or ACS, guidance, navigation and control (GNC or GN&C), communications, command and data handling (CDH or C&DH), power (EPS), thermal control (TCS), propulsion, and structures.

16.9 Scientific Rovers on the Surface of Mars

As of September 24, 2014, there were two scientific rovers on the surface of Mars beaming signals back to Earth. In addition, five orbiters are currently surveying the planet: Mars Odyssey, Mars Express, Mars Reconnaissance Orbiter, Mangalyaan, and MAVEN. Figure 16.6 reveals that the Curiosity rover checks its wheels in Gale crater.

NASA reported that current studies on Mars by the Curiosity and Opportunity rovers will give emphasis for evidence of ancient life, including a biosphere based on autotrophic, chemotrophic, and chemolithoautotrophic microorganisms as well as ancient water, including fluviolacustrine environments, that is, plains related to ancient rivers or lakes which may have been habitable [6–10].

16.9.1 Recent Mars Missions

In 2001, NASA's Mars Odyssey orbiter [10] entered in the orbit of Mars. Its Gamma Ray Spectrometer detected hydrogen in the upper meter or so of regolith on Mars which is assumed to be contain large deposits of water ice [11]. In 2003, the Mars Express mission of the European Space Agency (ESA) reached Mars that carried the Beagle 2 lander but was declared lost in February 2004 [12]. Mars Express Planetary Fourier Spectrometer team declared that the orbiter had detected methane in the Martian atmosphere. In addition, ESA announced in June 2006 the discovery of aurorae on Mars [13]. Curiosity's self-portrait

FIGURE 16.6
Curiosity rover checks its wheels in Gale crater. (Courtesy of NASA photos.)

FIGURE 16.7
Curiosity's self-portrait on the planet Mars at "Rocknest." (Courtesy of NASA photos.)

on the planet Mars at "Rocknest" (October 31, 2012) is shown in Figure 16.7. The opportunities of Mars expedition during 2013–2020 is shown in Table 16.1.

16.9.2 Mars Exploration Rover and Phoenix Lander

Mars Exploration Rover Mission (NASA—MER) is an ongoing robotic space mission which involves two rovers: (1) Spirit and (2) Opportunity. It began in 2003 with the launching of the two rovers, namely, MER-A Spirit and MER-B Opportunity. These probes mission was for exploring the Martian surface and geology and search for a wide range of rocks and

TABLE 16.1
Opportunities 2013–2020

Year	Launch	Spacecraft (Launched or Planned)
2013	Nov 2013	MAVEN, Mars Orbiter Mission
2016	Jan 2016–Apr 2016	InSight, ExoMars
2018	Apr 2018–May 2018	ExoMars, Inspiration Mars
2020	Jul 2020–Sep 2020	Mars 2020 Rover

FIGURE 16.8
Arctic surfaces as seen by Phoenix. (Courtesy of NASA photos.)

soils which hold clues to past water activity on Mars. Arctic surfaces as seen by Phoenix is shown in Figure 16.8.

16.9.2.1 Mars Orbiter Mission, ExoMars and Inspiration Mars

2013 Nov 2013 MAVEN, Mars Orbiter Mission 2016 Jan 2016—Apr 2016 InSight, ExoMars 2018 Apr 2018—May 2018 ExoMars, Inspiration Mars 2020 Jul 2020—Sep 2020 Mars 2020 Rover.

16.10 Future Missions

- ESA and the Russian Federal Space Agency have planned to send the Trace Gas Orbiter and lander to Mars in 2016, and the ExoMars rover in 2018 for searching past or present microscopic life on Mars
- NASA selected a $425 million lander mission for 2016, with a drill and seismometer for finding the interior structure of Mars

- The MetNet precursor was considered for a piggyback launch on Phobos-Grunt and on the planned 2016 and 2018 ExoMars landers
- The Mars-Grunt is a Russian mission concept for bringing a sample of Martian soil to Earth
- The ISRO has a plan to send a follow up mission to its Mars Orbiter Mission in the 2017–2020 timeframe
- China unveiled a prototype model of a rover including an orbiter, lander, and the rover which will be sent before 2020
- NASA Mars 2020 rover mission would be launched in 2020 and it would be based on the Mars Science Laboratory design. The information related to scientific payload will be focused on astrobiology
- United Arab Emirates announced that the first mission of its space agency will be for sending an unmanned mission to Mars by 2021

References

1. Mader, T. H. et al. (2011). Optic disc edema, globe flattening, choroidal folds, and hyperopic shifts observed in astronauts after long-duration space flight. *Ophthalmology* 118(10): 2058–2069.
2. McDougall, W. A. Shooting the duck, *American Heritage*, Winter 2010.
3. Swenson, L, Jr. et al. (1966). ...On October 4, 1957 Sputnik I shot into orbit and forcibly opened the space age. This New Ocean, A History of Project Mercury. Washington D.C.: National Aeronautics and Space Administration, Library of Congress Card No. 66-62424.
4. Vostok. Encyclopedia Astronautica. http://www.astronautix.com/v/vostok.html (retrieved August 06, 2016).
5. Mendis, D. A. (1988). A postencounter view of comets. *Annual Review of Astronomy and Astrophysics* 26(1): 11–49.
6. Grotzinger, J.P. (January 24, 2014). Introduction to Special Issue—Habitability, Taphonomy, and the Search for Organic Carbon on Mars. *Science* 343 (6169): 386–387. doi:10.1126/science.1249944.
7. Various (January 24, 2014). Special Issue—Table of Contents–Exploring Martian Habitability. *Science* 343 (6169): 345–452.
8. Various (January 24, 2014). Special Collection—Curiosity—Exploring Martian Habitability. *Science* 343 (6169): 386pp.
9. Grotzinger, J.P. et al. (January 24, 2014). A Habitable Fluvio-Lacustrine Environment at Yellowknife Bay, Gale Crater, Mars. *Science* 343 (6169). doi:10.1126/science.1242777.
10. NASA. NASA's Mars Odyssey Shifting Orbit for Extended Mission (October 9, 2008). http://www.nasa.gov/jpl/odyssey/nasas-mars-odyssey-orbiter-watches-comet-fly-near/ (retrieved November 15, 2008).
11. Britt, R. (March 14, 2003). Odyssey Spacecraft Generates New Mars Mysteries. Space.com. Archived from the original on March 15, 2006. http://web.archive.org/web/20060315103112/http://www.space.com/missionlaunches/odyssey_update_030314.html (retrieved June 13, 2006).
12. ESA News. ESA Media Relations Division (February 11, 2004). UK and ESA announce Beagle 2 inquiry. http://www.esa.int/Our_Activities/Space_Science/Mars_Express/UK_and_ESA_announce_Beagle_2_inquiry (retrieved June 28, 2011).
13. Bertaux, J.-L. et al. (June 9, 2005). Discovery of an aurora on Mars. *Nature* 435 (7043): 790–794. doi:10.1038/nature03603.

17

Radio Astronomy and Planetary Observations

17.1 Introduction

Radio astronomy [1] has played a major role in the area of planetary research. A few examples include

- *Improved Saturn Positions Help Spacecraft Navigation, Planet Studies, Fundamental Physics*—https://public.nrao.edu/news/pressreleases/vlba-measures-saturn-position

Scientists have used the National Science Foundation's very long baseline array (VLBA) [2] radio telescope system and NASA's Cassini spacecraft to measure the position of Saturn and its family of moons to within about a mile …

- *VLA Reveals a Close Pair of Potential Planetary Systems*—https://public.nrao.edu/news/pressreleases/potential-planetary-system

Planets apparently can form in many more binary-star systems than previously thought, according to astronomers who used the National Science Foundation's VLA [3] radio telescope to image protoplanetary disks around a close pair of stars.

17.2 Planetary Radio Astronomy—The Definition

Planetary radio astronomy is the study of the physical characteristics of the planets in the solar system by means of the electromagnetic radio radiation emitted by these objects.

Radio astronomy generally refers to the wavelength range from about 0.5 mm (600 GHz = 600 × 109 Hz) to a kilometer (300 kHz) and long ward. Observations from the ground cannot be carried out below a few megahertz because of the Earth's ionosphere is opaque to very low frequency radio waves. At millimeter and submillimeter wavelengths, the terrestrial atmosphere defines windows where radio emissions from cosmic sources can reach the ground. Radio emissions have been measured from all of the planets and some satellites, asteroids, and comets. The observed continuum emissions from the planets can be broadly classified as quasi-thermal (having the same general shape as a blackbody emitter) and nonthermal (i.e., cyclotron, synchrotron). Narrow spectral lines from molecules have been observed in the atmospheres. In addition, a number of solar system spacecraft have carried radio astronomical instrumentation.

17.3 Radio Astronomy History

Radio astronomy began in 1932 with the pioneering work of Karl G. Jansky [4]. While employed at Bell Labs, NJ, Jansky discovered radio emission from the Milky Way galaxy while studying the direction of radio bursts associated with thunderstorms. Ten years later (1942), during World War 2 while trying to track down the source of radio interference on a military antiaircraft radar system [5] in England, J.S. Hey [6] established the occurrence of radio emission from the Sun. Figure 17.1a and b shows early radar installations.

17.3.1 A Brief Radio Astronomy Lesson

Radio astronomy is the study of the invisible universe (celestial objects) that emit radio waves. We see that there have been many books written on this subject but, no matter what, we must describe and educate the reader before we go on. I believe that a good basic description of a subject is required for one to thoroughly understand the subject.

The author will not go into much heavy technical detail as to the internal workings of the entire system. After the reader has read the material and would like the more detailed version, they can refer to a "Suggested Readings" list at the end of this book.

Referring to Figure 17.2, the electromagnetic spectrum (EM) we see visible light which our eyes detect, a type of electromagnetic radiation. Objects on Earth and in space also emit other types of EM radiation that cannot be seen by the human eye, such as radio waves. The full range of radiation emitted by an object is called its EM spectrum.

(a) (b)

FIGURE 17.1
(a) This early radar device was used during World War II on the beach in Normandy. Photograph taken on June 22, 1944. (Courtesy of The Public Domain, The History of Radar, http://www.windows2universe.org/earth/Atmosphere/tornado/radar_history.html [retrieved August 22, 2016].) (b) Chain Home transmitter antenna, part of one of the first comprehensive radar systems. (Courtesy of Wikipedia, https://commons.wikimedia.org/wiki/File:CH_Radar_Mast_-_Stenigot_-_geograph.org.uk_-_150615.jpg [last modified on August 22, 2016].)

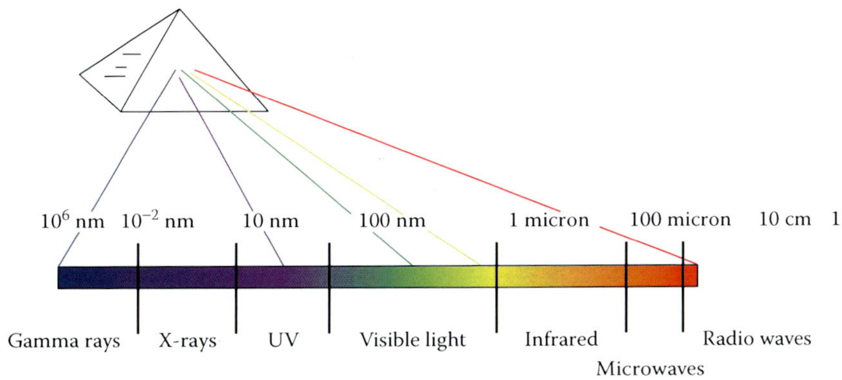

FIGURE 17.2
Electromagnetic (EM) spectrum. (Courtesy of Lichtman, Radio Astronomy Supplies, TX.)

With radio astronomy, scientists can study astronomical phenomena that are often invisible in other portions of the EM spectrum.

Whenever an electric charge changes speed or direction it gives off an EM wave. How fast the wave "wiggles" determines what kind of EM radiation is created. EM can be placed in order from lowest energy to highest energy as follows: radio, infrared, visible light, ultraviolet, X-rays, and gamma rays.

Using radio astronomy techniques, astronomers can observe the Cosmic Microwave Background (CMB) radiation, which is the remnant signal of the Big Bang. Radio astronomy can also probe the time before the onset of the first stars or galaxies, and study the earliest generation of galaxies. Radio astronomers can also analyze and explore the black holes that live at the center of most galaxies.

Since radio waves penetrate dust, scientists use radio astronomy techniques to study regions that cannot be seen in visible light, such as the dust-shrouded environments where stars and planets are born, the center of our galaxy, and other Milky Way. Radio waves also allow astronomers to trace the location, density, and motion of the hydrogen gas that constitutes three-fourths of the ordinary matter in the universe.

17.4 The Radio Telescope as a Tool

A radio telescope [7] is comprised of different blocks as seen in Figure 17.3. of a directional radio antenna (parabolic dish) is used to collect the incoming radio signals. The same types of antennas are also used in tracking and collecting data from satellites and space probes. In their astronomical role, they differ from optical telescopes in that they operate in the radio frequency portion of the EM spectrum where they can detect and collect data from radio sources. As mentioned above, radio telescopes are typically large parabolic ("dish") antennas used by themselves or in an array.

In most cases, radio observatories are located far from major centers of population to avoid radio frequency interference (RFI) from radio, TV, radar, and other interference emitting devices. This is similar to the locating of optical telescopes to avoid light pollution.

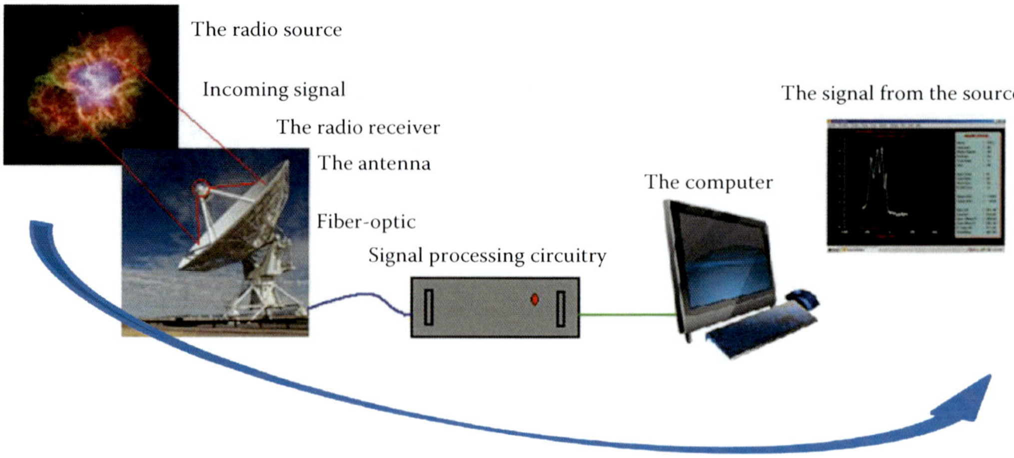

FIGURE 17.3
Radio telescope block diagram. (Courtesy of Lichtman, Radio Astronomy Supplies, TX.)

Radio telescopes receive radio waves (continuum and spectral) from objects within our universe such as stars, pulsars, quasars, galaxies, black holes, and some of our planetary neighbors like Jupiter and our Sun, all at different wavelengths. As seen above in Figure 17.3, the major parts of a radio telescope all contribute to a research radio telescope.

The cosmic radio waves are extremely weak (by many magnitudes) than those signals reaching our radio or televisions.

When the weak signal reach the antenna, it is amplified by a thermoelectrically cooled low noise amplifier (LNA). The amplified signal is routed to a receiver, also cooled (all parts of the receiver are housed at the focal point of the antenna). The amplified signal is sent via fiber-optic cable to further signal processing circuitry, then to the computer resulting in a signal footprint.

17.4.1 VLBI

One of the tools also used is VLBI (very-long-baseline interferometry) [8]. To explain the process, when a spacecraft or probe is launched into space and eventually orbits a planetary body, the data signal being transmitted back to earth is received by a radio telescope. In addition, signals (guidance, etc.) from ground-based control facilities are relayed to the satellite or probe. Figure 17.4 illustrates the basic VLBI operation.

Author's Note

In section 17.5, some images as well as text have been removed or replaced for clarity. By no means has the author/coauthor of this book deliberately tried to change the original content of Dr. De Patr or Dr. Gulkis. In some cases, images from the original document could not be located. The coauthor, Lichtman diligently worked to find closely related information. We do thank Dr. De Patr and Dr. Gulkis for their help and permission.

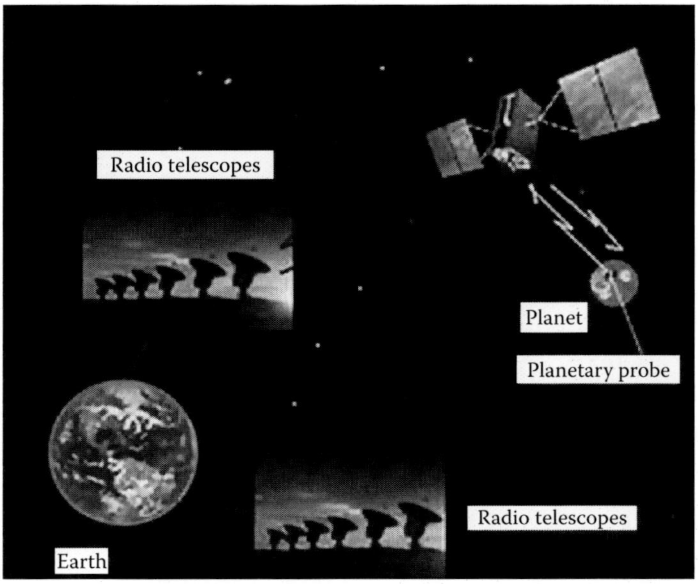

FIGURE 17.4
VLBI simplified.

17.5 Planetary Radio Observations

The following, courtesy of Dr. S. Gulkis, JPL and Dr. I. DePatr- R.H. Dicke [9] and R. Beringer working in the United States made the first intentional measurement of a solar system object in 1945. They observed radio emission from the moon at a wavelength of 1.25 cm, thus beginning the first scientific studies at radio wavelengths of the planets and satellites of the solar system. Subsequent observations revealed that the microwave emission from the moon varies with lunar phase but the amplitude of these variations is much smaller than that observed at infrared wavelengths. This result was interpreted in terms of emission originating below the surface of the moon, where the temperature variations are smaller than at the surface. Within a few years, it was widely recognized that the long wavelengths provided by radio measurements offered a new and important tool for solar system studies, namely the capability of probing into and beneath cloud layers and surfaces of the planets. However, thermal radio emissions [10] from the planets are exceedingly weak and nearly a decade elapsed before system sensitivity was sufficiently improved to enable the detection of planetary thermal emission.

Meanwhile, another unanticipated discovery was made. In June 1954, when the angular separation between the Sun and a supernova remnant known as the Crab Nebula was small, astronomers B. Burke and K. Franklin of the Carnegie Institution were attempting to study the effect of the solar corona on radio waves from the Crab Nebula. Occasionally, they observed bursts of radio interference, which they initially thought were due to the Sun. That hypothesis was discarded when it was discovered that the origin of the interference bursts was nearly fixed with respect to the background stars. Further observations and examinations of the data revealed that the emissions were in fact originating from Jupiter, which happened to be located in the same region of the sky as the Sun and the Crab Nebula. Because the intensity of the emissions was much too strong to be of thermal origin, Franklin and Burke concluded that Jupiter was a source of nonthermal radio emission.

The first successful measurements of thermal emission from the planets were made in 1956 at the Naval Research Laboratory in Washington, DC. C.H. Mayer, T.P. McCullough, and R.M. Sloanaker scanned Venus, Mars, and Jupiter with a 15 m parabolic antenna equipped with a new 3 cm wavelength radio receiver. They detected weak thermal emission from these three planets when each was observed at its closest distance to the Earth.

In the intervening years, thermal emission has been measured from all of the planets in the solar system. Because of the faintness of its radio emission, Pluto was the last planet to be detected. A few asteroids, satellites, and comets have been measured as well. Nonthermal radio emission has been measured from Jupiter, Saturn, Uranus, Neptune, and Earth. In this chapter we give an overview of the techniques used by planetary radio astronomers and discuss what has been learned from the measurements and what can be done in the future. In the interest of brevity, we do not discuss specific observations of asteroids and satellites, although they rightfully belong in any discussion of planetary radio astronomy.

17.5.1 Measurement Objectives

The primary objective of planetary exploration is to determine the physical characteristics of the planets, satellites, asteroids, and comets in order to obtain an understanding of the origin and evolution of the solar system, including the origin of life on Earth. One objective associated with this goal is to determine the composition and physical characteristics of these bodies and their atmospheres. Studies of the energy budget and redistribution of energy within solid surfaces and atmospheres are part of this work. Another objective is to investigate the magnetic fields and ionized plasmas that surround some of the bodies in the solar system and to understand the interaction of the magnetic fields with the solar wind and the cosmic environment.

Planetary research involves many scientific disciplines and requires a variety of instruments and techniques, including astronomical studies from the Earth, planetary spacecraft flybys, orbiters, and probes, and eventually manned landings. Each of the various approaches used has a particular strength that the experimenters try to exploit. Thus far, most planetary radio astronomy has been carried out from the ground, but the techniques carry over to spacecraft as well.

Planetary radio astronomy measurements provide complementary data to other observational techniques. They also provide some unique data. For example, radio measurements can be used to provide information about planetary atmospheres and planetary subsurface materials to much greater depth than other remote sensing techniques. The greater penetration is a result of neutral gases and solids being more transparent to radio waves than higher-frequency waves such as infrared or visible light. Also, the scattering from particulate materials in planetary atmospheres is generally less at radio wavelengths than at shorter wavelengths. The relative transparency of atmospheres, clouds, and surfaces to radio waves allows the planetary radio astronomer to measure thermal profiles of planetary atmospheres beneath the cloud layers in the atmosphere and to measure temperatures beneath the solid surface of the planet. Taking advantage of this property, radio astronomers were the first to measure the very high surface temperature of cloud covered Venus.

Very high-resolution spectroscopy [11] is another area in which radio astronomy provides a unique capability. It is possible to achieve very high spectral resolution in the radio and submillimeter spectral region by translating the frequency of the radio signal under study to a convenient place in the frequency spectrum where spectrum analysis is easier to achieve with either digital or analog techniques. This process takes place without disturbing the relation of the sidebands to the carrier frequency. The process of frequency translation is referred to by such names as heterodyne, mixing, or frequency conversion. The heterodyne technique makes it possible to measure absorption and emission line shapes in greater detail than has been possible at shorter wavelengths. Both temperature–altitude and composition–altitude profile distributions of absorbing chemical species (e.g., NH, H, O, CO) can be deduced from spectroscopic measurements.

Finally, we note that both synchrotron and cyclotron radiation [12] from (solar system) planets is confined to the radio region of the spectrum. This circumstance is due to the range of planetary magnetic field strengths and particle energies found in the solar system. The existence of Jupiter's strong magnetic field was first deduced from earth-based measurements of its polarized radio emission.

17.5.2 Physical Properties of the Planets

The physical characteristics of the planets are required in order to make qualitative estimates of the radio power flux densities expected from the planets. Table 17.1 presents physical data for the planets.

TABLE 17.1

Physical Data for the Planets

	Planetary Data[a]						
Planet	Mass (10^{24} kg)	Diameter (km)	Density (kg/m^3)	Length of Day[b] (hours)	Distance from Sun (10^6 km)	Orbital Period[c] (days)	Orbital Velocity[d] (km/s)
Mercury	0.330	4879	5427	4222.6	57.9	88.0	47.9
Venus	4.87	12,104	5243	2802.0	108.2	224.7	35.0
Earth	5.97	12,756	5515	24.0	149.6	365.2	29.8
Mars	0.642	6794	3933	24.7	227.9	687.0	24.1
Jupiter	1899	142,984	1326	9.9	778.6	4331	13.1
Saturn	568	120,536	687	10.7	1433.5	10,747	9.7
Uranus	86.8	51,118	1270	17.2	2872.5	30,589	6.8
Neptune	102	49,528	1638	16.1	4495.1	59,800	5.4
Pluto (dwarf)	0.0125	2390	1750	153.3	5870.0	90,588	4.7

Source: Courtesy of NASA, Planetary Fact Sheet—U.S. Units, http://nssdc.gsfc.nasa.gov/planetary/factsheet/planet_table_british.html (retrieved November 18, 2015). Planetary Fact Sheet Notes, http://nssdc.gsfc.nasa.gov/planetary/factsheet/planetfact_notes.html (retrieved November 18, 2015).

[a] Numerical data based on NASA information.

[b] This is the average time in hours that it takes for the Sun to move from the noon position in the sky at a point on the equator back to the same position.

[c] This is the time in Earth days that it takes for the planet to orbit the Sun.

[d] This is the average velocity, or speed, of the planet in kilometers per second as it orbits the Sun.

17.5.3 Planetary Atmospheres

MERCURY
42% Molecular oxygen
29.0% Sodium
22.0% Hydrogen
6.0% Helium
0.5% Potassium
Trace amounts of argon, nitrogen, carbon dioxide, water vapor, xenon, krypton, and neon

VENUS
~96.5% Carbon dioxide
~3.5% Nitrogen
0.015% Sulfur dioxide
0.007% Argon
0.002% Water vapor
0.0017% Carbon monoxide
0.0012% Helium
0.0007% Neon
Trace carbonyl sulfide
Trace hydrogen chloride
Trace hydrogen fluoride

EARTH
78.08% Nitrogen (N_2)
20.95% Oxygen (O_2)
0.93% Argon
0.038% Carbon dioxide
About 1% water vapor (varies with climate)

MARS
95.72% Carbon dioxide
2.7% Nitrogen
1.6% Argon
0.2% Oxygen
0.07% Carbon monoxide
0.03% Water vapor
0.01% Nitric oxide
2.5 ppm Neon
300 ppb Krypton
130 ppb Formaldehyde
80 ppb Xenon
30 ppb Ozone
10 ppb Methane

JUPITER
89.8% ± 2.0% Hydrogen (H_2)
10.2% ± 2.0% Helium
~0.3% Methane
~0.026% Ammonia

(*Continued*)

~0.003% Hydrogen deuteride (HD)
0.0006% Ethane
0.0004% Water

ICES
Ammonia
Water
Ammonium hydrosulfide (NH_4SH)

SATURN
~96% Hydrogen (H_2)
~3% Helium
~0.4% Methane
~0.01% Ammonia
~0.01% Hydrogen deuteride (HD)
0.0007% Ethane

ICES
Ammonia
Water
Ammonium hydrosulfide (NH_4SH)

URANUS
Composition (below 1.3 bar):
83% ± 3% Hydrogen (H_2)
15% ± 3% Helium
2.3% Methane
(0.007%–0.015%) Hydrogen deuteride (HD)

ICES
Ammonia
Water
Ammonium hydrosulfide (NH_4SH)
Methane (CH_4)

NEPTUNE
80% ± 3.2% Hydrogen (H_2)
19% ± 3.2% Helium
1.5% ± 0.5% Methane
~0.019% Hydrogen deuteride (HD)
~0.00015% Ethane

ICES
Ammonia
Water
Ammonium hydrosulfide (NH_4SH)
Methane

PLUTO (MINOR PLANET)
Nitrogen
Methane

17.5.4 The Role of Spacecraft

Earth-based radio observations of the planets have limitations which affect certain types of measurements. These limitations include (1) the inability to obtain spatial resolution on scales of a few meters or less, (2) restrictions on the viewing geometry of the planets, (3) limitations set by the opaqueness and variability of the terrestrial atmosphere, (4) an intrinsic faintness of the radio emissions from planetary bodies, and (5) interference from man-made radio noise. The opacity of the terrestrial atmosphere varies with frequency. The atmosphere is opaque at frequencies lower than about 5 MHz due to the terrestrial ionosphere. Attenuation due to the atmospheric gases, water vapor, and oxygen affects the centimeter, millimeter, and submillimeter bands but observations are possible from the ground by working in the transparent "windows" in the spectrum. Rain, fog, and clouds occasionally limit the usefulness of the centimeter and millimeter bands.

To overcome these difficulties, a number of spacecraft radio instruments have been proposed, and several have flown on US spacecrafts. The first planetary radio system on a US spacecraft was a two-channel microwave radiometer that operated at wavelengths of 13.5 and 19.0 mm, flown on the Mariner [13] (Figure 17.5) spacecraft to Venus in 1962. The microwave radiometer system weighed-10 kg and used an average power of 4 W. This early system was designed to take advantage of the high spatial resolution and sensitivity that could be achieved from a spacecraft. Another radio astronomy experiment was placed on the Voyager spacecraft that was launched in the late 1970s to the outer planets and targeted to fly by Neptune in 1989. This experiment measures the radio spectra of planetary emissions in the range 1.2 kHz to 40.4 MHz. The system was designed to measure planetary spectra below the frequency range that is cut off by the Earth's ionosphere and to take advantage of the unique viewing geometry provided by the spacecraft. The Magellan Venus orbiter spacecraft carried a 12.6 cm radio receiver, which allowed over 91% of its surface to be measured.

In the future, we expect to see many more radio astronomy spacecraft experiments. Spacecraft experiments will allow the submillimeter spectral range to be observed without hindrance from the terrestrial atmosphere. Planetary spectroscopy in the submillimeter spectral range is expected to reveal new information about the upper atmospheres of the planets.

FIGURE 17.5
Mariner spacecraft. (Courtesy of NASA and Wikipedia, https://en.wikipedia.org/wiki/Mariner_program#/media/File:Mariner_2.jpg [last modified on August 22, 2016].)

FIGURE 17.6
Mercury. (Courtesy of NASA, Mercury Fact Sheet, http://nssdc.gsfc.nasa.gov/planetary/factsheet/mercuryfact.html [accessed September 10, 2016].)

The ESA ROSETTA spacecraft [14] will carry a millimeter and submillimeter wave spectroscopic instrument (MIRO) to investigate the nucleus and coma of a comet.

17.5.4.1 Mercury

Radio measurements of Mercury (Figure 17.6) show no evidence of an atmosphere. Estimates based on other techniques including spacecraft flyby instruments show the atmosphere on Mercury to be extremely tenuous, with a surface pressure of ~5×10^{-15}″ bar. For the purpose of interpreting the radio data, it can safely be assumed that Mercury is an airless planet. The radio emission from Mercury is thermal in character, strongly controlled by the high eccentricity of Mercury's orbit and the synchronism between Mercury's spin period and its period of revolution. Solar tidal effects have caused the period of axial rotation of Mercury to be 58.642 days, precisely two-thirds of its orbital period of 87.97 days. One solar day on Mercury is equal to 3 stellar days or 2 Mercurian years. This period equals 176 mean Earth solar days. Because of the synchronism between spin period and revolution period, the Sun takes a curious diurnal path in the sky as seen from the surface of Mercury. At some longitudes, the sun rises and sets twice a Mercury day. At perihelion the insolation is approximately twice its value at aphelion. The insolation reaches a maximum value of -14×10^3 W/m^2, 10 times the value of Earth. The visual albedo of Mercury is similar to that of the moon. The spin–orbit coupling and eccentricity combine to cause the surface of Mercury to be heated very nonuniformly in longitude. A pair of longitudes 180″ apart alternatively faces the sun at perihelion. These longitudes are preferentially heated because their midday insolation occurs when the sun is nearly stationary on the meridian and the Sun–Mercury distance is at a minimum. At longitudes 90″ away from these hot longitudes, the heating is identical but much less than that received at the hot longitudes. The longitudinal temperature variations on Mercury are indicated schematically by the "hot," "warm," and "cold" regions. This pattern of temperature variations is fixed on Mercury because the spin–orbit coupling causes the heating to be cyclic at each longitude (i.e., the same pair of longitudes faces the sun at perihelion).

Radio spectra and images, together with Mariner 10 IR data of the planet have been used to infer Mercury's surface properties. At first glance, Mercury's surface is quite similar to that of the Moon. Due to the fact that Mercury, like the Moon, is continuously bombarded by small meteorites, one would expect the top few centimeters to have a very low density, while deeper layers are more compact. Observations indeed show that this is the case on both bodies. Researchers found that the microwave opacity on Mercury is roughly a factor of 2–3 smaller than that of most lunar samples. This suggests that the ilmenite content, which is the

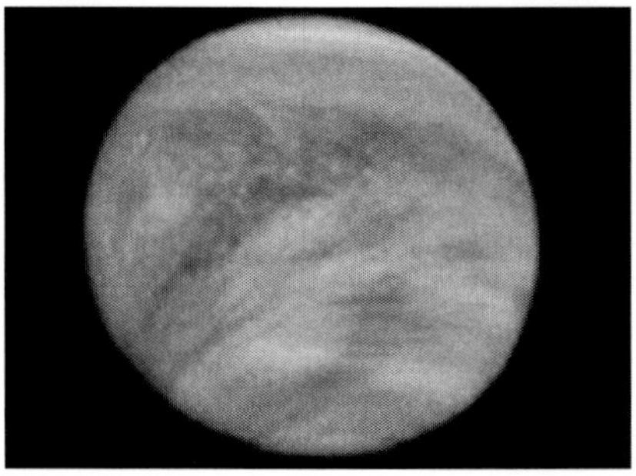

FIGURE 17.7
Venus.

most common titanium-bearing mineral on the Moon ((Fe, Mg)TiO), is much less abundant on Mercury than on the Moon. Ilmenite is also opaque at optical wavelengths and is largely responsible for the dark appearance of the moon's maria compared to its highlands. Its absence would explain why Mercury is brighter than the Moon at visible wavelengths.

17.5.4.2 Venus

Venus (Figure 17.7) has the densest atmosphere of all the terrestrial planets. The principal atmospheric constituents are carbon dioxide and nitrogen (N_2); their mixing ratios are approximately 96.5% and 3.5%, respectively, below 100 km altitude. Trace constituents below 100 km altitude are in the range of 0.1%. The Venus atmosphere is covered with thick clouds composed primarily of sulfuric acid and contaminants, making the surface invisible from above. The total pressure at the bottom of the cloud layer (–47 km) is approximately 1.3 atmospheres. Water is highly depleted throughout the atmosphere. The mean physical structure of the atmosphere (pressure and temperature profile) is reasonably well known from the data returned by a number of space probes. The surface pressure and temperature (on a mean surface) are approximately 94 atm and 737 K, respectively.

The effective temperature of Venus deduced from measurements in the infrared is about 240+. 8 K, corresponding to an altitude of approximately 60 km. Below this level, the temperature distribution generally follows that for an atmosphere that is in convective equilibrium, which implies that the temperature gradient in the atmosphere is close to the adiabatic value. The temperature gradient is approximately 8.6 Wkm. The high surface temperature is believed to be due to the greenhouse effect. The physical basis for this is that the visible light from the Sun is only partially absorbed by the clouds and atmosphere. Some of the light reaches the surface and warms it. The heated surface re-radiates in the infrared. The atmosphere is highly absorbent in the infrared spectral region to CO, and perhaps H, O. The atmospheric opacity traps the infrared radiation, thereby raising the surface temperature.

The atmosphere of Venus is opaque at millimeter and short centimeter wavelengths, gradually becoming transparent at longer wavelengths. Individual spectral lines are not observable below the clouds because of pressure broadening. The total vertical optical depth of the atmosphere of Venus at a wavelength of 1 cm is estimated to be slightly less than 20 and to vary approximately as (optical depth = 1 at –4 cm). A little more than half of the total opacity is due to collision-induced, nonresonant absorption in CO; the remaining opacity is produced by the minor constituents in the atmosphere. Other known or

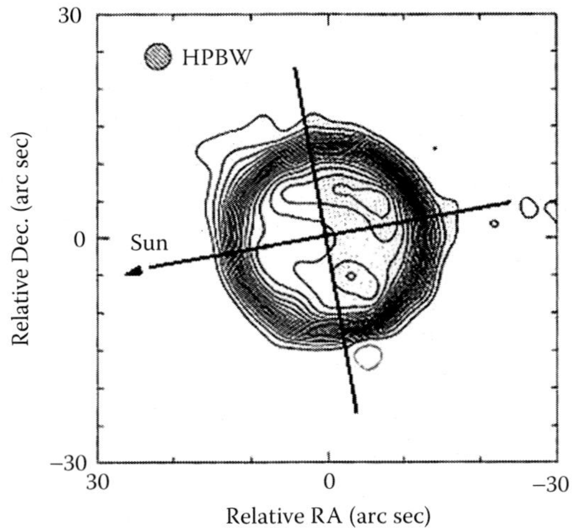

FIGURE 17.8
Radio map of Venus at 3 mm. (From de Pater, I. Radio images of the planets, *Ann. Rev. Astron. Astrophys.* 28, 347–399, 1990; Courtesy of SAO/NASA Astrophysics Data System, http://adsabs.harvard.edu/abs/1991Icar...90..282D [accessed September 10, 2016].)

suspected microwave absorbers in the atmosphere are H, O, SO, H, SO, and the sulfuric acid particles in the clouds. Near wavelengths of 6 cm and long ward, the atmosphere is sufficiently transparent that it is possible to measure the surface temperature of Venus using radio astronomical methods.

The brightness temperature is seen to rise from about 225 K and 3 mm to about 700 K near 6 cm. This increase in brightness temperature is due to the decreasing opacity of the atmosphere with increasing wavelength. The decreasing opacity allows radio waves to escape from deeper regions in the atmosphere, where it is warmer due to the adiabatic lapse rate. At wavelengths longer than –15 cm the brightness temperature decreases to −600 K. Figure 17.8 contain radio views of Venus at 3 mm.

Radio interferometer data, radar data, and spacecraft data have been used to study the surface of Venus, in particular to determine the dielectric constant. Radar reflectivity data place the dielectric constant in the range 4–5. The Magellan radiometer experiment observed the 12.6 cm wavelength radio emissivity of more than 91% of the Venus surface. With its 2° beam width, Magellan observations achieved surface resolutions that varied from 15 by 23 km at priapism (10°N latitude) to about 85 km at the North Pole. The global mean value of emissivity seen using horizontal linear polarization is 0.845, a value that corresponds to a dielectric permittivity of between 4.0 and 4.5, depending on the surface roughness. These values are considerably greater than the values for Mercury, Mars, and the moon, which range from 2.0 to 2.5. The higher values are suggestive of a surface composed of dry rock not unlike many rocks of the Earth's surface. These values are consistent with the dry basaltic minerals thought to compose the bulk of the Venus surface. The observations have confirmed earlier findings that a few regions on Venus, primarily located at high elevations, possess unexpectedly low values of radio thermal emissivity, occasionally reaching as low as 0.3.

Carbon monoxide is an important constituent of the upper atmosphere of Venus. It is formed primarily by the dissociation of carbon dioxide by solar ultraviolet radiation and is removed by chemical and transport processes in the atmosphere. The ground and first excited rotational states of CO (located at very high altitudes in Venus' atmosphere) have been observed to absorb the hot continuous background of the deeper atmosphere. Observed spectral lines have included $^{12}CO(0\text{-}1)$, $^{12}CO(1\text{-}2)$, $C^{18}(1\text{-}2)$, and $^{13}CO(1\text{-}2)$. These observations have been particularly useful for exploring the altitude range from 70 to 110 km.

It has been possible to derive the vertical temperature and mixing ratio profiles of CO in the upper atmosphere of Venus from such observations. The variability is believed to be the result of large-scale circulation in the upper atmosphere of the planet.

Detailed studies of this variability have been undertaken with arrays of telescopes, which operate at millimeter wavelengths. With such arrays the planet can be imaged in the CO lines. With this said, CO concentrations and temperature profiles at all solar phase angles can be measured simultaneously.

In addition, through Doppler shift measurements, the lines at various locations on Venus' disk, the winds in Venus' mesosphere can be studied. For the CO (0-1) transition, a mean wind speed of 100 m/s in the spectral line forming region produces a Doppler shift of 38 KHz. Such Doppler shifts have been observed at altitudes at least as high as 100 km. Continued observations of these spectral lines will lead to a better understanding of the large-scale circulation in Venus' upper atmosphere.

17.5.4.3 Mars

Mars moves in an orbit slightly larger than the Earth's, always turning its dayside toward the Earth as it approaches. Earth-based measurements of the night side are impossible and phase angle coverage is greatly restricted. The axis of rotation is tilted from the perpendicular to the plane of its orbit by 250, about the same as for the Earth. Radio observations of the disk brightness temperature of Mars show a nearly flat spectrum from about 1 mm to 21 cm. The mean disk brightness temperature is about 215 K. As seen from the Earth, the average surface disk temperature varies by +15 K as the sub-Earth point moves from afternoon to morning and from mid-latitudes to equatorial latitudes. Thus far, most observational data used in thermal modeling studies have come from infrared measurements made from spacecraft. The infrared measurements are limited to the near surface properties. Microwave images obtained with the VLA have extended the thermal models to the subsurface layers of the planet. Figure 17.9 is a radar image of Mars.

Microwave images in contrast to single dish observations also yield information on the spatial variations of the brightness temperature. For example, VLA images show the polar regions to be significantly colder than the disk-averaged temperature.

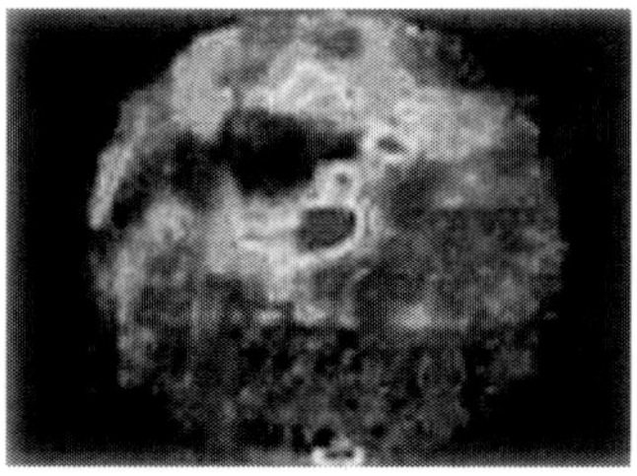

FIGURE 17.9
Radar image of Mars. (Courtesy of NRAO, http://www.gb.nrao.edu/visitors/gbtour/radiomars.jpg [accessed September 10, 2016].)

The centimeter radio brightness temperatures of Mars have been found to vary as a function of the central meridian longitude of the planet. Temperature differences as large as 5–10 K are observed over the full range of longitudes. The variations are believed to be due to nonconformity in the Martian surface properties; however, no completely satisfactory explanation of the observations exists.

The Martian atmosphere is very tenuous, having a surface pressure some 200 times less than on the Earth. The primary constituent of the lower atmosphere is CO. Photolysis of CO by solar ultraviolet radiation produces CO and O. The diatomic molecule CO plays an important role in determining the millimeter wave spectrum of Mars.

Both the ground state and first excited state transitions of CO have been observed in the Martian atmosphere. The altitude distribution of CO has been inferred by interpreting the observed line shape in terms of pressure broadening. A column abundance of CO equal to $2–5 \times 10^2$ molecules/cm^2 has been derived from the measurements. This number is very stable overtime, even though the CO line profiles show significant variability. This variability has been used to "monitor" Mars's thermal structure: the atmospheric temperature rises significantly during global dust storms, since the dust grains, heated by the Sun, warm up the atmosphere while at the same time preventing sunlight from penetrating down to the surface. Water vapor (from HDO observations at 226 GHz) has also been observed on Mars and used to determine water vapor distribution and atmospheric behavior. Clancy and collaborators used the VLA at a wavelength near 1.35 cm to image the water vapor in Mars's atmosphere. The water vapor shows up in emission around the planet, where the path length through the atmosphere is largest. The data clearly show the absence of emission over Mars' polar caps, where the atmospheric temperature is so low that all the water has been frozen out. We note that the ground-based detection of water vapor in Mars's atmosphere could only be achieved because of the VLA's high angular resolution, so that the atmosphere on the limb could be separated from the planet itself.

Author's Note

Radio Telescopes to Keep Sharp Eye on Mars Lander (2008)—As NASA's Phoenix Mars Lander descends through the Red Planet's atmosphere toward its landing on May 25, its progress will be scrutinized by radio telescopes from the National Radio Astronomy Observatory (NRAO). At NRAO control rooms in Green Bank, West Virginia, and Socorro, New Mexico, scientists, engineers, and technicians will be tracking the faint signal from the lander, 171 million miles from Earth [15].

17.5.4.4 Jupiter

Author's Note

Jupiter is the first planet detected at radio wavelengths. The discovery observations occurred in 1955, at the very low frequency of 22.2 MHz.

In 1955, Bernard Burke and Kenneth Franklin detected bursts of radio signals coming from Jupiter at 22.2 MHz [16]. The period of these bursts matched the rotation of the planet, and they were also able to use this information to refine the rotation rate. Radio bursts from Jupiter were found to come in two forms: long bursts (or L-bursts) which can last up to several seconds, and short bursts (or S-bursts) which can have a duration of less than a hundredth of a second. Figure 17.10 is an artist's conception of Jupiter and the Galileo space probe. Figure 17.11a and b show Jupiter radio bursts at 21.4 MHz.

Scientists discovered that there were three forms of radio signals transmitted from Jupiter. Decametric radio bursts (with a wavelength of tens of meters) vary with the rotation of Jupiter, and are influenced by interaction of Io with Jupiter's magnetic field.

Decametric radio emission (with wavelengths measured in centimeters) was first observed by Frank Drake and Hein Hvatum in 1959. The origin of this signal was from a torus-shaped belt around Jupiter's equator. This signal is caused by cyclotron radiation from electrons that are accelerated in Jupiter's magnetic field.

Jupiter revealed that its radio spectrum is extremely complicated in its workings complex, showing both thermal and nonthermal emission mechanisms. Thermal emission from the atmosphere is most prevalent in the Jovian spectrum on the short-wavelength side of 7 cm.

Observations of the Jovian show that the atmosphere is composed mostly of hydrogen and helium, in convective equilibrium. The principal source of opacity is ammonia (NH), which exhibits very strong absorption in the microwave spectral region.

Nonthermal synchrotron emission dominates the spectrum from –3 m to 7 cm; brightness temperatures exceed 105 K for the synchrotron component. Long ward of 7.5 m, Jupiter emits strong and sporadic nonthermal radiation. The radiation exhibits complex

FIGURE 17.10
Galileo spacecraft. (Courtesy of NASA, Galileo Legacy Site, http://science.nasa.gov/missions/galileo/ [retrieved March 17, 2016].)

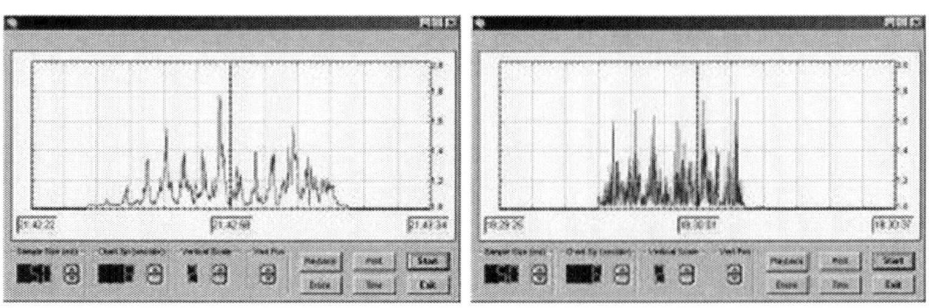

(a) (b)

FIGURE 17.11
Jupiter radio bursts: (a) L-burst and (b) S-burst. (Courtesy of RadioSky Publishing, Ocean View, HI, http://radiosky.com/.)

frequency, time, and polarization structure. The brightness temperature of this component exceeds 1017 K, suggesting a coherent source of emission.

Observations of Jupiter at high angular resolution with radio interferometers have been used to map the synchrotron radiation from Jupiter's radiation belts and to separate the thermal from the nonthermal synchrotron components. The nonthermal component is easily identifiable with a radio interferometer because it is greatly extended relative to the optical disk of Jupiter and is strongly linearly polarized.

In December 1995 the Galileo spacecraft released a probe into the atmosphere of Jupiter, which relayed its findings to the spacecraft via radio signals at a frequency of 1.4 GHz. By analyzing the attenuation of the probe radio signal, the ammonia abundance in Jupiter's deep atmosphere (at pressures over 8 bars) was derived to be a factor of –3.5 larger than the solar N value. Ground-based microwave measurements are most sensitive to layers where the clouds form (–0.5 bar) down to roughly 10–15 bar. The ground-based microwave measurements do not show as much ammonia as the probe. Apparently, the ammonia abundance in Jupiter's deep atmosphere is significantly decreased at higher altitudes, to roughly half the solar N value just below the upper cloud deck. Scientists do not (yet) understand why there is so much less ammonia in the upper regions of Jupiter's atmosphere compared to deeper layers.

Radio interferometric maps of Jupiter's synchrotron emission have been made at a number of different wavelengths. It has been possible to deduce a great deal of information about Jupiter's magnetosphere from the radio measurements. The radio astronomical measurements provided convincing proof that Jupiter has a strong magnetic field, and this information was used to design the first spacecraft sent to Jupiter. The radio measurements show that the magnetic field is primarily dipolar in shape with the dipole axis tilted about 10″ with respect to Jupiter's rotational axis. Using the well-developed theory of synchrotron emission, it has been possible to determine energies and densities of the high-energy electrons that are trapped in Jupiter's magnetic field. Figure 17.12a–f illustrates radio emission from Jupiter at 6, 13, 20, 21, and 90 cm.

The total radio intensity of Jupiter varies significantly over time (years), and appears to be correlated with solar wind parameters. Figure 17.13 shows the impact of comet Shoemaker-Levy 9 with Jupiter in July 1994 caused a sudden sharp increase in Jupiter's total flux density, by –20%. At the same time the brightness distribution of the radio flux density changed drastically. These observations suggest that the impact and associated phenomena significantly modified the electron distribution and possibly the magnetic field as well.

At frequencies below 40 MHz, Jupiter is a strong emitter of sporadic nonthermal radiation. A decameter wavelength (DAM) component observable from the ground is characterized by complex, highly organized structure in the frequency–time domain and dependent on the observer's position relative to Jupiter. The satellite Io modulates the DAM emission. The Voyager spacecraft added significantly to our knowledge of this low-frequency component when it flew by Jupiter in 1979. A kilometric wavelength (KOM) component was discovered at frequencies below 1 MHz and the observations of the DAM component were significantly improved. The extremely high brightness temperatures (K), narrow bandwidth emissions, and sporadic nature all suggest that the very low frequency emissions from Jupiter are generated by energetic particles acting coherently and interacting with the plasma that surrounds Jupiter. The details of the emission process are not well understood at present. Figure 17.14 shows the spectrum of Jupiter's radio emissions compared with spectra of four other magnetized planets.

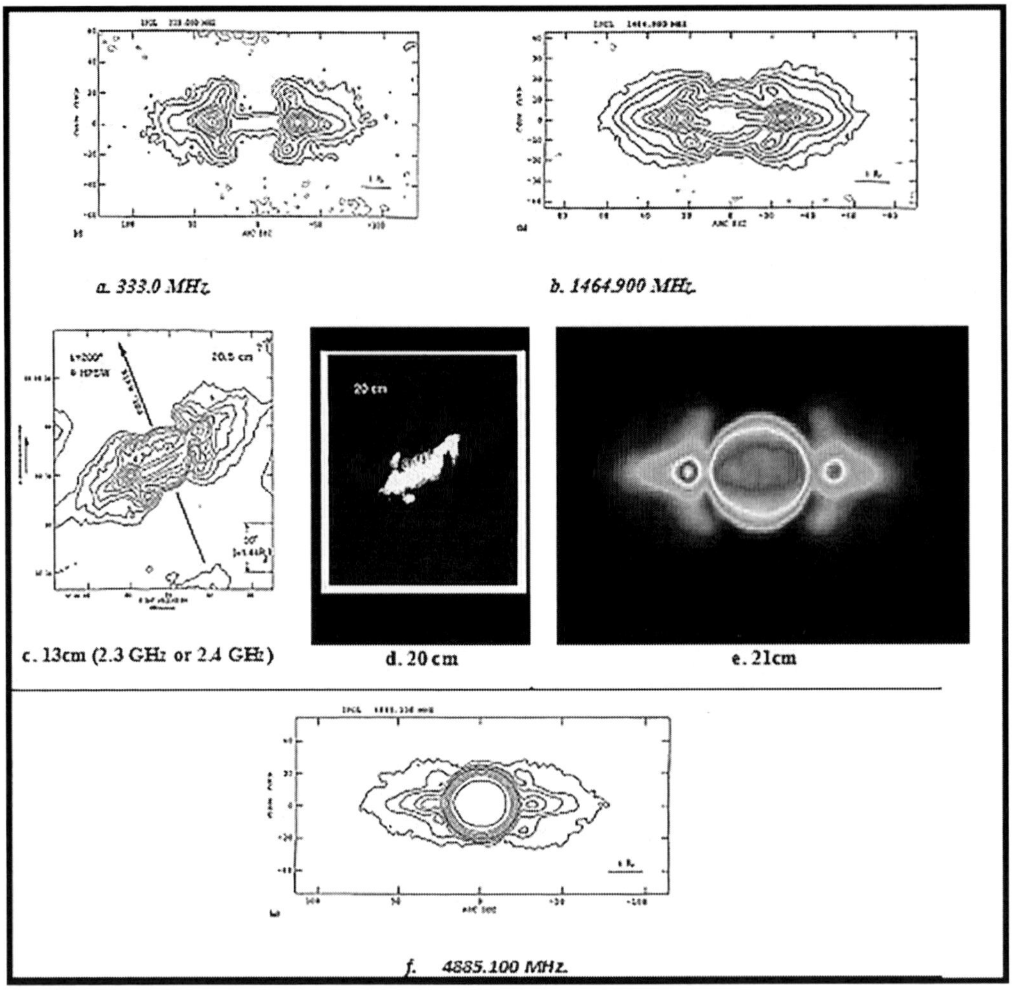

FIGURE 17.12
Radio images of Jupiter's synchrotron radiation at 6, 13, 20, 21, and 90 cm. (From de Pater, I., *Ann. Rev. Astron. Astrophys.*, 28, 347–399, 1990; Courtesy of NASA, Astrophysics Data System, http://adsabs.harvard.edu/abs/1991Icar...90..282D [retrieved September 10, 2016]; Courtesy of NRAO, http://images.nrao.edu/22 [retrieved September 10, 2016].)

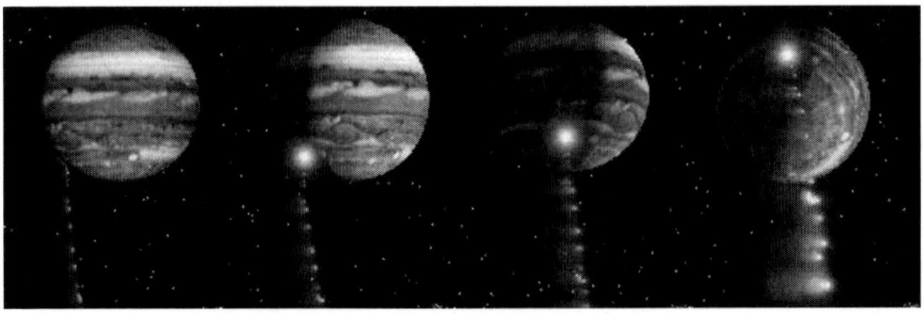

FIGURE 17.13
The impact of comet Shoemaker-Levy 9 with Jupiter. (Courtesy of NASA, http://www2.jpl.nasa.gov/sl9/ [retrieved August 23, 2016].)

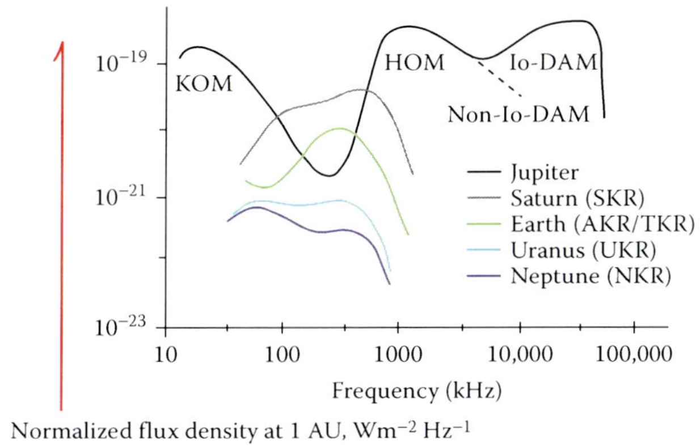

FIGURE 17.14
The spectrum of Jovian radio emissions compared with spectra of four other magnetized planets, where (N,T,S,U) KR means (Neptunian, Terrestrial, Saturnian and Uranian) kilometric radiation emissions. (Redrawn from Wikipedia, Magnetosphere of Jupiter, https://en.wikipedia.org/wiki/Magnetosphere_of_Jupiter [last modified on June 30, 2016].)

17.5.4.5 Saturn

Radio emission from Saturn has been observed from the Earth over a wavelength range from 1 mm to approximately 70 cm. The emission is thermal throughout this band, arising in both the atmosphere and the rings. The atmospheric emission is similar to that observed from Jupiter. Figure 17.15 shows radio observations of Saturn.

Interferometer observations of Saturn also detect the thermal emission from the ring particles. Most of the radio emission is due to scattering of Saturn's emission off the ring particles. In front of the planet, the rings are visible as an absorption feature; they block out Saturn's radio emission. The scattering characteristics of the rings contain information on the ring particle sizes and composition.

Interferometer observations of Saturn at centimeter and millimeter wavelengths have detected thermal emission from the ring particles. The rings have a low brightness temperature, approximately 10 K. The presence of ring particle at sizes larger than a few centimeters are suggested by the radio observations. The observations are consistent with the bulk properties of the ring particles being those of water ice.

The Voyager spacecraft detected two distinct classes of nonthermal emissions from Saturn at frequencies below 1 MHz. These emissions are not observable from the Earth because of the opaqueness of the Earth's ionosphere at frequencies below a few megahertz. The first class emission is called Saturn kilometric radiation, which is a relatively narrow band polarized emission. The second class is a broadband, impulsive emission, called Saturn electrostatic discharge [17].

17.5.4.6 Uranus

Uranus (Figure 17.16) is unique among the planets in having its rotation axis tilted close to the plane of the ecliptic. The north pole of Uranus is inclined –98″ to the ecliptic plane (8″ south of the plane), and the seasons on Uranus average 21 terrestrial years in length. The effect of this geometry on the large-scale circulation of the Uranium atmosphere is not yet fully understood, but it is expected to be significant.

As in the case of both Jupiter and Saturn, the disk brightness temperatures of Uranus significantly exceeds the expected equilibrium temperature. The observed temperatures are

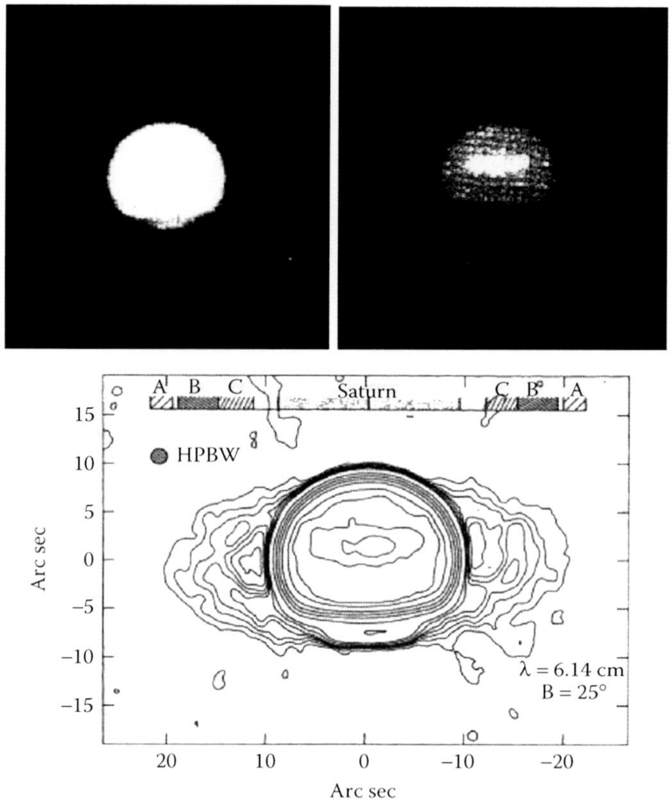

FIGURE 17.15
Radio images of Saturn at 2 and 6 cm and at different ring inclination angles B. The resolution in all images is 1.5″. (top left) Radio photos at λ = 2 cm, B = 26°; (top right) radio photo at λ = 6 cm, B = 20°; (bottom) contour map at λ = 6 cm, B = 25°. Contour values (in kelvins) are 1.9, 3.7, 5.6, 7.4, 9.3, 13.0, 18.6, 46.5, 4.3, 102.2, 130.0, 158.0, 167.0, 176.5, and 182.1. Negative contour values, with some absolute levels, are indicated as dashed contours. The images taken in December 1986, January 1984, and June 1984, respectively. The bar at the top shows schematically the A, B, and C rings at either side of Saturn. (From de Pater, I. Radio images of the planets, *Ann. Rev. Astron. Astrophys.* 28, 347–399, 1990; Courtesy of NASA Astrophysics Data System, http://adswww.harvard.edu/ [retrieved September 10, 2016].)

FIGURE 17.16
Uranus. (Courtesy of NASA, Uranus Fact Sheet, http://nssdc.gsfc.nasa.gov/planetary/factsheet/uranusfact.html [retrieved September 10, 2016].)

greater than 100 K at wavelengths greater than a few millimeters and long ward, whereas the predicted effective temperature is only about 55 K. The radio emission from Uranus is unpolarized within the measurement uncertainties of a few percent. Interferometer observations of Uranus show the emission to be confined to the solid angle of the visible disk, providing evidence that the excess emission is from the atmosphere and is not synchrotron emission. It is believed that the emission from Uranus is thermal, originating in the atmosphere of the planet. The radio emissions from Uranus arise from sufficient depths that collision-induced absorption by hydrogen is an important source of opacity at millimeter wavelengths. Ammonia is severely depleted in Uranus's atmosphere, at least at pressure levels less than 25 bar. Since, based upon planet formation theories, nitrogen must be present in at least solar proportions, it is believed that ammonia gas is abundant at deeper levels, but reacts with H and S to form a cloud of NH_4SH (NH_4SH as the Primary 3-Micron Absorber in Jupiter's Clouds) [18]. If indeed this process accounts for the "observed" depletion in NH, hydrogen sulfide should be enriched in Uranus's atmosphere by about an order of magnitude over solar S. Such an abundance of H, S itself will contribute to the radio opacity in Uranus's atmosphere, and actually help reconcile observed spectra with models.

Another interesting aspect of the radio emission from Uranus is that its total intensity varies slowly over time. With the help of interferometric (VLA) observations, it is well established by now that the Uranus pole is warmer than its equator. This will certainly lead to time variations during Uranus' orbit around the sun. It is not known, however, whether this indeed fully explains the observed time variability. Figure 17.17 shows radio images of Uranus.

The Voyager spacecraft detected a wide variety of radio emissions from Uranus during its encounter in January 1986. Most of the emissions were polarized and probably due to maser cyclotron emission. The emissions range in frequency from about 20–800 kHz, well below the frequency range that is observable from the Earth. The emissions suggest a magnetosphere rich in magnetohydrodynamic phenomena.

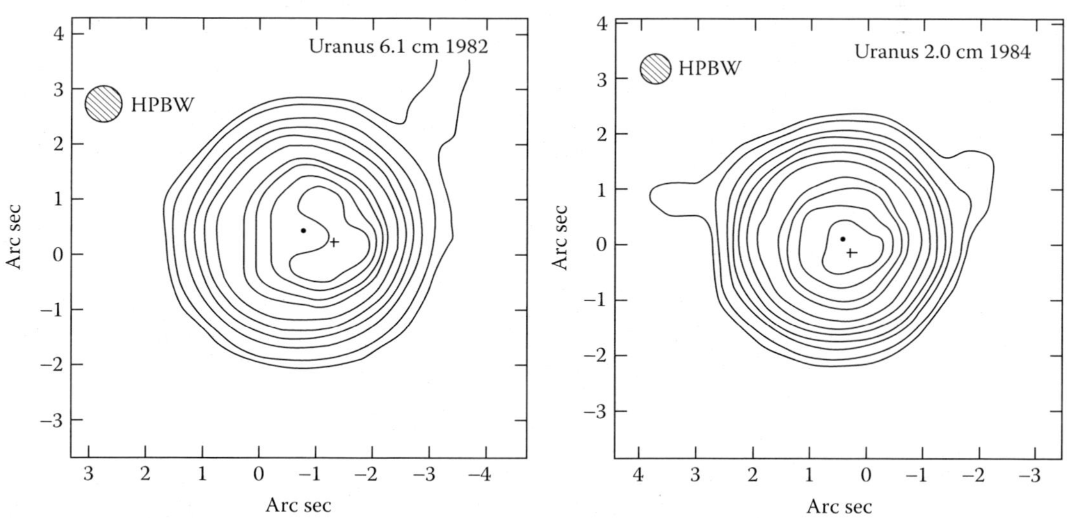

FIGURE 17.17
Radio images of Uranus at (left) 6 cm and (right) 2 cm. The cross indicates the position of the pole, and the dot is the position of the subpolar point. The resolution of the images is 0.65″ for the 6 cm and 0.5″ for the 2 cm image. The contour values (in kelvins) are (a) 14, 28, 69, 110, 151, 193, 234, 248, 261, and 270; and (b) 11, 22, 55, 88, 121, 154, 187, 198, 209, and 216. (From de Pater, I., Gulkis, S. VLA observations of Uranus at 1.3–20 cm, *Icarus*, 75, 306–323, 1988.)

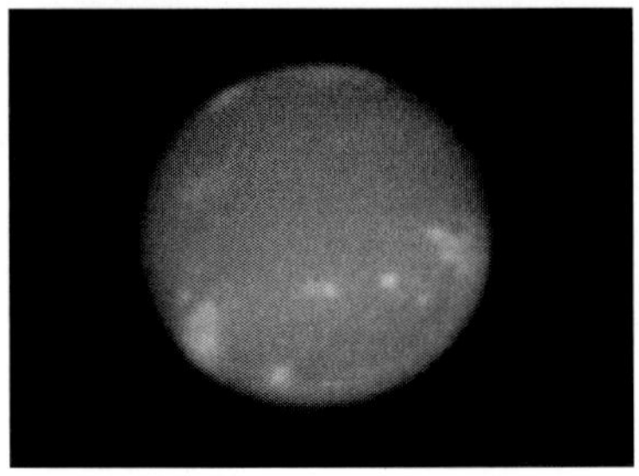

FIGURE 17.18
Neptune.

17.5.4.7 Neptune

Disk brightness temperature measurements of Neptune (Figure 17.18) at centimeter and millimeter wavelengths are sparse, but existing measurements suggest a spectrum is very similar to that of Uranus. The disk brightness temperatures exceed the predicted equilibrium temperature by 50 K or more. High angular resolution measurements obtained with the VLA show that the excess emission is not due to synchrotron emission. As in the case of Uranus, model studies suggest that ammonia must be depleted on Neptune (as on Uranus) by roughly two orders of magnitude compared to solar nitrogen values to explain the high brightness temperatures that are observed.

This depletion may be caused by a nearly complete removal of NH gas in the upper atmosphere through the formation of NH_4SH, the same process which depletes ammonia in the Uranus atmosphere. On Neptune, the H and S abundance must be even larger than on Uranus, about 30 times the solar sulfur/hydrogen ratio. Although ammonia gas may be close to the solar N value in Uranus deep atmosphere, it must be substantial subsolar throughout Neptune.

Here a problem arises as the subsolar value for nitrogen gas is inconsistent with theories on planet formation. If a planet forms directly from the primordial solar nebula, the elemental abundances must be equal to that measured on the Sun. Condensable materials accrete as solids, sublime in the protoplanet's atmosphere and therefore enhance the elemental abundances above solar values. Carbon, sulfur, and oxygen compounds present as CH, H, S, H, and O in the giant planets are therefore enhanced above solar values. Similarly, nitrogen is expected to be enhanced above solar N. A subsolar value on Neptune is simply not possible. This dilemma was solved, however, with the detection of emission lines in the 1–3 mm band of CO and HCN in Neptune's upper atmosphere.

Since both lines are seen in emission, CO and HCN must be located above the tropopause in Neptune's stratosphere. Although CO may be brought up from deep depths through rapid convection, HCN must be formed in the stratosphere from nitrogen and carbon products, since HCN, if it would exist in Neptune's deep atmosphere, would freeze out on its way up, well before it could reach the stratosphere. If, however, much of the nitrogen in Neptune's atmosphere would be present in the form of N, rather than NH, the N, might convect rapidly upward and form HCN in the stratosphere through chemical reactions with carbon products. Moreover, if nitrogen would exist in the form of N, rather than NH, the total nitrogen abundance in Neptune's atmosphere would be consistent with planet formation theories.

The Voyager spacecraft detected a variety of low frequency radio emissions from Neptune during its encounter in 1989. The emissions were similar to those observed from other major planets.

17.5.4.8 Pluto

In 1905, astronomer Percival Lowell noticed orbit variations of Neptune and Uranus. This led to further investigation and the thoughts of another planet might exist. The variations in the orbits were most likely due to the gravity of what was then to be known as planet x, which was asserting its presence. In 1915, Lowell predicted its location, but died without confirming its existence. In 1930, Clyde Tombaugh also at the Lowell Observatory, and using the research based on predictions from Lowell and other astronomers made the solid discovery.

17.6 Further Research

New Horizons is a NASA space probe [19] launched to study the dwarf planet Pluto, its moons, and one or two other Kuiper belt objects depending on which are in position to be explored. Part of the New Frontiers program, the mission was approved in 2001 after cancellation of Pluto Fast Flyby and Pluto Kuiper Express. The mission profile was proposed by a team led by principal investigator Alan Stern of Southwest Research Institute. After several delays on the launch site, New Horizons was launched on January 19, 2006, from Cape Canaveral, directly into an Earth-and-solar-escape trajectory with an Earth-relative speed of about 16.26 km/s (58,536 km/h; 36,373 mph); it set the record for the highest launch speed of a human-made object from Earth. After a brief encounter with asteroid 132524 APL, New Horizons proceeded to Jupiter, making its closest approach on February 28, 2007, at a distance of 2.3 million kilometers (1.4 million miles). The Jupiter flyby provided a gravity assist that increased New Horizons' speed by 4 km/s (14,000 km/h; 9000 mph). The encounter was also used as a general test of New Horizons' scientific capabilities, returning data about the planet's atmosphere, moons, and magnetosphere. After Jupiter, the probe continued toward Pluto. Much of the post-Jupiter voyage has been spent in hibernation mode to preserve onboard systems. On January 15, 2015, NASA reported the New Horizons spacecraft began its approach phase to Pluto, resulting in the first flyby of the dwarf planet on July 14, 2015 [20]. Figure 17.19 shows a radio picture of Pluto, Charon, and the New Horizons space probe on the pad, prior to launch (2006).

(a) (b)

FIGURE 17.19
(a) Pluto as seen in August 2014: The ALMA observes Pluto and Charon. (Courtesy of NRAO, ALMA Pinpoints Pluto to Help Guide NASA's New Horizons Spacecraft, https://public.nrao.edu/news/pressreleases/alma-pluto [retrieved August 23, 2016]). (b) the New Horizons space probe prior to launch. (Courtesy of Wikipedia, https://en.wikipedia.org/wiki/File:NewHorizons_Rocket_Bly.jpg [last modified on December 30, 2011].)

17.7 Radio Astronomy Contribution

Astronomers using the Atacama Large ALMA Pinpoints Pluto to Help Guide NASA's New Horizons Spacecraft. The millimeter/submillimeter Array (ALMA) are making high-precision measurements of Pluto's location and orbit around the Sun to help NASA's New Horizons spacecraft accurately home in on its target when it nears Pluto and its five known moons [21] in July 2015.

Since then the findings have been numerous. Further information can be found in [20].

17.8 Comets

Prior to the 1960s, comets were investigated primarily using visible wavelengths observations. The first infrared detection of a comet was achieved in the mid-1960s. Although radio astronomers had attempted to detect cometary emissions since the 1950s, the first widely accepted radio detection of a comet is that of the 18 cm OH transitions in Comet Kohoutek [22] in 1973. Radio observations of comets have been used to study all of the major components of comets—nucleus, dust, neutral gas, and the plasma.

Continuum observations of thermal emission from comets provides information on the nucleus temperature and on the large dust grains which surround the nucleus. The thermal emission from comets is very weak, and sensitive receiving systems are needed to make a detection. Short centimeter, millimeter, and submillimeter observations have been the most successful for the detection of thermal emission. In the 1990s sensitive detectors at millimeter and submillimeter wavelengths came on line, just in time to observe two very bright comets: Comet Hyakutake [23] in 1996 and Comet Hale–Bopp [24] in 1997. Radio spectra at submillimeter wavelengths were obtained for both comets. The thermal emission is clearly dominated by emission from dust grains, rather than the cometary nucleus, and the data have been used to derive the dust mass and dust mass production rate for each comet. Continuum observations of comets will be greatly improved when the new large millimeter and submillimeter arrays become available and when thermal emission measurements from spacecraft are possible.

Radio astronomy of comets has made its most significant advances in the area of spectroscopy. Many so-called "parent" molecules (molecules which sublimate directly off the nucleus, in contrast to "daughter" molecules which are products of the parents) have been observed at radio wavelengths through their rotation lines in the millimeter and submillimeter bands. Observations of these spectral lines give valuable information on the icy composition of the cometary nucleus, gas production rates, physical conditions in the coma, and on the variation of these in time and heliocentric distance.

The timely appearance of the bright comets Hyakutake and Hale–Bopp greatly expanded our knowledge of cometary volatiles. The number of identified parent molecules increased by more than a factor of two through observations of these comets; most of these new species were detected in the millimeter and submillimeter. Spectroscopic observations of comet Hale–Bopp, which could be followed in its orbit for several years, yielded invaluable data on composition and gas production rates. Gas production rates of different molecules could be followed over large heliocentric distances. At distances greater than 3 AU, the sublimation was dominated by gases more volatile than water. Carbon monoxide, the most

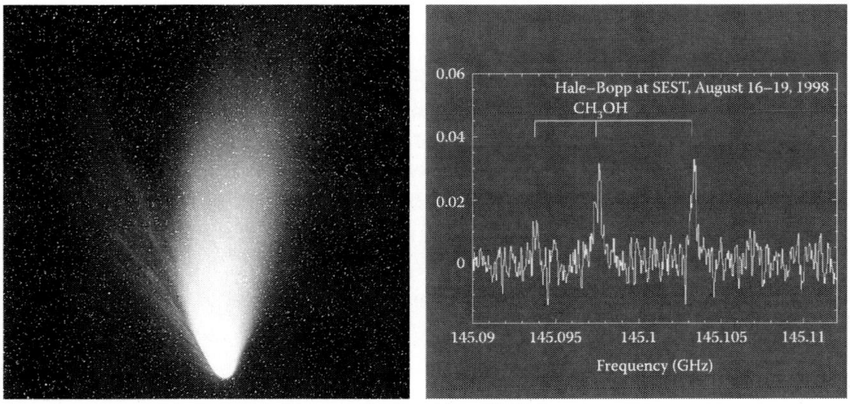

(a) (b)

FIGURE 17.20
(a) Comet Hale–Bopp. (Courtesy of Wikipedia, https://en.wikipedia.org/wiki/Comet_Hale%E2%80%93Bopp#/media/File:Comet_Hale-Bopp_1995O1.jpg [last modified April 07, 1997].) (b) Methanol and hydrogen cyanide detected. (Courtesy of ESO, New Observations of Comet Hale-Bopp from La Silla, http://www.eso.org/public/news/eso9849/ [retrieved September 10, 2016].)

volatile species observed, could be detected at a heliocentric distance of 7 AU. It displayed a steady increase in brightness while the comet moved in toward the Sun. This gas clearly must have been key to the observed visible brightness of the comet at these large heliocentric distances. At a heliocentric distance of 3 AU, OH became the most abundant species, confirming that the composition of the comet is dominated by water ice. Inside a heliocentric distance of 1.5 AU all species showed a dramatic increase in production rates. Radio images of various species have been obtained and have been used to derive the relative importance of sublimation directly from the nucleus or from grains in the coma. This varies from gas to gas, and probably differs from comet to comet. Figure 17.20a and b shows comet Hale–Bopp and chemical detection.

As mentioned above, the first radio detection of a comet was that of the 18 cm OH line. The study of this line has been an important component of cometary research for over 25 years. The 18 cm line is split into four transitions at 1612, 1665, 1667, and 1721 MHz. The observed intensity of the OH line is quite variable. The line is seen in emission or absorption depending on the comet's heliocentric velocity. The explanation of this effect is that it is due to pumping of the OH molecules by solar UV photons. If the comet's heliocentric velocity is such that Fraunhofer absorption lines from the Sun shift into the excitation frequency, the OH molecules are not excited. At all other times they are. This variation in excitation is known as the Swings effect. When OH is not excited, it is seen in absorption against the galactic background. If it is excited, the emission is actually maser emission rather than simple thermal emission.

17.9 Prospects for the Future

Planetary radio astronomical observations were initially limited to measurements of the disk-averaged brightness temperatures and to the strong nonthermal radio emission from Jupiter. With improved spatial resolution and high sensitivity receiving systems, it has been possible to map the planets (satellites and comets) and carry out high-resolution spectroscopic observations.

With the rapidly improving instrumentation now on the horizon, eventually it should be possible to study the planets using ground-based and spacecraft instruments in nearly the same detail in which the Earth is being studied with combined sensors on weather satellites and ground stations. As angular resolution improves with time, it should be possible to map the vertical and horizontal distribution of certain chemical species within planetary atmospheres, to measure wind speeds, and to search for local variations in subsurface properties. Both ground-based radio telescopes and spacecraft radio receivers will play a role in future observations. Large ground-based telescopes will have sufficient resolution to examine the global properties of the planets out to Neptune and to study satellites, asteroids, and comets. The advent of lighter-weight and lower-power radio receivers will enhance the possibilities of placing radio experiments on spacecraft. Radio telescopes working at submillimeter wavelengths are gradually becoming a reality as a few mountaintop observatories are nearing completion and plans for a submillimeter space telescope are being discussed in a number of countries. The shorter-wavelength region of the spectrum will provide new opportunities to study the upper atmospheres of the planets with spectroscopic techniques capable of detecting heretofore unobserved transitions that occur in the submillimeter and infrared. In addition, the shorter wavelengths will provide high angular resolution with only moderate-size telescopes. In summary, the future prospects for planetary radio astronomy are bright. This optimistic outlook is based on radio astronomical systems currently under development. The evolution of radio technology will undoubtedly lead to an even brighter future for planetary radio astronomy.

References

1. F. Ghigo. Pre-History of Radio Astronomy. National Radio Astronomy Observatory. http://www.nrao.edu/whatisra/hist_prehist.shtml (accessed April 9, 2010).
2. National Radio Astronomy Observatory (NRAO). Very long baseline array. http://www.vlba.nrao.edu/(accessed September 16, 2016).
3. National Radio Astronomy Observatory (NRAO). Very large array. http://www.vla.nrao.edu/ (accessed September 16, 2016).
4. National Radio Astronomy Observatory (NRAO). Karl Jansky and the discovery of cosmic radio waves. http://www.nrao.edu/whatisra/hist_jansky.shtml (accessed September 16, 2016).
5. D. K. van Keuren (1997). Science goes to war: the radiation laboratory, radar, and their technological consequences. *Reviews in American History* 25: 643–647.
6. A. Hewish (2002). James Stanley Hey, M.B.E. 3 May 1909–27 February 2000. *Biographical Memoirs of Fellows of the Royal Society* 48: 167.
7. J. M. Marr et al. (2015). *Fundamentals of Radio Astronomy: Observational Methods*. CRC Press, Boca Raton, FL. pp. 21–24.
8. Joint Institute for VLBI in Europe (JIVE). Astronomers demonstrate a global internet telescope. http://www.jive.nl/astronomers-demonstrate-global-internet-telescope (accessed May 6, 2011).
9. W. Happer et al. (September 1997). Obituary: Robert Henry Dicke. *Physics Today* 50 (9): 92–94.
10. S. Blundell and K. Blundell (2006). *Concepts in Thermal Physics*. Oxford University Press, Oxford, UK, p. 247.
11. P. V. Foukal (2004). *Solar Astrophysics*. Wiley-VCH Verlag, Weinheim, Germany, p. 69.
12. F. R. Elder (1947). Radiation from electrons in a synchrotron. *Physical Review* 71 (11): 829–830.

13. NASA. Mariner 1. http://nssdc.gsfc.nasa.gov/nmc/spacecraftDisplay.do?id=MARIN1. (accessed September 16, 2016).
14. European Space Agency (ESA). The Rosetta lander. http://www.esa.int/Our_Activities/Space_Science/Rosetta/The_Rosetta_lander (accessed September 16, 2016).
15. National Radio Astronomy Observatory (NRAO). http://search.nrao.edu/search?q=radio+astronomy+frequencies&btn=Search+NRAO (accessed August 29, 2016).
16. NASA Radio Jove. The discovery of Jupiter's radio emissions. http://radiojove.gsfc.nasa.gov/library/sci_briefs/discovery.html. (accessed September 16, 2016).
17. P. Zarka and W. S. Kurth (2005). Radio wave emissions from the outer planets before Cassini. *Space Science Reviews* 116 (1–2): 371–397.
18. L. A. Sromovsky and P. M. Fry (2010). Evidence for NH4SH as the primary 3-micron absorber in Jupiter's clouds. http://adsabs.harvard.edu/abs/2010DPS....42.3106S. (accessed September 16, 2016).
19. Wikipedia. New horizons. http://en.wikipedia.org/wiki/New_Horizons (last modified September 30, 2016). NASA. New horizons. http://science.nasa.gov/missions/new-horizons/ (accessed September 16, 2016).
20. NASA. New horizons. http://www.nasa.gov/mission_pages/newhorizons/main/index.html (accessed September 16, 2016).
21. National Radio Astronomy Observatory (NRAO). ALMA pinpoints Pluto to help guide NASA's new horizons spacecraft. https://public.nrao.edu/news/pressreleases/alma-pluto (accessed September 16, 2016).
22. C. A. Lundquist (1979). Observations of comet Kohoutek. *Skylab's Astronomy and Space Sciences.* Scientific and Technical Information Office, NASA, Washington, D.C. http://history.nasa.gov/SP-404/ch4.htm (accessed September 16, 2016).
23. NASA JPL. Comet C/1996 B2 Hyakutake. http://www2.jpl.nasa.gov/comet/hyakutake/ (accessed September 16, 2016).
24. NASA JPL. Comet Hale-Bopp. http://www2.jpl.nasa.gov/comet/(accessed September 16, 2016).
25. I. de Pater (1990). Radio images of the planets, *Ann. Rev. Astron. Astrophys.* 28: 347–399.
26. I. de Pater, S. Gulkis (1988). VLA observations of Uranus at 1.3-20 cm. *Icarus* 75: 306–323.

18

Planetary Radio Astronomy Projects

18.1 Planetary Radio Astronomy Curriculum Projects

There are simple projects that the student/class can develop in a classroom setup. Two of these projects are

> WARNING: To avoid eye damage or blindness, use protective eyewear/optics when viewing the sun.

- Radio and optical correlation of the Sun
- Radio observations of Jupiter

NOTE: Both solar and Jupiter observations are possible by using a single receiver along with some peripheral circuitry and modules described in this chapter.

18.2 Solar Radio Astronomy—An Introduction

The Sun is the closest star that we can study in great detail. We study the Sun with many different tools, that is, optical telescopes, spectrometers, space probes, and radio telescopes. Table 18.1 shows solar statistics and Figure 18.1, a photo of the Sun.

How do we study the Sun? Let's talk about the tools used in this area. We are all familiar with the optical telescope, which many of us have used. This, with solar filters, hydrogen alpha filters, and so on gives us wonderful visual views of the Sun. In addition, we also rely on radio telescopes.

18.3 Detection of Solar Flares

A flare [1] is defined as a sudden, rapid, and intense variation in brightness. A solar flare occurs when magnetic energy that has built up in the solar atmosphere is suddenly released. Radiation is emitted across virtually the entire electromagnetic spectrum, from radio waves at the long wavelength end, through optical emission to X-rays and gamma rays at the short wavelength end. The amount of energy released is the equivalent of millions of 100-megaton hydrogen bombs exploding at the same time! The first solar flare recorded in astronomical

TABLE 18.1

Solar Statistics

Solar Statistics (NASA)
Solar radius = 695,990 km = 432,470 mi = 109 Earth radii
Solar mass = 1.989 10^{30} kg = 4.376 10^{30} lb = 333,000 Earth masses
Solar luminosity (energy output of the Sun) = 3.846 10^{33} erg/s
Surface temperature = 5770°K = 10,400°F
Surface density = 2.07 10^{-7} g/cm^3 = 1.6 10^{-4} Air density
Surface composition = 70% H, 28% He, 2% (C, N, O,...) by mass
Central temperature = 15,600,000°K = 28,000,000°F
Central density = 150 g/cm^3 = 8 × Gold density
Central composition = 35% H, 63% He, 2% (C, N, O,...) by mass

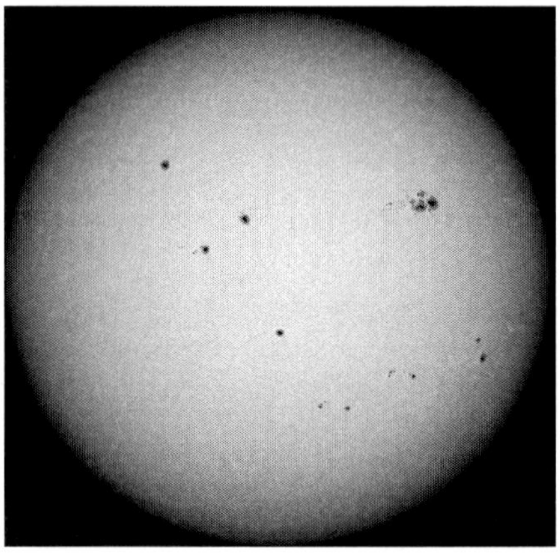

FIGURE 18.1
The Sun. (Courtesy of NASA, http://sdo.gsfc.nasa.gov/assets/gallery/preview/Nov_spots.jpg [retrieved August 22, 2016].)

literature was on September 1, 1859 [2]. Two scientists, Richard C. Carrington and Richard Hodgson, were independently observing sunspots at the time, when they viewed a large flare in white light. Figure 18.2 shows the Sun in hydrogen alpha emission.

As the magnetic energy is being released, particles, including electrons, protons, and heavy nuclei, are heated and accelerated in the solar atmosphere. The energy released during a flare is typically on the order of 1027 ergs (the erg is a unit of energy and work equal to 10^{-7} joules) per second. Large flares can emit up to 1032 ergs of energy. This energy is 10 million times greater than the energy released from a volcanic explosion. On the other hand, it is less than one-tenth of the total energy emitted by the Sun every second. There are typically three stages to a solar flare. First is the precursor stage, where the release of magnetic energy is triggered. Soft X-ray emission is detected in this stage. In the second or impulsive stage, protons and electrons are accelerated to energies exceeding 1 MeV. During the impulsive stage, radio waves, hard X-rays, and gamma rays are emitted. The gradual build up and decay of soft X-rays can be detected in the third, decay stage. The duration of these stages can be as short as a few seconds or as long as an hour. Solar flares extend out to the layer of the Sun called the corona. The corona is the outermost atmosphere of the Sun, consisting of highly rarefied gas. This gas normally

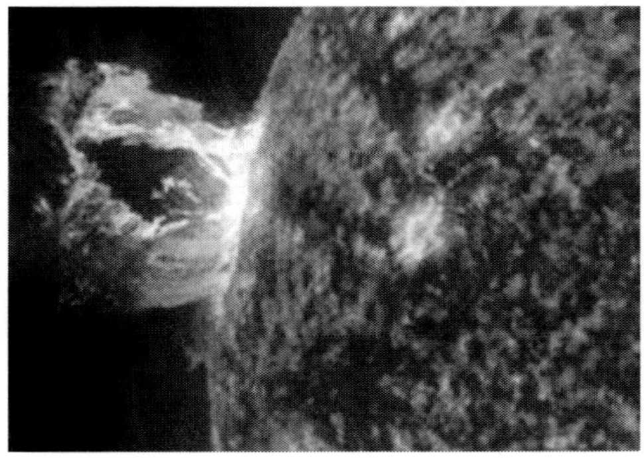

FIGURE 18.2
Sun in Hydrogen alpha filter. (Courtesy of NASA, SDO Gallery, http://sdo.gsfc.nasa.gov/gallery/main/ [retrieved August 22, 2016].)

has a temperature of a few million degrees Kelvin. Inside a flare, the temperature typically reaches 10 or 20 million degrees Kelvin, and can be as high as 100 million degrees Kelvin. The corona is visible in soft X-rays, as in Figure 18.2. Notice that the corona is not uniformly bright, but is concentrated around the solar equator in loop-shaped features. These bright loops are located within and connect areas of strong magnetic field called active regions. Sunspots are located within these active regions. Solar flares occur in active regions. The frequency of flares coincides with the Sun's 11-year cycle. When the solar cycle is at a minimum, active regions are small and rare and few solar flares are detected. These increase in number as the Sun approaches the maximum part of its cycle.

> WARNING: Never look directly at the sun or directly through a telescope unless proper filters have been installed! Eye damage/blindness can result.

A person cannot view a solar flare by simply staring at the Sun. Flares are in fact difficult to see against the bright emission from the photosphere. A Hydrogen alpha filter enables an observer to comfortably and safely view a solar flare in real time. Instead, specialized scientific instruments are used to detect the radiation signatures emitted during a flare. The radio and optical emissions from flares can be observed with telescopes on the Earth. Energetic emissions such as X-rays and gamma rays require telescopes located in space, since these emissions do not penetrate the Earth's atmosphere.

18.4 VLF Detection of the Sun

Our Sun is a variable star, G2 classification and close by (approximately, 93 million miles), which makes it a relatively easy study, in the radio and visual spectrum.

The Sun is also studied in the radio spectrum, it also emits from "DC to Daylight." In other words, throughout the spectrum. One part of the radio spectrum where observations are conducted is VLF (very low frequency). Receivers normally used in this area are called SID or SED (sudden ionospheric/enhancement of the D layer). Figure 18.3 shows a VLF receiver and Figure 18.4 shows VLF solar data.

FIGURE 18.3
VLF receiver at 40 KHz. (Courtesy of Lichtman, Radio Astronomy Supplies, TX.)

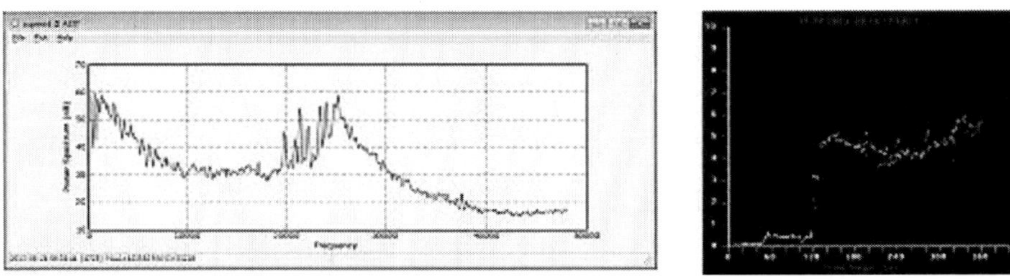

FIGURE 18.4
VLF Solar data. (Courtesy of Lichtman, Radio Astronomy Supplies, TX.)

18.4.1 Sudden Ionospheric Disturbance

A sudden ionospheric disturbance (SID) [3] is an abnormally high ionization/plasma. (In physics and chemistry, plasma is defined as a partially ionized gas, in which a certain proportion of electrons are free rather than being bound to an atom or molecule.) density in the D region of the ionosphere caused by a solar flare. The SID results in a sudden increase in radio-wave absorption that is most severe in the upper medium frequency (MF) and lower high frequency (HF) ranges (adapted from Federal Standard 1037C and from MIL-STD-188), and as a result often interrupts or interferes with telecommunications systems. When a solar flare occurs on the Sun a blast of intense ultraviolet and X-ray radiation hits the dayside of the Earth after a propagation time of about 8 min. This high-energy radiation is absorbed by atmospheric particles, raising them to excited states and knocking electrons free in the process of photo ionization. The low altitude ionospheric layers (D region and E region) immediately increase in density over the entire dayside. The ionospheric disturbance enhances VLF radio propagation. Scientists on the ground can use this enhancement to detect solar flares; by monitoring the signal strength of a distant VLF transmitter, SIDs are recorded and indicate when solar flares have taken place. Short wave radio waves (in the HF range) are absorbed by the increased particles in the low altitude ionosphere causing a complete blackout of radio communications. This is called a short wave fading. These fadeouts last for a few minutes to a few hours and are most severe in the equatorial regions where the Sun is most directly overhead. The ionospheric disturbance enhances long wave (VLF) radio propagation.

What other frequencies can we use? Many radio astronomers use frequencies allocated for radio astronomy "receive-only." Some of those are listed as follows:

- 13.36–13.41 MHz
- 25.55–25.67 MHz
- 73.00–74.60 MHz
- 150.05–153.00 MHz
- 406.10–410.00 MHz
- 1400.0–1427.0 MHz

18.4.2 VLF Detection of GRBs

Gamma ray bursts (GRBs) [4] are flashes of gamma rays associated with extremely energetic explosions in distant galaxies. They are the most luminous electromagnetic events occurring in the universe. Bursts can last from milliseconds to nearly an hour, although a typical burst lasts a few seconds. The initial burst is usually followed by a longer-lived "afterglow" emitting at longer wavelengths (X-ray, ultraviolet, optical, infrared, and radio). Most observed GRBs are believed to be a narrow beam of intense radiation released during a supernova event, as a rapidly rotating, high-mass star collapses to form a black hole. A subclass of GRBs (the "short" bursts) appears to originate from a different process, possibly the merger of binary neutron stars. The sources of most GRBs are billions of light years away from Earth, implying that the explosions are both extremely energetic (a typical burst releases as much energy in a few seconds as the Sun will in its entire 10 billion year lifetime) and extremely rare (a few per galaxy per million years). All observed GRBs have originated from outside the Milky Way galaxy, although a related class of phenomena, soft gamma repeater flares, are associated with magnetars [5] within the Milky Way. It has been hypothesized that a GRB in the Milky Way could cause a mass extinction on Earth. GRBs were first detected in 1967 by the Vela satellites, a series of satellites designed to detect covert nuclear weapons tests. Hundreds of theoretical models were proposed to explain these bursts in the years following their discovery, such as collisions between comets and neutron stars. Little information was available to verify these models until the 1997 detection of the first X-ray and optical afterglows and direct measurement of their redshifts using optical spectroscopy. These discoveries, and subsequent studies of the galaxies and supernovae associated with the bursts, clarified the distance and luminosity of GRBs, definitively placing them in distant galaxies and connecting long GRBs with the deaths of massive stars. Radio astronomers who are looking at the data from radio telescope may have seen this type of event before. Early on, we referred to them as HEPs (high energy pulses), a term first used by the author in the 1970s. HEPs showed a similarity to what were believed to be the first direct link possibly to GRBs. Software design engineer James Van Prooyen has been very successful in detecting GRBs.

Figure 18.5 verifies the Swift spacecraft reports a GRB at about the same time giving the event it official name, GRB080919.

The event was observed on September 19, 2008, using the radio telescopes of Grand Rapids Radio Observatory (GRRO), located in Grand Rapids, Michigan. The 40 kHz radio telescope receiver was built by Radio Astronomy Supplies (http://www.radioastronomy-supplies.com). The feed is a dipole of several hundred turns of wire, mounted on a 3 in. diameter, 10 ft long PVC pipe.

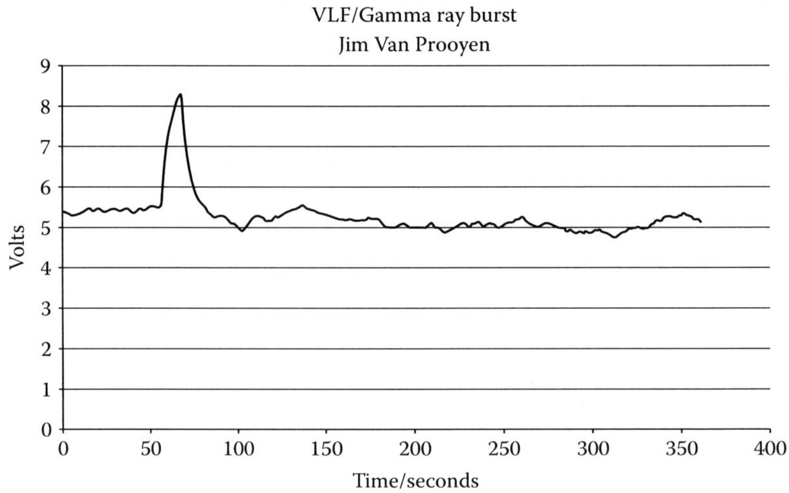

FIGURE 18.5
VLF GRB. (Courtesy of James Van Prooyen, Radio Astronomy Supplies, TX.)

The time line of VLF events:

- The plots both start at 19:48:31 Eastern Time or 23:48:31 universal time
- The VLF event start is 23:49:28 universal
- High point of the VLF event is at 23:49:39
- End of the VLF event is at 23:50:00. Figure 18.6 shows GRB profiles. Notice the close similarities between the NASA plots and the Van Prooyen plot

18.4.3 Satellite GRB Investigation

The NASA/Swift mission [7] with international participation, scientists have a tool dedicated to answering these questions and solving the GRB mystery. Its three instruments give scientists the ability to scrutinize GRBs like never before. Within seconds of detecting a burst, Swift relays its location to ground stations, allowing both ground- and space-based telescopes around the world the opportunity to observe the burst's afterglow. Swift is part of NASA's medium explorer (MIDEX) program and was launched into a low-Earth orbit on a Delta 7320 rocket on November 20, 2004. The principal investigator is Dr. Neil Gehrels (NASA-GSFC).

18.5 Radio and Optical Correlation of the Sun

While studying astronomy in the early 1970's (Suffolk Community College—Vanderbilt Planetarium, NY), I embarked on a semester project for the study of radio and optical correlation of the sun. Figures 18.7 is a block diagram and Figure 18.8 shows a similar receivers used by the author.

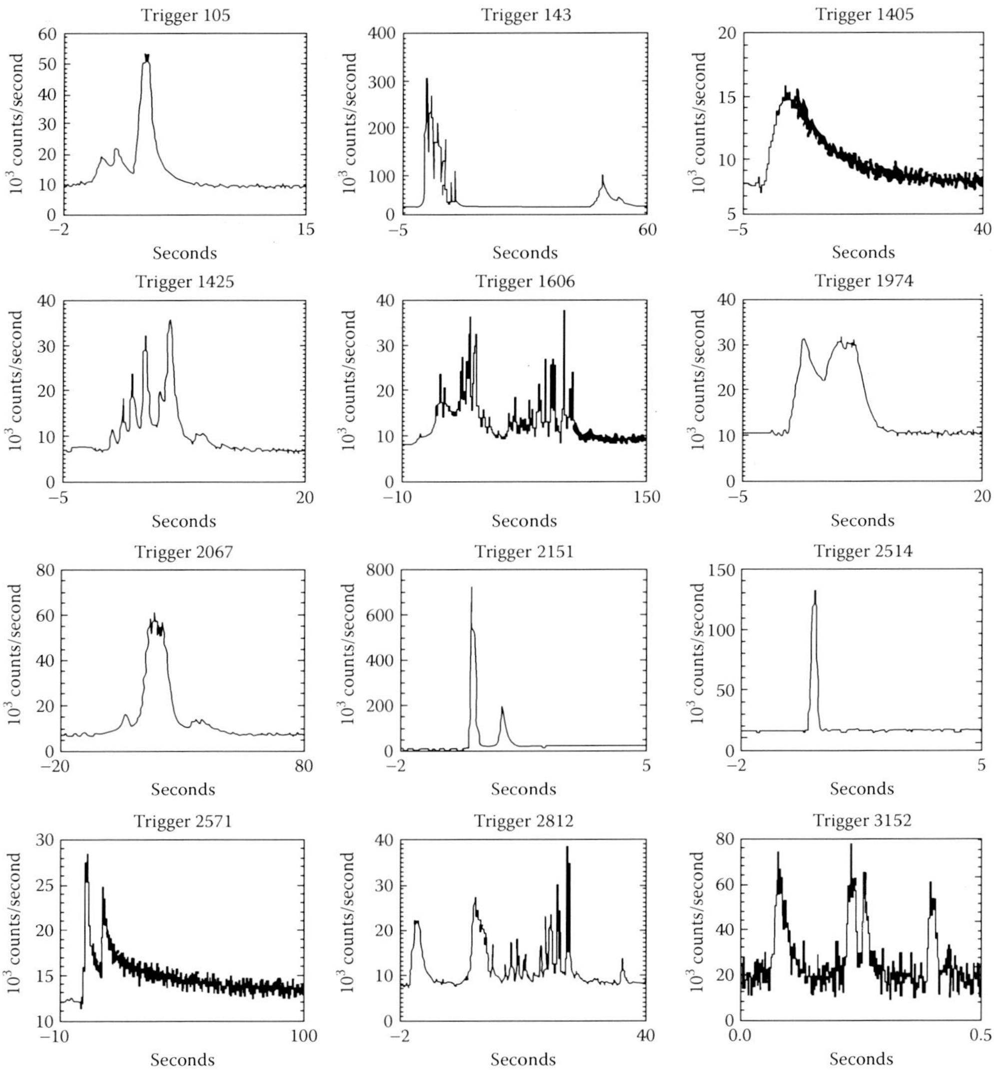

FIGURE 18.6
NASA GRB light curves. (Courtesy of NASA, http://radiojove.gsfc.nasa.gov/telescope/ [retrieved August 23, 2014].)

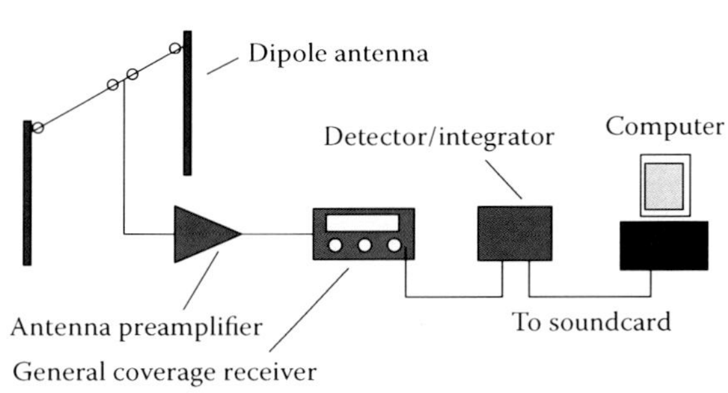

FIGURE 18.7
Solar/Jupiter radio telescope block diagram general coverage (Radio Shack) Short wave receiver Jove receiver. (Courtesy of Lichtman, Radio Astronomy Supplies, TX.)

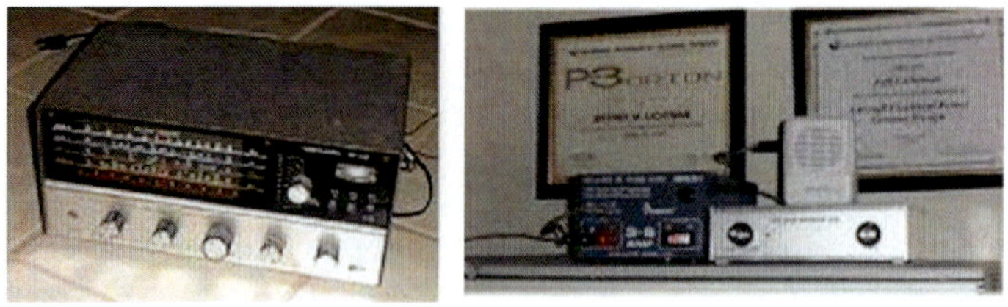

General coverage (Radio shack) short wave receiver

Jove receiver

FIGURE 18.8
Solar and Jupiter receivers (Courtesy of Lichtman, Radio Astronomy Supplies, TX.)

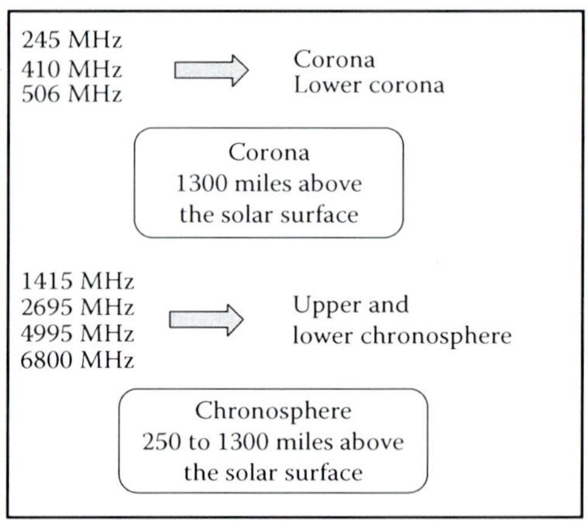

FIGURE 18.9
Solar frequency spectrum. (Courtesy of Lichtman, Radio Astronomy Supplies, TX.)

The frequency range that I would be working in was in direct correlation to the solar corona region. As per the Figure 18.9, the ranges of frequencies connect to certain portions of the solar atmosphere.

18.5.1 The Electronic Package

The "original" electronic package consisted of the following:

- Eico Sky Ranger short wave receiver
- Dow Key antenna preamplifier
- Home-made detector/integrator
- Esterline Angus chart recorder

NOTE: To "modernize," you can use any general coverage receiver such as

Radio Shack DX-394 General Coverage Receiver, Kenwood R-1000 Communication Receiver or similar and a 20dB Preamplifier Kit General Purpose Broadband Amplifier

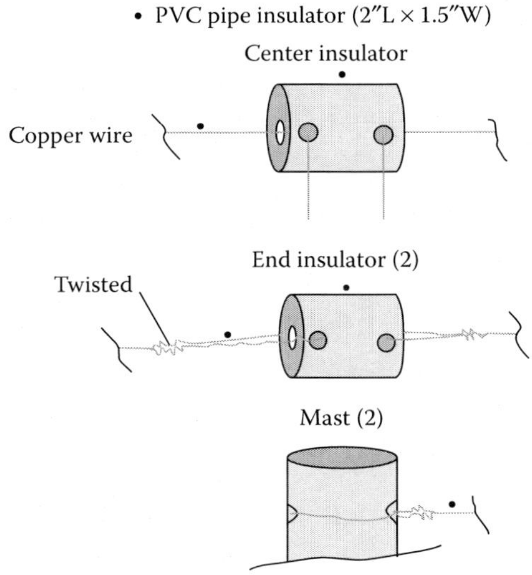

FIGURE 18.10
Fabricated insulator. (Courtesy of Lichtman, Radio Astronomy Supplies, TX.)

AADE HF 20dB Preamplifier Kit General Purpose Broadband Amplifier*

To replace the chart recorder, Radio Skypipe software and Spectrum Lab (Freeware)†

18.5.2 Putting It Together

As shown previously in Figure 18.7, the dipole antenna is the first in line and the most important portion of the system. Without this, no signals will get to the receiver and the data retrieval area.

18.5.3 Antenna

To figure out the length of the dipole, one may use a simple calculation. First, you must decide on a frequency for observation. Other frequencies may be used with the same formula. Thirty MHz was used as an example. The formula is as follows:

$$\tfrac{1}{2} \text{ Wave at 30 MHz } L_{ft} = 468/\text{F MHz or } 468/30 = 15.6 \text{ ft } (4.81 \text{ m})$$
$$\tfrac{1}{4} \text{ Wave at 30 MHz } L_{ft} = 234/\text{F MHz or } 234/30 = 7.8 \text{ ft } (2.3 \text{ m}) \text{ Dipole Calculator [8]}$$

NOTE: Formulae assumes velocity of propagation equal to 95% the velocity of light. The antenna wire may be a standard copper wire. In addition, you will require insulators. In the event that these insulators are not available, you may construct these from pieces from PVC pipe, as shown in Figure 18.10.

* Available at: http://www.mtechnologies.com/aade/hfpreamp.htm.
† Available at: http://www.radiosky.com/skypipeishere.html and http://www.qsl.net/dl4yhf/spectral.html.

Then antenna supports may be trees or 2 in. diameter PVC pipe. The pipe should be placed in a 1-ft-deep hole, allowing approximately 8 ft above ground.

18.5.4 Antenna Preamplifier

This General Purpose Broadband Amplifier module may be built or purchased from Ramsey Electronics, (http://www.ramseyelectronics.com/). The preamplifier serves as an amplifier of weak signals as well as a type of band pass filter. Generally, this module is placed right at the antenna. The output of the preamplifier is connectorized and connected to the length of coaxial cable going to the receiver antenna terminals. The output from the antenna preamplifier enters the receiver via the antenna connections shown in Figure 18.11.

18.5.5 The Receiver

This portion of the system acts as an additional antenna amplifier as well as a tuner, intermediate frequency (IF) section, and an audio section. The receiver originally used for this research was an Eico standard tabletop short-wave receiver. These receivers, as well as similar ones can be found at local Ham radio flea markets or EBay. Prices have been seen at around $25 and up. In the event that one is not obtainable, you may also purchase a new type of receiver used for Ham radio applications. These may be found online or at local Ham radio supply stores.

- Radio Shack DX-394 General Coverage Receiver
- Kenwood R-1000 Communication Receiver
- Jove Receiver (http://radiojove.gsfc.nasa.gov/telescope/)
- Any similar General Coverage Receivers

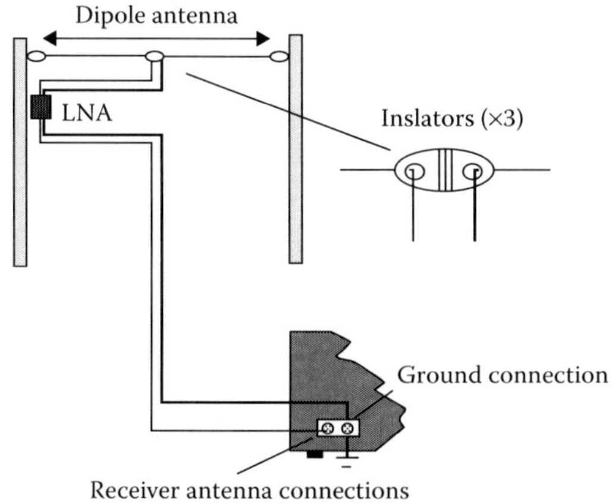

FIGURE 18.11
Antenna to receiver connections.

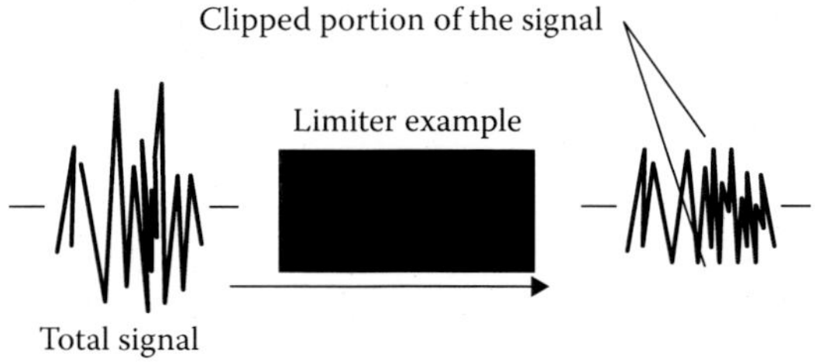

FIGURE 18.12
AM versus FM circuitry. (Courtesy of Lichtman, Radio Astronomy Supplies, TX.)

In any case, you will want to get a radio with "AM-type detection!" Most of the newer radios will have a selector switch for continuous wave (CW) and amplitude modulation (AM). It is most important to use an AM-type detection system in place of an FM type. Without going into much technical detail, an FM detector has circuitry know as a limiter. This does exactly that. Let's take a look at Figure 18.12.

To further clarify the description, the AM signal is a sine wave, which carries the signal information on the envelope. When your home FM system is on, you have noticed that it is much quieter than an AM-type radio, this is due to the FM circuitry which includes a limiter section. The limiter shown in simple form above, acts like a pipe. If you try to put 10 lb of signal through a pipe that allows only 5 lb, guess what! The results are clearly shown above. Your valuable signal information is cut away.

NOTE: FM receivers may be used for radio astronomy but the detection circuitry must be modified to an AM type.

18.5.6 Detector/Integrator

The detector/integrator module serves as a type of signal translator. To further explain, the signal that you are receiving via the audio portion of the receiver is an AM sine-wave signal. Most recording devices, such as the original chart recorder used, were a direct current (DC) recorder. Basically, the signal must be translated or converted from the AC component to a DC component, via a full-wave rectifier (shown below). Since the old ink type chart recorders have slowly disappeared over the years, the new method of recording is via software to soundcard such as RadioSkyPipe software. The integrator section (Figure 18.13) acts as a signal smoothing section. As seen below, the integrator portion has three capacitors of different values. The values teamed with the impedance of the actual recording device will determine the time constant (TC) which is the time a capacitor charges and discharges. For example, if your output recording device had an impedance of 1 Meg Ohm, and the value of your capacitor was 4 microfarads, your time constants will be 4.7 s.

The formula is as follows [9]:

$$T = R \times C$$

T = Time, R = Resistance (ohms)
C = Capacitance (farads)

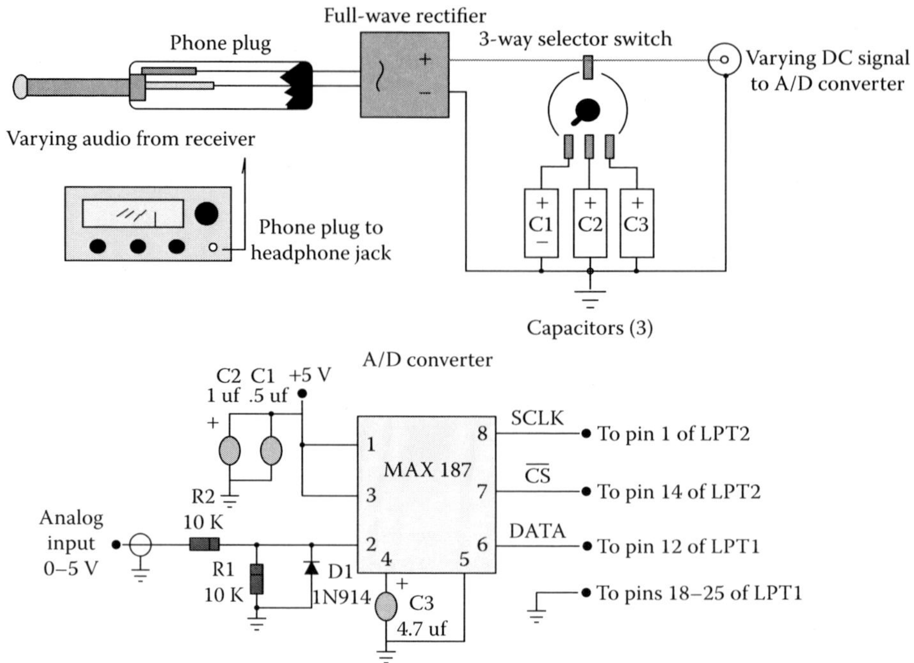

FIGURE 18.13
Detector/integrator circuit.

18.6 Optical Instrumentation Rear Projection Solar Viewing Screen

WARNING: Never view the sun directly through the eyepiece.
Serious eye damage or blindness will occur.

In 1973, I was experimenting with different versions of solar observation. To observe the Sun, one must be aware of danger to the eyes so, my thoughts revolved around ease of viewing, cost, and safety factors for this design. Of course, the design had to be safe and exhibit a good solar disk resolution. There are a few ways of observing the Sun (Figure 18.14). The following are a few of the most popular:

- Projected front side image
- Rear projected image
- Hershel wedge and filter
- Hydrogen alpha filter system

After considering all of the above methods, I chose the rear projection method. When choosing a material for the housing, I determined that a lightweight plastic or PVC tube would serve very well. I obtained a piece of 12 in. long by 4 in. wide piece of tubing. The next step was to cut out a portion of the wall for access to the focusing knob. Once this was completed, I went to the art supply store and purchased a piece of frosted Mylar and a piece of red ruby (used for PC board designs, etc.). At this point, I cut each piece to fit the inside diameter of

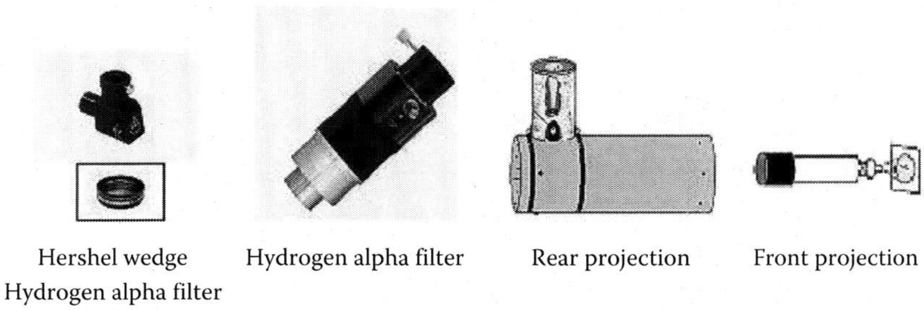

FIGURE 18.14
Solar observation methods.

the pipe. In order to hold the disks in place and keep them rigid, I also cut two thin pieces of Plexiglas to sandwich the Mylar and ruby (or red acetate available at Hobby Lobby or any art supply store). This so-called filter was held in place by some silicon material.

NOTE: To hold the projection screen in place, L brackets worked well. Two were screwed to the walls at the base of the tube and then placed on the telescope body. To secure the tube to the telescope, large screwdriver adjustable rings were used. Refer to Figure 18.15.

> WARNING: Never view the sun directly through the eyepiece. Serious eye Damage or blindness will occur.

The project was accompanied by a daily viewing and sketching of the sun. Notes were taken on the phenomenon occurring at the time. In some cases, it was closely correlated to what

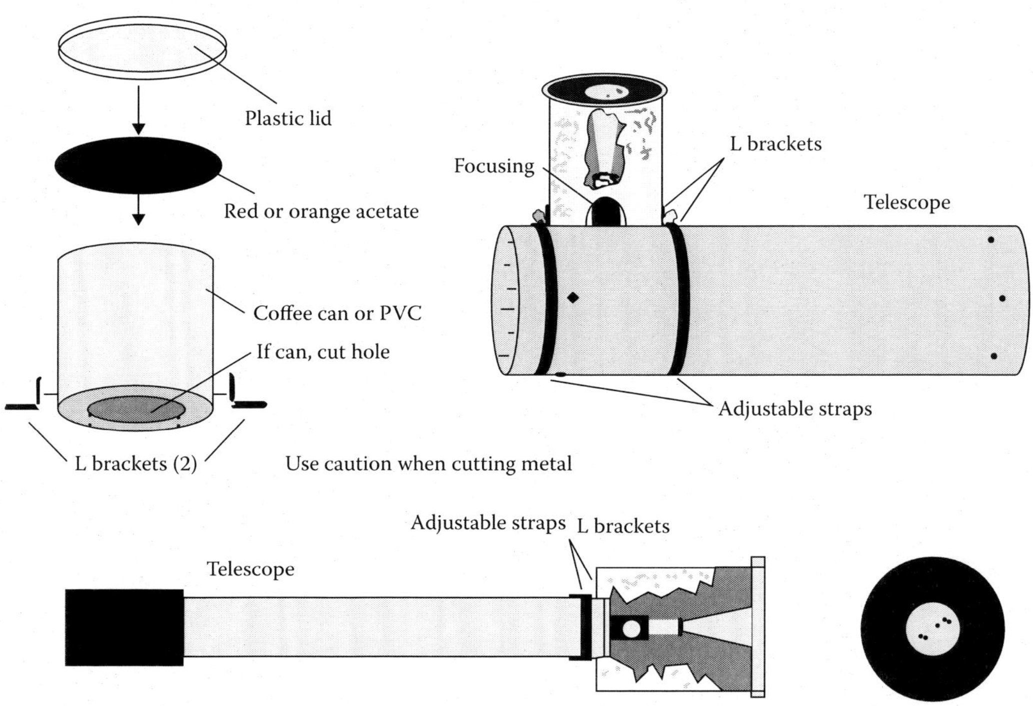

FIGURE 18.15
Rear projection solar viewer instructions.

FIGURE 18.16
Esterline Angus chart recorders.

was happening with the radio telescope. Figure 18.16 shows old type Esterline Angus Chart Recorders, similar to the ones used by the author. Figure 18.17 exhibits some actual observations.

18.7 Jupiter Observations

Jupiter, perhaps one of the most interesting planets in our solar system, is known as a giant. The planet offers both radio and optical mysteries, such as the great red spot, a storm which has continued for hundreds of years. In addition, it has many moons.

In 1955, Bernard Burke and Kenneth Franklin detected bursts of radio signals coming from Jupiter at 22.2 MHz. [10]. The period of these bursts matched the rotation of the planet, and they were also able to use this information to refine the rotation rate. Radio bursts from Jupiter were found to come in two forms: long bursts (or L-bursts) lasting up to several seconds and short bursts (or S-bursts) that had a duration of less than a hundredth of a second. Figure 18.18 illustrates an L-burst and an S-burst [11].

Scientists discovered that there were three forms of radio signals transmitted from Jupiter. Decametric radio bursts (with a wavelength of tens of meters) vary with the rotation of Jupiter, and are influenced by interaction of Io with Jupiter's magnetic field.

Decimetric radio emission (with wavelengths measured in centimeters) was first observed by Frank Drake and Hein Hvatum in 1959. The origin of this signal was from a torus-shaped belt around Jupiter's equator. This signal is caused by cyclotron radiation from electrons that are accelerated in Jupiter's magnetic field.

18.8 Components of the Decametric Telescope

For a bird's eye one may consult the following site [12].

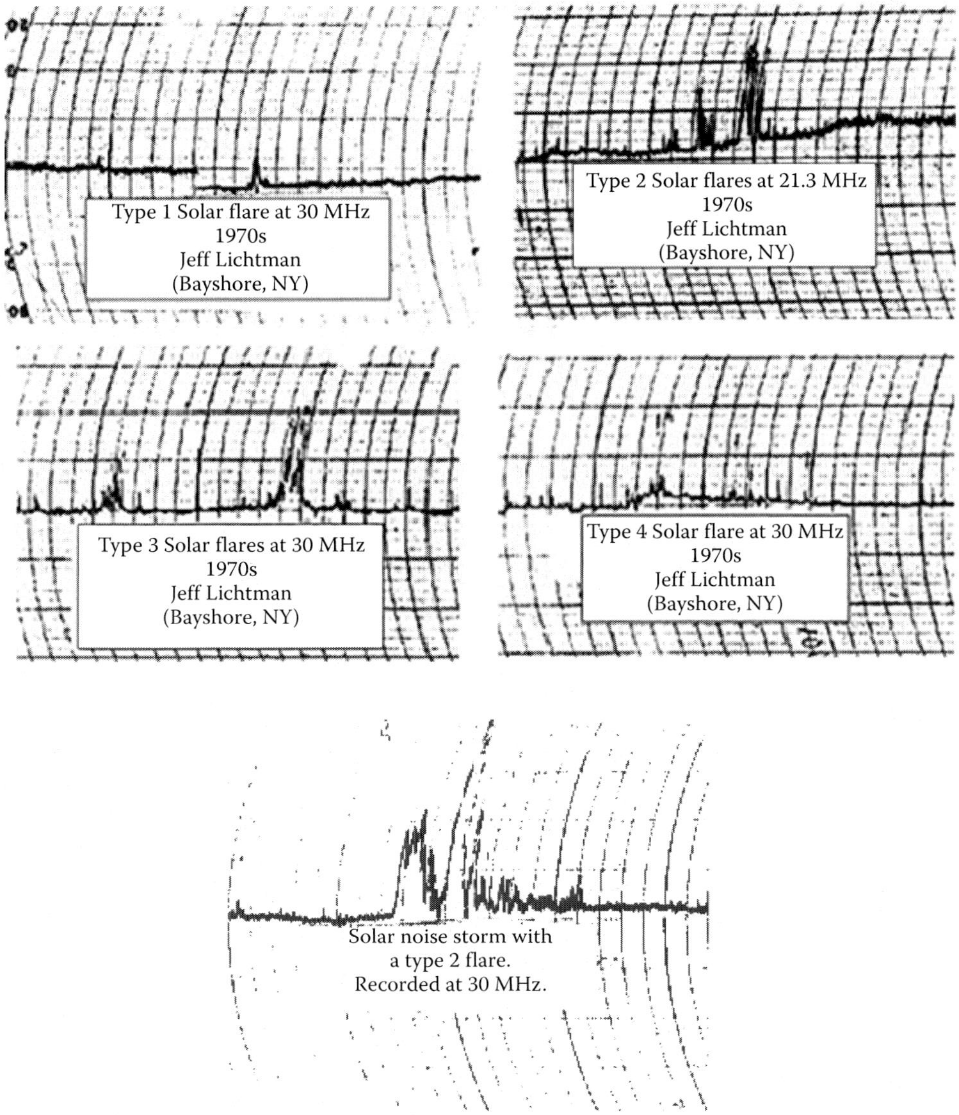

FIGURE 18.17
Actual observations. (Courtesy of Lichtman, Radio Astronomy Supplies, TX.)

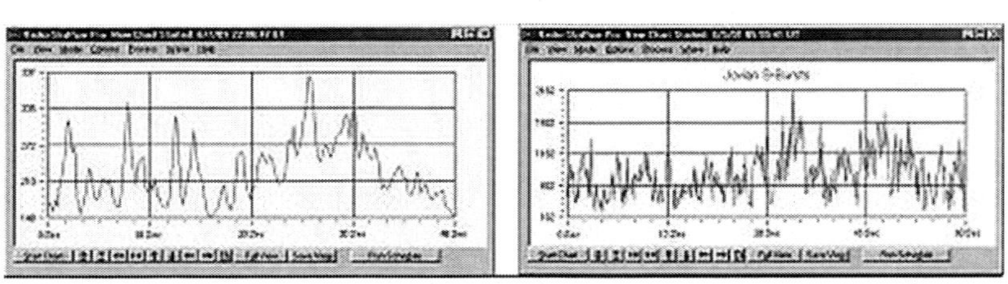

FIGURE 18.18
L and S radio bursts captured by Radio Skypipe software. (Courtesy of RadioSky Publishing, Ocean View, HI, http://radiosky.com/.)

18.8.1 Radio Receiver

The "standard" receiver for the Radio JOVE project is the RJ 1.1 (Figure 18.19). As described earlier, other receivers will work fine for this project, that is, Icom 75, Radio Shack DX-394 General Coverage Receiver or a Kenwood R-1000 Communication Receiver.

18.8.2 Jupiter Antenna Array

A two element phased dipole array (Figure 18.20) is recommended for Jupiter observations. Refer to [13] for construction details.

FIGURE 18.19
Radio Jove receiver. (Courtesy of Lichtman, Radio Astronomy Supplies, TX.)

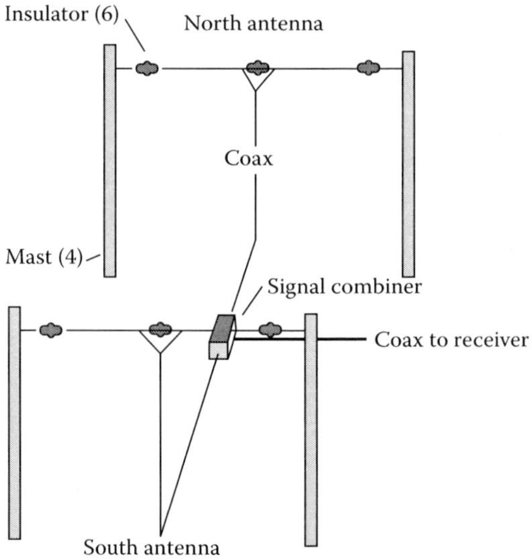

FIGURE 18.20
Two-Element phased dipole array. (Courtesy of Lichtman, Radio Astronomy Supplies, TX.)

18.9 Data Collection and Analysis

Gathering and measuring information on variables of interest, in an established systematic fashion that enables one to answer stated research questions, test hypotheses, and evaluate outcomes. The data collection component of research is common to all fields of study including astronomy (optical and radio), physical and social sciences, humanities, business, and so on. While methods vary by discipline, the emphasis on ensuring accurate and honest collection remains the same. Each field of study has its preferred set of data collection instruments. The hallmark of laboratory sciences is the meticulous documentation of the lab notebook. Regardless of the discipline, comprehensive documentation of the collection process before, during, and after the activity is essential to preserving data integrity. Regardless of the field of study or preference for defining data (quantitative, qualitative), accurate data collection is essential to maintaining the integrity of research. Both the selection of appropriate data collection instruments (existing, modified, or newly developed) and clearly delineated instructions for their correct use reduce the likelihood of errors occurring.

Consequences from improperly collected data include:

- Inability to answer research questions accurately
- Inability to repeat and validate the study
- Distorted findings resulting in wasted resources
- Misleading other researchers to pursue fruitless avenues of investigation
- Compromising decisions for public policy
- Causing harm to human participants and animal subjects

While the degree of impact from faulty data collection may vary by discipline and the nature of investigation, there is the potential to cause disproportionate harm when these research results are used to support public policy recommendations [14].

18.9.1 Record Keeping

As mentioned above, there are different requirements for different areas of research or study. The way you determine what points of data are necessary for your particular area will be determined by guidelines established by the director of the project. For radio astronomy, this is also the same. In this area as well as others, precise data records must be taken in order to prove your observation or experiment. Repeatable results are important and others reading or trying to duplicate the work will require good data. Figure 18.21 shows a sample data/observation record form used for radio astronomy. The fields used were chosen by the author for his own record keeping.

DATA OBSERVATION RECORDS
Date
Time
Object
Observer
Location
Frequency
RA and DEC
Antenna Type
Drift
Tracking
System Description
Weather Conditions

Notes: Supporting Data (attach recording or place in an envelope to be included).

FIGURE 18.21
Sample data record form.

References

1. G. Kopp (2005). The Total Irradiance Monitor (TIM): Science Results. *Solar Physics* 20 (1–2): 129–139.
2. P. Tony (January 21, 2009). Severe Space Weather—Social and Economic Impacts. NASA *Science: Science News*. http://science.nasa.gov/science-news/science-at-nasa/2009/21jan_severespaceweather/ (accessed February 16, 2011).
3. Federal Standard 1037C Glossary of Telecommunications Terms. http://www.its.bldrdoc.gov/fs-1037/fs-1037c.htm (accessed December 15, 2011).
4. B. Gendre et al. (2013). The Ultra-Long Gamma-Ray Burst 111209A: The Collapse of a Blue Supergiant?. *The Astrophysical Journal* 766: 30.
5. A. L. Melott et al. (2004). Did a gamma-ray burst initiate the late Ordovician mass extinction?. *International Journal of Astrobiology* 3: 55–61.
6. *NASA*, Radio Telescope for Observing Jupiter and the Sun. http://radiojove.gsfc.nasa.gov/telescope/ (retrieved August 23, 2016).
7. *NASA*, Highlights from NASA's Structure and Evolution of the Universe Theme - 2005. http://www.nasa.gov/centers/goddard/news/topstory/2006/2005seu_resource.html (accessed February 23, 2008).
8. Angelfire, Amandx presents dipole antenna. http://www.angelfire.com/mb/amandx/dipole.html (accessed August 26, 2016).
9. I. C. Purdie. VK2TIP. http://www.electronics-tutorials.com/basics/capacitance.htm (accessed September 16, 2016).
10. B. F. Burke and K. L. Franklin (1955). Observations of a variable radio source associated with the planet Jupiter. *Journal of Geophysical Research* 60: 213–217.
11. L. N Garcia. The Jovian Decametric Radio Emission. NASA. http://radiojove.gsfc.nasa.gov/library/sci_briefs/decametric.htm (accessed February 18, 2007).
12. NASA. Jupiter L-Bursts. http://radiojove.gsfc.nasa.gov/observing/sample_data.htm (accessed June 27, 2013).
13. The Radio JOVE Project Team. Antenna Kit and Manual Developed for NASA Radio JOVE Project. http://radiojove.gsfc.nasa.gov/telescope/ant_manual.pdf (accessed August 23, 2016).
14. Data Collection, Responsible Conduct in Data Management. http://ori.hhs.gov/education/products/n_illinois_u/datamanagement/dctopic.html (accessed August 22, 2016).

19

The Telescope—The Essential Tool

19.1 Introduction

In this chapter, we discuss the telescope evolution, their properties, and function of telescopes from Galileo to NASA's Great Observatories [1–3]. In addition, we talk about some of the supporting instrumentation as well as some new programs and discoveries. Figure 19.1 illustrates a rough timeline of telescope technology.

Space science researchers of NASA control some of the most sophisticated space probes and orbiting telescopes to get images of objects in space which we have also summarized.

19.2 Lippershey and Galileo's Refracting Telescope

In 1608—Hans Lippershey applies for a patent for the first recorded functional refractor telescope. A year later in 1609, Galileo refractor used two lenses for the purpose of concentrating the light from celestial objects. The system was designed in a manner so that it can deliver more light to the human eye than it can gather on its own. The light was refracted through a lens of spherical shape, forming an image. The spherical primary lens made the images blurry. The lens also split light into colors, producing a fringe of color around bright objects. A simple arrangement of the refracting telescope is shown in Figure 19.2.

As shown in Figure 19.2, the parallel light rays from the star are collected by the primary lens which is a convex spherical lens. After passing through the lens light is being concentrated on a small area on a concave eyepiece (parallel light rays) lens and then to the observer's eye.

19.2.1 Kepler's Design for an Improved Refractor

A spherical-type lens brings light rays to different focal points, causing blurry images. Kepler constructed a less curved spherical "primary lens" with a larger "focal length" than Galileo had used (1611). The less curved lens would make light rays move farther before they crossed thus bringing the focal points closer together. This improvement made the fuzzy image sharper.

Parallel light rays from star appear on a less curved primary lens. The eyepiece lens is also convex which would give the telescope the additional benefit of a larger field of view. As shown in Figure 19.3, the focal points would be closer together. As a result, the image would be clearer. It may be mentioned that Kepler actually never made this telescope, but this proposed design largely influenced future refractors, like those of Hevelius and Huygens.

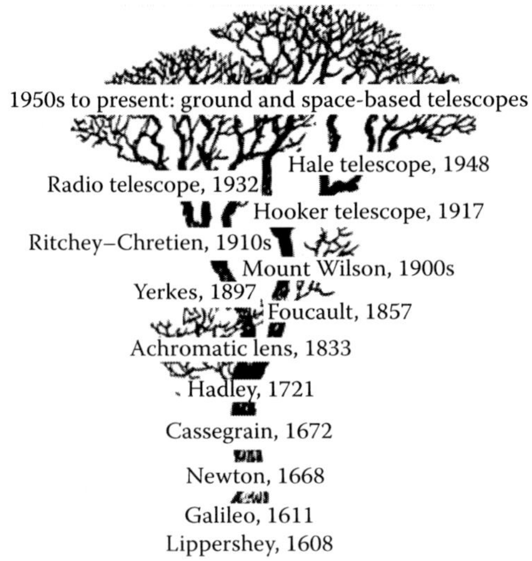

FIGURE 19.1
Telescope history. (Courtesy of Lichtman, Radio Astronomy Supplies, TX.)

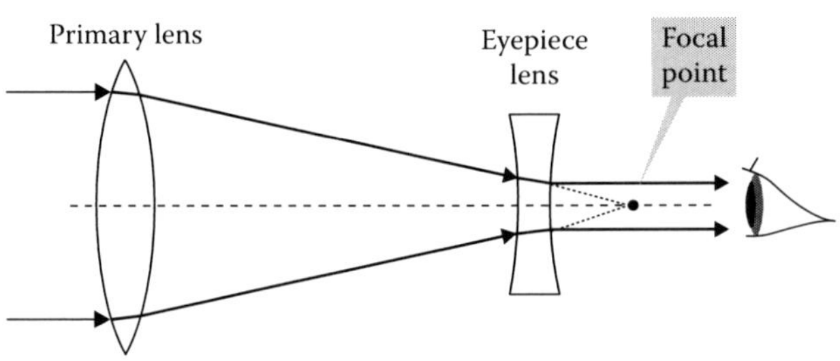

FIGURE 19.2
Galileo's refracting telescope.

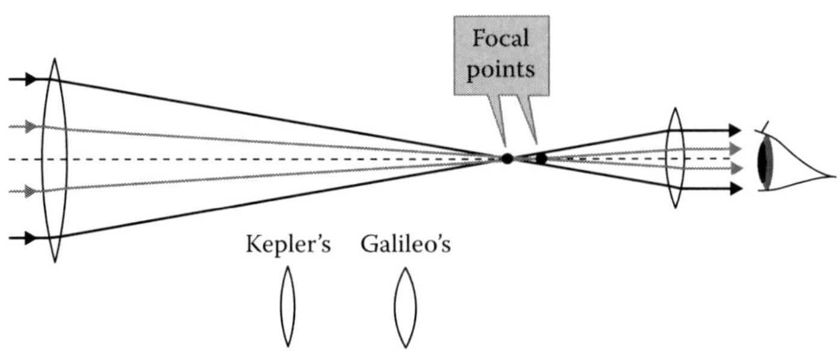

FIGURE 19.3
Kepler's design for an improved refractor.

19.2.2 The Achromatic Refractor

The great refractors took Kepler's basic telescope design and used a doublet lens as the primary lens; thus, creating the achromatic refractor.

The doublet lens arrangement consists of a pair of two different shaped lenses. In addition, two different types of glass are used in the doublet. When used together, the two different shapes and types of glass make the light rays meet at a single "focal point." This eliminates both chromatic and spherical aberration. Figure 19.4 shows the achromatic refractor.

19.2.3 Gregorian Reflecting Telescope

In 1633, James Gregory constructed a telescope with a "parabolic" primary mirror. The images of the telescope would have been free of both chromatic and spherical aberration. By making use of a mirror rather than a lens, Gregory eliminated the effect of chromatic error. The shape of the mirror was parabolic and not "spherical" and thus eliminating the spherical aberration. The Gregorian reflecting telescope is shown in Figure 19.5.

19.2.4 Newton's Reflecting Telescope

Newton used a metal "primary mirror" for capturing light and a "secondary mirror" for directing the light through the telescope (1668). Like Gregory, he avoided the problem of chromatic error by using a mirror instead of a lens. Newton was unable to clear up the blurry images caused by the "spherical" shape of the mirror. "Spherical aberration" occurs in both spherical mirrors and lenses. Figure 19.6 shows the Newton's reflecting telescope.

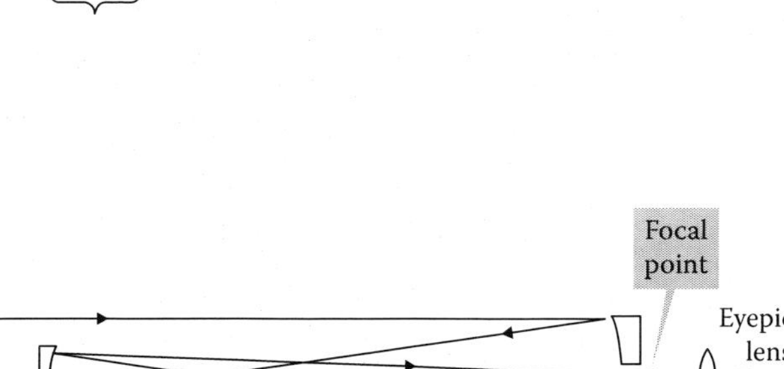

FIGURE 19.4
Achromatic refractor.

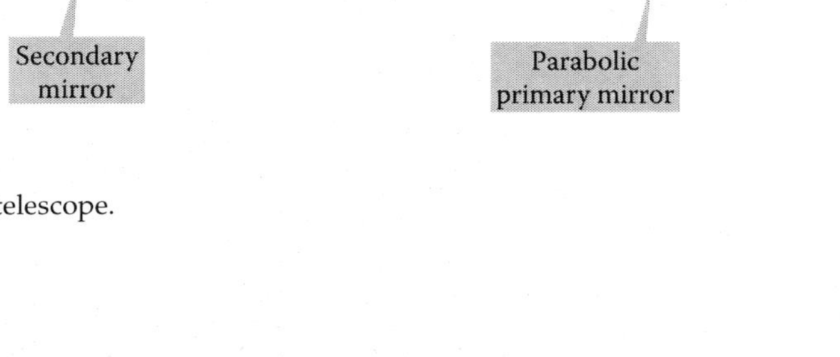

FIGURE 19.5
Gregorian reflecting telescope.

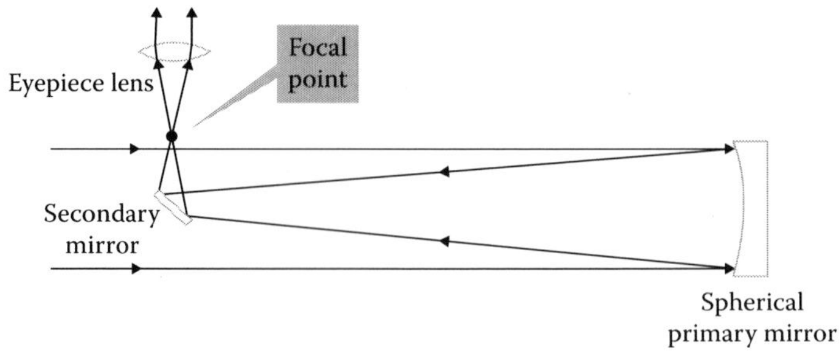

FIGURE 19.6
Newton's reflecting telescope.

19.3 The Multiple Mirror Telescope

The multiple mirror concept/use was first examined in the 1970s. In Arizona, one group of astronomers built a telescope using six individual mirrors held by one mounting. All the six mirrors created individual images, which the astronomers lined up by using joystick like controls. Subsequently, computers were used to align the images (Figure 19.7) [4,5].

A group of astronomers at the Keck Observatory pieced together 36 individual mirror segments into a very big mirror for making the Keck I telescope, as shown in Figure 19.8. Subsequently, Keck I got a twin sister Keck II. The two separate mirrors of the telescope can be used together for producing a single image.

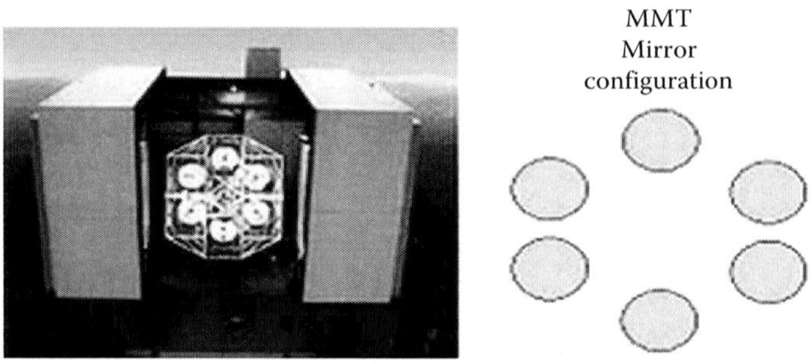

FIGURE 19.7
(a) Multiple mirror telescope's original mirrors and (b) visibility of the six separate mirrors of the original multiple mirror telescope.

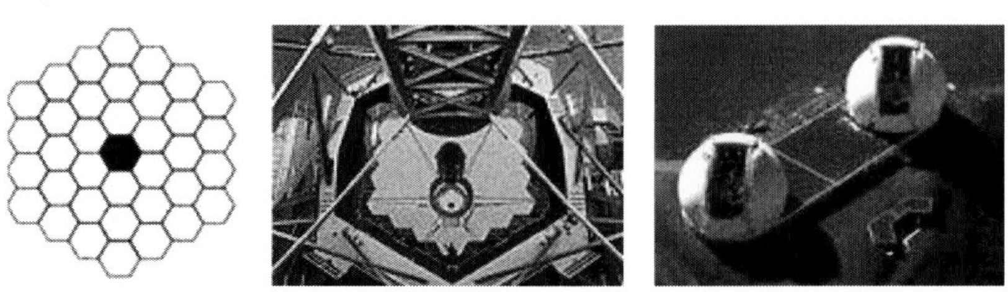

FIGURE 19.8
Keck I and twin sister Keck II (36 mirror).

19.4 Design of the Snow Solar Telescope

The proximity of the Sun makes observing relatively easy such as there is no need for huge mirrors. One such telescope does this very well. The Snow Telescope is a unique solar telescope in that it was laid out horizontally and it does not have a tube (Figure 19.9). There are four major parts of the telescope: (1) "Coelostat" is a tracking mirror that collects the Sun's light; it rotates from east to west to follow the Sun; (2) "Flat mirror" reflects the image from the tracking mirror to the primary mirror; (3) "Primary mirror": is parabolic in shape; it focuses the light/image; and (4) "Spectroheliograph" makes portraits of the Sun using only a narrow range of wavelengths; the pictures help astronomers to study the Sun's composition and structure.

19.5 Space Telescopes

Putting telescopes in space, astronomers could break free the distorting and shielding effects of the atmosphere of the Earth. In order to maintain these space observatories, we need significant advances in telescopes, instruments, computers, and the spacecraft used to launch them. Many of these technological advances came about as a direct result of the rivalry between the United States and the Soviet Union during the Cold War between World War II and the late 1980s. Scientists and engineers have created telescopes which can study the universe in various types of light and record information not available from Earth: space telescopes [6].

A plus for space telescopes are that they are the better option to avoid the blurriness caused by the atmosphere. NASA's Great Observatories program constructed four orbiting telescopes. These are (1) the Hubble Space Telescope, (2) the Compton Gamma Ray Observatory, (3) the Chandra X-ray Observatory, and (4) the Spitzer Space Telescope.

Space telescopes have their own drawbacks. They are more expensive to build and launch than ground-based (GB) observatories. Since they must be launched into space, they can't be as large as GB observatories. Moreover, as space telescopes remain in orbit, fixing and upgrading them can be either difficult or impossible. When they break down,

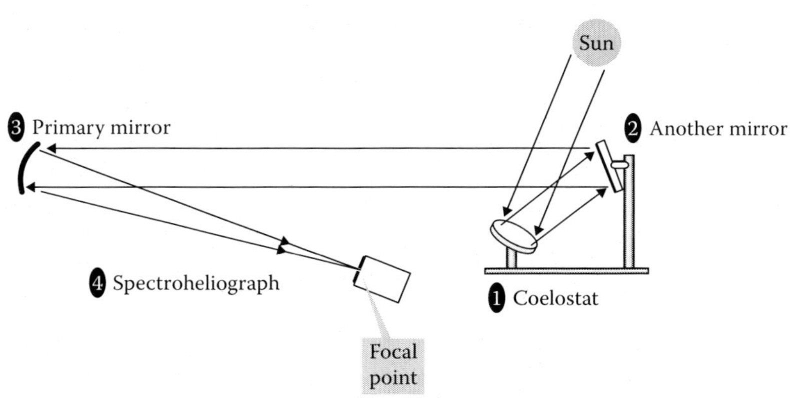

FIGURE 19.9
Snow solar telescope.

they are gone forever. A big advantage with the space-based telescopes is, however, keeping telescopes in orbit above the Earth, astronomers were finally free to view the universe in all wavelengths of light [5] without atmospheric limitations. In addition, space observatories have higher resolution and can look at tiny regions of the sky in detail. By observing the images from a variety of space- and GB telescopes, astronomers obtain a more complete view of an object.

19.6 Characteristics of a Good Telescope

A good quality telescope is judged by considering the following qualities:

Its light-collecting ability: Faint objects are hard to find. Objects appear faint as they are far away, and also because they are much dimmer. The more light a telescope can collect, the better it can observe faint objects. Large mirrors permit telescopes to collect more light.

Note: A mirror collects more light than a lens. This is the difference between the reflector- and refractor-type telescopes.

Its resolution: Optical resolution describes the ability of an imaging system to resolve detail in the object that is being imaged. A telescope with greater resolution will be able to see two points of light as being separate from one another while a telescope with low resolution will blur the two points together into a single point of light. Figure 19.10 reveals lower and higher resolution images, respectively.

Its magnifying power: Except for solar telescopes, magnification is the least important element. Magnification depends on focal length; as the magnification increases, the telescope focuses on a smaller piece of the sky. Telescopes are usually designed to operate at the smallest magnification possible, for wide field viewing. The distance and details they observe depends more on their light-collecting ability and resolution rather than their magnification. Solar telescopes can rely on magnification mainly as they do not have to look deep into space or gather much light for clearly viewing the Sun.

Instrument quality: A research telescope is only as good as the cameras and other instruments which record and analyze the light that it captures. Instruments are judged by factors like the quality of their images, how effectively they spread out light, and how much light they capture.

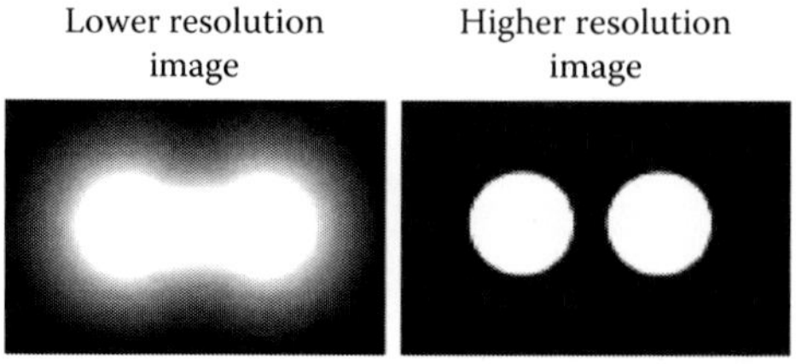

FIGURE 19.10
Lower and higher resolution images.

19.7 NASA's Great Observatories

The four Great Observatories are a series of space telescopes with a purpose to give the complete picture of objects at different wavelengths. Each observatory investigates a particular wavelength region in a greater detail [6–8].

The telescopes, in order of launch, are (1) the Hubble Space Telescope (1990); (2) Compton Gamma Ray Observatory, popularly called CGRO (1991); (3) Chandra X-ray Observatory (1999); and (4) Spitzer Space Telescope (2003). Figure 19.11 shows each telescope including the wavelength region it was built to observe. All except for CGRO are providing information at present.

Figure 19.12 exhibits the images showing the remains of an exploded star (Kepler's supernova), as seen by three telescopes of the Great Observatories.

19.7.1 Hubble Space Telescope's Design

The Hubble Space Telescope was deployed from the space shuttle Discovery during STS-31 on April 25, 1990. Since then, there have been five servicing missions that continued to upgrade the telescope's scientific instruments and operational systems. Hubble reached a major milestone, its twentieth anniversary in orbit, on April 24, 2010.

As seen in Figure 19.13, when light hits the "concave" primary mirror, it is reflected to the convex secondary mirror, then returns through a hole in the center of the "primary mirror." The light comes to the "focal point" and then passes to one of Hubble's "instruments." Telescopes of this design are called Cassegrain telescopes, after the person who first designed the same. Figure 19.13a shows the design of the telescope [9–11].

Hubble is one of the largest and most versatile and is well known as both a vital research tool and a public relations boon for astronomy. The telescope (Figure 19.13b) was built by the United States space agency NASA with contributions from the European Space Agency. This is operated by the Space Telescope Science Institute [9]. The telescope is operating largely since 2015 and may last until 2030–2040. Its scientific successor, the James Webb Space Telescope, is scheduled for launch in 2018.

19.7.2 Chandra X-Ray Observatory

X-rays are difficult to observe, the reason being, instead of bouncing the X-rays aimed directly at a telescope mirror, they will penetrate it. Chandra's mirrors have to be placed

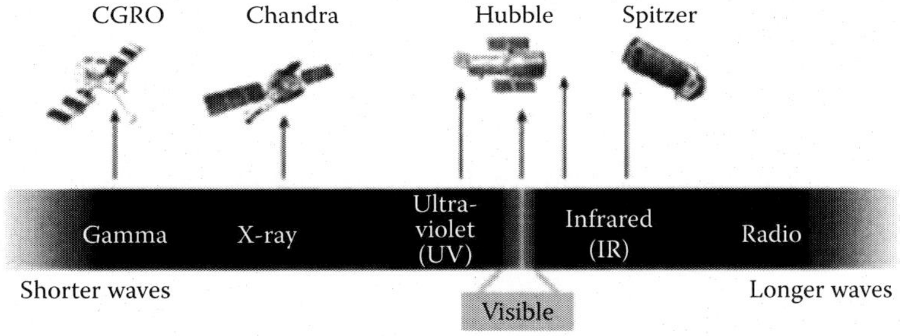

FIGURE 19.11
Space telescopes of NASA's Great Observatories.

FIGURE 19.12
The images of an exploded star as found by three telescopes of the great observatories. (Courtesy of NASA Hubble, Caught in the act: Hubble captures first-ever predicted exploding star, http://hubblesite.org/newscenter/archive/releases/star/supernova/2015/46/ [retrieved August 12, 2016].)

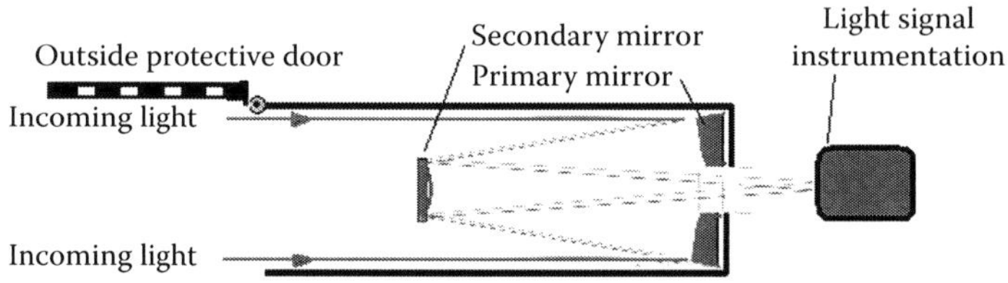

(a)

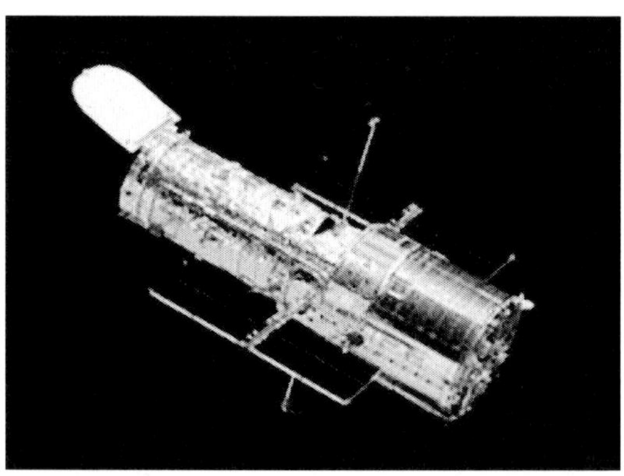

(b)

FIGURE 19.13
(a) Design of the Hubble Space Telescope (Courtesy of Lichtman, Radio Astronomy Supplies, TX.) and (b) photograph of the Hubble Space Telescope. (Courtesy of NASA, http://spaceflight.nasa.gov/gallery/images/shuttle/sts-119/hires/s125e011848.jpg [retrieved August 12, 2016].)

at just the right angle for X-rays to glance off their surfaces. In principle, Chandra's mirrors are positioned in a manner so that they are almost parallel to the entering X-rays. The mirrors look like open cylinders or barrels. The X-rays skip across the mirrors almost like stones skip across the surface of a pond. Figure 19.14 illustrates the principle while Figure 19.15 shows how the incoming X-rays after passing through the pair of mirrors are moving towards the detector.

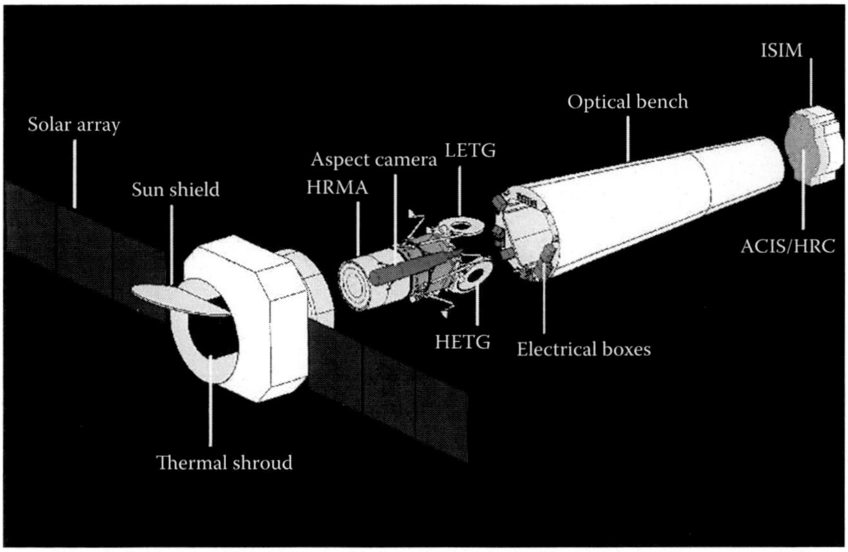

FIGURE 19.14
Chandra X-ray observatory instruments. (Courtesy of NASA, https://www.nasa.gov/mission_pages/chandra/astronomy/index.html [retrieved August 12, 2016].)

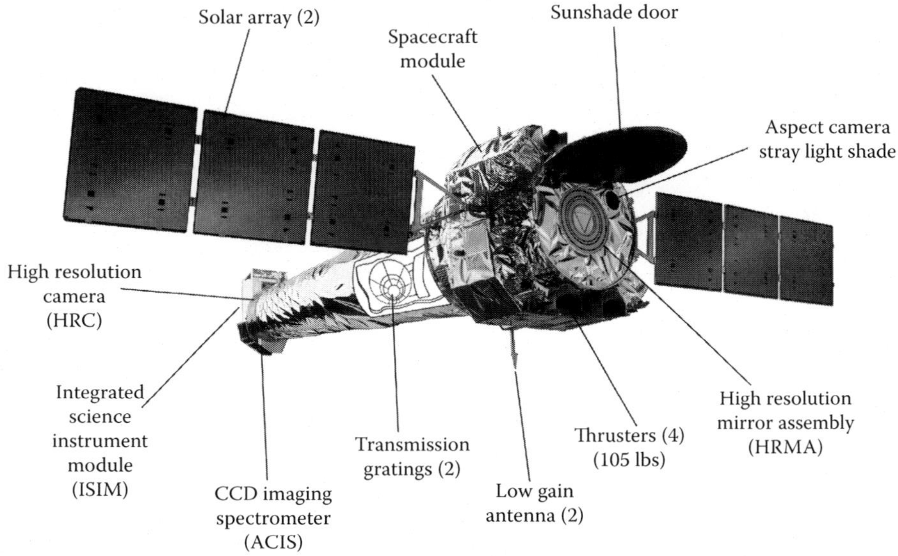

FIGURE 19.15
Chandra spacecraft and significant instrument locations (Courtesy of NASA, https://www.nasa.gov/sites/default/files/images/704250main_chandra-telescope_full.jpg [retrieved August 12, 2016].)

19.7.3 Webb Space Telescope—Distance from Earth

The James Webb Space Telescope, previously known as Next Generation Space Telescope, is a space observatory under construction and scheduled to launch in October 2018. It has been planned that the Webb telescope will be located much farther away from Earth than other space telescopes [11,13]. It will be placed in a location known as the L2 (Lagrange point The "Lagrangian points" also "Lagrange points," "L-points," or "libration points" are positions in an orbital configuration of two large bodies where a small object affected only by gravity can maintain a stable position relative to the two large bodies. The Lagrange points mark positions; there are five, labeled L1 to L5, positions in the orbital plane of the two large bodies. The first three are on the line connecting the two large bodies, which is far beyond the orbit of the Moon. In fact, the telescope will be placed nearly four times as far away from Earth as the Moon (Figure 19.16).

19.7.4 Keeping a Telescope at the Lagrange Point

By the term Lagrange points we mean areas in space where the gravity and orbital motions of nearby bodies come into balance [15,16]. There are altogether five Lagrange points around the Earth. The Webb telescope will be located in the L2 point which is 940,000 miles (1.5 million km) from Earth. At this L2 point, the Earth's gravity will pull Webb along, so it can "keep up" with Earth in its path around the Sun. This is essential since Webb has to be kept in a very specific position, sunshield protection against light from the Sun, Earth, and Moon at the same time. At this Lagrange point, the Sun and Earth will always be lined up on the other side of Webb's sunshield. Figure 19.17 shows that an object at the Earth's L2 point will always stay in line with the Sun and Earth.

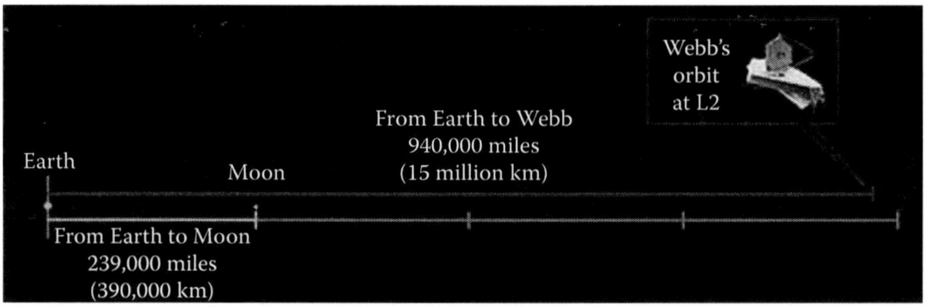

FIGURE 19.16
Webb Space Telescope distance from Earth.

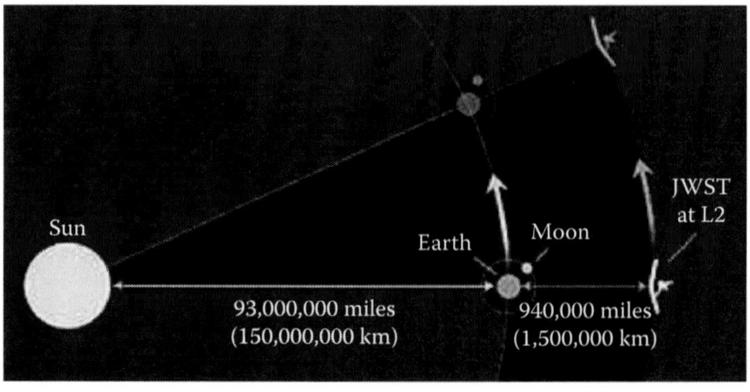

FIGURE 19.17
Object at the Earth's L2 point will stay in line with the Sun and Earth.

19.8 Mirror Sizes—A Comparison of Space Telescope and GB Mirrors

The size of a space telescope primary mirror is limited mainly by the size of the vehicle carrying it into orbit [12–14]. It is easier to build large primary mirrors into GB telescopes. We have made a comparison of mirror sizes in Table 19.1. A size comparison of primary mirrors is shown in Figure 19.18.

19.8.1 Linking Telescopes Together

Images from the multimirror telescope Keck I and its twin, Keck II, are combined by a computer and software. To some extent, the effect is to get one gigantic mirror which can collect more light and resolve information in a greater detail, than either Keck I or Keck II could alone. An enhancement of light-collecting ability as well as increase of resolution is demonstrated in Figures 19.19 and 19.20, respectively.

If we carefully examine Figure 19.19, we find that the light-collecting ability increases in the system, in a manner as if surface areas of all the segments were being added together.

TABLE 19.1
Comparison of Telescope Mirror Sizes

Spitzer	Hubble	Webb	GB Telescopes
Spitzer is presently NASA's largest infrared telescope in space. Its mirror is sufficiently small to fit inside the rocket that launched it into orbit.	Hubble has a large, single mirror that has the right size to fit inside the hold of the space shuttle.	Webb, the next infrared space telescope, has a mirror designed for folding up inside the rocket which will launch it. Once in space, it will unfurl into the biggest mirror of any telescope in orbit.	Telescopes on Earth, that is, Keck telescopes in Hawaii, have fewer technological limitations and have much larger mirrors than the space telescopes.

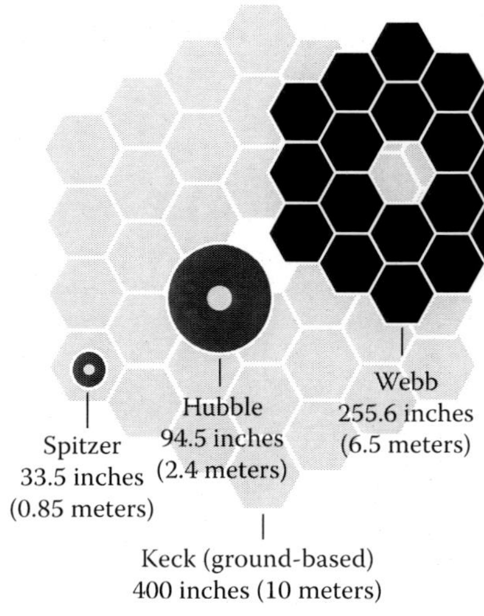

FIGURE 19.18
Size of primary mirrors.

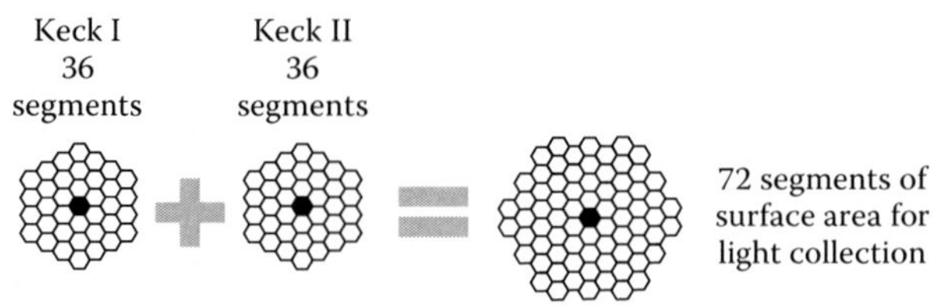

FIGURE 19.19
Enhancement of light collecting ability.

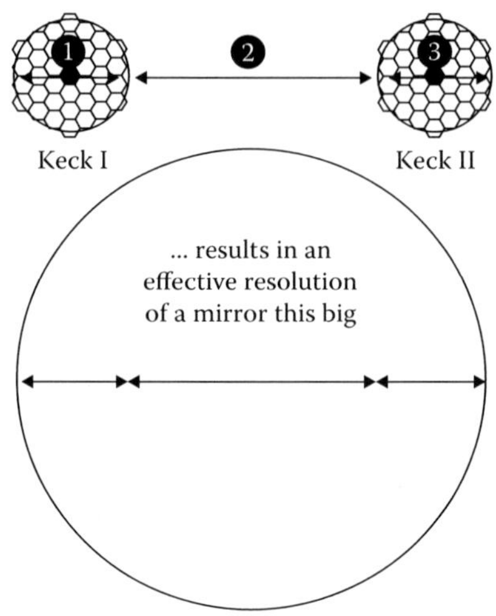

FIGURE 19.20
Increase of resolution.

On the other hand, in Figure 19.20, resolution increases in a manner as if there were a mirror with a diameter as big as the diameters of both Kecks in addition to the distance between them.

19.9 The Radio Telescope

Radio telescopes are basically reflecting telescopes (described in detail in Chapter 17). A large "parabolic" dish collects the incoming radio waves to focus them at the receiver, which is mounted at the "focal point" of the dish (Figure 19.21). Figure 19.22 illustrates the range of sizes. These specially designed telescopes observe the longest wavelengths of light, ranging from 1 mm to over 10 m long. For comparison, visible light waves are only a few hundred nanometers long.

19.9.1 VLBI—Multiple Dish Arrays

Very long baseline interferometry (VLBI) [15] is a type of astronomical interferometry used in radio astronomy. In VLBI a signal from an astronomical radio source, such as a

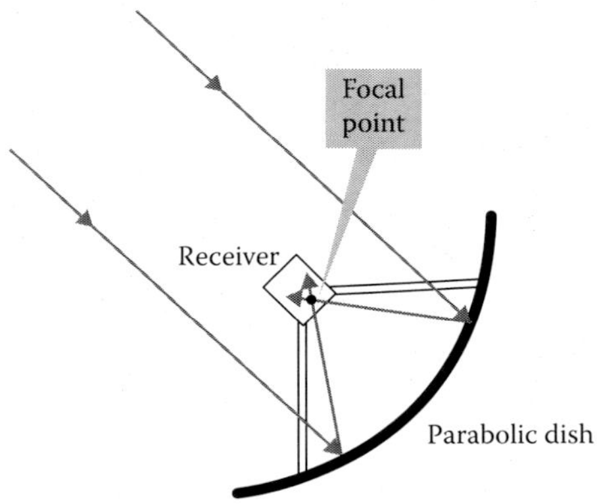

FIGURE 19.21
"Parabolic" dish collecting the incoming radio waves.

FIGURE 19.22
The Green Bank Telescope (GBT) and the very large array. (Courtesy of NRAO, http://www.nrao.edu [retrieved September 10, 2016].)

quasar, is collected at multiple radio telescopes on Earth. The distance between the radio telescopes is then calculated using the time difference between the arrivals of the radio signal at different telescopes. This allows observations of an object that are made simultaneously by many radio telescopes to be combined, emulating a telescope with a size equal to the maximum separation between the telescopes. The dishes are moveable and can be moved closer together or farther apart for getting different resolutions. When the dishes are at the farthest apart, the array has a resolution equal to a single dish which is 36 km in diameter (Figures 19.23 and 19.24).

The arrangement of the array can be illustrated by considering Figure 19.24. Gray circles represent the arrangement of array. In the upper figure, the 27 dishes have the resolution of a dish as large as this circle. In the lower figure, they are spaced slightly farther apart and in this case also the 27 dishes have the resolution of a dish as large as this circle. It may be mentioned that the resolution of the array depends on whether the dishes are positioned near each other or spread out. However, the amount of light the array can collect depends on the size and number of dishes and never on the space between the dishes. It is clear that if instead of a single dish, 27 dishes have 27 times the surface area causing a collection of 27 times more light (Figure 19.25).

FIGURE 19.23
Very large array with 27 individual radio dishes laid out in a Y-shape.

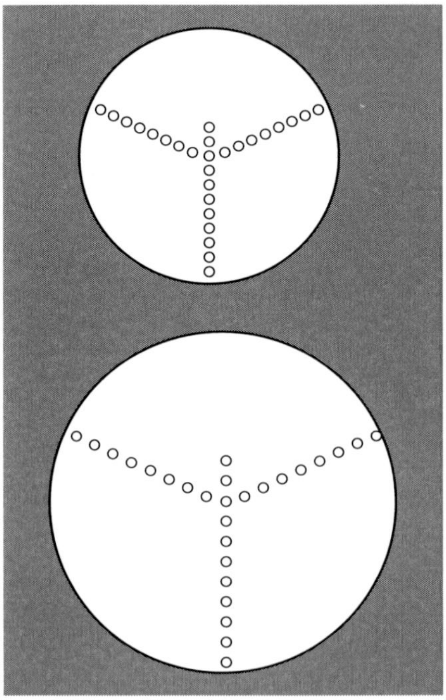

FIGURE 19.24
Arrangement of the array by two different means.

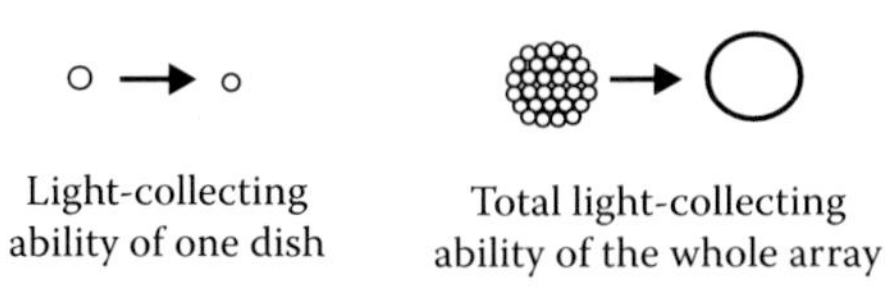

FIGURE 19.25
Variation of light collecting ability with the number of dishes.

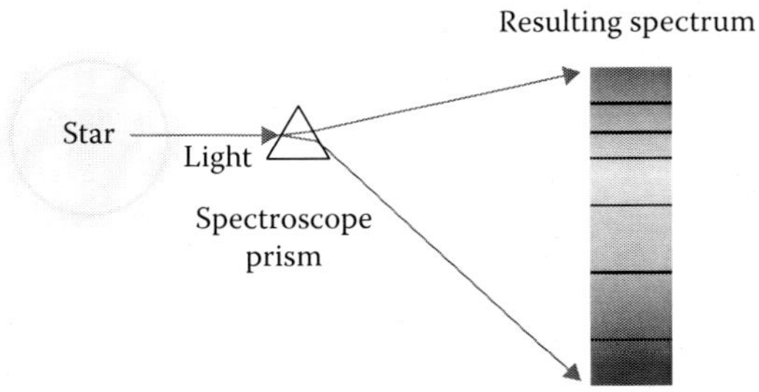

FIGURE 19.26
Spectroscope output.

19.10 Spectrometer Contribution

A spectrometer [16], in particular, shows the intensity of light as a function of wavelength or of frequency. The deflection is produced either by refraction in a prism or by diffraction in a diffraction grating. This instrument is often used to investigate the makeup of a star. (It separates or "disperses" white light from a star into a wide spectrum of colors.) This is, in fact, wider than normal prism displays. Interestingly, when spread very wide, black lines appear in the spectrum. Once scientists understood the element responsible for a certain pattern, they knew which elements were present in the star and thus it was indeed a revolutionary discovery (Figure 19.26).

19.11 Making Primary Mirrors for Modern Reflecting Telescopes

Modern reflecting telescopes use a parabolic-shaped primary mirror coated with a thin film of aluminum [11,13,14,17,18]. There are three main steps in the mirror-making process. The steps followed are outlined below:

Casting the glass disk: Molten glass is poured into a mold for forming disk with a honeycomb-like structure. This makes the disk both stronger and lighter.

Grinding the disk: The top of the glass disk is made into perfect concave, parabolic shape.

Applying the mirrored surface: A thin, shiny metal coating, like silver or aluminum, is applied at the top of the disk.

In Figure 19.27, from top to bottom, we have shown, respectively, (1) casting the glass disk, (2) grinding the disk, and (3) mirror coatings.

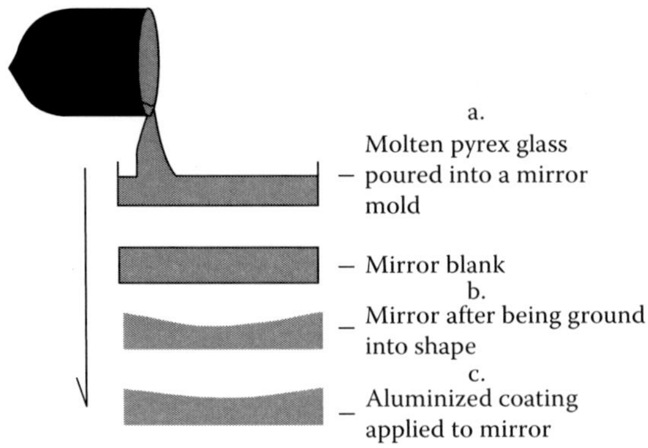

FIGURE 19.27
The three steps in mirror making.

19.12 Reducing the Effects of Atmospheric Distortion

A computer-controlled mirror with a surface that can change shape reacts continuously to incoming rays constantly shifting angles. The computer samples the distortions of the atmosphere and tells the mirror what shape to take to keep all the rays aimed at one focal point. The mirror is in constant motion.

Figure 19.28 illustrates the mirror in motion. One single, stable point of focus is obtained causing a distinct image. It is important to note from the figure that the single focal point stays in one place.

19.13 Space Telescopes versus Earth-Based Telescopes

The Hubble Space Telescope has beamed thousands of images of the stars and galaxies back to Earth where scientists have been able to resolve questions or gain knowledge of the universe and shed light on some mysteries, that is, the existence of dark energy. All these

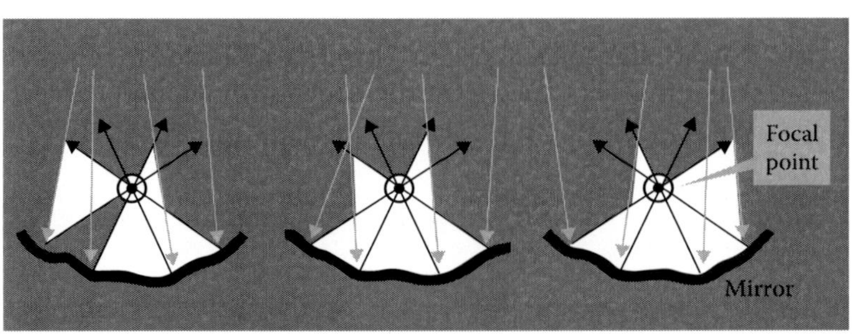

FIGURE 19.28
Three sample snapshots of the mirror in motion.

extraordinary advancements have been possible due to the Hubble images. These images surpass those taken by Earth-based telescopes [7,13].

While GB observatories are generally located in highly elevated areas with minimal light pollution, they must contend with atmospheric turbulence which limits the sharpness of images taken from this vantage point. But with space telescopes we are able to obtain clearer images of distant objects.

A second disadvantage for GB telescopes is that the Earth's atmosphere absorbs much of the infrared and UV light which passes through it. Space telescopes can conveniently detect these waves. Newer GB telescopes are using technological advances like adaptive optics to correct or limit atmospheric distortion.

One drawback of the space telescopes like the Hubble is that they are too difficult to maintain and upgrade plus very expensive. In fact, Hubble is the first telescope designed to be repaired in space to some extent while other space telescopes cannot be serviced at all.

19.14 "Super-Earth" Alien Planet Spotted by GB Telescope

To observe Earth-like planets beyond our solar system has largely been the work of space-based telescopes [9,14,18]. However, new observations from a remote island suggest something else. The Nordic Optical Telescope on La Palma, one of the Canary Islands off the west coast of Africa, observed 55 Cancri e, a planet which is twice the size of Earth. First identified in 2004 by a space-based telescope, 55 Cancri e has a diameter of nearly 26,000 km, about twice that of Earth. This alien world is eight times as massive as Earth, making it a so-called super-Earth but smaller than gas giants like Neptune and Uranus. While not habitable, the size of the planet and position around a sun-like star makes it similar to planets that might support life [19]. A GB telescope in Spain was able to identify 55 Cancri e, which also suggests that telescopes on the ground may help in the search for habitable planets around other stars. An artist's conception exhibits the size of super-Earth 55 Cancri e in comparison to Earth (Figure 19.29).

FIGURE 19.29
An artist's concept showing the size of super-Earth 55 Cancri e compared to Earth. (Courtesy of M. Kornmesser and NASA/ESA/Hubble, Super-Earths: Exoplanets Close to Earth's Size, http://www.space.com/30231-super-earth.html [retrieved August 12, 2016].)

19.15 STEREO

STEREO is the abbreviation of Solar TErrestrial RElations Observatory [20–22]. This consists of two space-based observatories: one ahead of Earth in its orbit and the other trailing behind. With this pair of viewpoints, scientists will be able to observe the structure and evolution of solar storms as they blast from the Sun and move out through space. Thus, it is a solar observation mission.

In 2006, two almost identical spacecraft were launched into orbits around the Sun which cause them to respectively pull farther ahead of and fall behind the Earth gradually. This helps to get stereoscopic imaging of the Sun and solar phenomena, such as coronal mass ejections (CMEs) [23–25]. It is the third mission in NASA's Solar Terrestrial Probes (STP) program. Some particulars of STEREO are given in Table 19.2.

19.15.1 STEREO Spacecraft and Its Scientific Objectives

The two STEREO observatories are almost identical with selective redundancy. The building of the spacecraft bus and the integration of the instruments were made by the Johns Hopkins University Applied Physics Laboratory (APL). The two solar-powered observatories with three-axis stabilization each had a mass at launch of nearly 620 km, including

TABLE 19.2

Some Particulars of STEREO

Mission Type	Solar Observation
Operator	NASA
COSPAR ID	STEREO A: 2006-047A STEREO B: 2006-047B
SATCAT No	STEREO A: 29510 STEREO B: 29511
Website	http://stereo.gsfc.nasa.gov/ http://stereo.jhuapl.edu/
Mission duration	Planned: >3 years Elapsed: 8 years, 9 months, and 16 days
Spacecraft properties	
Manufacturer	Johns Hopkins University Applied Physics Laboratory
Launch mass	640 kg (1412 lb)
Dimensions	7.5 × 8.7 × 5.9 m
Power	475 W
Start of mission	
Launch date	October 26, 2006, 00:52 UTC
Rocket	Delta II 7925-10L
Launch site	Cape Canaveral SLC-17B, Florida
Contractor	United Launch Alliance
Orbital parameters	
Reference system	Heliocentric
Period	STEREO A: 346 days STEREO B: 388 days

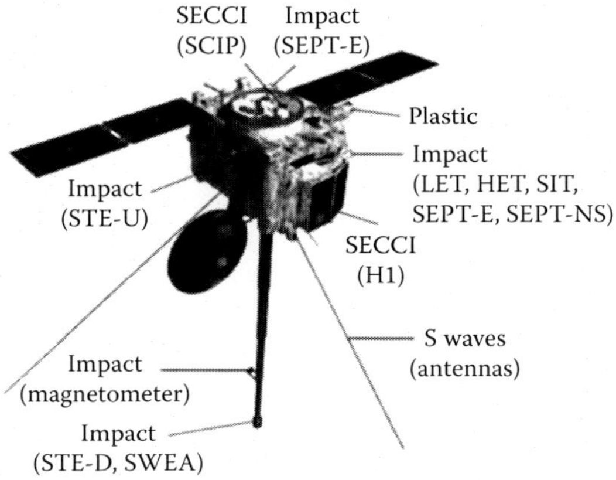

FIGURE 19.30
The different parts of the STEREO spacecraft.

the weight of the propellant. The spacecraft communicate with the APL-based Mission Operations Center through NASA's Deep Space Network. The challenge in spacecraft design is the large number and extent of the instrument fields of view, coupled with different instruments [26,27]. The main design drivers for supporting the performance of the instruments are a conductive outer surface for the energetic particle experiments, stringent electromagnetic compatibility and interference for the radio burst tracker and particulates for the imager experiment. The different parts of the spacecraft are shown in Figure 19.30.

The spacecraft bus consists of six operational subsystems which support two instruments and two instrument suites [28]. The combination gives a total of 16 instruments per observatory. The subsystems include command and data handling, radio frequency communications, guidance and control, propulsion, power, and thermal.

Key characteristics of twin observatories

- Mass: 1364 lb (620 km)
- Dimensions:
  - 3.75 ft (1.14 m)
  - 4.00 ft (1.22 m) wide (launch configuration)
  - 21.24 ft (6.47 m) wide (solar arrays deployed)
  - 6.67 ft (2.03 m) deep
- Power consumption: 475 W
- Data downlink: 720 kilobits per second
- Memory: 1 gigabyte
- Attitude:
  - Control: within 7 arcseconds (0.0019 degrees)
  - Knowledge: within 0.1 arcsecond (0.000028 degrees)

NOTE: 1 arcsecond = 1/3600 of a degree

Scientific Objectives

The scientific objectives are as follows:

- Understand the mechanisms of CME initiation and characterize the propagation through the heliosphere
- Discover the mechanisms and sites of energetic particle acceleration in the low corona and the interplanetary medium
- Improve the investigation of the structure of the ambient solar wind

19.15.2 The Spacecraft Launch Information

The twin STEREO spacecraft were launched in October 2006 on a Delta II 7925-10L rocket (Figure 19.31) from Cape Canaveral Air Force Station in Florida.

The following four instrument packages are mounted on each of the two STEREO spacecraft: an extreme UV imager, two white-light coronagraphs, and a heliospheric imager [29–31]. These instruments study the 3D evolution of CMEs from the very beginning at the Sun's surface through the corona and interplanetary medium to its eventual impact at Earth. Positions of STEREO A and B for August 11, 2015, 08:00 UT, are shown in Figure 19.32.

The figure plots the current positions of the STEREO spacecraft relative to the Sun. The dotted lines reveal the angular displacement from the Sun (units are in AU). When

FIGURE 19.31
STEREO launch.

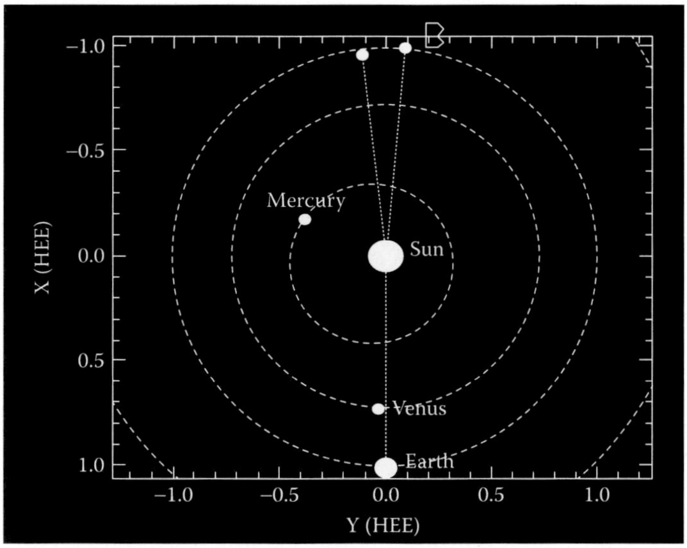

FIGURE 19.32
Current positions of the STEREO spacecraft relative to the Sun.

the two spacecraft come close to Earth, an expanded view of the region around Earth will appear on the right, in the same orientation as the Sun-centered view [32].

Communications with the STEREO "behind" spacecraft were interrupted on October 1, 2014, just after a planned reset of the spacecraft performed as part of a test of solar conjunction operations. The spacecraft telemetry indicated an anomaly in the control system. The STEREO mission achieved its prime science goals within the two-year prime mission.

19.15.3 Studying Space Weather

Space weather happens when a solar storm from the Sun moves through the space and impacts the Earth's magnetosphere. Studying space weather is important to national economy since solar storms can affect the advanced technology such as the electrical grid or communication systems [29,33]. Energy and radiation from solar flares and CMEs can harm astronauts in space, damage sensitive electronics on orbiting spacecraft, and cause colorful auroras, often observed in the higher latitudes. Figure 19.33 shows a solar storm, aurora from space, and aurora on Earth.

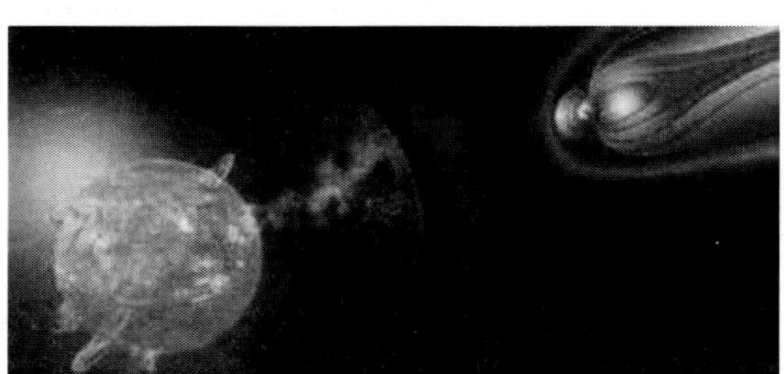

FIGURE 19.33
A solar storm, aurora from space, and aurora on Earth.

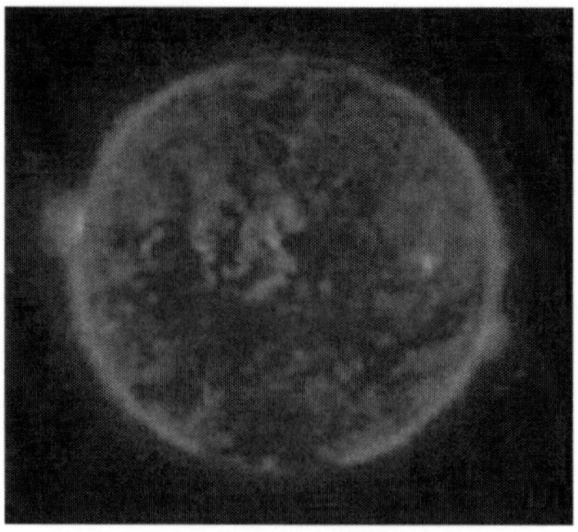

FIGURE 19.34
A time-for-space wiggle image obtained by STEREO. (Courtesy of NASA, 3D Left Right Images, http://www.nasa.gov/mission_pages/stereo/multimedia/LeftRightImages.html [retrieved August 12, 2016].)

In order to understand the changing Sun and its effects on the solar system and consequences to life and society, NASA's Heliophysics research program is actively engaged. NASA missions also focus on the Sun and its interactions with Earth [30,34]. Current missions include STEREO, SDO, SOHO, ACE, SORCE, and the Van Allen Probes. The STEREO Science Center serves the following functions: (1) archive for STEREO telemetry, mission support data, and higher level instrument data; (2) processing center for STEREO space weather beacon data; (3) focal point for science coordination; and (4) focal point for education and public outreach activities. Figure 19.34 shows a time-for-space wiggle image obtained by STEREO.

19.16 Other Space Observatories

Besides NASA's Great Observatories, there have been other small space-based observatories which have focused on particular objects or sections of the electromagnetic spectrum. NASA's Cosmic Background Explorer (COBE) spacecraft was launched in 1989 on a mission to find Cosmic Microwave Background radiation, light left over from shortly after the Big Bang. COBE's instruments measured small variations in the background, to give the proof for the Big Bang model of the universe. NASA also launched a new mission, the Microwave Anisotropy Probe (MAP), in June 2000 to determine the variations in the microwave background in a greater detail. It must be pointed out that NASA is not the only space agency to launch space observatories. The European Space Agency (ESA) has launched a good number of its own observatories to gather information of the universe. The Infrared Space Observatory provided astronomers with views of the universe at infrared wavelengths in the mid- and late 1990s. In 1999, ESA launched XMM-Newton, an orbiting X-ray observatory similar to NASA's Chandra spacecraft. XMM-Newton and Chandra jointly provided lot of information. Chandra is designed to take detailed X-ray images of objects and XMM-Newton focuses to measure the spectra of those objects at X-ray wavelengths. Japan has also contributed some small space observatories. The Advanced Satellite for Cosmology and Astrophysics spacecraft was launched in 1993 which continues to operate in the early twenty-first century, studying the universe at X-ray wavelengths. The Yohkoh spacecraft was launched in 1991 to investigate the Sun in X-rays. The Halca spacecraft, launched in 1997, conducted joint observations with radio telescopes on Earth. The Soviet Union also flew a few space observatories, including the Gamma ray observatory and the Granat X-ray observatory.

19.17 Future Space Observatories

The success of past and present space observatories and progress of other space agencies have encouraged to develop a new series of larger and complex spacecrafts which will be able to examine many fine points. These future observatories are under the umbrella of the Next Generation Space Telescope, the successor to the Hubble Space Telescope. NASA is also supporting the improvement of other new space observatories, including GLAST, a gamma-ray observatory launched in 2006. ESA is also developing some space observatories that will observe the universe at different wavelengths. One mission, called Terrestrial Planet Finder, has the opportunity to combine images from several telescopes. NASA is also developing Constellation-X to use several X-ray telescopes with the purpose to create a virtual telescope 100 times more powerful than existing ones. In the more distant future, astronomers have planned to develop large telescopes and arrays on the surface of the Moon.

Review Questions

19.1 An achromatic telescopic objective with a focal length 1.5 m, consists of two thin lenses in contact, having dispersive powers 0.050 and 0.075, respectively. Find the focal length of each lens, stating its nature.

Solution

We find $f_c/f_f = -(2/3)(1/f) = (1/f_1) - (3/2f_f)$ or $f_f = -1.5$ m and $f_c = +1.0$ m. The flint glass is concave and the other convex.

19.2 An astronomical telescope has an angular magnification of 5× for distant objects. The separation between the objective and the eyepiece is 36 cm. The final image is formed at infinity. What are the focal lengths of the objective and the eyepiece?

Solution

Angular magnification is given by

$$M = \frac{f_0}{f_e} = 5 \text{ (given)}$$

(Note that we have ignored the negative sign because it simply means inverted image.)

Also

$$f_0 + f_e = 36$$

Solving we get to $f_0 = 30$ cm, $f_e = 6$ cm.

19.3 A telescope consisting of two equiconvex lenses of refractive index 3/2 is focused for infinity for an eye adjusted for parallel rays. If the tube is now filled with a liquid of refractive index 4/3 and focused as before, show that the distance between the lenses will have to be doubled.

Solution

If f_0 be the second focal length of objective and f_e be the first focal length of eyepiece then the tube length of telescope adjusted for normal vision is $l = |f_0| + |f_e|$. (1) When there is air between the lenses, the focal length f of lens is given by $1/f = (n-1)[(1/r_1) - (1/r_2)]$. Taking $n = 3/2$ for both the lenses; $r_1 = +r$ and $r_2 = -r$ for objective; while $r_1 = +r'$ and $r_2 = -r'$ for eyepiece, we get $f_0 = r$ and $f_e = r'$. Thus $1 = r + r'$.

(2) When there is liquid of refractive index 4/3 within the lenses we get,

$$\frac{n_3}{v} - \frac{n_1}{u} = \frac{n_2 - n_1}{r_1} - \frac{n_3 - n_2}{r_2} \tag{19.1}$$

For objective, $n_1 = 1$, $v = +f_0'$, $n_3 = 4/3$, $u = \infty$, $r_1 = +r$, $r_2 = -r$, and $n_2 = 3/2$. Putting these values in Equation 19.1 we get $f_0' = 2r$.

For eyepiece, $n_1 = 4/3, v = \infty, n_3 = 1, u = f_e', r_1 = +r', r_2 = -r'$, and $n_2 = 3/2$. Putting these values in Equation (19.1) we get $f_e' = 2r'$.

Thus $l = |f_0'| + |f_e'| = 2(r + r')$; which is double the distance between the lenses when air was present between them.

19.4 The objective lens of a telescope has a focal length of 32 cm and a diameter of 4 cm. When focused at infinity, the exit pupil is 2.5 cm in diameter. Find the angular magnification of the system and the focal length of the eyepiece.

Solution

$$\text{Magnification} = \frac{\text{Diameter of the objective (entrance pupil)}}{\text{Diameter of the exit pupil}}$$

$$= \frac{4\text{ cm}}{0.25\text{ cm}} = 16\text{x}$$

Again magnification, $|M| = f_0/f_e$; or, $16 = 32\text{ cm}/f_e$

Therefore, $f_e = 2$ cm.

19.5 Give an outline of the telescopes used at the early days.

19.6 Describe Galileo's refracting telescope, Explain Kepler's design for an improved refractor and what is an achromatic refractor?

19.7 Explain briefly the Gregorian reflecting telescope and Newton's reflecting telescope.

19.8 What are multimirror telescopes?

19.9 Describe the design of the snow solar telescope.

19.10 What are the needs of space telescopes?

19.11 Mention the characteristics of a good telescope.

19.12 Describe the design of Hubble Space Telescope.

19.13 Make a comparison of a space telescope and ground-based mirror sizes.

19.14 Explain how to make primary mirrors for modern reflecting telescopes.

19.15 Discuss how to reduce the effects of atmospheric distortion.

19.16 What is STEREO spacecraft? What are its scientific objectives?

19.17 Make a brief note on the spacecraft launch information.

19.18 What is space weather?

19.19 Give an outline of future space observatories.

19.20 For an astronomical telescope, the focal lengths of the objective and the eyepiece are respectively given as 120 cm and 0.5 cm. Find the tube length and the magnification when the instrument is focused for (a) infinity and (b) near point.

Hint: (a) For infinity focusing, length of the tube, $L = f_0 + f_e = 120 + 0.5 = 120.5$ cm.

$$m = f_0 / f_e = 120/0.5 = 240$$

(b) For near point focusing, length of the tube, $L = f_0 + [D f_e / (D + f_e)] = 120.5$ cm (approx.)

$$m = (f_0 / f_e)\,(1 + f_e/D) = 244.8$$

19.21 An objective of an astronomical telescope has diameter of 12 cm and a focal length of 180 cm. If the focal length of the eyepiece is 3 cm, find the magnification when focused for infinity and also the length of the instrument. [**Ans.** 60, 183 cm]

19.22 A telescope with an objective of focal length 50 cm is used to bring into view an object at 150 cm distant. When the eyepiece is adjusted to form the image at infinity, the magnifying power is 5. Calculate (i) the focal length of the eyepiece and (ii) the magnifying power of the telescope, if the eyepiece is adjusted to view objects at infinity. [**Ans.** (i) 15 cm and (ii) 3.33]

Definition of Terms	
Chromatic aberration	In optics, chromatic aberration (CA, also called achromatism, chromatic distortion, and spherochromatism) is an effect resulting from dispersion in which there is a failure of a lens to focus all colors to the same convergence point. It occurs because lenses have different refractive indices for different wavelengths of light. The refractive index of transparent materials decreases with increasing wavelength in degrees unique to each. Chromatic aberration manifests itself as "fringes" of color along boundaries that separate dark and bright parts of the image, because each color in the optical spectrum cannot be focused at a single common point. Since the focal length f of a lens is dependent on the refractive index n, different wavelengths of light will be focused on different positions.

(*Continued*)

Definition of Terms	
Spherical aberration	Spherical aberration is an optical effect observed in an optical device (lens, mirror, etc.) that occurs due to the increased refraction of light rays when they strike a lens or a reflection of light rays when they strike a mirror near its edge, in comparison with those that strike nearer the center. It signifies a deviation of the device from the norm, that is, it results in an imperfection of the produced image.
Lagrangian points, Lagrange points, L-points, or libration points	Positions in an orbital configuration of two large bodies where a small object affected only by gravity can maintain a stable position relative to the two large bodies.

References

1. Stevens, D. J. and Scott, G. B. (2013). A posteriori transit probabilities. *Publications of the Astronomical Society of the Pacific* 125 (930): 933–950.
2. Thalmann, C. et al. (2009). Discovery of the coldest imaged companion of a Sun-like star. *The Astrophysical Journal* 707 (2): L123–L127.
3. Deming, D. et al. (2005). Infrared radiation from an extrasolar planet. *Nature* 434 (7034): 740–743.
4. SpaceTelescope.org. Fact Sheet. http://www.spacetelescope.org/about/general/fact_sheet/ (retrieved October 22, 2013).
5. Ryba, J. STS-31. NASA. http://www.nasa.gov/mission_pages/shuttle/shuttlemissions/.../sts-31.html (retrieved March 7, 2012).
6. Harwood, W. (May 30, 2013). Four years after final service call, Hubble Space Telescope going strong. *CBS News*. http://www.cbsnews.com/network/news/space/home/spacenews/files/1ae7cac0d167055e41e1f0da7b0ac6a3-588.html (retrieved June 3, 2013).
7. N2YO.Com. HST Satellite details 1990-037B NORAD 20580. (January 27, 2015). http://www.n2yo.com/satellite/?s=20580 (retrieved January 27, 2015).
8. Laidler, V. et al. (2005). *Synphot User's Guide* (PDF). Version 5.00. Space Telescope Science Institute, Baltimore, MD. p. 27. http://y7.4g99.eu.org/__efa_16_0_fgfqnf.fgfpv.rqh/stsci_python_epydoc/SynphotManual.pdf (retrieved November 3, 2012).
9. Canright, S. NASA's Great Observatories. http://www.nasa.gov/audience/forstudents/postsecondary/features/F_NASA_Great_Observatories_PS.html (retrieved April 26, 2008).
10. Oberth, H. (1923). *Die Rakete zu den Planetenräumen*. Feucht, Germany: R. Oldenbourg-Verlag. p. 85.
11. Spitzer, L. Jr. (1990). Report to Project Rand: Astronomical Advantages of an Extra-Terrestrial Observatory, reprinted in *NASA SP-2001-4407: Exploring the Unknown*, Chapter 3, Document III-1, p. 546ff.
12. Spitzer, L. Jr., Caltech. Archived from the original on March 27, 2008. https://web.archive.org/web/20080327091202/http://www.spitzer.caltech.edu/about/spitzer.shtml (retrieved April 26, 2008).
13. Baum, W. A. et al. (November 1946). Solar ultraviolet spectrum to 88 kilometers. *Physical Review* 70(9–10): 781–782 (American Physical Society). doi:10.1103/PhysRev.70.781.
14. NASA Goddard Space Flight Center. Orbiting Solar Observatories (OSO 1), June 26, 2003. https://heasarc.gsfc.nasa.gov/docs/heasarc/missions/oso1.html (retrieved September 25, 2011).
15. Joint Institute for VLBI in Europe (JIVE). Astronomers demonstrate a global internet telescope. http://www.jive.nl/astronomers-demonstrate-global-internet-telescope (retrieved May 6, 2011).
16. Wikipedia. Spectrometer. https://en.wikipedia.org/wiki/Spectrometer (last modified on August 2, 2016).

17. NASA. Orbiting Astronomical Observatory 2. http://science.nasa.gov/missions/oao/ (retrieved April 26, 2008).
18. Spitzer, L. Jr. (1979). History of the space telescope. *Quarterly Journal of the Royal Astronomical Society* 20: 29–36.
19. MIT Technology Review (2011). Rogue planets could harbor life in interstellar space, say astrobiologists. http://www.technologyreview.com/view/422659/rogue-planets-could-harbor-life-in-interstellar-space-say-astrobiologists/ (retrieved August 29, 2016).
20. NASA Missions. NASA Launch Schedule. (September 20, 2006). http://www.nasa.gov/launchschedule/ (retrieved September 20, 2006).
21. NASA. First Ever STEREO Images of the Entire Sun. http://www.nasa.gov/mission_pages/stereo/news/entire-sun.html (retrieved August 6, 2016).
22. NASA (November 7, 2014). STEREO - Behind Status. http://stereo-ssc.nascom.nasa.gov/behind_status.shtml (retrieved December 7, 2014).
23. Sun bears all for twin space probes. *CBC News*. February 7, 2011. http://www.cbc.ca/technology/story/2011/02/07/science-sun-nasa-stereo.html (retrieved February 8, 2011).
24. Lemonick, M. (February 6, 2011). NASA Images the Entire Sun, Far Side and All. *TIME*. http://content.time.com/time/health/article/0,8599,2046569,00.html (retrieved February 8, 2011).
25. Winter, M. (February 7, 2011). Sun shines in twin probes' first 360-degree images. *USA Today*. http://content.usatoday.com/communities/ondeadline/post/2011/02/sun-shines-in-twin-probes-first-360-degree-images/1#.V5nl2kt961s (retrieved February 8, 2011).
26. Stereo satellites move either side of Sun. *BBC News*. February 6, 2011. http://www.bbc.co.uk/news/science-environment-12365083 (retrieved February 8, 2011).
27. Royal Astronomical Society, United Kingdom (April 19, 2011). STEREO turns its steady gaze on variable stars. *Astronomy*. http://www.astronomy.com/news-observing/news/2011/04/stereo%20turns%20its%20steady%20gaze%20on%20variable%20stars (retrieved April 19, 2011).
28. NASA (July 23, 2014). Near Miss: The Solar Superstorm of July 2012. http://science.nasa.gov/science-news/science-at-nasa/2014/23jul_superstorm/ (retrieved July 24, 2014).
29. NASA Missions. STEREO Spacecraft & Instruments. March 8, 2006. http://www.nasa.gov/mission_pages/stereo/spacecraft/index.html (retrieved May 30, 2006).
30. Howard, R. A. et al. (2002). Sun Earth Connection Coronal and Heliospheric Investigation (SECCHI). *Solar Variability and Solar Physics Missions Advances in Space Research* 29(12): 2017–2026. doi:10.1016/S0273-1177(02)00147-3.
31. Luhmann J. G. et al. (2005). IMPACT: Science goals and firsts with STEREO. *Solar Encounter, Solar-B and Stereo Advances in Space Research* 36(8): 1534–1543. doi:10.1016/j.asr.2005.03.033.
32. Compton, G. and Reavis, B. (2005). Honeywell to provide miniature inertial measurement units for STEREO spacecraft. Honeywell International. https://web.archive.org/web/20051125023147/http://www.honeywell.com:80/en/aerospace/announcement_details.jsp?rowID=115&docID=3995&catID=2 (retrieved October 25, 2006).
33. Driesman, A. et al. (2008). The STEREO Observatory. *Space Science Reviews* 136:17–44.
34. Mewaldt, R. A. et al. (2008). The low-energy telescope (LET) and SEP central electronics for the STEREO mission, in *The STEREO Mission*, C. Russell (ed.), pp. 285–362. New York: Springer Science+Business Media.

20

Space Telescopes—An In-Depth View

20.1 Introduction

The space telescopes are categorized by the task that they were designed for. In most cases, these tasks are for the research of gamma rays, X-rays, ultraviolet light, visible light, infrared, microwave emission, and radio waves. Space telescopes which collect particles, such as cosmic ray nuclei, electrons, and instruments used to detect gravitational waves, are also among the categories listed.

Two values are given for the dimensions of the initial orbit. For telescopes in Earth orbit, the minimum and maximum altitude are given in kilometers. For telescopes in solar orbit, the minimum distance (periapsis) and the maximum distance (apoapsis) between the telescope and the center of mass of the Sun are given in astronomical units (AU).

20.2 Gamma Ray Space Telescopes

Gamma ray telescopes (Table 20.1) are used to collect and measure individual, high-energy gamma rays from different astrophysical sources. These are absorbed by the atmosphere, requiring that observations are made by high-altitude balloons or space missions. Gamma rays are generated by supernovae, pulsars, neutron stars, and black holes. Gamma ray bursts, with sufficiently high energies, have also been detected but have yet to be identified [1].

20.3 X-Ray Telescopes

The purpose of X-ray telescopes (Table 20.2) are to measure high-energy photons termed as X-rays. X-rays cannot travel long distances through the atmosphere. They can only be seen in the upper altitudes of our atmosphere or in space. There are a variety of astrophysical objects that emit X-rays, from galaxy clusters, through black holes, supernova remnants, stars, and binary stars having white dwarf and neutron stars. Some solar system bodies emit X-rays through the process of reflection. A combination of many unresolved X-ray sources is assumed to produce the observed X-ray background.

TABLE 20.1

Particulars of Gamma Ray Space Telescopes

Name	Space Agency	Launch Date	Terminated	Location	Ref(s)
3rd High Energy Astronomy Observatory (HEAO 3)	NASA	September 20, 1979	May 29, 1981	Earth orbit (486.4–504.9 km)	[2,3]
Astrorivelatore Gamma ad Immagini LEggero (AGILE)	ISA	April 23, 2007	–	Earth orbit (524–553 km)	[4,5]
Compton Gamma Ray Observatory (CGRO)	NASA	April 5, 1991	June 4, 2000	Earth orbit (362–457 km)	[6–8]
Cos-B	ESA	August 9, 1975	April 25, 1982	Earth orbit (339.6–99,876 km)	[9–11]
Gamma	USSR, CNES, RSA	July 1, 1990	1992	Earth orbit (375 km)	[12]
Fermi Gamma-ray Space Telescope	NASA	June 11	–	Earth orbit (555 km)	[13]
Granat	CNRS and IKI	December 1, 1989	May 25, 1999	Earth orbit (2,000–200,000 km)	[14–16]
High Energy Transient Explorer 2 (HETE 2) HETE-1, (Lost) November 4, 1996	NASA	October 9, 2000	2007	Earth orbit (590–650 km)	[17–19]
International Gamma Ray Astrophysics Laboratory (INTEGRAL)	ESA	October 17, 2002	–	Earth orbit (639–153,000 km)	[20,21]
Low Energy Gamma Ray Imager (LEGRI)	INTA	May 19, 1997	2002	Earth orbit (600 km)	[22,23]
Second Small Astronomy Satellite (SAS 2)	NASA	November 15, 1972	June 8, 1973	Earth orbit (443–632 km)	[24,25]
Swift Gamma Ray Burst Explorer	NASA	November 20, 2004	–	Earth orbit (585–604 km)	[26,27]

20.4 Ultraviolet Telescopes

Ultraviolet space telescopes (Table 20.3) study the ultraviolet wavelength spectrum which lie in between approximately 10 and 320 nm. At these wavelengths light is absorbed by the Earth's atmosphere. Hence, observations at these wavelengths are performed from the upper atmosphere or from space [89]. Objects emitting ultraviolet radiation include the Sun, other stars, and galaxies [90].

20.5 Optical (Visible Light) Space Telescope

Optical or visible light astronomy is the oldest astronomy which extends from 400 to 700 nm approximately [116]. Positioning is critical for an optical telescope in space. The position is important to eliminate any atmospheric effects as to provide higher resolution images. Optical telescopes are utilized for looking at stars, galaxies, planetary nebulae, and proto-planetary disks [117]. Table 20.4 lists the optical space telescopes in use since 1989.

TABLE 20.2

Particulars of X-Ray Space Telescopes

Name	Space Agency	Launch Date	Terminated	Location	Ref(s)
1st High Energy Astronomy Observatory (HEAO 1)	NASA	August 12, 1977	January 9, 1979	Earth orbit (445 km)	[28,30]
3rd High Energy Astronomy Observatory (HEAO 3)	NASA	9/2079	May 29, 1981	Earth orbit (486.4–504.9 km)	[2,3]
A Broadband Imaging X-Ray All-Sky Survey (ABRIXAS)	DLR	April 28, 1999	July 1, 1999	Earth orbit (549–598 km)	[31,33]
Advanced Satellite for Cosmology and Astrophysics (ASCA)	ISAS and NASA	February 20, 1993	March 2, 2001	Earth orbit (523.6–615.3 km)	[34,35]
AGILE	ISA	April 23, 2007	–	Earth orbit (524–553 km)	[4,5]
Ariel V	SRC and NASA	October 15, 1974	March 14, 1980	Earth orbit (520 km)	[36,37]
Array of Low Energy X-Ray Imaging Sensors (Alexis)	LANL	April 25, 1993	2005	Earth orbit (749–844 km)	[38,40]
Aryabhata	ISRO	April 19, 1975	April 23, 1975	Earth orbit (563–619 km)	[41]
Astron	IKI	March 23, 1983	6/89	Earth orbit (2,000–200,000 km)	[42–44]
Astronomical Netherlands Satellite (ANS)	SRON	August 30, 1974	6/76	Earth orbit (266–1176 km)	[45,46]
BeppoSAX	ASI	April 30, 1996	April 30, 2002	Earth orbit (575–594 km)	[47–49]
Broad Band X-Ray Telescope/Astro 1	NASA	December 2, 1990	December 11, 1990	Earth orbit (500 km)	[50,51]
Chandra X-Ray Observatory	NASA	July 23, 1999	–	Earth orbit (9,942–140,000 km)	[52,53]
Cos-B	ESA	August 9, 1975	April 25, 1982	Earth orbit (339.6–99,876 km)	[9–11]
Cosmic Radiation Satellite (CORSA)	ISAS	February 6, 1976	February 6, 1976	Failed launch	[54,55]
Einstein Observatory (HEAO 2)	NASA	November 13, 1978	April 26, 1981	Earth orbit (465–476 km)	[56,57]
EXOSAT	ESA	May 26, 1983	April 8, 1986	Earth orbit (347–191,709 km)	[58–60]
Ginga (Astro-C)	ISAS	February 5, 1987	November 1, 1991	Earth orbit (517–708 km)	[61–63]
Granat	CNRS and IKI	December 1, 1989	May 25, 1999	Earth orbit (2,000–200,000 km)	[14–16]
Hakucho	ISAS	February 21, 1979	April 16, 1985	Earth orbit (421–433 km)	[64–66]
High Energy Transient Explorer 2 (HETE 2)	NASA	October 9, 2000	–	Earth orbit (590–650 km)	[17–19]
International Gamma Ray Astrophysics Laboratory (INTEGRAL)	ESA	October 17, 2002	–	Earth orbit (639–153,000 km)	[20,21]
Nuclear Spectroscopic Telescope Array (NuSTAR)	NASA	June 13, 2012	–	Earth orbit (603.5 km)	[67,68]
ROSAT	NASA & DLR	June 1, 1990	February 12, 1999	Re-entry October 23, 2011 [69] Formerly Earth orbit (580 km)	[70–72]
Rossi X-Ray Timing Explorer (RXTE)	NASA	December 30, 1995	January 3, 2012	Earth orbit (409 km)	[73–75]
Suzaku (ASTRO-E2)	JAXA and NASA	July 10, 2005	–	Earth orbit (550 km)	[76,77]
Swift Gamma Ray Burst Explorer	NASA	November 20, 2004	–	Earth orbit (585–604 km)	[26,27]
Tenma	ISAS	February 20, 1983	January 19, 1989	Earth orbit (489–503 km)	[78–80]
Third Small Astronomy Satellite (SAS-C)	NASA	May 7, 1975	4/79	Earth orbit (509–516 km)	[81–83]
Uhuru	NASA	December 12, 1970	3/73	Earth orbit (531–572 km)	[84–86]
XM M-Newton	ESA	December 10, 1999	–	Earth orbit (7,365–114,000 km)	[87,88]

TABLE 20.3

Particulars of Ultraviolet Space Telescopes

Name	Space Agency	Launch Date	Terminated	Observing Location	Ref(s)
Astro 2	NASA	March 2, 1993	March 18, 1993	Earth orbit (349–363 km)	[91,92]
Astron	IKI	March 23, 1983	6/89	Earth orbit (2000–200,000 km)	[42–44]
Far Ultraviolet Camera/Spectrograph (UVC)	NASA	April 16, 1972	April 23, 1972	Descartes Highlands on Lunar surface	[93]
Astronomical Netherlands Satellite (ANS)	SRON	August 30, 1974	6/76	Earth orbit (266–1176 km)	[45,46]
Broad Band X-ray Telescope/Astro 1	NASA	December 2, 1990	December 11, 1990	Earth orbit (500 km)	[50,51]
Cosmic Hot Interstellar Spectrometer (CHIPS)	NASA	January 13, 2003	April 11, 2008	Earth orbit (578–594 km)	[94,95]
Extreme Ultraviolet Explorer (EUVE)	NASA	June 7, 1992	January 31, 2001	Earth orbit (515–527 km)	[96,97]
Far Ultraviolet Spectroscopic Explorer (FUSE)	NASA and CNES and CSA	June 24, 1999	July 12, 2007	Earth orbit (752–767 km)	[98,99]
Galaxy Evolution Explorer (GALEX)	NASA	April 28, 2003	June 28, 2013	Earth orbit (691–697 km)	[100–102]
Hisaki (SPRINT-A)	JAXA	September 24, 2013	–	–	[103]
Hubble Space Telescope	NASA and ESA	April 24, 1990	–	Earth orbit (586.47–610.44 km)	[104]
Interface Region Imaging Spectrograph (IRIS)	NASA	June 27, 2013	–	Earth orbit	[105,106]
International Ultraviolet Explorer (IUE)	ESA and NASA & SERC	January 26, 1978	September 30, 1996	Earth orbit (32,050–52,254 km)	[107,108]
Korea Advanced Institute of Science and Technology Satellite 4 (Kaistsat 4)	KARI	September 27, 2003	2007?	Earth orbit (675–695 km)	[109,110]
OAO-2 (Stargazer)	NASA	December 7, 1968	1/73	Earth orbit (749–758 km)	[111,112]
OAO-3 *Copernicus*	NASA	August 21, 1972	2/81	Earth orbit (713–724 km)	[111]
Orion 1 and Orion 2 Space Observatories	USSR	(Orion 1) April 19, 1971 (Orion 2) December 18, 1973	1971, 1973	Earth orbit (Orion 1: 200–222 km; Orion 2: 188–247 km)	[113,114]
Swift Gamma Ray Burst Explorer (Swift)	NASA	November 20, 2004	–	Earth orbit (585–604 km)	[26,27]
Venus	NASA	November 26, 2013	Reusable	Suborbital to 300 km	[115]

TABLE 20.4

Particulars of Optical Space Telescopes

Name	Space Agency	Launch Date	Terminated	Location	Ref(s)
BRITE constellation	Austria, Canada, Poland	2013–2014	–	Earth orbit	[118]
COROT	CNES & ESA	December 27, 2006	2013	Earth orbit (872–884 km)	[119,120]
Hipparcos	ESA	August 8, 1989	3/93	Earth orbit (223–35,632 km)	[121–123]
Hubble Space Telescope	ESA & NASA	April 24, 1990	–	Earth orbit (586.47–610.44 km)	[104]
Kepler Mission	NASA	March 6, 2009	–	Earth-trailing heliocentric orbit	[124–126]
MOST	CSA	June 30, 2003	–	Earth orbit (819–832 km)	[127,128]
Swift Gamma Ray Burst Explorer	NASA	November 20, 2004	–	Earth orbit (585–604 km)	[26,27]
Gaia mission (astrometry)	ESA	December 19, 2013	–	Sun-Earth L_2 Lagrange point	[129]

20.6 Infrared Space Telescopes

The primary source of infrared (IR) radiation is heat or thermal radiation. In comparison to visible light IR light can only be detected by IR viewers. This is due to the part of the spectrum in which our eyes do not see. Table 20.5 shows a list of IR satellites in orbit since 1995. These satellites can view the following: cool stars including brown dwarves, nebulae, and red-shifted galaxies [130]. Table 20.5 lists the particulars of IR space telescopes.

Note: IR is invisible radiant energy, electromagnetic radiation with longer wavelengths than those of visible light, extending from the nominal red edge of the visible spectrum at 700 nm (frequency 430 THz) to 1 mm (300 GHz).

20.7 Microwave Space Telescopes

These telescopes (Table 20.6) have mainly been used to determine cosmological parameters from the Cosmic Microwave Background. They also find synchrotron radiation, free emission, and spinning dust from our galaxy, in addition to extragalactic compact sources and galaxy clusters through the Sunyaev–Zel'dovich effect (https://en.wikipedia.org/wiki/Sunyaev%E2%80%93Zel%27dovich_effect).

TABLE 20.5

Particulars of Infrared Space Telescopes

Name	Space Agency	Launch Date	Terminated	Location	Ref(s)
AKARI	JAXA	February 21, 2006	November 24, 2011 [131]	Earth orbit (586.47–610.44 km)	[132,133]
Herschel Space Observatory	ESA and NASA	May 14, 2009 [134]	April 29, 2013 [135]	Sun–Earth L_2 Lagrange point	[136–138]
IRAS	NASA	January 25, 1983	November 21, 1983	Earth orbit (889–903 km)	[139,140]
Infrared Space Observatory (ISO)	ESA	November 17, 1995	May 16, 1998	Earth orbit (1000–70500 km)	[141,142]
Infrared Telescope in Space	ISAS and NASDA	March 18, 1995	April 25, 1995	Earth orbit (486 km)	[143,144]
Midcourse Space Experiment (MSX)	USN	April 24, 1996	February 26, 1997	Earth orbit (900 km)	[145]
Spitzer Space Telescope	NASA	August 25, 2003	–	Solar orbit (0.98–1.02 AU)	[146,147]
Submillimeter Wave Astronomy Satellite (SWAS)	NASA	December 6, 1998	Last used in 2005	Earth orbit (638–651 km)	[148,149]
Wide Field Infrared Explorer (WIRE)	NASA	March 5, 1999	No observations	Re-entered May 10, 2011 [150]	[151]
Wide-Field Infrared Survey Explorer (WISE)	NASA	December 14, 2009	(hibernation) February 2011–August 2013	Earth (500 km)	[152–154]

TABLE 20.6

Particulars of Microwave Space Telescopes

Name	Space Agency	Launch Date	Terminated	Location	Ref(s)
Cosmic Background Explorer (COBE)	NASA	November 18, 1989	December 23, 1993	Earth orbit (900 km)	[155,156]
Odin	Swedish Space Corporation	February 20, 2001	–	Earth orbit (622 km)	[157,158]
Planck	ESA	May 14, 2009	10/13	Sun–Earth L_2 Lagrange point (mission) Heliocentric (Derelict)	[137,159,160]
WMAP	NASA	June 30, 2001	10/10	Sun–Earth L_2 Lagrange point	[161]

20.8 Very-Long-Baseline Interferometry—Radio Telescopes in Space

Radio telescopes in space are largely used for very-long-baseline interferometry. These space-based satellites work in conjunction with ground-based radio telescopes. This combination of space systems and ground-based systems allows for observations of an

TABLE 20.7

Particulars of Radio Telescopes in Space for Very-Long-Baseline Interferometry

Name	Space Agency	Launch Date	Terminated	Location	Ref(s)
Highly Advanced Laboratory for Communications and Astronomy (HALCA, or VSOP)	ISAS	February 12, 1997	November 30, 2005	Earth orbit (560–21,400 km)	[162–164]
RadioAstron	ASC LPI	5/11	–	Earth orbit (10,000–390,000 km)	[165–167]

object that are made simultaneously by many radio telescopes to be combined, emulating a telescope with a size equal to the maximum separation between the telescopes. Then by correlating their signals to simulate a radio telescope the size of the separation between the two telescopes can be done. Various types of observations like supernova remnants, masers, gravitational lenses, and starburst galaxies are possible. Table 20.7 lists satellites that are performing this type of research.

20.9 Particle Detection

Spacecraft and space-based modules can be used successfully for particle detection, looking for cosmic rays and electrons. These particles are emitted by the Sun, that is, by solar energetic particles, galactic cosmic rays, and extragalactic cosmic rays. There are also ultrahigh-energy cosmic rays from active galactic nuclei (Table 20.8).

TABLE 20.8

Particulars of Particle Detection, Looking for Cosmic Rays

Name	Space Agency	Launch Date	Terminated	Location	Ref(s)
3rd High Energy Astrophysics Observatory (HEAO 3)	NASA	September 20, 1979	September 25, 1981	Earth orbit (486.4–504.9 km)	[2,3]
Alpha Magnetic Spectrometer 01 (AMS-01)	NASA	June 2, 1998	June 12, 1998	Earth orbit (296 km)	[168]
Alpha Magnetic Spectrometer 02 (AMS-02)	NASA	May 16, 2011	–	Earth orbit (353 km)	[169]
IBEX	NASA	October 19, 2008	–	Earth orbit	[170]
Payload (PAMELA)	ISA, INFN, RSA, DLR, and SNSB	May 15, 2006	–	Earth orbit (350–610 km)	[171,172]
SAMPEX	NASA/DE	July 3, 1992	June 30, 2004	Earth orbit	[173]

20.10 Gravitational Waves

A new type of telescope, under consideration, is one that detects gravitational waves; ripples in space–time generated by colliding neutron stars and black holes (Table 20.9).

20.11 List of Proposed Space Observatories

List of proposed space observatories (Table 20.10), plans, concepts, and new proposals including (1) active; (2) other; and (3) merged, cancelled, and superseded are available in the literature, as outlined below.

20.12 Requirement for Further Investigation

The universe transmits light all across the electromagnetic spectrum. But much of this radiation does not reach the Earth's surface as our atmosphere blocks out certain types of radiation while letting other types through. With this said, a requirement for further investigation is needed. For life on Earth, we are very fortunate that our atmosphere blocks out harmful high-energy radiation like X-rays, gamma rays, and most of UV rays. The atmosphere also absorbs a major part of the IR radiation. On the contrary, our atmosphere is transparent to visible light, most radio waves, and small windows within the IR region. Ground-based optical and IR observatories are placed, in general, near the summits of dry mountains to get above much of the atmosphere. Radio telescopes can operate round-the-clock from the Earth's surface. These ground-based observatories supply valuable long-term studies of objects in space, but can only detect the light which passes through our atmosphere. Figure 20.1 shows a plot of opacity in the atmosphere corresponding to different wavelengths.

TABLE 20.9

Particulars of Proposed Telescope Related to Gravitational Waves

Name	Space Agency	Launch Date	Terminated	Location	Ref(s)
eLISA (see also LISA Pathfinder)	ESA	2034?	–	Heliocentric	[174]
To Be Launched					
Astrosat	ISRO	2015	–	Earth orbit (650 km)	[175,176]
James Webb Space Telescope	NASA/ ESA/CSA	October 2018	–	On Earth. Planned for Sun–Earth L_2 Lagrange point	[177]
Tel Aviv University Ultraviolet Explorer (TAUVEX)	Israeli Space Agency	TBA	–	–	[178]
Hard X-Ray Modulation Telescope (HXMT)	CNSA	2015–2016	–	–	[179]
Dark Matter Particle Explorer (DAMPE)	CNSA	2015–2016	–	–	[180]

TABLE 20.10
Particulars of Proposed Space Observatories

Name	Agency	Planned Launch Date	Fate	Planned Location	Ref(s)
Advanced technology large aperture space telescope	Space telescope	–	–	Sun–Earth L_2 Lagrange point	[181]
X-ray evolving universe (XEUS)	ESA	Merged into IXO		–	[182]
Constellation-X	NASA	Merged into IXO		–	
International X-ray observatory (IXO)	NASA and ESA and JAXA	No funding 2011		–	[183]
Spectrum-X-gamma	IKI & ESA	–	Cancelled	–	[184]
Astrosat	ISRO	TBA	–	Earth orbit (650 km)	[175,176]
SIM lite astrometric observatory	NASA	2015	No funding 2010	–	[185]
Darwin mission	ESA	–	–	Sun–Earth L_2 Lagrange point	[186]
Euclid	ESA	2019	–	Sun–Earth L_2 Lagrange point	
Terrestrial planet finder	NASA	TBA	No funding 2011	–	[187]
Dark universe	NASA	Superseded by WFIRST		Earth orbit (600 km)	[188,189]
Dark energy space telescope	NASA and DOE	Superseded by WFIRST		–	[190]
SPHEREx	NASA	TBA	–	–	[191]
Wide field infrared survey telescope (WFIRST) (IR)	NASA and DOE	2020	–	Sun–Earth L_2 Lagrange point	[192,193]
Astromag free-flyer (particles)	NASA	January 1, 2005	–	Earth orbit (500 km)	[194,195]
VSOP-2 (Astro-G) (Radio)	JAXA	2012	Cancelled 2011	–	[196]
Public space telescope (PST)	astrofactum	2019	–	Earth orbit (800 km)	[197]

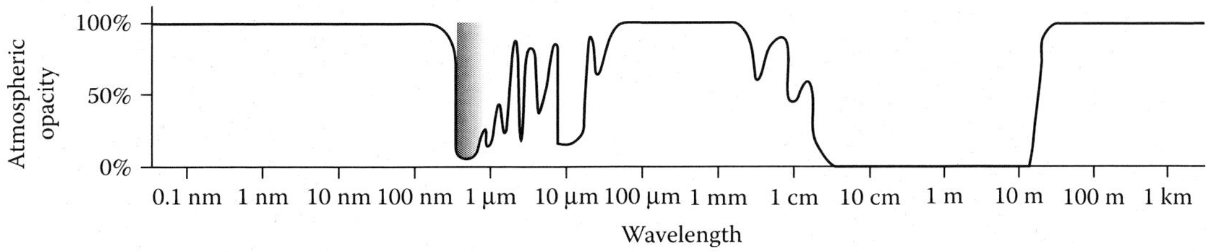

FIGURE 20.1
Plot of atmospheric opacity versus wavelength.

Review Questions

20.1 Briefly discuss the working principle and operation of (a) gamma ray telescopes, (b) X-ray telescopes, (c) ultraviolet telescopes, (d) optical (visible light) space telescope, (e) infrared space telescopes, and (f) microwave space telescopes.

20.2 Describe the radio telescopes in space for very-long-baseline interferometry and also give the working principle in brief.

20.3 What are gravitational waves?

20.4 Discuss the requirement for sending ultraviolet, optical, and infrared telescopes in space and make a comparison among them.

Definition of Terms

Term	Definition
Cosmic rays	Cosmic rays are immensely high-energy radiation, mainly originating outside the solar system. They may produce showers of secondary particles that penetrate and impact the Earth's atmosphere and sometimes even reach the surface. Composed primarily of high-energy protons and atomic nuclei, they are of mysterious origin. Data from the Fermi space telescope (2013) has been interpreted as evidence that a significant fraction of primary cosmic rays originate from the supernovae of massive stars. However, this is not thought to be their only source. Active galactic nuclei probably also produce cosmic rays.
Gamma ray bursts	Gamma ray bursts (GRBs) are flashes of gamma rays associated with extremely energetic explosions that have been observed in distant galaxies. They are the brightest electromagnetic events known to occur in the universe. Bursts can last from ten milliseconds to several hours. The initial burst is usually followed by a longer-lived "afterglow" emitted at longer wavelengths (X-ray, ultraviolet, optical, infrared, microwave, and radio).
Sunyaev–Zel'dovich	The Sunyaev–Zel'dovich effect (often abbreviated as the SZ effect) is the result of high-energy electrons distorting the Cosmic Microwave Background (CMB) radiation through inverse Compton scattering, in which the low-energy CMB photons receive an average energy boost during collision with the high-energy cluster electrons. Observed distortions of the CMB spectrum are used to detect the density perturbations of the universe. Using the Sunyaev–Zel'dovich effect, dense clusters of galaxies have been observed.
X-rays	X-radiation (composed of X-rays) is a form of electromagnetic radiation. Most X-rays have a wavelength ranging from 0.01 to 10 nanometers, corresponding to frequencies in the range 30 petahertz to 30 exahertz (3×10^{16} Hz to 3×10^{19} Hz) and energies in the range 100 eV to 100 keV. X-ray wavelengths are shorter than those of UV rays and typically longer than those of gamma rays. In many languages, X-radiation is referred to with terms meaning Röntgen radiation, after Wilhelm Röntgen [1], who is usually credited as its discoverer, and who had named it X-radiation to signify an unknown type of radiation [2]. Spelling of X-ray(s) in the English language includes the variants X-ray(s), Xray(s), and x-ray(s).

References

1. Gamma rays. NASA. http://missionscience.nasa.gov/ems/12_gammarays.html (retrieved March 1, 2008).
2. NASA – NSSD – Spacecraft – Trajectory Details (HEAO 3). NASA. http://nssdc.gsfc.nasa.gov/nmc/spacecraftOrbit.do?id=1972-091A (retrieved February 27, 2008).
3. NASA – NSSDC – Spacecraft – Details (HEAO 3). NASA. http://nssdc.gsfc.nasa.gov/nmc/spacecraftDisplay.do?id=1972-091A (retrieved February 27, 2008).
4. NASA – NSSD – Spacecraft – Trajectory Details (AGILE). NASA. http://nssdc.gsfc.nasa.gov/nmc/spacecraftOrbit.do?id=1999-022A (retrieved February 27, 2008).
5. NASA – NSSDC – Spacecraft – Details (AGILE). NASA. http://nssdc.gsfc.nasa.gov/nmc/spacecraftOrbit.do?id=1999-022A (retrieved February 27, 2008).
6. NASA – NSSD – Spacecraft – Trajectory Details (Compton Gamma Ray Observatory). NASA. http://nssdc.gsfc.nasa.gov/nmc/spacecraftOrbit.do?id=1991-027B (retrieved February 27, 2008).
7. NASA – NSSDC – Spacecraft – Details (Compton Gamma Ray Observatory). NASA. http://nssdc.gsfc.nasa.gov/nmc/spacecraftDisplay.do?id=1991-027B (retrieved February 27, 2008).
8. CGRO Science Support Center. NASA. http://heasarc.gsfc.nasa.gov/docs/cgro/index.html (retrieved February 27, 2008).
9. NASA – NSSD – Spacecraft – Trajectory Details (COS-B). NASA. http://nssdc.gsfc.nasa.gov/nmc/spacecraftOrbit.do?id=1975-072A (retrieved February 27, 2008).
10. NASA – NSSDC – Spacecraft – Details (COS-B). NASA. http://nssdc.gsfc.nasa.gov/nmc/spacecraftDisplay.do?id=1975-072A (retrieved February 27, 2008).
11. ESA – Space Science – Cos-B overview. ESA. http://www.esa.int/Our_Activities/Space_Science/Cos-B_overview2 (retrieved February 28, 2008).
12. The Gamma Satellite. NASA. http://heasarc.gsfc.nasa.gov/docs/heasarc/missions/gamma.html (retrieved February 29, 2008).
13. GLAST Overview – The GLAST Mission. NASA. December 19, 2007. http://www.nasa.gov/mission_pages/GLAST/main/index.html (retrieved February 27, 2008).
14. NASA – NSSD – Spacecraft – Trajectory Details (GRANAT). NASA. http://nssdc.gsfc.nasa.gov/nmc/spacecraftOrbit.do?id=1989-096A (retrieved February 27, 2008).
15. NASA – NSSDC – Spacecraft – Details (GRANAT). NASA. http://nssdc.gsfc.nasa.gov/nmc/spacecraftDisplay.do?id=1989-096A (retrieved February 27, 2008).
16. 1999 Reentries. The Aerospace Corporation, Center for Orbital and Reentry Debris Studies. http://www.aerospace.org/ (retrieved February 26, 2008).
17. NASA – NSSD – Spacecraft – Trajectory Details (HETE 2). NASA. http://nssdc.gsfc.nasa.gov/nmc/spacecraftOrbit.do?id=2000-061A (retrieved February 27, 2008).
18. NASA – NSSDC – Spacecraft – Details (HETE 2). NASA. http://nssdc.gsfc.nasa.gov/nmc/spacecraftDisplay.do?id=2000-061A (retrieved February 27, 2008).
19. The High Energy Transient Explorer (HETE-2). Massachusetts Institute of Technology. March 28, 2007. http://space.mit.edu/HETE/ (retrieved February 27, 2008).
20. NASA – NSSD – Spacecraft – Trajectory Details (INTEGRAL). NASA. http://nssdc.gsfc.nasa.gov/nmc/spacecraftOrbit.do?id=2002-048A (retrieved February 27, 2008).
21. NASA – NSSDC – Spacecraft – Details (INTEGRAL). NASA. http://nssdc.gsfc.nasa.gov/nmc/spacecraftDisplay.do?id=2002-048A (retrieved February 27, 2008).
22. Instrumentation: Low Energy Gamma Ray Imager (LEGRI). Birmingham University. January 24, 2006. http://www.sr.bham.ac.uk/instrument/legri.html (retrieved February 27, 2008).
23. LEGRI. NASA. December 4, 1997. http://imagine.gsfc.nasa.gov/features/satellites/index.html (retrieved February 28, 2008).
24. NASA – NSSD – Spacecraft – Trajectory Details (SAS-B). NASA. http://nssdc.gsfc.nasa.gov/nmc/spacecraftOrbit.do?id=1972-091A (retrieved February 27, 2008).

25. NASA – NSSDC – Spacecraft – Details (SAS-B). NASA. http://nssdc.gsfc.nasa.gov/nmc/spacecraftDisplay.do?id=1972-091A (retrieved February 27, 2008).
26. NASA – NSSD – Spacecraft – Trajectory Details (Swift). NASA. http://nssdc.gsfc.nasa.gov/nmc/spacecraftOrbit.do?id=2004-047A (retrieved February 27, 2008).
27. NASA – NSSDC – Spacecraft – Details (Swift). NASA. http://nssdc.gsfc.nasa.gov/nmc/spacecraftDisplay.do?id=2004-047A (retrieved February 27, 2008).
28. NASA – NSSDC – Spacecraft – Details (HEAO 1). NASA. http://nssdc.gsfc.nasa.gov/nmc/spacecraftDisplay.do?id=1977-075A (retrieved February 27, 2008).
29. The High Energy Astrophysics Observatory-1 – Overview. NASA. http://heasarc.gsfc.nasa.gov/docs/heao1/heao1.html (retrieved February 27, 2008).
30. The High Energy Astrophysics Observatory-1 – Mission Overview. NASA. http://heasarc.gsfc.nasa.gov/docs/heao1/heao1_about.html (retrieved February 27, 2008).
31. NASA – NSSD – Spacecraft – Trajectory Details (ABRIXAS). NASA. http://nssdc.gsfc.nasa.gov/nmc/spacecraftOrbit.do?id=1999-022A (retrieved February 27, 2008).
32. NASA – NSSDC – Spacecraft – Details (ABRIXAS). NASA. http://nssdc.gsfc.nasa.gov/nmc/spacecraftDisplay.do?id=1999-022A (retrieved February 27, 2008).
33. ABRIXAS. Astronautix.com. http://www.astronautix.com/craft/abrixas.htm (retrieved February 28, 2008).
34. NASA – NSSD – Spacecraft – Trajectory Details (ASCA). NASA. http://nssdc.gsfc.nasa.gov/nmc/spacecraftOrbit.do?id=1993-011A (retrieved February 27, 2008).
35. NASA – NSSDC – Spacecraft – Details (ASCA). NASA. http://nssdc.gsfc.nasa.gov/nmc/spacecraftDisplay.do?id=1993-011A (retrieved February 27, 2008).
36. The Ariel V Satellite. NASA. http://nssdc.gsfc.nasa.gov/nmc/spacecraftDisplay.do?id=1993-011A (retrieved February 29, 2008).
37. The Ariel V Satellite – About. NASA. http://heasarc.nasa.gov/docs/ariel5/ariel5_about.html (retrieved February 29, 2008).
38. NASA – NSSD – Spacecraft – Trajectory Details (Alexis). NASA. http://nssdc.gsfc.nasa.gov/nmc/spacecraftOrbit.do?id=1993-026A (retrieved February 27, 2008).
39. NASA – NSSDC – Spacecraft – Details (Alexis). NASA. http://nssdc.gsfc.nasa.gov/nmc/spacecraftDisplay.do?id=1993-026A (retrieved February 27, 2008).
40. AeroAstro Returning to Space With Launch of STPSat-1 This Fall. Space.com. June 5, 2006. https://web.archive.org/web/20060608025316/http://www.space.com/spacenews/businessmonday_060605.html (retrieved February 28, 2008).
41. The Aryabhata Satellite. NASA. http://heasarc.gsfc.nasa.gov/docs/heasarc/missions/aryabhata.html (retrieved February 29, 2008).
42. NASA – NSSD – Spacecraft – Trajectory Details (ASTRON). NASA. http://nssdc.gsfc.nasa.gov/nmc/spacecraftOrbit.do?id=1983-020A (retrieved February 27, 2008).
43. NASA – NSSDC – Spacecraft – Details (ASTRON). NASA. http://nssdc.gsfc.nasa.gov/nmc/spacecraftDisplay.do?id=1983-020A (retrieved February 27, 2008).
44. The Astron Satellite. NASA. http://heasarc.gsfc.nasa.gov/docs/heasarc/missions/astron.html (retrieved February 28, 2008).
45. NASA – NSSD – Spacecraft – Trajectory Details (ANS). NASA. http://nssdc.gsfc.nasa.gov/nmc/spacecraftDisplay.do?id=1974-070A http://nssdc.gsfc.nasa.gov/nmc/spacecraftDisplay.do?id=1974-070A (retrieved February 27, 2008).
46. NASA – NSSDC – Spacecraft – Details (ANS). NASA. http://nssdc.gsfc.nasa.gov/nmc/spacecraftOrbit.do?id=1974-070A (retrieved February 27, 2008).
47. NASA – NSSD – Spacecraft – Trajectory Details (SAX). NASA. http://nssdc.gsfc.nasa.gov/nmc/spacecraftOrbit.do?id=1996-027A (retrieved February 27, 2008).
48. NASA – NSSDC – Spacecraft – Details (SAX). NASA. http://nssdc.gsfc.nasa.gov/nmc/spacecraftDisplay.do?id=1996-027A (retrieved February 27, 2008).
49. HEASARC: BeppoSAX Guest Observer Facility. NASA. http://heasarc.gsfc.nasa.gov/docs/sax/saxgof.html (retrieved February 28, 2008).

50. NASA – NSSD – Spacecraft – Trajectory Details (Astro 1). NASA. http://nssdc.gsfc.nasa.gov/nmc/spacecraftOrbit.do?id=ASTRO-1 (retrieved February 27, 2008).
51. NASA – NSSDC – Spacecraft – Details (Astro 1). NASA. http://nssdc.gsfc.nasa.gov/nmc/spacecraftDisplay.do?id=ASTRO-1 (retrieved February 27, 2008).
52. NASA – NSSD – Spacecraft – Trajectory Details (Chandra). NASA. http://nssdc.gsfc.nasa.gov/nmc/spacecraftDisplay.do?id=ASTRO-1 (retrieved February 27, 2008).
53. NASA – NSSDC – Spacecraft – Details (Chandra). NASA. http://nssdc.gsfc.nasa.gov/nmc/spacecraftDisplay.do?id=1999-040B (retrieved February 27, 2008).
54. NASA – NSSD – Spacecraft – Trajectory Details (CORSA). NASA. http://nssdc.gsfc.nasa.gov/nmc/spacecraftOrbit.do?id=CORSA (retrieved February 27, 2008).
55. NASA – NSSDC – Spacecraft – Details (CORSA). NASA. http://nssdc.gsfc.nasa.gov/nmc/spacecraftDisplay.do?id=CORSA (retrieved February 27, 2008).
56. NASA – NSSD – Spacecraft – Trajectory Details (Einstein Observatory). NASA. http://nssdc.gsfc.nasa.gov/nmc/spacecraftOrbit.do?id=1978-103A (retrieved February 27, 2008).
57. NASA – NSSDC – Spacecraft – Details (Einstein Observatory). NASA. http://nssdc.gsfc.nasa.gov/nmc/spacecraftDisplay.do?id=1978-103A (retrieved February 27, 2008).
58. NASA – NSSD – Spacecraft – Trajectory Details (Exosat). NASA. http://nssdc.gsfc.nasa.gov/nmc/spacecraftOrbit.do?id=1983-051A (retrieved February 27, 2008).
59. NASA – NSSDC – Spacecraft – Details (Exosat). NASA. http://nssdc.gsfc.nasa.gov/nmc/spacecraftDisplay.do?id=1983-051A (retrieved February 27, 2008).
60. ESA Science & Technology: Exosat. ESA. http://sci.esa.int/exosat/ (retrieved February 28, 2008).
61. NASA – NSSD – Spacecraft – Trajectory Details (Astro-C). NASA. http://nssdc.gsfc.nasa.gov/nmc/spacecraftOrbit.do?id=1987-012A (retrieved February 27, 2008).
62. NASA – NSSDC – Spacecraft – Details (Astro-C). NASA. http://nssdc.gsfc.nasa.gov/nmc/spacecraftDisplay.do?id=1987-012A (retrieved February 27, 2008).
63. The Ginga Observatory. NASA. http://heasarc.gsfc.nasa.gov/docs/ginga/ginga.html (retrieved February 28, 2008).
64. NASA – NSSD – Spacecraft – Trajectory Details (Hakucho). NASA. http://nssdc.gsfc.nasa.gov/nmc/spacecraftOrbit.do?id=1979-014A (retrieved February 27, 2008).
65. NASA – NSSDC – Spacecraft – Details (Hakucho). NASA. http://nssdc.gsfc.nasa.gov/nmc/spacecraftDisplay.do?id=1979-014A (retrieved February 27, 2008).
66. The Hakucho (CORSA-B) Satellite. NASA. http://heasarc.gsfc.nasa.gov/docs/hakucho/hakucho.html (retrieved February 28, 2008).
67. W. Harwood (June 13, 2012). NuSTAR X-ray telescope launched on mission to search for black holes. *CBS News*. Retrieved July 26, 2012. NuSTAR Quickfacts. California Institute of Technology. Archived from the original on February 3, 2008. http://www.cbsnews.com/network/news/space/home/spacenews/files/nustar_launch.html (retrieved February 28, 2008).
68. F. Harrison and Y. Kim (May 2010). NusTAR Newsletter (PDF). NASA JPL. http://www.nustar.caltech.edu/uploads/updates/NuSTAR_2010may_update.pdf (retrieved June 24, 2010).
69. ROSAT – latest news. *DLR Portal*. October 25, 2011. http://www.dlr.de/dlr/en/desktopdefault.aspx/tabid-10424 (retrieved October 25, 2011).
70. NASA – NSSD – Spacecraft – Trajectory Details (ROSAT). NASA. http://nssdc.gsfc.nasa.gov/nmc/spacecraftOrbit.do?id=1990-049A (retrieved February 27, 2008).
71. NASA – NSSDC – Spacecraft – Details (ROSAT). NASA. http://nssdc.gsfc.nasa.gov/nmc/spacecraftDisplay.do?id=1990-049A (retrieved February 27, 2008).
72. The Roentgen Satellite. NASA. http://heasarc.gsfc.nasa.gov/docs/rosat/rosat.html (retrieved February 28, 2008).
73. The RXTE Mission is Approaching the End of Science Operations. http://heasarc.nasa.gov/docs/xte/xhp_new.html#endofoperations (retrieved January 4, 2012).
74. NASA – NSSD – Spacecraft – Trajectory Details (X-Ray Timing Explorer). NASA. http://nssdc.gsfc.nasa.gov/nmc/spacecraftOrbit.do?id=1995-074A (retrieved February 27, 2008).

75. NASA – NSSDC – Spacecraft – Details (X-Ray Timing Explorer). NASA. http://nssdc.gsfc.nasa.gov/nmc/spacecraftDisplay.do?id=1995-074A (retrieved February 27, 2008).
76. NASA – NSSD – Spacecraft – Trajectory Details (Suzaku). NASA. http://nssdc.gsfc.nasa.gov/nmc/spacecraftOrbit.do?id=2005-025A (retrieved February 27, 2008).
77. NASA – NSSDC – Spacecraft – Details (Suzaku). NASA. http://nssdc.gsfc.nasa.gov/nmc/spacecraftOrbit.do?id=2005-025A (retrieved February 27, 2008).
78. NASA – NSSD – Spacecraft – Trajectory Details (Tenma). NASA. http://nssdc.gsfc.nasa.gov/nmc/spacecraftOrbit.do?id=1983-011A (retrieved February 27, 2008).
79. NASA – NSSDC – Spacecraft – Details (Tenma). NASA. http://nssdc.gsfc.nasa.gov/nmc/spacecraftDisplay.do?id=1983-011A (retrieved February 27, 2008).
80. Institute of Space and Astronautical Science – JAXA – Tenma. JAXA. https://web.archive.org/web/20070208103229/http://www.isas.jaxa.jp:80/e/enterp/missions/complate/tenma.shtml (retrieved February 28, 2008).
81. NASA – NSSD – Spacecraft – Trajectory Details (SAS-C). NASA. http://nssdc.gsfc.nasa.gov/nmc/spacecraftOrbit.do?id=1975-037A (retrieved February 27, 2008).
82. NASA – NSSDC – Spacecraft – Details (SAS-C). NASA. http://nssdc.gsfc.nasa.gov/nmc/spacecraftDisplay.do?id=1975-037A (retrieved February 27, 2008).
83. The Third Small Astronomy Satellite (SAS-3). NASA. http://heasarc.nasa.gov/docs/sas3/sas3.html (retrieved February 28, 2008).
84. NASA – NSSD – Spacecraft – Trajectory Details (Uhuru). NASA. http://nssdc.gsfc.nasa.gov/nmc/spacecraftOrbit.do?id=1970-107A (retrieved February 27, 2008).
85. NASA – NSSDC – Spacecraft – Details (Uhuru). NASA. http://nssdc.gsfc.nasa.gov/nmc/spacecraftDisplay.do?id=1970-107A (retrieved February 27, 2008).
86. The Uhuru Satellite. NASA. http://heasarc.gsfc.nasa.gov/docs/uhuru/uhuru.html (retrieved February 28, 2008).
87. NASA – NSSD – Spacecraft – Trajectory Details (XMM-Newton). NASA. http://nssdc.gsfc.nasa.gov/nmc/spacecraftOrbit.do?id=1999-066A (retrieved February 25, 2008).
88. NASA – NSSDC – Spacecraft – Details (XMM-Newton). NASA. http://nssdc.gsfc.nasa.gov/nmc/spacecraftDisplay.do?id=1999-066A (retrieved February 25, 2008).
89. A. N. Cox, ed. (2000). *Allen's Astrophysical Quantities*. New York: Springer-Verlag.
90. Ultraviolet Waves. NASA. http://science.hq.nasa.gov/kids/imagers/ems/uv.html (retrieved February 28, 2008).
91. NASA – NSSD – Spacecraft – Trajectory Details (Astro 2). NASA. http://nssdc.gsfc.nasa.gov/nmc/spacecraftOrbit.do?id=ASTRO-2 (retrieved February 27, 2008).
92. NASA – NSSDC – Spacecraft – Details (Astro 2). NASA. http://nssdc.gsfc.nasa.gov/nmc/spacecraftDisplay.do?id=ASTRO-2 (retrieved February 27, 2008).
93. Experiment Operations during Apollo EVAs. http://ares.jsc.nasa.gov/HumanExplore/Exploration/EXLibrary/docs/ApolloCat/Part1/UVC.htm (retrieved May 15, 2013).
94. NASA – NSSD – Spacecraft – Trajectory Details (CHIPS). NASA. http://nssdc.gsfc.nasa.gov/nmc/spacecraftOrbit.do?id=2003-002B (retrieved February 27, 2008).
95. NASA – NSSDC – Spacecraft – Details (CHIPS). NASA. http://nssdc.gsfc.nasa.gov/nmc/spacecraftDisplay.do?id=2003-002B (retrieved February 27, 2008).
96. NASA – NSSD – Spacecraft – Trajectory Details (EUVE). NASA. http://nssdc.gsfc.nasa.gov/nmc/spacecraftOrbit.do?id=1992-031A (retrieved February 27, 2008).
97. NASA – NSSDC – Spacecraft – Details (EUVE). NASA. http://nssdc.gsfc.nasa.gov/nmc/spacecraftDisplay.do?id=1992-031A (retrieved February 27, 2008).
98. NASA – NSSD – Spacecraft – Trajectory Details (FUSE). NASA. http://nssdc.gsfc.nasa.gov/nmc/spacecraftOrbit.do?id=1999-035A (retrieved February 27, 2008).
99. NASA – NSSDC – Spacecraft – Details (FUSE). http://nssdc.gsfc.nasa.gov/nmc/spacecraftDisplay.do? id=1999-035A (retrieved February 27, 2008).
100. NASA – NSSD – Spacecraft – Trajectory Details (GALEX). NASA. http://nssdc.gsfc.nasa.gov/nmc/spacecraftOrbit.do?id=2003-017A (retrieved February 27, 2008).

101. NASA – NSSDC – Spacecraft – Details (GALEX). NASA. http://nssdc.gsfc.nasa.gov/nmc/spacecraftDisplay.do?id=2003-017A (retrieved February 27, 2008).
102. NASA Decommissions Its Galaxy Hunter Spacecraft. Spacecraft. http://www.galex.caltech.edu/newsroom/glx2013-03r.html (retrieved August 22, 2016).
103. Spectroscopic Planet Observatory for Recognition of Interaction of Atmosphere. JAXA. http://global.jaxa.jp/projects/sat/sprint_a/index.html (retrieved August 8, 2013).
104. NASA – NSSD – Spacecraft – Trajectory Details (Hubble Space Telescope). NASA. http://nssdc.gsfc.nasa.gov/nmc/spacecraftOrbit.do?id=1990-037B (retrieved February 27, 2008).
105. NASA's Newest Solar Satellite Arrives at Vandenberg AFB for Launch. NASA. http://www.nasa.gov/mission_pages/iris/news/arrives-vafb.html (retrieved April 17, 2013).
106. NASA's IRIS Mission Updates. NASA. http://www.nasa.gov/rss/iris_main_update.xml (retrieved June 25, 2013).
107. ESA Science & Technology: IUE. ESA. http://sci.esa.int/iue/ (retrieved February 29, 2008).
108. NASA – NSSDC – Spacecraft – Details (IUE). NASA. http://nssdc.gsfc.nasa.gov/nmc/spacecraftDisplay.do?id=1978-012A (retrieved February 27, 2008).
109. NASA – NSSD – Spacecraft – Trajectory Details (Kaistsat 4). NASA. http://nssdc.gsfc.nasa.gov/nmc/spacecraftOrbit.do?id=2003-042G (retrieved February 27, 2008).
110. NASA – NSSDC – Spacecraft – Details (Kaistsat 4). NASA. http://nssdc.gsfc.nasa.gov/nmc/spacecraftDisplay.do?id=2003-042G (retrieved February 27, 2008).
111. D. Darling. Orbiting Astronomical Observatory (OAO) in the Internet Encyclopedia of Science. http://www.daviddarling.info/encyclopedia/O/OAO.html (retrieved February 28, 2008).
112. Orbiting Astronomical Observatory OAO-2. University of Wisconsin–Madison. http://www.sal.wisc.edu/OAO/ (retrieved February 28, 2008).
113. G. A. Gurzadyan and J. Ohanesyan (September 8, 1972). Observed Energy Distribution of α Lyra and β Cen at 2000–3800 Å. Nature.com. http://www.nature.com/nature/journal/v239/n5367/abs/239090a0.html (retrieved September 4, 2012).
114. G. A. Gurzadyan (July 19, 1974). Ultraviolet spectra of Capella. Nature.com. http://www.nature.com/nature/journal/v250/n5463/abs/250204a0.html (retrieved September 4, 2012).
115. What is VeSpR? Center for Space Physics, Boston University. November 2013. http://www.bu.edu/csp/PASS/science/venusfaq.html (retrieved November 28, 2013).
116. P. Moore (1997). *Philip's Atlas of the Universe*. Great Britain: George Philips Limited.
117. HubbleSite – The Telescope – Hubble Essentials. NASA. http://hubblesite.org/the_telescope/hubble_essentials/ (retrieved March 1, 2008).
118. BRITE-Constellation. http://www.brite-constellation.at/ (retrieved March 9, 2015).
119. NASA – NSSD – Spacecraft – Trajectory Details (COROT). NASA. http://nssdc.gsfc.nasa.gov/nmc/spacecraftOrbit.do?id=2006-063A (retrieved February 27, 2008).
120. NASA – NSSDC – Spacecraft – Details (COROT). NASA. http://nssdc.gsfc.nasa.gov/nmc/spacecraftOrbit.do?id=2006-063A (retrieved February 27, 2008).
121. NASA – NSSD – Spacecraft – Trajectory Details (Hipparcus). NASA. http://nssdc.gsfc.nasa.gov/nmc/spacecraftOrbit.do?id=1989-062B (retrieved February 27, 2008).
122. NASA – NSSDC – Spacecraft – Details (Hipparcus). NASA. nssdc.gsfc.nasa.gov/nmc/spacecraftDisplay.do?id=1989-062B (retrieved February 27, 2008).
123. The Hipparcos Space Astrometry. ESA. www.rssd.esa.int/index.php?project=HIPPARCOS (retrieved February 28, 2008).
124. Staff writers (March 7, 2009). NASA launches Earth. *BBC News*. news.bbc.co.uk/2/hi/science/nature/7926277.stm (retrieved March 14, 2009).
125. Kepler Mission. NASA. http://www.kepler.arc.nasa.gov (retrieved February 28, 2008).
126. Kepler – About – Mission News. NASA. https://web.archive.org/20070726070942/http://kepler.nasa.gov/about/news.html (retrieved February 28, 2008).
127. NASA – NSSD – Spacecraft – Trajectory Details (MOST). NASA. http://nssdc.gsfc.nasa.gov/nmc/spacecraftOrbit.do?id=2003-031D (retrieved February 27, 2008).

128. NASA – NSSDC – Spacecraft – Details (MOST). NASA. http://nssdc.gsfc.nasa.gov/nmc/spacecraftDisplay.do?id=2003-031D (retrieved February 27, 2008).
129. ESA – Space Science – Gaia overview. ESA. http://www.esa.int/esaSC/120377_index_0_m.html (retrieved February 28, 2008).
130. Cool Cosmos. California Institute of Technology. http://coolcosmos.ipac.caltech.edu/cosmic_classroom/ir_tutorial/importance.html (retrieved March 1, 2008).
131. 赤外線天文衛星「あかり」ASTRO-Fの運用終了について (in Japanese). JAXA. November 24, 2011. http://www.jaxa.jp/press/2011/11/20111124_akari_j.html (retrieved November 24, 2011).
132. NASA – NSSD – Spacecraft – Trajectory Details (Akari). NASA. http://nssdc.gsfc.nasa.gov/nmc/spacecraftOrbit.do?id=2006-005A (retrieved February 27, 2008).
133. NASA – NSSDC – Spacecraft – Details (Akari). NASA. http://nssdc.gsfc.nasa.gov/nmc/spacecraftDisplay.do?id=2006-005A (retrieved February 27, 2008).
134. Jet Propulsion Laboratory Herschel Mission News. Herschel.jpl.nasa.gov. http://herschel.jpl.nasa.gov/newsLatest.shtml (retrieved August 19, 2009).
135. Amos, Jonathan (April 29, 2013). Herschel space telescope finishes mission. *BBC News*. http://www.bbc.co.uk/news/science-environment-21934520 (retrieved April 29, 2013).
136. NASA – NSSDC – Spacecraft – Details (Herschel Space Observatory). NASA. http://nssdc.gsfc.nasa.gov/nmc/spacecraftDisplay.do?id=FIRST (retrieved February 27, 2008).
137. Planck – Home Page. ESA. March 6, 2008. http://www.rssd.esa.int/index.php?project=Planck (retrieved March 15, 2008).
138. Herschel Science Centre – A short Herschel mission overview. ESA. November 20, 2007. http://herschel.esac.esa.int/overview.shtml (retrieved March 15, 2008).
139. NASA – NSSD – Spacecraft – Trajectory Details (IRAS). NASA. http://nssdc.gsfc.nasa.gov/nmc/spacecraftOrbit.do?id=1983-004A (retrieved February 27, 2008).
140. NASA – NSSDC – Spacecraft – Details (IRAS). NASA. http://nssdc.gsfc.nasa.gov/nmc/spacecraftDisplay.do?id=1983-004A (retrieved February 27, 2008).
141. NASA – NSSD – Spacecraft – Trajectory Details (ISO). NASA. http://nssdc.gsfc.nasa.gov/nmc/spacecraftOrbit.do?id=1995-062A (retrieved February 27, 2008).
142. ESA Science & Technology: ISO. ESA. http://sci.esa.int/science-e/www/area/index.cfm?fareaid=18 (retrieved February 28, 2008).
143. IRTS Home Page. Institute of Space and Astronautical Science. http://www.ir.isas.ac.jp/irts/irts_E.html (retrieved February 29, 2008).
144. H. Murakami et al. (1996). The IRTS (Infrared Telescope in Space) Mission. *Publications of the Astronomical Society of Japan* 48(5): L41–L46. doi:10.1093/pasj/48.5.l41.
145. MSX Project Page. Infrared Processing and Analysis Center. http://www.ipac.caltech.edu/ipac/msx/msx.html (retrieved February 29, 2008).
146. NASA – NSSD – Spacecraft – Trajectory Details (Spitzer). NASA. http://nssdc.gsfc.nasa.gov/nmc/spacecraftOrbit.do?id=2003-038A (retrieved February 27, 2008).
147. NASA – NSSDC – Spacecraft – Details (Spitzer). NASA. http://nssdc.gsfc.nasa.gov/nmc/spacecraftDisplay.do?id=2003-038A (retrieved February 27, 2008).
148. NASA – NSSD – Spacecraft – Trajectory Details (SWAS). NASA. http://nssdc.gsfc.nasa.gov/nmc/spacecraftOrbit.do?id=1998-071A (retrieved February 27, 2008).
149. NASA – NSSDC – Spacecraft – Details (SWAS). NASA. http://nssdc.gsfc.nasa.gov/nmc/spacecraftDisplay.do?id=1998-071A (retrieved February 27, 2008).
150. WIRE Web Team. Caltech: WIRE. Ipac.caltech.edu. http://www.ipac.caltech.edu/wire/ (retrieved September 4, 2012).
151. NASA – NSSDC – Spacecraft – Details (WIRE). NASA. http://nssdc.gsfc.nasa.gov/nmc/spacecraftDisplay.do?id=1999-011A (retrieved February 27, 2008).
152. WISE: Wide-Field Infrared Survey Explorer. University of California, Los Angeles, CA. http://wise.ssl.berkeley.edu/ (retrieved March 3, 2008).

153. WISE – Mapping the Infrared Sky (PDF). NASA. http://wise.ssl.berkeley.edu/documents/2007-Fact_Sheet.pdf (retrieved March 3, 2008).
154. Wide-field Infrared Survey Explorer. NASA. http://www.nasa.gov/mission_pages/WISE/main/index.html (retrieved November 30, 2009).
155. NASA – NSSD – Spacecraft – Trajectory Details (COBE). NASA. http://nssdc.gsfc.nasa.gov/nmc/spacecraftOrbit.do?id=1989-089A (retrieved February 27, 2008).
156. NASA – NSSDC – Spacecraft – Details (COBE). NASA. http://nssdc.gsfc.nasa.gov/nmc/spacecraftDisplay.do?id=1989-089A (retrieved February 27, 2008).
157. NASA – NSSD – Spacecraft – Trajectory Details (Odin). NASA. http://nssdc.gsfc.nasa.gov/nmc/spacecraftOrbit.do?id=2001-007A (retrieved February 27, 2008).
158. NASA – NSSDC – Spacecraft – Details (Odin). NASA. http://nssdc.gsfc.nasa.gov/nmc/spacecraftDisplay.do?id=2001-007A (retrieved February 27, 2008).
159. NASA – NSSDC – Spacecraft – Details (PLANCK). NASA. http://nssdc.gsfc.nasa.gov/nmc/spacecraftDisplay.do?id=PLANCK (retrieved February 27, 2008).
160. Planck – Home Page. ESA. http://www.rssd.esa.int/index.php?project=Planck (retrieved February 28, 2008).
161. NASA – NSSDC – Spacecraft – Details (WMAP). NASA. http://nssdc.gsfc.nasa.gov/nmc/spacecraftDisplay.do?id=2001-027A (retrieved February 27, 2008).
162. NASA – NSSD – Spacecraft – Trajectory Details (HALCA). NASA. http://nssdc.gsfc.nasa.gov/nmc/spacecraftOrbit.do?id=1997-005A (retrieved February 27, 2008).
163. NASA – NSSDC – Spacecraft – Details (HALCA). NASA. http://nssdc.gsfc.nasa.gov/nmc/spacecraftDisplay.do?id=1997-005A (retrieved February 27, 2008).
164. Minutes of VSOP-2 Tracking Station Meeting, October 31–November 1, 2006. National Astronomical Observatory of Japan. http://vsop.mtk.nao.ac.jp/VSOP2TSMAPL07/VSOP2-link-Nov06.pdf (retrieved February 28, 2008).
165. Description of the RadioAstron project. Russian Space Research Institute. http://www.asc.rssi.ru/radioastron/description/status_eng.htm (retrieved February 28, 2008).
166. Description of the RadioAstron project – Orbit. Russian Space Research Institute. http://www.asc.rssi.ru/radioastron/description/orbit_eng.htm (retrieved February 28, 2008).
167. RadioAstron Newsletter # 1, January 1, 2010. Russian Astro Space Center. January 1, 2010. http://www.asc.rssi.ru/radioastron/newsletter/index.html (retrieved June 5, 2010).
168. AMS Collaboration et al. (August 2002). The Alpha Magnetic Spectrometer (AMS) on the International Space Station: Part I – results from the test flight on the space shuttle. *Physics Reports* 366(6): 331–405. doi:10.1016/S0370-1573(02)00013-3.
169. Alpha Magnetic Spectrometer – 02 (AMS-02). NASA. December 18, 2009. http://www.nasa.gov/mission_pages/station/science/experiments/AMS-02.html (retrieved December 24, 2009).
170. Ray and Justin (October 19, 2008). Mission Status Center: Pegasus/IBEX. *Spaceflight Now*. http://www.spaceflightnow.com/pegasus/ibex/status.html (retrieved November 27, 2009).
171. PAMELA Mission Official Website. Istituto Nazionale di Fisica Nucleare. http://pamela.roma2.infn.it/index.php?option=com_content&task=view&id=64&Itemid=129 (retrieved March 9, 2008).
172. PAMELA Mission Official Website – Partners. Istituto Nazionale di Fisica Nucleare. http://pamela.roma2.infn.it/index.php (retrieved March 9, 2008).
173. SAMPEX Data Center. Srl.caltech.edu. http://www.srl.caltech.edu/sampex/DataCenter/ (retrieved September 4, 2012).
174. eLISA Consortium. eLISA Mission Description. Max Planck Institute for Gravitational Physics. https://www.elisascience.org/articles/elisa-mission/elisa-mission-gravitational-universe (retrieved January 4, 2015).
175. ASTROSAT – An Indian Multiwave Astronomy Satellite. Inter-University Centre for Astronomy & Astrophysics. http://www.gktoday.in/astrosat/ (retrieved February 28, 2008).

176. National Institute of Aeronautics and Space – LAPAN Republic of Indonesia (2010). Indonesia's equatorial orbit twin satellites for space-based safety application in the disaster mitigation and relief effort, *16th Session of Asia Pacific Regional Space Agency Forum*, Bangkok, Thailand, January 26–29, 2010. http://www.aprsaf.org/data/aprsaf16_data/D3-1130_AP16_CR_LAPAN.pdf (retrieved August 12, 2016).
177. The James Webb Space Telescope. NASA. http://www.jwst.nasa.gov (retrieved February 28, 2008).
178. TAUVEX: UV Astronomy Mission. Indian Institute of Astronomy. https://prints.iiap.res.in/handle/2248/4706 (retrieved February 28, 2008).
179. Institute of High Energy Physics, Chinese Academy of Sciences. Space missions - hard X-ray modulation telescope (HXMT). http://english.ihep.cas.cn/rs/fs/sm/SM/ (retrieved October 03, 2016).
180. Nuclear and Corpuscular Physics Department (DPNC), Geneva, Switzerland. DArk Matter Particle Explorer (DAMPE). http://dpnc.unige.ch/dampe/ (retrieved August 22, 2016).
181. Advanced Technology Large-Aperture Space Telescope. Space Telescope Science Institute. https://arxiv.org/ftp/arxiv/papers/0904/0904.0941.pdf (retrieved October 8, 2010).
182. KEUS – The X-Ray Evolving Universe. ESA. http://www.rssd.esa.int/index.php (retrieved February 28, 2008).
183. Official NASA IXO Home Page. NASA. http://ixo.gsfc.nasa.gov/ (retrieved February 28, 2008).
184. NASA – NSSDC – Spacecraft – Details (Spectrum-X-Gamma). NASA. http://nssdc.gsfc.nasa.gov/nmc/spacecraftDisplay.do?id=1966-027A (retrieved February 27, 2008).
185. SIM Lite JPL. NASA. http://www.northropgrumman.com/Capabilities/SIMLite/Pages/default.aspx (retrieved March 19, 2009).
186. ESA Science & Technology: Darwin. ESA. http://sci.esa.int/darwin/40844-darwin-mission-summary-status/ (retrieved February 28, 2008).
187. Planet Quest: Missions – Terrestrial Planet Finder. NASA. http://www.nasa.gov/multimedia/imagegallery/image_feature_706.html (retrieved March 3, 2008).
188. Dark Universe Observatory. Sonoma State University. ftp://ftp.mpe.mpg.de/pub/publication/Predehl/DUO_Proposal_DLR_2004.pdf (retrieved February 29, 2008).
189. Dark Universe Observatory – About the Launch Vehicle and Orbit. Sonoma State University. ftp://ftp.mpe.mpg.de/pub/publication/Predehl/DUO_Proposal_DLR_2004.pdf (retrieved February 2, 2008).
190. Destiny JDEM Mission Public Page. National Optical Astronomy Observatory. http://www.noao.edu/noao/staff/lauer/destiny.htm (retrieved February 28, 2008).
191. O. Doré et al. (2014). SPHEREx: An All-Sky Spectral Survey. arXiv:1412.4872 [astro-ph.CO].
192. $1.6 Billion Telescope Would Search Alien Planets and Probe Dark Energy. SPACE. http://www.space.com/8944-1-6-billion-telescope-search-alien-planets-probe-dark-energy.html (retrieved October 9, 2010).
193. Astro2010 Report Release Presentation (PDF). NASA. http://sites.nationalacademies.org/bpa/bpa_049810 (retrieved October 9, 2010).
194. NASA – NSSD – Spacecraft – Trajectory Details (Astromag FF). NASA. http://nssdc.gsfc.nasa.gov/nmc/spacecraftDisplay.do?id=ASTRMAG (retrieved February 27, 2008).
195. NASA – NSSDC – Spacecraft – Details (Astromag-FF). NASA. http://nssdc.gsfc.nasa.gov/nmc/spacecraftDisplay.do?id=ASTRMAG (retrieved February 27, 2008).
196. VSOP-2 project. JAXA. http://www.vsop.isas.jaxa.jp/vsop2/ (retrieved February 28, 2008).
197. Public Telescope. http://www.publictelescope.org/language/en/ (retrieved August 29, 2016).

21

The Pluto Missions—History and Findings

21.1 Introduction

In January 2006, the New Horizons mission was launched as an interplanetary space probe as part of the NASA New Frontiers program. Johns Hopkins University Applied Physics Laboratory (APL) and the Southwest Research Institute (SwRI) engineered the spacecraft. It was launched mainly to study Pluto, its moons, and the Kuiper belt, and also to perform flybys of the Pluto system and other Kuiper belt objects. The spacecraft is the result of many years of work on missions to send a spacecraft to Pluto.

The Dawn mission initially had a projected cost of US$373 million, but the cost overruns inflated the final cost to US$446 million in 2007 [1]. The Dawn mission team started work under the leadership of Christopher T. Russell.

The New Frontiers program was established for missions that fit in between the big budgets of the Flagship Program and the low budgets of the Discovery Program.

21.2 Dawn Spacecraft

The Dawn mission is the first NASA exploratory mission to use ion propulsion, which helped it to enter and leave the orbit of multiple celestial bodies. The Dawn mission studied two of the three known protoplanets of the asteroid belt, Vesta and Ceres. Currently, it is in orbit around its second target, the dwarf planet Ceres. Dawn is the first spacecraft to orbit two extraterrestrial bodies [2]. It entered the orbit of Vesta on July 16, 2011, and completed a 14-month survey mission before leaving for Ceres in late 2012 [3,4]. Dawn entered Ceres orbit on March 6, 2015 [5,6]. Table 21.1 presents Dawn mission particulars.

The Dawn mission was designed for investigating two large bodies in the asteroid belt with a view to get an idea about the formation of the solar system, and additionally, to test the performance of its ion drive in deep space. Ceres and Vesta were selected as two contrasting protoplanets, the first one apparently "wet" that is, icy and cold while the other "dry" that is, rocky, whose accretion was terminated by the formation of Jupiter. The two bodies bridge the understanding of the science between the formation of rocky planets and the icy bodies of the solar system, and under what conditions a rocky planet can hold water [12]. Figure 21.1 shows scale comparison of Vesta, Ceres, and Earth's moon.

The International Astronomical Union (IAU) adopted a new terminology defining a planet on August 24, 2006, which used the name "dwarf planet." This term refers to bodies that are too small to qualify for planetary status. Dawn arrived at Ceres a few months

TABLE 21.1

Dawn Mission Particulars

Mission Type	Multitarget Orbiter
Operator	NASA
COSPAR ID	2007-043A
Website	http://dawn.jpl.nasa.gov/
Mission duration	~9 years [7]
Spacecraft Properties	
Manufacturer	Orbital Sciences · JPL (Jet Propulsion Laboratory) · UCLA (University of California, Los Angeles, CA)
BOL mass	1240 kg (2730 lb) (wet) [8]
Power	1300 W (Solar array) at 3 AU [8]
Start of Mission	
Launch date	September 27, 2007, 11:34:00 UTC [9] (7 years, 10 months, and 15 days ago)
Rocket	Delta II 7925H
Launch site	Space Launch Complex 17B Cape Canaveral Air Force Station, Florida
Flyby of Mars (Gravity Assist)	
Closest approach	February 4, 2009 (6 years, 6 months, and 7 days ago)
Distance	549 km (341 mi)
4 Vesta Orbiter	
Orbital insertion	July 16, 2011, 04:47 UTC [10] (4 years and 26 days ago)
Departed orbit	September 5, 2012 (2 years, 11 months, and 6 days ago)
1 Ceres Orbiter	
Orbital insertion	March 6, 2015 [11]

FIGURE 21.1
Scale comparison of Vesta, Ceres, and Earth's moon. (Courtesy of Gregory H. Revera, NASA/JPL-Caltech/UCLA/MPS/DLR/IDA and Wikipedia, https://commons.wikimedia.org/wiki/File:Ceres,_Earth_%26_Moon_size_comparison.jpg [last modified on December 13, 2015].)

FIGURE 21.2
Dawn image of Ceres from 13,600 km, May 4, 2015.

before the arrival of the New Horizons probe at Pluto in July 2015. Figure 21.2 shows a Dawn image of Ceres from 13,600 km as observed on May 4, 2015.

Data has shown that Ceres spectral characteristics suggest a composition similar to that of a water-rich carbonaceous chondrite. Vesta, a smaller, water-poor achondritic asteroid comprising a tenth of the mass of the asteroid belt, has experienced significant heating and differentiation. It shows signs of a metallic core, a Mars-like density, and lunar-like basaltic flows [13]. Available information suggests that both bodies formed very early in the history of the solar system and processes from the time of the formation of the terrestrial planets. Radionuclide dating of pieces of meteorites assumed to come from Vesta indicates that Vesta differentiated quickly, in three million years or less. Investigations of thermal evolution suggest that Ceres had formed some time later, more than three million years after the formation of CAIs (calcium–aluminum-rich inclusions), the oldest known objects of solar system origin [14]. Further, Vesta appears to be the source of many smaller objects in the solar system. Most V-type near-Earth asteroids and some outer main-belt asteroids have spectra similar to Vesta, and are called "vestoids." Five percent of the meteoritic samples seen on Earth, the howardite–eucrite–diogenite (HED) meteorites, are assumed to be the result of collisions with Vesta.

It is assumed that Ceres may have a differentiated interior [15]. Its oblateness appears very small for an undifferentiated body, which suggests that it consists of a rocky core overlain with an icy mantle [16]. There is a huge collection of potential samples from Vesta accessible to scientists, in the form of over 1400 HED meteorites [17], giving insight into Vesta's geologic history and structure. Vesta is considered to consist of a metallic iron–nickel core, an overlying rocky olivine mantle, and crust [18–20].

21.2.1 Objectives

The Dawn mission's target is to characterize the conditions and methods of the solar system by studying two of the largest protoplanets remaining intact since their formation [21]. The mission addresses the role played by size and water as to determining the evolution of the planets [21,22]. Dawn's journey to date is shown in Figure 21.3.

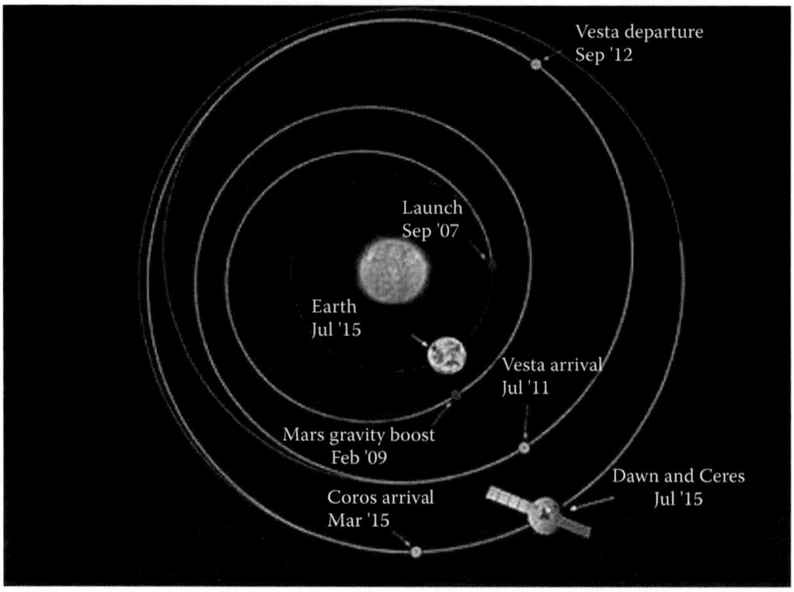

FIGURE 21.3
Dawn's Journey. (Courtesy of NASA/JPL and Marc Rayman, http://www.planetary.org/image/dawn_trajectory_20090900.png [retrieved August 12, 2016].)

FIGURE 21.4
A Dawn image of Vesta from orbit, taken on July 17, 2011. (Courtesy of NASA, Dawn Spacecraft Snaps Close-up Image of Asteroid Vesta, http://science.nasa.gov/science-news/science-at-nasa/2011/18jul_dawn4/ [retrieved August 12, 2016].)

The contrasting characteristics of Ceres and Vesta are considered to have resulted from them forming in two different regions of the early solar system [22]. A Dawn image of Vesta from orbit, taken on July 17, 2011, is presented in Figure 21.4.

There are three principal scientific drivers for the mission. First, the Dawn mission can capture the earliest moments in the origin of the solar system, the second is to determine the nature of the building blocks from which the terrestrial planets formed and the third it contrasts the formation and evolution of two small planets that followed very different evolutionary paths, permitting us to find what factors control that evolution [22].

FIGURE 21.5
Dawn prior to encapsulation at its launch pad on July 1, 2007.

21.2.2 Dimensions

With its solar array in the retracted launch position, it is 2.36 m long while with its solar arrays is 19.7 m [23]. The solar arrays have a total area of 36.4 m^2 [24]. Dawn prior to encapsulation at its launch pad on July 1, 2007, is shown in Figure 21.5.

21.3 Propulsion System

The Dawn spacecraft is propelled by three xenon ion thrusters based on an evolution of the NSTAR technology utilized by the Deep Space 1 spacecraft [25]. They have a specific impulse of 3100 s and create a thrust of 90 mN [26]. The whole spacecraft is powered by a 10 kW triple-junction gallium arsenide photovoltaic solar array [27,28]. *Dawn* was allocated 275 kg of xenon for its Vesta approach and carried another 110 kg to reach Ceres [29], out of a total capacity of 425 kg of on-board propellant [30]. With the propellant it carries, Dawn can perform a velocity change of more than 10 km/s [29].

21.3.1 Payload

The Jet Propulsion Laboratory provided overall planning and management of the mission, flight system, and payload development. It also provided the ion propulsion system. A camera view of the Ceres bright spots is shown in Figure 21.6.

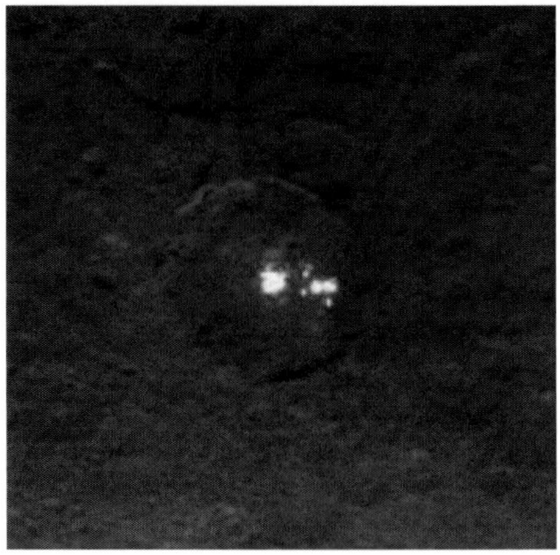

FIGURE 21.6
A camera view of the Ceres bright spots.

21.4 Transit from Earth to Vesta

After preliminary testing, during which the ion thrusters accumulated more than 11 days, 14 h of thrust, Dawn started long-term cruise on December 17, 2007 [31]. On October 31, 2008, Dawn covered its first thrusting phase sending it on to Mars for a gravity assist flyby. During this first interplanetary cruise phase, Dawn spent 270 days, or 85% of this phase, using its thrusters. It expended nearly 72 kg of xenon propellant for a total change in velocity of 1.81 km/s. On November 20, 2008, Dawn made its first trajectory correction maneuver (TCM1), firing its number 1 thruster for 2 h, 11 min. Dawn made its closest approach of 549 km to Mars on February 17, 2009, during a successful gravity assist [32,33]. This flyby slowed the orbital speed of Mars by about 2.3 cm over 180 million years. A greyscale NIR image of Mars taken by Dawn during its 2009 flyby is shown in Figure 21.7.

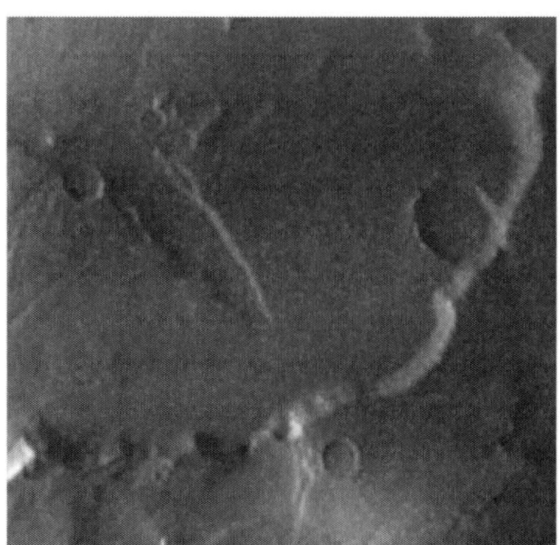

FIGURE 21.7
Grayscale NIR image of Mars taken by Dawn during its 2009 flyby.

21.5 Vesta Approach

As Dawn approached Vesta, the on-board camera took high-resolution images (Figure 21.8). These photographs were published online and at NASA and MPI news conferences.

On May 3, 2011, Dawn acquired its first targeting image, 1,200,000 km from Vesta and started its approach phase to the asteroid [34]. On June 12, 2011, Dawn's speed relative to Vesta was slowed in preparation for its orbital insertion 34 days later [35,36]. NASA later confirmed that it received telemetry from Dawn suggesting that the spacecraft entered orbit around Vesta [37].

21.5.1 Vesta Orbit

After being captured by the gravity of Vesta and entering its orbit on July 16, 2011 [38], Dawn moved to a lower, closer orbit by running its xenon-ion engine using solar power. On August 2, it paused its spiraling approach to enter a 69-h survey orbit at an altitude of 2750 km. It assumed a 12.3-h high-altitude mapping orbit at 680 km on September 27 and ultimately entered a 4.3-h low-altitude mapping orbit at 210 km on December 8 [39–41]. After being captured by Vesta's gravity and entering its orbit on July 16, 2011, Dawn moved to a lower, closer orbit and took photographs as shown for July 17, 18, 23, and 24, 2011 (Figure 21.9).

In May 2012, NASA released the preliminary results derived from Dawn's investigation of Vesta. In addition, an estimate of the size of Vesta's metal-rich core, which is theorized to be 220 km across. It was also stated that the Vesta is the "last of its kind"—the only remaining example of the large planetoids which came together to form the rocky planets during the formation of the solar system [42–44]. In October 2012, NASA further stated that the data from Dawn had shown the origin of anomalous dark spots and streaks on Vesta's surface, which were likely deposited by ancient asteroid impacts [45–47]. In December 2012, it

FIGURE 21.8
Vesta. (Courtesy of NASA, http://www.nasa.gov/mission_pages/dawn/multimedia/pia15678.html [retrieved August 12, 2016].)

FIGURE 21.9
Closer and closer to the Vesta surface (high resolution). (Courtesy of NASA, http://photojournal.jpl.nasa.gov/target/Vesta?start=200 [retrieved September 10, 2016].)

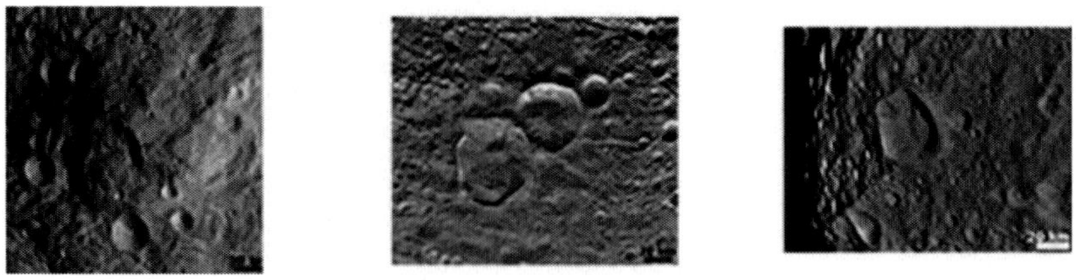

Central mound at the S-pole on Vesta — Snowman-shaped craters on Vesta — Craters and ridges on Vesta

FIGURE 21.10
Physical features of Vesta's surface.

was reported that Dawn had found gullies on the surface of Vesta which were interpreted to have been eroded by transiently flowing liquid water [48,49].

Dawn was initially scheduled to depart Vesta and start its two and a half year journey to Ceres on August 26, 2012 [4]. However, due to a problem with one of the wheels of the spacecraft, Dawn's departure from Vesta's gravity was forcibly delayed until September 5, 2012 [50–53]. Some scientific discoveries at asteroid Vesta showing different physical conditions are presented in Figure 21.10.

21.5.2 Transit from Vesta to Ceres

During its period in orbit around Vesta, the probe experienced many failures of its reaction wheels. Scientists planned to change their activities upon arrival at Ceres for close range geographical survey mapping. The Dawn team described that they would orient the probe using a "hybrid" mode, which would conserve fuel using both reaction wheels and ion thrusters. On November 13, 2013, during the transit Dawn completed a 27-h-long series of exercises of the said hybrid mode [54]. In order to avoid a lapse in propulsion, the mission team exchanged the active ion engine and electrical controller with another (all manipulations were done remotely from earth). The team had a plan in place to revive this disabled component later in 2014. Upon exiting the safe mode on September 15, 2014, the ion thruster of the probe resumed normal operation [55]. Dawn investigators also noticed that after the propulsion issue, it could not aim its main communications antenna toward Earth. To correct the problem, the probe's computer was reset and the aiming mechanism of the main antenna was restored. Imaging dates (2014–2015) and corresponding distances, diameter, resolution, and portion of disk illuminated [54,55] have shown in Table 21.2.

21.6 Approach to Ceres

The initiation for photographing the extended disk of Ceres began on December 1, 2014 [56], with images of partial rotations on January 13 and 25, 2015 (first released as animations). Images obtained of Ceres after January 26, 2015, exceeded the resolution of comparable images from the Hubble Space Telescope [57]. Progression of images of Ceres taken between January and March 2015 can be seen in Figure 21.11.

TABLE 21.2
Imaging Dates (2014–2015) and Corresponding Distances, Diameter, Resolution, and Portion of Disk Illuminated

Date	Distance (km)	Diameter (px)	Resolution (km/px)	% of Disk Illuminated
December 1, 2014	1,200,000	9	112	94%
January 13, 2015	383,000	27	36	95%
January 25, 2015	237,000	43	22	96%
February 2, 2015	146,000	70	14	97%
February 12, 2015	83,000	122	7.8	98%
February 19, 2015	46,000	222	4.3	87%
February 25, 2015	40,000	255	3.7	44%
March 1, 2015	49,000	207	4.6	23%
April 10, 2015	33,000	306	3.1	17%
April 15, 2015	22,000	453	2.1	49%

Source: NASA/JPL, Dawn fills out its Ceres dance card, December 3, 2013. http://www.jpl.nasa.gov/news/news.php?release=2013-347 (retrieved March 18, 2015); NASA/JPL, Dawn operating normally after safe mode triggered, September 16, 2014. https://web.archive.org/20141225012242/http://dawn.jpl.nasa.gov/feature_stories/Dawn_operated_normally.asp (retrieved March 18, 2015).

FIGURE 21.11
Photographic images of Ceres.

Owing to the failure of two reaction wheels, Dawn made fewer camera observations of Ceres during its approach phase than it made during the Vesta approach. For camera observations of the spacecraft, we use thrusters, resulting in further hydrazine fuel consumption.

21.6.1 Ceres Orbit

Dawn entered Ceres orbit on March 6, 2015, 4 months before the arrival of New Horizons at Pluto and thus became the first mission to observe a dwarf planet at close range [58,59]. Initially it entered a polar orbit around Ceres and continued to refine its orbit. It achieved its first full topographic map of Ceres during this period [60]. Mapping orbits (2015) and resolution are shown in Table 21.3.

During the period of April 23–May 9, Dawn entered a "rotation characterization 3" (RC3) orbit at an altitude of 13,500 km. This orbit lasted 15 days, during which Dawn alternated between taking pictures and sensor measurements which were then relayed to Earth [62]. On May 9, 2015, Dawn powered up its ion engines and started a month long spiral descent down to its second mapping point at an altitude of 4430 km, (three times closer to Ceres than the previous orbit). On June 30, 2015, Dawn experienced a software glitch because of an anomaly in its orientation system. Engineers identified the cause of the anomaly to be related to the mechanical gimbal system connected with one of Dawn's ion engines. Switching to a separate ion engine and conducting tests from July 14

TABLE 21.3
Mapping Orbits (2015) and Resolution

Orbit Phase	No.	Dates [62]	Altitude (km)	Orbital Period	Resolution (km/px)	Improvement Over Hubble
RC3	1st	April 23–May 9, 2015	13,500	15 days	1.3	24×
Survey	2nd	June 6–30, 2015	4,400	3.1 days	0.41	73×
HAMO	3rd	August 17–October 23	1,450	19 h	0.14 (140 m)	217×
LAMO	4th	December 16–end of mission	375	5.5 h	0.035 (35 m)	850×

Source: NASA, *Dawn journal*, March 31, 2015. http://dawnblog.jpl.nasa.gov/2016/03/31/dawn-journal-march-31-3/ (retrieved April 20, 2015).

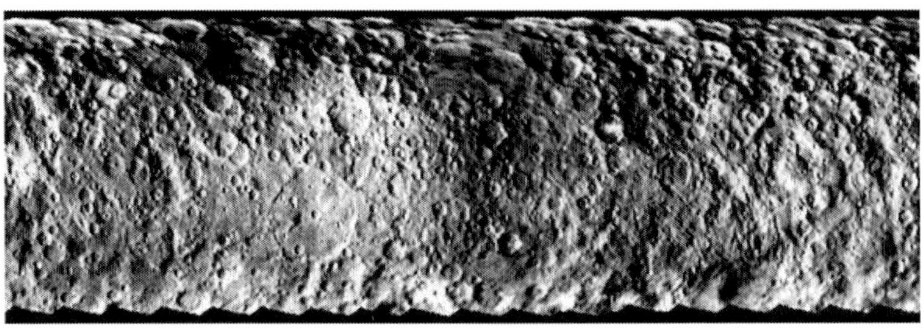

FIGURE 21.12
The first map of Ceres by Dawn (exaggerated color, March 2015).

to July 16, engineers certified the ability to continue the mission. As of July 17, 2015, Dawn maneuvered to its third science orbit phase [63]. The first color map of Ceres by Dawn is shown in Figure 21.12.

It is likely that Dawn will become a perpetual satellite of Ceres when the mission is over, due to its highly stable orbit. A flyby of the asteroid 2 Pallas after the completion of the Ceres mission was suggested but would not be possible for Dawn, because of the high inclination of Pallas' orbit relative to Ceres [64,65]. Two important features on Ceres are shown in Figures 21.13 and 21.14. Out of the two features, Figure 21.13 reveals bright "Spot 5"

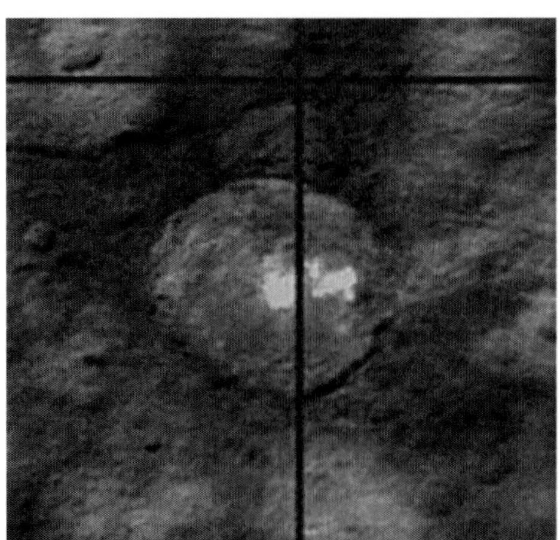

FIGURE 21.13
Reveals prominent "Spot 5" in Occator (crater) (July 28, 2015).

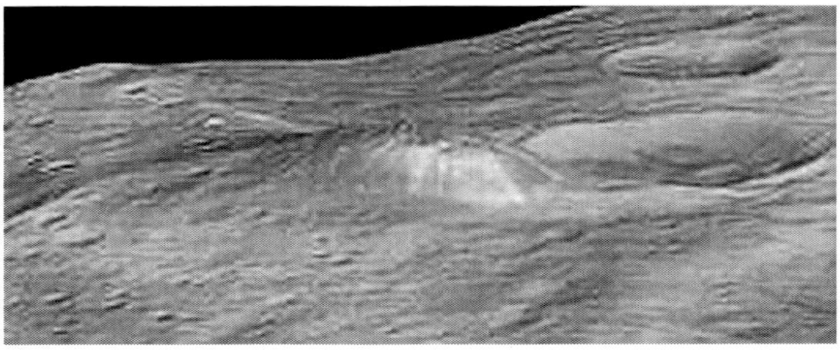

FIGURE 21.14
Reveals elevation map of the mountain and the impact crater seen at the bottom of the preceding image. (Courtesy of NASA, http://www.nasa.gov/image-feature/jpl/pia19976/topographic-view-of-ceres-mountain [retrieved August 12, 2016].)

in Occator (crater). On the other hand, Figure 21.14 reveals the elevation map of the mountain and the impact crater seen at the bottom of the preceding image.

21.7 New Horizons—Interplanetary Space Probe

New Horizons is an interplanetary space probe which was launched as part of New Frontiers program of NASA [66]. Johns Hopkins University APL and the SwRI engineered the spacecraft. It was launched mainly to study Pluto, its moons, and the Kuiper belt, and also to perform flybys of the Pluto system and other Kuiper belt objects [67–70]. The spacecraft is the result of many years of work on missions to send a spacecraft to Pluto.

The proposed mission named "Pluto Kuiper Express" would be a lightweight, cost-effective spacecraft for the further study of Pluto and the Kuiper belt. Figure 21.15 shows a photograph of the New Horizons space probe. Some particulars of Pluto flyby are shown in Table 21.4.

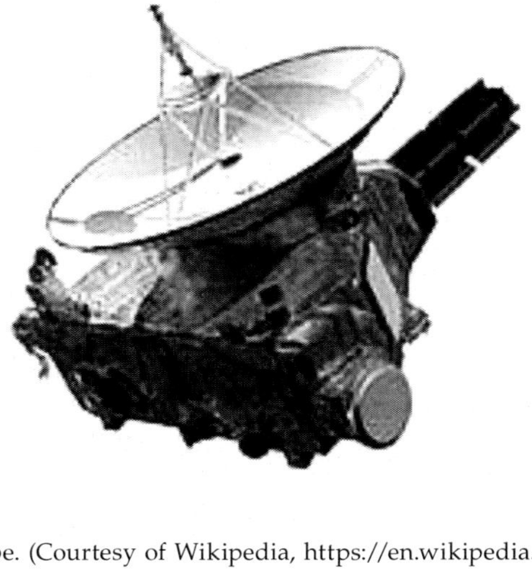

FIGURE 21.15
The New Horizons space probe. (Courtesy of Wikipedia, https://en.wikipedia.org/wiki/New_Horizons [last modified on August 26, 2016].)

TABLE 21.4

Particulars of Pluto Flyby

Identification	
Operator	NASA
COSPAR ID	2006-001A
SATCAT No	28928
Website	pluto.jhuapl.edu www.nasa.gov
Mission duration	Primary mission: 9.5 years
Spacecraft Properties	
Manufacturer	APL · Southwest Research Institute
Launch mass	478 kg (1054 lb)
Power	228 W
Start of Mission	
Launch date	January 19, 2006, 19:00 UTC (9 years, 6 months, and 23 days ago)
Rocket	Atlas V 551
Launch site	Cape Canaveral SLC-41
Contractor	International Launch Services
Flyby of Moon	
Closest approach	January 20, 2006, 04:00 UTC (9 years, 6 months, and 22 days ago)
Distance	189,916 km (118,008 mi)
Flyby of (132524) APL (Incidental)	
Closest approach	June 13, 2006 04:05 UTC (9 years, 1 month, and 29 days ago)
Distance	101,867 km (63,297 mi)
Flyby of Jupiter (Gravity Assist)	
Closest approach	February 28, 2007, 05:43:40 UTC (8 years, 5 months, and 14 days ago)
Distance	2,300,000 km (1,400,000 mi)
Flyby of Pluto	
Closest approach	July 14, 2015, 11:49:57 UTC (28 days ago)
Distance	12,500 km (7,800 mi)

After a short encounter with asteroid 132524 APL, New Horizons proceeded to Jupiter, making its closest approach on February 28, 2007, at a distance of 2.3 million kilometers. The Jupiter flyby provided a gravity assist which increased New Horizons speed by 14,000 km/h. The encounter was used to examine the capabilities of the spacecraft, returning data about the planet's atmosphere, moons, and magnetosphere.

On January 15, 2015, the New Horizons spacecraft started its approach phase to Pluto. On July 14, 2015, it flew 12,500 km above the surface of Pluto [71,72], making it the first spacecraft to explore the dwarf planet [73]. Thirteen hours later [74] NASA received the first communication from the probe following a flyby at the time expected. Data showed that the flyby was successful and the probe operated in all respects as thought [75–77].

Note: Dawn is the first spacecraft to orbit two extraterrestrial bodies, the first spacecraft to visit either Vesta or Ceres, and also the first to visit a dwarf planet, arriving at Ceres in March 2015, a few months before New Horizons flew by Pluto in July 2015.

21.8 Earlier Pluto Missions—The Pluto Underground

In May 1989, a group of scientists and engineers formed an alliance called the "Pluto Underground." The group began a letter writing campaign which aimed to bring to attention Pluto as a viable target for exploration [78]. In 1990, engineers at NASA decided to look into concepts for a mission to Pluto. It was thought that the atmosphere of Pluto would freeze and fall to the surface during winter and so a lightweight spacecraft was necessary as it would be able to reach Pluto before such an event would occur. One of the many early concepts for a mission to Pluto was to send a Mariner Mark II spacecraft which, however, later ruled out in favor of a smaller, less expensive spacecraft similar to the Pluto 350 concept (Figure 21.16).

The Pluto 350 was a smaller spacecraft, weighing 35–50 kg [79]. The minimalistic design of the spacecraft would allow it to move faster and be more cost-effective, in comparison to most other big-budget projects NASA were developing at the time, such as Galileo and Cassini. However, Pluto 350 would soon become controversial among NASA mission planners, who considered the project to be too small and too risky. An alternate plan was also considered utilizing a different configuration, namely, the Mariner Mark II spacecraft, which would weigh 2000 kg and cost US$3.2 billion [80]. When both the projects competed for approval, Pluto 350 was more favored by NASA mission planners, who adopted smaller missions like Mars Pathfinder and NEAR Shoemaker [79].

In 1992, Staehle with the help of JPL engineers and students from the California Institute of Technology formed the Pluto Fast Flyby project. The mission heralded the same ideology as the Pluto 350 concept: small in size and cost-effective in scope, so that the spacecraft would be able to get to Pluto faster and be affordable to develop and launch. Described as a "radical" mission concept, the mission would see two spacecraft being sent to Pluto. Both spacecraft were to weigh only around 35–50 kg each (including 7 kg worth of scientific instruments), and the project would cost less than US$500 million to develop, excluding launch costs. Described by Staehle as a "faster, better, and cheaper" approach than the

FIGURE 21.16
Pluto 350 concept.

FIGURE 21.17
New Horizons at the Kennedy Space Center in 2005. (Courtesy of NASA, http://mediaarchive.ksc.nasa.gov/detail.cfm?mediaid=27362 [retrieved August 12, 2016].)

Pluto 350 and Mariner Mark II projects, it caught the attention of then-NASA administrator Daniel S. Goldin, who ordered all work on both Pluto 350 and Mariner Mark II to cease and shift all resources to the new Pluto Fast Flyby project instead.

21.8.1 Design and Construction of the Spacecraft Subsystems

The body of the New Horizons forms a triangle of nearly 0.76 m thick. It may be mentioned that the Pioneers have hexagonal bodies while the Voyagers, Galileo, and Cassini–Huygens have decagonal, hollow bodies. A 7075 aluminum alloy tube forms the main structural column, between the launch vehicle adapter ring at the "rear," and the 2.1 m radio dish antenna affixed to the "front" flat side [80]. The titanium fuel tank lies in this tube. The structure is larger than strictly required, with empty space inside. The structure is designed for shielding and reducing damage to electronics due to radiation. The mass distribution for a spinning spacecraft demands a wider triangle. Figure 21.17 shows the spacecraft at the Kennedy Space Center in 2005.

The interior structure is painted black for equalizing temperature by radiative heat transfer. Unlike the Pioneers and Voyagers, the radio dish of the spacecraft is also enclosed in blankets that extend to the body.

21.9 Pluto System Encounter

As mentioned earlier, the closest approach of the New Horizons spacecraft to Pluto occurred on July 14, 2015, at a range of 12,472 km from the surface [81] and 13,658 km from the center of Pluto. Figure 21.18 shows a concept drawing of the Pluto encounter.

Telemetry data confirming a successful flyby were received on Earth from the vicinity of the Pluto system on July 15, 2015 [82] (after 22 h of planned radio silence owing to the spacecraft being pointed toward the Pluto system). The download of all the data and analysis of the same took longer time.

Mission managers determined a 1 in 10,000 chance that debris could have destroyed it during the course of flyby, preventing it from sending data to Earth [83]. Figure 21.19a shows Pluto as viewed by New Horizons on July 13, 2015, while Figure 21.19b shows Charon as viewed by New Horizons on July 13, 2015.

FIGURE 21.18
Artist's impression of New Horizons' close encounter with the Pluto–Charon system. (Courtesy of NASA/JHU APL/SwRI and Steve Gribben, http://www.nasa.gov/sites/default/files/15-011a.jpg [retrieved August 12, 2016].)

(a) (b)

FIGURE 21.19
(a) Pluto viewed by New Horizons on July 13, 2015. (Courtesy of NASA, http://www.nasa.gov/sites/default/files/thumbnails/image/071215_pluto_alone_0.png [retrieved August 12, 2016].) (b) Charon viewed by New Horizons on July 13, 2015. (Courtesy of NASA/Johns Hopkins University Applied Physics Laboratory/Southwest Research Institute and Alan Stern, Principal investigator, http://pluto.jhuapl.edu/News-Center/Press-Conferences/July-24-2015.php [retrieved August 12, 2016].)

21.9.1 Mission Objectives

The objectives of the mission are grouped in three distinct priorities as summarized below:

- *Primary objectives (required)*
  - Characterize the global geology and morphology of Pluto and Charon
  - Characterize the neutral (nonionized) atmosphere of Pluto and its escape rate
  - Map chemical compositions of Pluto and Charon surfaces
- *Secondary objectives (expected)*
  - Characterize the time variability of Pluto's surface and atmosphere
  - Characterize Pluto's ionosphere (upper layer of the atmosphere) and its interaction with the solar wind

- Search for neutral species such as molecular hydrogen, hydrocarbons, hydrogen cyanide, and other nitriles in the atmosphere
- Search for any Charon atmosphere
- Determine bolometric bond albedos for Pluto and Charon
- Map surface temperatures of Pluto and Charon
- Map any additional surfaces of outermost moons: Nix, Hydra, Kerberos, and Styx
- Image select Pluto and Charon areas in stereo
- Map the terminators (day/night border) of Pluto and Charon with high resolution
- Map the chemical compositions of select Pluto and Charon areas with high resolution

- *Tertiary objectives (desired)*
  - Characterize the energetic particle environment at Pluto and Charon
  - Search for additional moons and any rings
  - Refine bulk parameters (radii, masses) and orbits of Pluto and Charon

21.10 Additional Mission Facts

The New Horizons spacecraft mission was a success. Here are some of the highlights:

- Fastest launch ever recorded when it blasted off in January 2006
- It passed the Moon after just 9 h and reached Jupiter the following year
- Took only three minutes to cross the diameter of Pluto
- The first images beamed back reveal that the diameter of Pluto was 2370 km

Pluto is slightly larger than previously thought. Pluto is less dense than thought; meaning it probably contains more ice beneath the surface. A photo of Pluto obtained by the New Horizons spacecraft is shown in Figure 21.20.

Sensors on New Horizons have identified a thin nitrogen atmosphere extending far out into space and the surface coloring indicates a patchwork of different concentrations of frozen methane and nitrogen. These perhaps sublime into the atmosphere when the planet is closer to the sun and drift back to the ground as snowfall during colder periods.

FIGURE 21.20
Photo of Pluto obtained by the New Horizons spacecraft. (Courtesy of NASA/Johns Hopkins University Applied Physics Laboratory/Southwest Research Institute, http://pluto.jhuapl.edu/News-Center/Press-Conferences/July-24-2015.php [retrieved August 12, 2016].)

21.11 Pluto Flyby: Complete Coverage

The New Horizons probe has lifted the veil on Pluto. On July 14, 2015, it performed the first-ever flyby of the faraway dwarf planet, zooming within 12,500 km of its frigid surface. The close encounter is giving researchers their first up-close looks at Pluto, which has remained mysterious since its 1930 discovery by Clyde Tombaugh. The spacecraft will continue beaming flyby data home for months to come and it may cruise past the Kuiper belt in 2019, if NASA approves and funds a proposed mission extension. Figure 21.21 shows the Pluto flyby trajectories on July 14, 2015, while Figure 21.22 reveals icy mountains on Pluto on the same date.

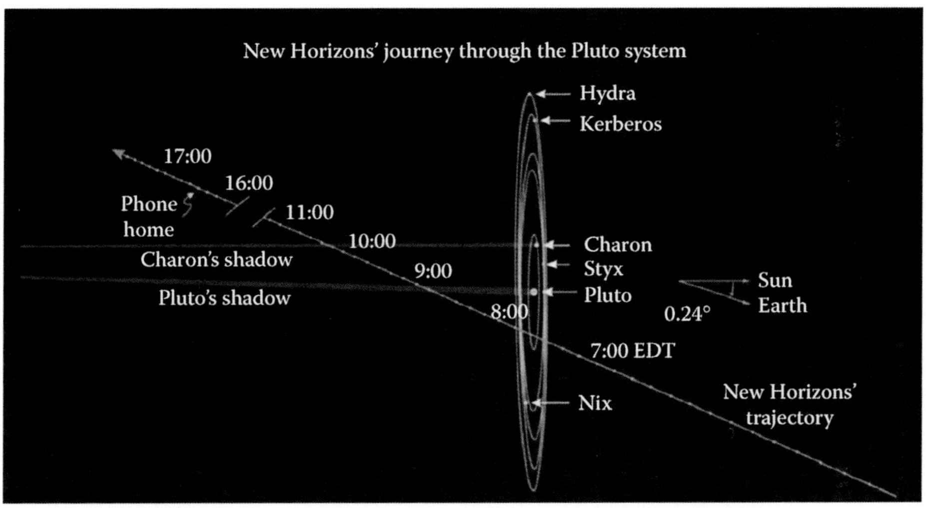

FIGURE 21.21
Pluto trajectory flyby on July 14, 2015. (Courtesy of NASA/JHUAPL, http://pluto.jhuapl.edu/Mission/Where-is-New-Horizons/index.php [retrieved August 12, 2016].)

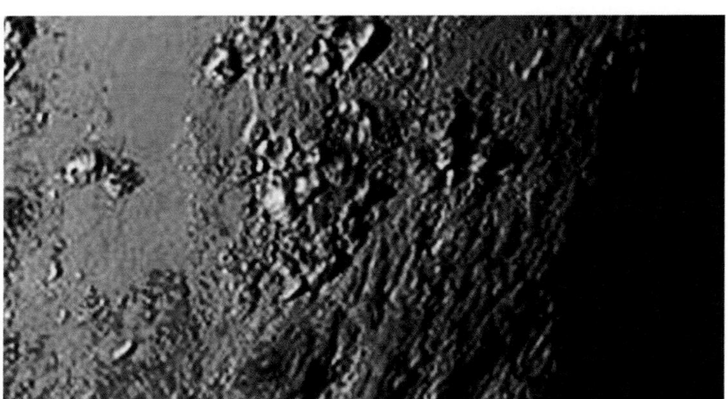

FIGURE 21.22
Icy mountains on Pluto, July 14, 2015. (Courtesy of NASA, https://www.nasa.gov/image-feature/the-icy-mountains-of-pluto [retrieved August 12, 2016].)

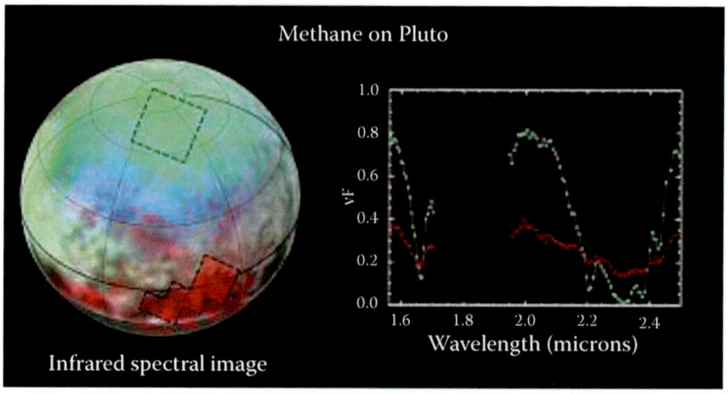

FIGURE 21.23
Distribution of methane ice on Pluto (July 12, 2015).

Reflected sunlight from Charon will permit some imaging observations of the night side. Also backlighting by the Sun will highlight any rings or atmospheric hazes. REX (Radio Science Experiment) will perform radiometry of the night side. Distribution of methane ice on Pluto (July 12, 2015) is shown in Figure 21.23.

Review Questions

21.1 Write a brief technical description on the *Dawn* spacecraft. What are its objectives? Mention the dimensions of the *Dawn* spacecraft.

21.2 Describe the propulsion system and write a brief description on the payload.

21.3 Discuss the transit of the Dawn spacecraft from Earth to Vesta. Also explain the orbits of Vesta and Ceres.

21.4 Describe the interplanetary space probe, New Horizons.

21.5 Make a comparison of the Pluto Underground, Pluto 350, and Mariner Mark II.

21.6 Write a brief technical description of the design and construction of the spacecraft subsystems in Problem 21.5.

21.7 Briefly explain the Pluto System Encounter.

Definition of Terms	
Achondritic	An achondrite is a stony meteorite that does not contain chondrules. It consists of materials similar to terrestrial basalts or plutonic rocks and has been differentiated and reprocessed to a lesser or greater degree due to melting and recrystallization on or within meteorite parent bodies. As a result, achondrites have distinct textures and mineralogies indicative of igneous processes.

(Continued)

Definition of Terms	
	Achondrites account for about 8% of meteorites overall, and the majority (about two-thirds) of them are HED meteorites, possibly originating from the crust of asteroid 4 Vesta. Other types include Martian, Lunar, and several types thought to originate from as-yet unidentified asteroids. These groups have been determined on the basis of, for example, the Fe/Mn chemical ratio and the $^{17}O/^{18}O$ oxygen isotope ratios, thought to be characteristic "fingerprints" for each parent body.
Howardite–eucrite–diogenite (HED)	HED meteorites are a clan (subgroup) of achondrite meteorites. HED stands for "howardite–eucrite–diogenite." These achondrites came from a differentiated parent body and experienced extensive igneous processing not much different from the magmatic rocks found on Earth and for this reason they closely resemble terrestrial igneous rocks.

References

1. Dawn. NASA – National Space Science Data Center. http://nssdc.gsfc.nasa.gov/planetary/text/dawn_pr_20110716.txt (retrieved July 16, 2011).
2. Rayman, M. (April 8, 2015). *Now Appearing at a Dwarf Planet Near You: NASA's Dawn Mission to the Asteroid Belt* (Speech). Silicon Valley Astronomy Lectures. Foothill College, Los Altos, CA.
3. NASA's Dawn Spacecraft Hits Snag on Trip to 2 Asteroids. (August 15, 2012). Space.com. http://www.space.com/17119-nasa-dawn-asteroid-spacecraft-vesta.html (retrieved August 27, 2012).
4. Dawn Gets Extra Time to Explore Vesta. (April 18, 2012). NASA. http://www.jpl.nasa.gov/news/news.php?release=2012-107 (retrieved April 24, 2012).
5. Landau, E. and Brown, D. (March 6, 2015). NASA Spacecraft becomes First to Orbit a Dwarf Planet. NASA. http://www.jpl.nasa.gov/news/news.php?feature=4503 (retrieved March 6, 2015).
6. Rayman, M. (March 6, 2015). Dawn Journal: Ceres Orbit Insertion! Planetary Society. http://www.planetary.org/blogs/guest-blogs/marc-rayman/20150306-dawn-journal-ceres-orbit-insertion.html (retrieved March 6, 2015).
7. GSpace Topics: Dawn. Planetary Society. http://planetary.org/explore/space-topics/space-missions/missions-beyond-mars.html#dawn (retrieved November 9, 2013).
8. Rayman, M. et al. (April 5, 2006). Dawn: A mission in development for exploration of main belt asteroids Vesta and Ceres *Acta Astronautica* 58(11):605–616. doi:10.1016/j.actaastro.2006.01.014.
9. Dawn Spacecraft Successfully Launched. (September 27, 2007). NASA. http://www.nasa.gov/mission_pages/dawn/news/dawn-20070927.html (retrieved November 9, 2013).
10. NASA's Dawn Spacecraft Begins Science Orbits of Vesta. (August 1, 2011). NASA. http://www.nasa.gov/mission_pages/dawn/news/dawn20110801.html (retrieved August 6, 2011).
11. Boyle, A. (March 6, 2015). Dawn Spacecraft Slips Quietly into Orbit Around Dwarf Planet Ceres. *NBCNews.com* (NBCUniversal Media, LLC). http://www.nbcnews.com/science/space/dawn-spacecraft-slips-quietly-orbit-around-dwarf-planet-ceres-n318371 (retrieved March 11, 2015).
12. DAWN Mission Objectives. NASA. http://dawn.jpl.nasa.gov/mission/goal_index.asp (retrieved March 2010).
13. Hamilton, C.J. Vesta. http://solarviews.com/eng/vesta.htm (retrieved January 6, 2013).
14. McCord, T.B. and Sotin, C. (2005). Ceres: Evolution and current state. *Journal of Geophysical Research* 110(E5):E05009. doi:10.1029/2004JE002244.
15. Thomas, P.C. et al. (2005). Differentiation of the asteroid Ceres as revealed by its shape. *Nature* 437(7056):224–226. doi:10.1038/nature03938.
16. Malik, T. (March 27, 2006). NASA Reinstates Cancelled Asteroid Mission. Space.com. http://www.space.com/2205-nasa-reinstates-cancelled-asteroid-mission.html (retrieved March 27, 2006).
17. Meteoritical Bulletin Database. http://www.lpi.usra.edu/meteor/metbull.php (retrieved November 27, 2014).
18. Ghosh, A. and McSween, H.Y. (1998). A Thermal Model for the Differentiation of Asteroid 4 Vesta, Based on Radiogenic Heating. *Icarus* 134(2):187. doi:10.1006/icar.1998.5956.

19. Sahijpal, S. et al. (2007). Numerical simulations of the differentiation of accreting planetesimals with ^{26}Al and ^{60}Fe as the heat sources. *Meteoritics & Planetary Science* 42(9):1529–1548. doi:10.1111/j.1945-5100.2007.tb00589.x.
20. Gupta, G. and Sahijpal, S. (2010). Differentiation of Vesta and the parent bodies of other achondrites. *Journal of Geophysical Research: Planets* 115(E8):E08001. doi:10.1029/2009JE003525.
21. Mission Objectives. (2006) JPL – NASA. http://dawn.jpl.nasa.gov/mission/goal_index.asp (retrieved March 23, 2013).
22. Geveden, R. (2006). Dawn Mission Reclama. http://www.nasa.gov/pdf/145511main_Dawn_letter.pdf (retrieved March 27, 2006).
23. Dawn Mission Overview. NASA. http://www.nasa.gov/mission_pages/dawn/spacecraft/index.html (retrieved August 17, 2014).
24. Benson, S.W. (November 8, 2007). Solar Power for Outer Planets Study. NASA. http://www.lpi.usra.edu/opag/nov_2007_meeting/presentations/solar_power.pdf (retrieved November 27, 2014).
25. Dawn Mission. NASA – JPL. http://dawn.jpl.nasa.gov/mission/spacecraft.asp (retrieved July 18, 2011).
26. Dawn, Ion Propulsion. NASA. http://dawn.jpl.nasa.gov/mission/ion_prop.asp (retrieved September 28, 2007).
27. Dawn, Spacecraft. NASA. http://dawn.jpl.nasa.gov/mission/spacecraft.asp (retrieved September 28, 2007).
28. Dawn Solar Arrays. Dutch Space. http://www.airbusdefenceandspacenetherlands.nl/ (retrieved July 18, 2011).
29. Dawn: Mission description. (October 17, 2006). UCLA Space Physics Center. http://www.nasa.gov/mission_pages/dawn/mission/ (retrieved September 28, 2007).
30. Watanabe, S. (July 5, 2007). Dawn: Spacecraft & Instruments. NASA. http://www.nasa.gov/mission_pages/dawn/spacecraft/ (retrieved August 10, 2006).
31. Rayman, M.D. Dawn Journal: December 17, 2007. JPL. http://dawn.jpl.nasa.gov/mission/journal_12_17_07.asp (retrieved November 9, 2013).
32. Rayman, M.D. Dawn Journal: Aiming away from a bull's eye at Mars. The Planetary Society. Archived from the original on July 16, 2011. http://www-ssc.igpp.ucla.edu/dawn/DawnJournal_24August08.html (retrieved November 9, 2013).
33. Malik, T. (February 18, 2009). Asteroid-Bound Probe Zooms Past Mars. Space.com. http://www.space.com/4293-asteroid-bound-probe-zooms-mars.html (retrieved November 9, 2013).
34. NASA's Dawn Captures First Image of Nearing Asteroid. (May 11, 2011). NASA. http://www.nasa.gov/mission_pages/dawn/news/dawn20110511.html (retrieved September 1, 2012).
35. NASA'S Dawn Spacecraft Begins Science Orbits Of Vesta. (August 1, 2011). NASA. http://www.nasa.gov/home/hqnews/2011/aug/HQ_11-254_Dawn_Image.html (retrieved August 7, 2014).
36. View of Vesta from Dawn. NASA/JPL MYSTIC simulator (updated periodically). http://neo.jpl.nasa.gov/orbits/fullview4.jpg (retrieved September 1, 2012).
37. Amos, J. (July 17, 2011). Dawn probe orbits asteroid Vesta. *BBC News*. http://www.bbc.com/news/science-environment-14160135 (retrieved January 22, 2015).
38. Russell, C.T. et al. (May 11, 2012). Dawn at Vesta: Testing the Protoplanetary Paradigm. *Science* 336(6082):684–686. doi:10.1126/science.1219381.
39. Rayman, M. (2011). Mission status updates 2011. DAWN, NASA JPL. http://dawn.jpl.nasa.gov/mission/status_2011.html (retrieved September 28, 2016).
40. Russell, C.T. and Raymond, C.A. (2011). The Dawn Mission to Vesta and Ceres. *Space Science Reviews* 163(1–4):3–23. doi:10.1007/s11214-011-9836–2.
41. 2011: The Dawn of Vesta Science (September 2011). Spacegrant.org. http://erps.spacegrant.org/uploads/images/images/iepc_articledownload_1988-2007/2011index/IEPC-2011-326.pdf (retrieved November 9, 2013).
42. Vega, P. and Brown, D. (July 16, 2011). NASA's Dawn Spacecraft Enters Orbit Around Asteroid Vesta. NASA. http://www.nasa.gov/mission_pages/dawn/news/dawn20110716.html (retrieved July 17, 2011).

43. Asteroid Vesta is 'last of a kind' rock. (May 11, 2012). BBC. http://www.bbc.co.uk/news/science-environment-18027933 (retrieved January 20, 2015).
44. Incredible video 'fly-over' by Nasa's Dawn probe reveals huge rippled asteroid Vesta is more like a small planet. (May 15, 2012). *Daily Mail*. http://www.dailymail.co.uk/sciencetech/article-2144652/Asteroid-Vesta-Nasas-Dawn-probe-reveals-huge-rippled-asteroid-like-small-planet.html?ito=feeds-newsxml (retrieved May 23, 2012).
45. Reddy, V. et al. (November–December 2012). Delivery of dark material to Vesta via carbonaceous chondritic impacts. *Icarus* 221(2):544–559. doi:10.1016/j.icarus.2012.08.011.
46. McCord, T.B. et al. (November 1, 2012). Dark material on Vesta from the infall of carbonaceous volatile-rich material. *Nature* 491(7422):83–86. doi:10.1038/nature11561.
47. NASA's Dawn Spacecraft Probes Proto-Planet Vesta, Discovers Deposits That Give Scientists Insight into the Origins of the Solar System. *Latino Post*. October 31, 2012. http://www.latinospost.com/articles/6285/20121031/nasas-dawn-spacecraft-probes-proto-planet-vesta.htm (retrieved November 28, 2012).
48. Scully, J.E.C. et al. (2014). Sub-Curvilinear Gullies Interpreted As Evidence For Transient Water Flow On Vesta. *45th Lunar and Planetary Science Conference*. Universities Space Research Association. http://www.hou.usra.edu/meetings/lpsc2014/pdf/1796.pdf (retrieved August 7, 2014).
49. Dawn probe spies possible water-cut gullies on Vesta. BBC. December 6, 2012. http://www.bbc.co.uk/news/science-environment-20582704 (retrieved December 8, 2012).
50. Dawn departs Vesta to become first asteroid hopper. *New Scientist*. September 6, 2012. https://www.newscientist.com/blogs/shortsharpscience/2012/09/asteroid-hopping-spacecraft-ma.html (retrieved November 9, 2013).
51. Dawn Engineers Assess Reaction Wheel. NASA/JPL. August 18, 2012. http://www.nasa.gov/mission_pages/dawn/news/dawn20120905.html (retrieved January 22, 2015).
52. 'Dawn has departed the giant asteroid Vesta'. NASA/JPL. https://www.yahoo.com/news/nasas-dawn-spacecraft-says-goodbye-giant-asteroid-vesta-205041936.html?ref=gs (retrieved September 5, 2012).
53. SPACE.com Staff (September 06, 2012). NASA's Dawn Spacecraft says goodbye to giant asteroid Vesta. Yahoo.com. https://www.yahoo.com/news/nasas-dawn-spacecraft-says-goodbye-giant-asteroid-vesta-205041936.html?ref=gs (retrieved September 28, 2016).
54. Dawn Fills out its Ceres Dance Card. NASA/JPL. December 3, 2013. http://www.jpl.nasa.gov/news/news.php?release=2013-347 (retrieved March 18, 2015).
55. Dawn Operating Normally After Safe Mode Triggered. NASA/JPL. September 16, 2014. https://web.archive.org/20141225012242/http://dawn.jpl.nasa.gov/feature_stories/Dawn_operated_normally.asp (retrieved March 18, 2015).
56. Dawn Journal. NASA/JPL. December 29, 2014. http://dawnblog.jpl.nasa.gov/2014/12/29/dawn-journal-december-29/ (retrieved January 21, 2015).
57. Dawn Journal October 31. NASA. October 31, 2014. http://dawnblog.jpl.nasa.gov/2014/10/31/dawn-journal-october-31/#sthash.OPRlufJs.dpuf (retrieved January 18, 2015).
58. NASA's Dawn Fills out its Ceres Dance Card. NASA. December 3, 2013. http://www.jpl.nasa.gov/news/news.php?release=2013-347 (retrieved December 8, 2013).
59. McCord, T.B. et al. (2011). Ceres: Its Origin, Evolution and Structure and Dawn's Potential Contribution. In *The Dawn Mission to Minor Planets 4 Vesta and 1 Ceres*. p. 63.
60. LPSC 2015: First results from Dawn at Ceres: provisional place names and possible plumes. Planetary Society. March 19, 2015. http://www.planetary.org/blogs/emily-lakdawalla/2015/03191629-lpsc-2015-dawn-at-ceres.html (retrieved March 21, 2015).
61. Dawn Journal March 31. NASA. March 31, 2015. http://dawnblog.jpl.nasa.gov/2016/03/31/dawn-journal-march-31-3/ (retrieved April 20, 2015).
62. Dawn Journal February 28. NASA. February 28, 2014. http://dawnblog.jpl.nasa.gov/category/dawn-journal/ (retrieved February 13, 2015).
63. Dawn Maneuvering to Third Science Orbit. NASA. July 17, 2015. http://www.jpl.nasa.gov/news/news.php?feature=4659 (retrieved July 20, 2015).

64. Dawn Journal December 29. NASA/JPL. January 11, 2015. http://dawnblog.jpl.nasa.gov/2014/12/29/dawn-journal-december-29/#comment-7392 (retrieved March 10, 2015).
65. Ambrosiano, N. (March 28, 2006). NASA's Dawn mission is a go. Los Alamos National Laboratory. http://dawn.jpl.nasa.gov/ (retrieved October 1, 2007).
66. Chang, K. (July 18, 2015). The Long, Strange Trip to Pluto, and How NASA Nearly Missed It. *New York Times*. http://www.nytimes.com/2015/07/19/us/the-long-strange-trip-to-pluto-and-how-nasa-nearly-missed-it.html?_r=0 (retrieved July 19, 2015).
67. New Horizons to Pluto, Mission Website. (July 2, 2015). US National Aeronautics and Space Administration (NASA). https://www.nasa.gov/mission_pages/newhorizons/main/ (retrieved July 7, 2015).
68. Chang, K. (July 13, 2015). A Close-Up for Pluto After Spacecraft's 3-Billion-MileTrip. *New York Times*. http://www.nytimes.com/2015/07/14/science/a-close-up-for-pluto-after-spacecrafts-3-billion-mile-trip.html (retrieved July 13, 2015).
69. Chang, K. (July 6, 2015). Almost Time for Pluto's Close-Up. *New York Times*. http://www.nytimes.com/2015/07/07/science/space/almost-time-for-plutos-close-up.html (retrieved July 6, 2015).
70. Overbye, D. (July 6, 2015). Reaching Pluto, and the End of an Era of Planetary Exploration. *New York Times*. http://www.nytimes.com/2015/07/07/science/space/reaching-pluto-and-the-end-of-an-era-of-planetary-exploration.html (retrieved July 7, 2015).
71. Chang, K. (July 14, 2015). NASA's New Horizons Spacecraft Completes Flyby of Pluto. *New York Times*. http://www.space.com/29850-new-horizons-pluto-flyby-complete-coverage.html (retrieved July 14, 2015).
72. Dunn, M. (July 14, 2015). Pluto close-up: Spacecraft makes flyby of icy, mystery world. *AP News*. http://www.mercurynews.com/science/ci_28480927/pluto-close-up-spacecraft-makes-flyby-icy-mystery (retrieved July 14, 2015).
73. Brown, D. et al. (July 14, 2015). 15-149 NASA's Three-Billion-Mile Journey to Pluto Reaches Historic Encounter. NASA. http://www.nasa.gov/press-release/nasas-three-billion-mile-journey-to-pluto-reaches-historic-encounter/ (retrieved July 14, 2015).
74. Staff (July 14, 2015). NASA News Briefing on New Horizons 09:30 pm–10:30 pm. NASA. http://www.nasa.gov/newhorizons/pr-archives/ (retrieved July 14, 2015).
75. Staff (July 14, 2015). NewHorizons 'phoned home' & is healthy! *Twitter @NASA*. http://www.newsweek.com/new-horizons-phones-home-pluto-flyby-went-smoothly-353925 (retrieved July 14, 2015).
76. Brown, D. et al. (July 14, 2015). 15-150 - NASA's New Horizons 'Phones Home' Safe after Pluto Flyby. NASA. http://www.nasa.gov/press-release/nasas-new-horizons-phones-home-safe-after-pluto-flyby (retrieved July 15, 2015).
77. Chang, K. (July 14, 2015). NASA's New Horizons Spacecraft Sends Signal from Pluto to Earth. *New York Times*. http://www.nytimes.com/2015/07/15/science/space/nasa-new-horizons-spacecraft-reaches-pluto.html (retrieved July 14, 2015).
78. Betz, E. (June 24, 2015). How'd we get New Horizons? You can thank the Pluto Underground. *Astronomy*. Kalmbach Publishing. http://www.astronomy.com/year-of-pluto/2015/06/howd-we-get-new-horizons-you-can-thank-the-pluto-underground (retrieved July 8, 2015).
79. Stern, A. and Christopher, R. (2009). *New Horizons: Reconnaissance of the Pluto-Charon System and the Kuiper Belt*. Springer Publishing (e-book). pp. 6, 7.
80. Davis, J. (July 7, 2015). Pushing Back the Frontier: How the Planetary Society Helped Send a Spacecraft to Pluto. *The Planetary Society*. http://www.planetary.org/blogs/jason-davis/2015/20150707-pushing-back-frontier.html (retrieved July 8, 2015).
81. NASA's Three-Billion-Mile Journey to Pluto Reaches Historic Encounter. Johns Hopkins University Applied Physics Laboratory. July 14, 2015.
82. Lakdawalla, E. (July 15, 2015). New Horizons 'phones home' after Pluto flyby. The Planetary Society.
83. Phipps, C. (July 15, 2015). Pluto: New Horizons probe makes contact with Earth. *The Guardian*.

Appendix A

The Sun and Planets of the Solar System with Some Summary Data

This section exhibits sampling of solar system planets, selected for size and quality of imagery, and also they are sorted by volume. The distances shown in Figure A.1 are not to scale.

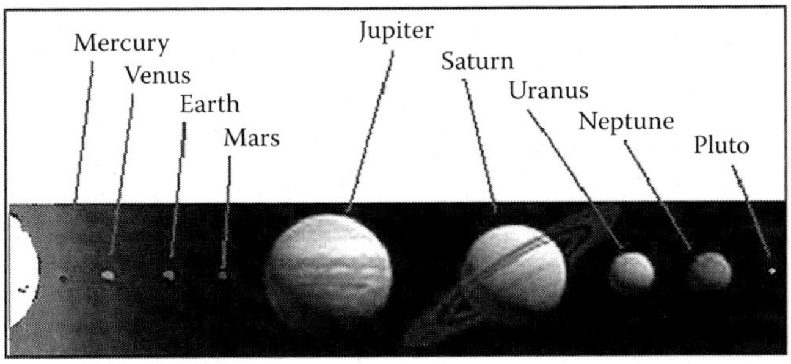

FIGURE A.1
The Sun and planets of the solar system.

Appendix B

Timeline of Solar System Evolution

The time frame of the formation of solar system has been obtained using radiometric dating. Estimation by scientists reveals that the solar system is 4.6 billion years old as has already been pointed out before. It may be pointed out that the oldest known mineral grains on Earth are nearly 4.4 billion years old. Earth's surface is constantly changing its shape by erosion, volcanism, and plate tectonics.

In order to estimate the age of the solar system, scientists use meteorites that were produced during the early condensation of the solar nebula. Most of the meteorites are found to have an age of 4.6 billion years, which indicates that the solar system must be at least this old. Careful studies of the disks around other stars have established a time frame for solar system formation. Stars that are between 1 and 3 million years old possess disks rich in gas, whereas disks around stars more than 10 million years old have little to no gas. Chronology of formation and evolution of the solar system is shown in Figure B.1. The life cycle of the Sun presented in Figure B.1 is not to scale.

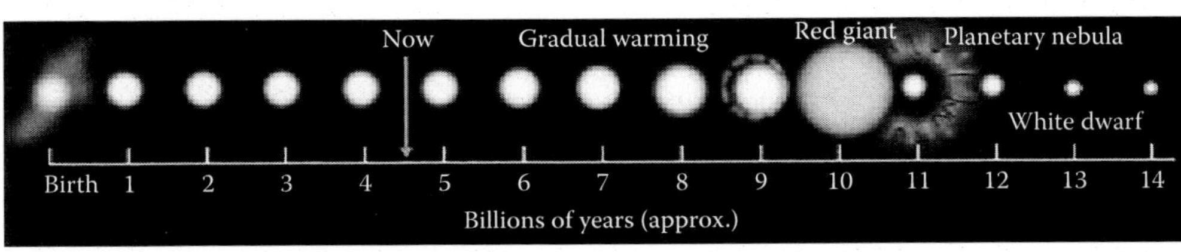

FIGURE B.1
Chronology of formation and evolution of the solar system.

Appendix C

Graduate Programs for Planetary Education

Institution	Department
Arizona State University	School of Earth and Space Exploration
Boston University	Astronomy
Brown University	Planetary Geosciences
California Institute of Technology	Geological and Planetary Sciences
Catholic University of America	Institute for Astrophysics and Computational Science
Clemson University	Physics and Astronomy
Colorado School of Mines	Geophysics
Cornell University	Astronomy
Florida Institute of Technology	Physics and Space Sciences
George Mason University	Physics and Astronomy
Georgia Institute of Technology	School of Earth and Atmospheric Sciences
Hampton University	Atmospheric and Planetary Sciences
Harvard University	Earth and Planetary Sciences
Howard University	Physics and Astronomy
Institutet för rymdfysik Swedish	Institute of Space Physics
Johns Hopkins University	Earth and Planetary Sciences
Kansas University	Physics
Keele University	Astrophysics Group
Kyung Hee University	Astronomy and Space Science
Leiden University	Leiden Observatory
Massachusetts Institute of Technology	Earth, Atmospheric, and Planetary Sciences
New Mexico State University	Astronomy and Space Science
Northern Arizona University	Physics and Astronomy
Northwestern University	Earth and Planetary Sciences
Princeton University	Astrophysical Sciences
Purdue University	Earth and Atmospheric Sciences
Rice University	Physics and Astronomy
SUNY Stony Brook	Geosciences
SUNY Stony Brook	Physics and Astronomy
Texas Tech University	Geosciences
The Australian National University	Research School of Astronomy and Astrophysics
Universidad de Chile	Astronomia
University College	London Astrophysics Group
University of Alaska Fairbanks	Geophysical Institute
University of Arizona	Lunar and Planetary Laboratory
University of Arkansas	Center for Space and Planetary Sciences
University of British Columbia	Institute of Planetary Science
University of California Berkeley	Astronomy

(*Continued*)

Institution	Department
University of California Berkeley	Earth and Planetary Science
University of California Los Angeles	Earth and Space Sciences
University of California Los Angeles	Astronomy and Astrophysics
University of California Los Angeles	Atmospheric and Oceanic Sciences
University of California Santa Cruz	Earth and Planetary Sciences
University of Cambridge	Institute for Astronomy
University of Canterbury	Physics and Astronomy
University of Central Florida	Planetary Sciences Group
University of Chicago	Geophysical Sciences
University of Chicago	Astronomy and Astrophysics
University of Colorado	Astrophysical and Planetary Sciences, Physics, Geological Sciences, Atmospheric and Oceanic Sciences, Aerospace Engineering, Chemistry
University of Florida	Gainsville Astronomy
University of Hawaii	Institute for Astronomy
University of Hawaii	Institute of Geophysics and Planetology
University of Idaho	Physics and Astronomy
University of Illinois-Urbana Champaign	Geology
University of Iowa	Physics and Astronomy
University of Manchester Jodrell Bank	Centre for Astrophysics
University of Manitoba	Physics and Astronomy
University of Maryland	Astronomy
University of Massachusetts	Amherst Astronomy
University of Michigan	Atmospheric, Oceanic, and Space Sciences
University of Minnesota	School of Physics and Astronomy
University of New Hampshire	Institute for the Study of Earth, Oceans and Space
University of Oxford	Atmospheric, Oceanic and Planetary Physics
University of Pittsburgh	Geology and Planetary Science
University of St. Andrews	School of Physics and Astronomy
University of Tennessee	Knoxville Earth and Planetary Sciences
University of Texas, Arlington	Physics and Astronomy
University of Texas, Austin	Astronomy Program
University of Texas, San Antonio/Southwest Research	Institute Astronomy/Space Science
University of Tokyo	Earth and Planetary Science
University of Toledo	Physics and Astronomy
University of Toronto	Astronomy and Astrophysics
University of Virginia	Astronomy
University of Wales, Aberystwyth	Institute of Mathematics and Physics
University of Washington	Earth and Space Sciences
University of Washington	Astronomy
University of Western Ontario Earth Sciences	Physics and Astronomy
Vanderbilt University	Physics and Astronomy
Washington University in St. Louis	Earth and Planetary Sciences
Wesleyan University	Planetary Sciences Group
Yale University	Astronomy

Glossary

Radio Astronomy/Astrophysics Glossary

absolute magnitude: a scale to measure the actual brightness of a celestial object without accounting the distance. Absolute magnitude determines how bright an object would appear if it were 10 parsecs away from Earth. In this scale, the Sun has an absolute magnitude of +4.8 while it has an apparent magnitude of −26.7.

accretion: the process through which dust and gas accumulated into larger bodies such as stars and planets.

albedo: it is the reflective property of a nonluminous object. A perfect mirror have an albedo of 100% while a black hole have an albedo of 0%.

antenna temperature: measure of the noise power collected by the antenna and delivered to the radio receiver. Specifically, the temperature at which a resistor, substituted for the antenna, would have to be maintained in order to deliver the same noise power to the receiver in the same frequency bandwidth.

aphelion: it represents the point in the orbit of a planet or other celestial body where it is farthest from the Sun.

apogee: it represents the point in the orbit of the Moon or other satellite where it is farthest from the Earth.

asteroid: it is a small planetary body in orbit around the Sun. This is larger than a meteoroid but smaller than a planet. Asteroids are usually observed in a belt between the orbits of Mars and Jupiter.

asteroid belt: it is a region of space between Mars and Jupiter where majority of asteroids are found.

astronomical unit (AU): a unit equal to the average distance between the Earth and the Sun, which is 93 million miles approximately.

atmosphere: gaseous layer surrounding a planet, moon, or star. The Earth's atmosphere is 120 miles thick and is composed of nitrogen, oxygen, carbon dioxide, and a few other trace gases.

aurora: a glow in the ionosphere of a planet caused by the interaction between the planet's magnetic field and charged particles from the Sun. This phenomenon is termed as the aurora borealis in the Earth's northern hemisphere and aurora australis in the Earth's southern hemisphere.

Aurora Australis: this is alternately called as the southern lights. It is an atmospheric phenomenon which displays a diffuse glow in the sky in the southern hemisphere and is caused by charged particles from the Sun as they interact with the Earth's magnetic field.

Aurora Borealis: alternately called as the northern lights. It is an atmospheric phenomenon that displays a diffuse glow in the sky in the northern hemisphere and is caused by charged particles from the Sun as they interact with the Earth's magnetic field.

Big Bang: the theory which suggests that the universe was produced from a single point in space during a cataclysmic explosion nearly 13.7 billion years ago. This is the presently accepted theory for the origin of the universe and is supported by measurements of background radiation and the observed expansion of space.

blackbody: idealized object that absorbs all electromagnetic radiation that is incident on it. The radiation properties of blackbody radiators are described by the Planck function. Planetary radio astronomers use the properties of blackbody radiators to describe the radiation from planets.

black hole: it is the collapsed core of a massive star. Stars which are very massive will collapse under their own gravity when their fuel is exhausted. The collapse continues so long after all matter is crushed out of existence into what is known as a singularity. The gravitational pull is so strong that not even light can escape.

blueshift: a shift in the lines of an spectrum of an object toward the blue end. Blueshift suggests that an object is moving toward the observer. The greater the blueshift, the faster the object is moving.

brightness temperature: the definition is not unique; great care is needed to decipher the intentions of the authors. The temperature at which a blackbody radiator would radiate an intensity of electromagnetic radiation identical to that of the planet for a specific frequency, frequency bandwidth, and polarization under consideration is one definition of brightness temperature. A second definition is that it is the intensity of radiation under consideration divided (normalized) by the factor $(h^2/2k)$. The normalization factor dimensionally scales the intensity to have units of temperature. The two definitions show the largest departures at low temperatures and high frequencies.

celestial poles: the North and South poles of the celestial sphere.

chaos: this represents a distinctive area of broken terrain.

comet: a ball of rock and ice generally referred to as a "dirty snowball." It is few kilometers in diameter. Comets orbit the Sun in paths that either permit them to pass by the Sun only once or that repeatedly brings them through the solar system.

constellation: a grouping of stars which make an imaginary picture in the sky.

corona: the outer part of the atmosphere of the Sun. The corona can be observed from Earth during a total solar eclipse. It is the bright glow found during the period of most solar eclipse.

coronagraph: this is a special telescope that blocks light from the disk of the Sun to study the faint solar atmosphere.

coronal hole: regions in the corona wherefrom the high-speed solar wind originates. Coronal holes found near the Sun's poles are large regions in the corona that are less dense and cooler than the surrounding region.

cosmic ray: atomic nuclei which are seen to strike the Earth's atmosphere with extremely high energy.

cosmological constant: it is a mathematical device used to give space-time an inbuilt tendency to expand.

cosmology: a branch of science which deals with studying the origin, structure, and nature of the universe.

crater: a bowl-shaped depression produced by a comet or meteorite colliding with the surface of a planet, moon, or asteroid. On geologically active moons and planets, craters can result from volcanic activity.

dark matter: the terminology describes matter in the universe which cannot be seen but can be detected by its gravitational effects on other bodies.

debris disk: this is a ring-shaped circumstellar disk of dust and debris in orbit around a star. Debris disks are the next phase in planetary system development following the protoplanetary disk phase. They can also be produced by collisions between planetesimals.

declination: the angular distance of an object in the sky from the celestial equator.

double asteroid: two asteroids which revolve around each other and are held together by the gravity between them. This is also named as a binary asteroid.

double star: it is a grouping of two stars which can be apparent, where the stars seem close together or physical, such as a binary system.

dwarf planet: a celestial body orbiting the Sun which is sufficiently massive to be rounded by its own gravity but has not cleared its surrounding region of planetesimals and is not a satellite. It has enough mass to overcome rigid body forces and achieve hydrostatic equilibrium. Pluto is a dwarf planet.

effective area: equivalent cross section or collecting area of a radio telescope to an incident radio wave; a measure of a radio telescope's capability to detect weak radio signals.

effective temperature: temperature at which a blackbody radiator would radiate over all frequencies an intensity of electromagnetic radiation identical to that radiated from a planet.

elliptical galaxy: a galaxy whose structure looks like an ellipse and is smooth and lacks complex structures like spiral arms.

equivalent blackbody disk temperature: temperature of a blackbody radiator with the same solid angle as the planet that gives the same radiation intensity at the Earth as observed from the planet at a specified frequency and bandwidth.

evolved star: a star which is near the end of its life cycle because most of its fuel were used up. At this point the star begins to lose mass in the form of stellar wind.

extragalactic: a term which means outside of or beyond our own galaxy.

extraterrestrial: a term used to describe anything which does not originate on Earth.

fireball: it represents an extremely bright meteor, also known as bolides. Fireballs can be few times brighter than the full moon.

flare star: a faint red star which appears to change in brightness due to explosions on its surface.

flux density: power per unit area and per unit frequency of an electromagnetic wave crossing an imaginary plane surface from one side to the other. In observational radio astronomy, the mks system of units is generally used, and the unit's f flux density is watts per square meter per hertz.

flux unit or jansky: commonly used unit of flux density equal to 1 × W m-Hz. The size of the unit is suited to planetary radio emissions, which are very weak.

Galilean moons: the name is related to Jupiter's four largest moons, namely, Io, Europa, Callisto, and Ganymede.

gravitational lens: a concentration of matter such as a galaxy or cluster of galaxies which bends light rays from a background object.

gravitational waves: ripples in space which travel at the speed of light, produced by the movement of very massive bodies.

great red spot: a circulating storm situated in Jupiter's upper atmosphere, which rotates around the planet in 6 days and has a width of two to three Earths.

greenhouse effect: it is an increase in temperature caused when incoming solar radiation is allowed in but outgoing thermal radiation is blocked by atmosphere. Carbon dioxide and water vapor are two major gases responsible for this effect.

habitable zone: a zone around a star where planets with liquid water may be present. Planets on the near edge of the habitable zone have a surface temperature slightly lower than the boiling point of water. A planet on the distant edge of the habitable zone would have a surface temperature slightly larger than the freezing point of water.

hadron: particles made of quarks which are influenced by strong nuclear force.

Hertzsprung–Russell diagram: it is a graph whose horizontal axis plots star color against a vertical axis plotting stellar luminosity.

Hubble's law: it states that the farther a galaxy is from us, the faster it is moving away from us.

inferior conjunction: a conjunction of an inferior planet which occurs when the planet is lined up directly between the Earth and the Sun.

inferior planet: a planet that orbits between the Earth and the Sun. In our solar system, Mercury and Venus are the only two inferior planets.

International Astronomical Union (IAU): an international organization which unites national astronomical societies from around the world. It acts as the internationally recognized authority for assigning designations to celestial bodies and their surface features.

interplanetary magnetic field: the magnetic field carried along with the solar wind.

interstellar medium: the gas and dust which exists in open space between the stars.

Io: it is the innermost of Jupiter's four large moons. Owing to Jupiter's gravitational might, Io is geologically active and its surface is peppered with volcanoes that send sulfurous eruptions into its thin atmosphere. Io has the most active volcanoes in the solar system.

ionosphere: a region of charged particles in the upper atmosphere of any planet. In Earth's atmosphere, the ionosphere begins at an altitude of about 60 km.

irregular galaxy: it is the galaxy with no spiral structure and no symmetric shape. Irregular galaxies are filamentary or clumpy in shape.

Jansky: it is the unit used in radio astronomy for indicating the flux density of electromagnetic radiation received from outer space. A typical radio source has a spectral flux density of about 1 Jy.

jet: it is the narrow stream of gas or particles ejected from an accretion disk surrounding a star or black hole.

Jovian atmosphere: it is the atmosphere surrounding the giant, massive planet Jupiter. The Jovian atmosphere is mainly composed of hydrogen (about 90%) and helium (about 10%). Other minor ingredients are water, hydrogen sulfide, methane, and ammonia.

Jovian planets: the planets Jupiter, Saturn, Uranus, and Neptune are called Jovian planets due to similarities in their composition and location. This group is also called as the "giant planets," the "gas planets," and when grouped with the planet Pluto, the "outer planets."

Kuiper belt: it is the large ring of icy, primitive objects beyond the orbit of Neptune. Kuiper belt objects are considered to be remnants of the original material which formed the solar system. Astronomers believe Pluto and Charon are Kuiper belt objects.

limb: it is the outer edge or border of a planet or other celestial body.

luminosity: it is the amount of light emitted by a star.

Magellanic Clouds: two small, irregular galaxies observed just outside our own Milky Way galaxy. The Magellanic Clouds can be found in the skies of the southern hemisphere.

magnetosphere: it represents the area around a planet most affected by its magnetic field. The boundary of this field is set by the solar wind.

major planet: a planet that is larger and more massive than the Earth and contains large amount of hydrogen and helium. Jupiter and Neptune are two examples of major planets.

meteor: this represents small particles of rock or dust which burns away in the Earth's atmosphere. Meteors are also referred to as shooting stars.

meteor shower: it is the event where a large number of meteors enter the Earth's atmosphere from the same direction in space at nearly the same time. Most of the meteor showers take place when the Earth passes through the debris left behind by a comet.

meteorite: it is an object, usually a chunk or metal or rock, which survives entry through the atmosphere to reach the Earth's surface. Meteors become meteorites when they reach the ground.

meteoroid: it is a small, rocky object in orbit around the Sun, smaller than an asteroid.

nebula: it is a cloud of dust and gas in space, illuminated by one or more stars. It represents the raw material the stars are made of.

neutrino: a fundamental particle produced by the nuclear reactions in stars.

neutron star: a compressed core of an exploded star made up of neutrons. Neutron stars have a strong gravitational field and emit pulses of energy along their axis called pulsars.

nonthermal radio emission: radio emission produced by processes other than thermal emission. Cyclotron and synchrotron radio emission are two examples of nonthermal radio emission.

nova: a star which flares up to several times its original brightness for some time before returning to its original state.

nuclear fusion: the nuclear process by which several small nuclei are combined to make a larger one. Nuclear fusion is the reaction which fuels the Sun, where hydrogen nuclei are fused to form helium.

Oort cloud: it is a theoretical shell of comets which is believed to exist at the outermost regions of our solar system.

open cluster: a collection of young stars that formed together. Some of the youngest open clusters are still embedded in the gas and dust from which they were produced.

opposition: the position of a planet when it is exactly opposite to the Sun in the sky as seen from Earth.

optical depth: atmospheric attenuation is usually expressed by giving the dimensionless quantity "optical depth" along a specified path. A signal that passes through an atmosphere whose optical depth T is attenuated by the factor $\boldsymbol{e - T}$.

orbit: it represents the path of a celestial body as it moves through space.

parallax: the apparent change in position of two objects seen from different locations.

parsec: a large distance used in astronomy. A parsec is equal to 3.26 light-years.

penumbra: the area of partial illumination surrounding the darkest part of a shadow produced by an eclipse.

perigee: the point in the orbit of the Moon or other satellite at which it is nearest to the Earth.

perihelion: the point in the orbit of a planet or other body where it is nearest to the Sun.

planetary nebula: it indicates a shell of gas surrounding a small, white star. The gas is illuminated by the star producing a variety of colors and shapes, in general.

planetesimal: a solid object which is assumed to exist in protoplanetary disks and in debris disks. Planetesimals are produced from small dust grains that collide and stick together and are the building blocks which form planets in new planetary systems.

prograde orbit: a prograde orbit means that the satellite orbits the planet in the same direction as the planet's rotation. A planet is considered to have a prograde orbit if the direction of its orbit is the same as that of the majority of other planets in the system.

protostar: dense regions of molecular clouds where stars are produced.

pulsar: a spinning neutron star which emits energy along its gravitational axis. This energy is received as pulses as the star moves.

quasar: an unusually bright object seen in the remote areas of the universe. Quasars release incredible amounts of energy.

quasi-stellar object: alternately named as quasi-stellar source. This is a star-like object with a large redshift, which gives off a strong source of radio waves. They are highly luminous and presumed to be extragalactic.

radio galaxy: a galaxy which gives off large amounts of energy in the form of radio waves.

red giant: a stage in the evolution of a star when the fuel starts to exhaust and the star expands to about 50 times its normal size.

red shift: a shift in the lines of the spectrum toward the red end. Red shift indicates that an object is moving away from the observer.

regolith: the layer of loose rock resting on bedrock found on the Earth, the Moon, or a planet. Regolith is made up of soils, sediments, weathered rocks, and hard near-surface crusts. On the surface of the Moon, regolith is a fine rocky layer of fragmentary debris.

retrograde orbit: the orbit of a satellite where the satellite moves in a direction opposite to the direction of the planet's rotation.

ring galaxy: a galaxy which has a ring-like appearance which contains luminous blue stars. Ring galaxies are assumed to have been produced by collisions with other galaxies.

Roche limit: the smallest distance from a planet or other body at which purely gravitational forces can hold together a satellite or secondary body of the same mean density as the primary.

Seyfert galaxy: a main-sequence star which rotates rapidly, resulting in a loss of matter to an ever-expanding shell.

short-period comet: a comet which orbits in the inner solar system. Halley's Comet is an example of a short-period comet.

sidereal period: the period of revolution of a planet around the Sun.

solar constant: the average amount of solar radiation reaching a planet; usually expressed in watts per square meter. For Earth, the solar constant equals 1372 W/m^2.

solar cycle: the periodic changing of the Sun's magnetic field that determines the number of sunspots and the amount of particles emitted in the solar wind. The period of the cycle is around 11 years.

solar eclipse: a phenomenon when the Moon's disk passes in front of the Sun, blocking sunlight. A total eclipse occurs when the Moon totally obscures the Sun's disk, leaving only the solar corona visible.

solar flare: a bright eruption of hot gas in the photosphere of the Sun.

solar maximum: the midpoint in the solar cycle where the amount of sunspot activity and the output of cosmic particles and solar radiation is highest.

solar minimum: the beginning and the end of a sunspot cycle when only a few sunspots are found and the output of particles and radiation is normal.
solar nebula: the cloud of dust and gas out of which the solar system was assumed to have formed about 5 billion years ago.
solar wind: a flow of charged particles which travels from the Sun out into the solar system.
spiral galaxy: a galaxy which contains a prominent central bulge and luminous arms of gas, dust, and young stars that wind out from the central nucleus in a spiral formation.
sprites: gamma-ray flashes produced in Earth's atmosphere by lightning storms and upper atmospheric events.
star: a giant ball of hot gas which creates and emits its own radiation through nuclear fusion.
stellar wind: the ejection of gas from the surface of a star.
sunspot: areas of the Sun's surface which are cooler than the surrounding areas.
superior planet: a planet which exists outside the orbit of the Earth.
super massive black hole: a black hole located at the center of a galaxy. This is formed by material falling onto the galaxy's core.
supernova: a supernova is a cataclysmic explosion produced when a star exhausts its fuel and ends its life.
supernova remnant: an expanding shell of gas ejected at high speeds by a supernova explosion. Supernova remnants are visible as diffuse gaseous nebulae with a shell-like structure.
synchronous rotation: a period of rotation of a satellite about its axis which is the same as the period of its orbit around its primary. This causes the satellite to always keep the same face to the primary. Our Moon is in synchronous rotation around the Earth.
terrestrial planet: a planet composed mainly of rock and iron, similar to that of Earth.
thermal radio emission: continuous radio emission from an object that results from the object's temperature. Blackbody radiation is a form of thermal radio emission. Planetary radio astronomy is the study of the physical characteristics of the planets in the solar system by means of the electromagnetic radio radiation emitted by these objects. The term is also used more generally to include the study of planetary ring systems, the moon, asteroids, satellites, and comets in the solar system. Radio astronomy generally refers to the (vacuum) wavelength range from about 0.5 mm (600 GHz = 600×10^9 Hz) to a kilometer (300 kHz) and long ward. There is no wide acceptance of the upper limit frequency. Observations from the ground cannot be carried out below a few megahertz because of the Earth's ionosphere that is opaque to very low frequency radio waves. At millimeter and submillimeter wavelengths, the terrestrial atmosphere defines windows where radio emissions from cosmic sources can reach the ground. Radio emissions have been measured from all of the planets and some satellites, asteroids, and comets. The observed continuum emissions from the planets can be broadly classified as quasi-thermal (having the same general shape as a blackbody emitter) and nonthermal (i.e., cyclotron, synchrotron). Narrow spectral lines from molecules have been observed in the atmospheres of planets, satellites, and comets. Planetary radio emissions originate in the solid mantles, atmospheres, and magnetospheres of the planets. A number of solar system spacecraft have carried radio astronomical instrumentation.
tidal force: the differential gravitational pull exerted on any extended body within the gravitational field of a second body.
transit: the passage of a celestial body across an observer's meridian.

ultraviolet: electromagnetic radiation at wavelengths shorter than the violet end of visible light.

umbra: the area of total darkness in the shadow produced by an eclipse.

Van Allen belts: radiation zones of charged particles which surround the Earth. The shape of the Van Allen belts is found by the Earth's magnetic field.

variable star: a star which fluctuates in brightness. These include eclipsing binaries.

white dwarf: a very small white star produced when an average-sized star uses up its fuel supply and collapses.

X-ray astronomy: the field of astronomy which is concerned with the celestial objects by the X-rays they emit.

X-ray star: a bright celestial object which gives off X-rays as a major portion of its radiation.

yellow dwarf: An ordinary star such as the Sun at a stable point in its evolution.

zodiac: an imaginary belt across the sky where the Sun, moon, and all of the planets can always be found.

Planetary Glossary[*†]

'A'a: blocky, angular, and rough type of lava flow.

accretion: the growth of planetary bodies from smaller objects by impact; one impact at a time. After formation, the bodies are said to have "accreted" from small objects.

achondrite: a class of stony meteorites that crystallized from magmas. The term means without chondrules.

AGB stars: cool, luminous, and pulsating red giant stars. Most stars in the universe that have left the main sequence will reach their final evolutionary stage as stars on the asymptotic giant branch (AGB).

agglutinates: common particle type in lunar "regolith"; agglutinates consist of small rock, mineral, and glass fragments bonded together with glass—a glass that is formed by flash heating when micrometeorite impacts melt the lunar regolith. The heat can also release solar-wind-implanted hydrogen and helium from the lunar regolith, causing vesicles (bubbles) in the quickly quenched glass. Agglutinates are typically tens of micrometers to a few millimeters in size.

albedo: the brightness of an object or surface; it is the percentage of incoming radiation (light) that the object or surface reflects.

Amazonian epoch: the geologic history of Mars has been divided into three broad time periods, or epochs. From oldest to youngest, these are "Noachian," "Hesperian," and Amazonian (named after places on Mars). These epochs are defined by the number of impact craters on the ground surface; older surfaces have more craters. The Amazonian epoch extends from about 2.9–3.3 billion years ago to the present day. The actual timing of the epochs is not known because we have different models of the rates of meteorite falls on Mars.

amino acids: molecules that are essential components of life as the chemical building units of proteins and enzymes. Amino acids are made of carbon, hydrogen, oxygen, and nitrogen. In the context of planetary science, amino acids are found in carbonaceous "chondrite" meteorites. There are ongoing research studies of the ramifications of

---

[*] Content was retrieved from NASA's SMD Planetary Science Division and Hawai'i Space Grant Consortium, Planetary Science Research Discoveries. http://www.psrd.hawaii.edu/PSRDglossary.html (accessed August 23, 2016).

[†] Supported by NASA's SMD Cosmochemistry Program and Hawai'i Space Grant Consortium.

prebiotic chemical reactions in early solar system environments and delivery of organic material from asteroids and comets to the early Earth.

ammonia: a gas molecule composed of one part nitrogen and three parts hydrogen; NH_3.

amorphous: lacking long-range crystalline structure; disordered.

andesite: dark-colored, fine-grained extrusive igneous rock with about 52–63 weight percent silica (SiO_2). Andesite consists mainly of plagioclase and one or more mafic minerals. The word andesite is derived from the Andes Mountains, located along the western edge of South America, where andesite rock is common.

angrite: a type of achondrite meteorite consisting of pyroxene, olivine, and plagioclase.

angstrom (Å): a unit of length equal to 10–8 cm.

anomaly: a deviation from the common rule, type, or form; something abnormal or inconsistent.

anorthosite: an "igneous" rock made up almost entirely of "plagioclase feldspar." This rock dominates the lunar highlands.

ANSMET (The Antarctic Search for Meteorites): an exploration program to collect meteorites in Antarctica supported by grants from the Office of Polar Programs of the U.S. National Science Foundation and by NASA. The work of the ANSMET teams provides the research community with a steady source of new extraterrestrial materials, including lunar and Martian samples [*ANSMET website*].

antenna: a conductor by which electromagnetic waves are transmitted or received.

antipode: a point on the surface of a sphere that is located diametrically opposite another point. For example, on the Earth the antipode of Hawaii is Botswana. The term is used commonly in lunar geology to describe locations on the Moon in relation to impact basins. Researchers have hypothesized that accumulations of converged ejecta materials, production of impact shock stresses, anomalies in lunar crustal magnetic fields, and geochemical anomalies are associated with antipodes of basins.

aphelion: point in the elliptical orbit of a comet, or asteroid, or planet about the Sun in which it is at its farthest distance from the Sun. The opposite orbital position is called "perihelion."

Apollo: U.S. Space Program that included six piloted lunar landings between 1969 and 1972. Apollo astronauts collected and returned 382 kg of rock and sediment samples from the Moon.

apparition: an appearance; in astronomy, the period of observation of a returning periodic comet.

argument of perihelion: angle along the orbit of a planet or other solar system object as measured from the ascending node (analogous to right ascension and longitude).

ascending node: the plane of a planet's orbit (or other solar system object) is usually tilted with respect to the plane defined by the solar system (the ecliptic). These two great circles intersect at two points. As the planet travels around its orbit, at one of the intersection points it will pass from below the ecliptic plane to above it (i.e., northward). This point is called the ascending node.

asteroid: a small, mostly rocky body orbiting the Sun. Asteroids range in size from 1000 km in diameter to tiny objects you could hold in your hand. Most asteroids orbit the Sun between Mars and Jupiter (the asteroid belt), and are the source of most meteorites. Principal asteroids are 1 Ceres (a dwarf planet), 2 Pallas, 3 Juno, and 4 "Vesta." The majority of asteroids fall into three types: C-type (carbonaceous), S-type (silicate or stony), and M-type (metallic).

asteroid belt: the region of space between the orbits of Mars and Jupiter, from 2.1 to 3.3 AU, where most asteroids are found.

asteroid family: group of asteroids with similar orbital elements (distance from the Sun, inclination of the orbital plane to the Earth's orbital plane, and eccentricity of the orbit) thought to be remnants of one larger asteroid parent.

astrobiology: study of the origin, distribution, and destiny of life in the universe. These are interdisciplinary studies involving astronomers, astrophysicists, biologists, chemists, geneticists, geologists, paleontologists, planetary scientists, and specialists from other scientific fields. There are three broad themes of astrobiology: Is there life elsewhere in the universe? How did life begin and evolve? What is the future for life on Earth and beyond?

astronaut: person engaged in or trained for spaceflight.

astronomical twilight: the time when the Sun is 12°–18° below the horizon. When the Sun moves from 19° to 18° below the horizon, astronomers can begin to detect an increase in the brightness of the sky, although to the naked eye the sky still looks dark.

atmophile: (gas-loving) element. Geochemists divide elements into four basic types (based on the Goldschmidt classification). Atmophile elements remain largely in the gas phase. The other types are "chalcophile," "lithophile," and "siderophile" elements.

atmosphere: mixture of gases that surround and are gravitationally attached to a planet.

AU (Astronomical Unit): the mean Earth–Sun distance, equal to 149.5 million kilometers or 93 million miles.

bacteria: single-celled microorganisms whose cells lack a nucleus. Bacteria comprise a major domain of life called prokaryotes. In contrast, the cells in eukaryotes have a well-defined nucleus; eukarya include molds, plants, and animals. Another major domain is called archaea, microorganisms with genetic features distinct from prokarya and eukarya. Most methane-producing bacteria are actually archaea, but for simplicity many biologists informally refer to all noneukaryotes as "bacteria."

basalt: fine-grained, dark-colored extrusive igneous rock with less than about 52 weight percent silica (SiO_2). Because of its low silica content, basalt has a low viscosity (resistance to flow). Basalt is composed primarily of plagioclase feldspar and pyroxene; other minerals such as olivine and ilmenite are usually present. Basalt is the most common rock on the surfaces of terrestrial planets and "differentiated" asteroids such as "4 Vesta." Basaltic magmas are made by the partial melting of "peridotite."

basin: see *lunar basin*.

biosphere: the part of Earth's crust, water, and atmosphere where living organisms can survive.

breccia: consolidated, fragmental rock consisting of rock, mineral, and/or glass fragments (also called clasts) embedded in a fine-grained matrix. (See also *monomict* and *polymict*.)

CAI: calcium–aluminum-rich inclusions (CAIs) are found in "chondritic" meteorites. CAIs are primitive objects that are formed in the "solar nebula" before the planets formed. CAIs are light-colored objects rich in "refractory" elements (that condense at a high temperature). Besides calcium and aluminum, this includes magnesium, titanium, and rare earth elements. CAIs range in size from about a millimeter to a centimeter. Meteoriticists have identified several distinct varieties of CAIs, but all share a high temperature origin. Some might be condensates from the solar nebula. Other CAIs might be evaporation residues. See also *FUN CAI*.

caldera: a large, basin-shaped volcanic depression caused by collapse after withdrawal of magma and eruption of lava.

canonical: the generally accepted, or established, procedure or quantity.
carbon: an element with atomic number 6; symbol: C. Graphite is a low-pressure form of carbon. Diamond is a high-pressure form of carbon. (Carbon is one of the four elements essential for life. The others are hydrogen, oxygen, and nitrogen.)
carbon dioxide: a compound formed by combining one carbon atom with two oxygen atoms, making the molecule CO_2. Carbon dioxide is an important part of the atmospheres of Venus and Mars. Carbon dioxide gas condenses to a solid below –78°C. This solid is commonly known as dry ice. The polar ice caps on Mars are made of frozen water and carbon dioxide.
carbon star: a rare class of red giant stars unusually rich in carbon or carbon compounds.
carbonaceous chondrite: type of "chondritic" meteorite that is classified by bulk chemical composition (into groups CI, CM, CR, CO, CV, CK) and by the amount of aqueous alteration (into types 1 and 2) or heating without much water (types 3–6). They are chemically primitive meteorites containing water-bearing minerals and carbon compounds including a variety of organic molecules such as "amino acids."
carbonate: common rock-forming mineral containing the carbonate ion, CO_3^{2-}. Common carbonate minerals are calcite, $CaCO_3$; siderite, $FeCaCO_3$; magnesite, $MgCO_3$; dolomite, $CaMg(CO_3)_2$; and ankerite, $Ca(Fe, Mg, Mn)(CO_3)_2$.
celestial body: a solid object found in space.
celestial equator: an imaginary great circle in the sky drawn concentric to the Earth's equator.
celestial sphere: an imaginary sphere centered on the Earth, or arbitrary large radius on the surface of which the stars are considered to be fixed.
Celsius: a temperature scale that assigns the value 0°C to the freezing point of water and the value of 100°C to the boiling point of water at standard pressure. To convert Celsius to "Fahrenheit": multiply by 9, then divide by 5, then add 32.
Ceres: the largest of the known asteroids; a dwarf planet. Also called 1 Ceres because it was the first asteroid to be discovered (by Piazzi in 1801).
chalcophile: (ore-loving) element. Geochemists divide elements into four basic types (based on the Goldschmidt classification). Chalcophile elements follow sulfur into the sulfide minerals. The other types are "lithophile," "siderophile," and "atmophile" elements.
chassignites: one of the three types of meteorites from Mars (the *SNC* meteorites). Chassignites are dunites, a rock composed almost entirely of the mineral olivine. The signature stone, Chassigny, fell in France in 1815.
chirality: in chemistry, the term describing a molecule that, due to its geometry, lacks an internal plane of symmetry and has a nonsuperposable mirror image. These mirror images can be designated L and D. Many amino acids are chiral. Life on Earth uses L-amino acids to make proteins.
chondrite: an abundant class of stony meteorites with chemical compositions similar to that of the Sun and characterized by the presence of chondrules (see definition below). Chondrites come from asteroids that did not melt when formed and are designated as H, L, LL, E, or C depending on chemical compositions. The H, L, and LL types are called ordinary chondrites. The L chondrites are composed of silicate minerals (mostly olivine and pyroxene, but feldspar as well), metallic nickel iron, and iron sulfide (called troilite). Most L chondrites are severely shock damaged, probably by a large impact on the asteroid in which they formed. The E type are called enstatite chondrites, a rare type that formed under very reducing conditions and are composed primarily of a magnesium silicate called enstatite.

They are subdivided into the low-iron (EL) chemical group and the high-iron (HL) group. The C—"carbonaceous chondrites"—contain water-bearing minerals and carbon compounds including a variety of organic molecules such as "amino acids." Carbonaceous chondrites are the most primitive meteorites—primitive in a chemical way. For example, the CI group of carbonaceous chondrites are closest in composition to the photosphere (visible surface) of the Sun.

chondrule: roughly spherical objects found in a type of meteorite called chondrites. Most chondrules are 0.5–2 mm in size and are composed of olivine and pyroxene, with smaller amounts of glass and iron-nickel metal. Two main chondrule types have been identified. Type I contain only small amounts of oxidized iron (FeO); olivine crystals in them contain only about 2 mole percent of the iron-rich olivine fayalite (Fe_2SiO_4) end member. Type II chondrules contain much more FeO; olivine crystals in them typically contain 10–30 mole percent fayalite. The shapes of the mineral grains in them indicate that chondrules were once molten droplets floating freely in space.

clementine: a global mapping mission to the Moon launched in 1994 by the U.S. Department of Defense, with science support from NASA.

clinopyroxene: monoclinic "pyroxene" mineral with a general formula of $XY(Si, Al)_2O_6$, where X can be Na^+, Ca^{+2}, Mg^{+2}, Fe^{+2}, Mn^{+2} and Y can be Mn^{+2}, Fe^{+3}, Mg^{+2}, Al^{+3}, Cr^{+3}, Ti^{+4}. Common clinopyroxenes include $CaMgSi_2O_6$, diopside; (Ca, Na)(Mg, Fe, Al)$(Si, Al)_2O_6$, augite; and $CaFeSi_2O_6$, hedenbergite.

closure temperature: the temperature low enough so that there is no measureable "diffusion" of radioactive decay products (e.g., 40Ar) out a rock or of elements between co-existing minerals (e.g., between two minerals such as "spinel" and "olivine"). Closure temperature depends on cooling rate and diffusion rate, hence specific minerals, isotopes, and elements are involved.

coma: a large cloud of dust and gas which escapes from the nucleus of an active comet.

comet: a small body (a "dirty iceball," typically 1 km across, with dust and plasma tails) that circulates the Sun with a highly elliptical orbit.

condensation: defined in cosmochemical terms, the process that transforms a gas into a solid. For example, corundum (Al_2O_3) is one of the earliest solid phases predicted to condense from a gas of solar composition. In classical geochemical terms, condensation describes a phase change of a gas into a liquid. But liquids are not stable at low pressures, such as those calculated for the "solar nebula" as being one-thousandth to one-millionth of the atmospheric pressure at Earth's surface. At these low pressures, minerals condense at temperatures below their melting points, so they condense as solids.*

convection: movements of "mantle" material, laterally or in upward–downward directions, due to heat variations.

Copernican: the geologic history of Earth's Moon has been divided into five broad time periods, or systems. From oldest to youngest, these are "pre-Nectarian," "Nectarian," "Imbrian," "Eratosthenian," and "Copernican." These systems are defined by the number of impact craters on the ground surface (more craters = older surface) correlated to the absolute ages determined for the lunar rock samples returned to Earth by the U.S. "Apollo" missions and unpiloted Soviet "Luna" missions. The Copernican system extends from 1.1 billion years ago to the present day.

---

* For more details, we refer you to the textbook, McSween Jr, H. Y. and Huss, G. R. (2010). *Cosmochemistry*. Cambridge University Press, Cambridge, UK. pp. 192–227.

core: the central region of a planet or moon frequently made of denser materials than the surrounding regions (mantle and crust). Earth and the Moon are thought to have cores of iron and nickel.

cosmic ray exposure age: the time interval that represents how long a surface of an airless body or a meter-sized or smaller asteroid traveled in space; based on measurements of accumulated radionuclides in the meteorite that were produced by exposure to the cosmic rays in space.

cosmic rays: extremely high-energy subatomic particles that continuously bombard Earth and other planetary bodies from all directions. Most cosmic rays hit and break up atomic nuclei in Earth's upper atmosphere. Airless bodies are not protected and their surfaces (asteroids and the Moon, for example) are exposed. When cosmic rays collide with the atoms of the elements in soils and rocks on a planetary surface, the nuclei of the atoms emit "gamma rays"—by measuring these gamma rays with a gamma ray spectrometer it is possible to calculate the abundance and distribution of the chemical elements on the planet's or moon's surface.

cosmochemistry: interdisciplinary science that overlaps with geochemistry, geology, astronomy, astrophysics, and geophysics to discover the materials and fundamental processes in the "solar nebula" and our "solar system." These sciences give us complementary ways of looking at our origins by addressing questions like How did the Sun and planets form? Where did we come from? Cosmochemistry, because it is a careful examination of the building blocks of the cosmos, also plays an important role in developing local resources on the Moon, Mars, and asteroids, essential to sustained human presence in space.

crater: (see *Impact*, *Transient crater*) a hole or depression. Most are roughly circular or oval in outline. On Earth most natural craters are of volcanic origin. On the Moon most are of impact origin.

crater chain: several craters along a general line.

crust: the outermost layer of a planet or moon, above the mantle.

cryptomaria (singular: cryptomare): buried mare basalt lavas on the Moon. Many light plains in the lunar highlands have impact craters surrounded by dark piles of ejecta, nicknamed "dark-haloed craters." The dark-haloed craters formed when mare basalt lava flows were covered with ejecta from large impact craters and basins, and then small craters punctured through the ejecta to toss out mare basalt.

dark mantle deposits: deposits of dark glass on the Moon, possibly products of volcanic fire fountaining.

declination: celestial coordinate which is equivalent to the latitude of an object on Earth. This is an angular measure of how far above the celestial equator an object lies.

density: mass per unit volume; a measure of how much material is in a given space. Bulk density refers to the entire volume of a sample, including "pore" space whereas grain density is based on the solid material, excluding the pore space.

density current: water movement due to differences in the density of the water from place to place usually caused by changes in the amount of material held in suspension or changes in temperature or variations in salinity.

Descartes: lunar highlands site of Apollo 16 landing on April 21, 1972.

deuterium: an isotope of hydrogen; also called heavy water (not radioactive). Its nucleus contains a proton and a neutron; hydrogen's nucleus contains only a proton. As a reference, the measured deuterium/hydrogen ratio (D/H) in Earth's standard mean ocean water (*SMOW*) is 1.558×10^{-4}. In our solar system, the water D/H ratio is a kind of fingerprint of where icy "planetesimals" formed; the ratio increases with increasing formation distance from the Sun.

differentiation: chemical zonation caused by differences in the densities of minerals; heavy materials sink, less dense materials float. Differentiation transforms an initially homogeneous body into a body with compositionally distinct "core," "mantle," and "crust."

diffusion: the random movement (spreading) and mass transport of molecules within a phase from regions of highest concentrations to regions of lowest concentrations.

dissociation: the splitting apart of a molecule into its constituent atoms. For example, the splitting of carbon monoxide into carbon and oxygen or the splitting of water into hydrogen and oxygen when struck by ultraviolet radiation.

diurnal: a single daily cycle.

earthquake: sudden motion or trembling of Earth caused by the abrupt release of slowly accumulated elastic energy in rocks.

eccentricity: the measure of the degree of elongation of an ellipse. For example, a circle has an eccentricity of 0, and a parabola (an open figure) an eccentricity of 1. The deviation of an orbit from a perfect circle.

ecliptic: the plane of the solar system.

ejecta: material thrown out from and deposited around an impact crater.

electromagnet spectrum: the full range of frequencies, from radio waves to gamma rays, that characterizes light. The electromagnetic spectrum can be expressed in terms of energy, wavelength, or frequency (where the wavelength equals the speed of light divided by the frequency and energy equals Planck's constant times the frequency.)

electron microprobe: an instrument that analyzes the chemistry of very small spots by bombarding the sample with electrons and measuring the X-rays produced.

electron volt: the energy given to an electron by accelerating it through 1 volt of electric potential difference.

emissivity: the ratio of radiant flux (total power of electromagnetic radiation including infrared, ultraviolet, and visible per unit of area) from a body to that from a black body at the same kinetic temperature.

enstatite: $MgSiO_3$, The magnesium-rich version of a group of minerals called "pyroxenes."

ephemeris: a table of positions in right ascension (celestial coordinate corresponding to longitude on Earth) and declination (celestial coordinate corresponding to latitude on Earth), as a function of time for a moving celestial object.

eratosthenian: the geologic history of Earth's Moon has been divided into five broad time periods, or systems. From oldest to youngest, these are "pre-Nectarian," "Nectarian," "Imbrian," "Eratosthenian," and "Copernican." These systems are defined by the number of impact craters on the ground surface (more craters = older surface) correlated to the absolute ages determined for the lunar rock samples returned to Earth by the U.S. "Apollo" missions and unpiloted Soviet "Luna" missions. The Eratosthenian system extends from 3.2 billion to 1.1 billion years ago.

erosion: removal of weathered rocks by moving water, wind, ice, or gravity.

eruption: (see *Source*) a break out or burst of volcanic matter.

eucrite: a common class of achondrite meteorites composed of pigeonite and plagioclase. These meteorites formed as basaltic flows on a parent body, probably asteroid 4 Vesta. (Data collected by NASA's Dawn Mission, in orbit around Vesta from 2011–2012, strengthened the association between Vesta and eucrite meteorites.)

exsolution: the segregation, during cooling, of a homogeneous solid solution into two or more different minerals.

extrusive: refers to igneous rocks erupted on the surface.

Fahrenheit: a temperature scale with the freezing point of water assigned the value 32°F and the boiling point of water 212°F at standard pressure. To convert Fahrenheit to "Celsius": deduct 32, then multiply by 5, then divide by 9.

fall: a meteorite that was observed as it fell through Earth's atmosphere and was retrieved. Since most falls are collected soon after hitting the ground, these meteorites are not badly degraded or weathered.

farside: the side of the Moon that always faces away from Earth or never faces the Earth.

fault: A fracture along which there has been movement or displacement.

feldspar: see *plagioclase feldspar.*

felsic: term describing the chemical composition of an igneous rock composed of >65 percent silica. (See also *mafic* and *ultramafic.*)

find: a meteorite that was not observed falling through the Earth's atmosphere, but was found at some later date.

fireball: the streak of light and loud noise of a large meteor passing through the Earth's atmosphere.

Fischer–Tropsch reaction: production of organic molecules by the addition of hydrogen to carbon monoxide aided by a suitable catalyst.

fissure: crack extending far into a planet or moon through which magma travels to and erupts onto the surface.

Fra Mauro: landing site of Apollo 14 on the Moon on February 5, 1971.

fractional crystallization: geochemical and physical processes that result in the sequential formation and separation of minerals from a melt. The residual magma changes in composition and becomes enriched in "incompatible elements."

FUN CAI: a rare type of *CAI.* F stands for fractionation, UN stands for unidentified nuclear isotope properties. FUN CAIs are characterized by 26Al/27Al ratios much lower than the "canonical" value of $\sim 5 \times 10^{-5}$ (at the time of our solar system's formation); they also can have large isotopic anomalies in many elements.

fusion crust: the glassy, melted rind on a meteorite that forms when the rock passes through the Earth's atmosphere.

gabbro: a course-grained igneous rock rich in the minerals olivine, pyroxene, and plagioclase feldspar.

Galilean moons: Jupiter's four largest moons: Io, Europa, Ganymede, and Callisto, discovered independently by Galileo Galilei and Simon Marius in 1609–1610.

Galileo mission: U.S. mission to study Jupiter's atmosphere, moons, and magnetosphere. The Galileo spacecraft was carried into space and released in 1989 by the shuttle Atlantis. It used its own rocket booster to leave Earth's orbit. The Galileo spacecraft arrived at Jupiter in 1995 and completed its two-year primary mission before going on to explore Europa. The 14-year mission ended in 2003. *Project Galileo* website.

gamma ray: gamma rays are like the light we see with our eyes, the X-rays used to probe our bodies in search of problems, and the radio waves that our television sets translate into inane shows. The only difference among these types of electromagnetic radiation is the wavelengths of the waves and how fast they vibrate. Gamma rays are the most energetic (and vibrate the most); radio waves are the least energetic and vibrate the least. When "cosmic rays" collide with the atoms of the elements in soils and rocks on a planetary surface, the nuclei of the atoms emit gamma rays—by measuring these gamma rays with a gamma ray spectrometer it is possible to calculate the abundance and distribution of the chemical elements on the planet's or moon's surface.

genesis mission: U.S. mission launched in August 2001 to observe and collect solar wind samples and return them to Earth. Sample collections were obtained at the L1 point, approximately one million miles from Earth where the gravities of Earth and Sun are balanced. The sample capsule landed back on Earth in September 2004 in an unexpected crash after its deceleration sensor failed and the parachutes did not open. Nevertheless, solar wind samples were recovered, cleaned, and stored successfully and are now being studied meticulously. *Genesis Mission* webpage.

geocentric distance: the distance from Earth.

geologist: scientist who studies Earth, its structure, its materials, the physical and chemical processes and changes that occur on the surface and in the interior, and the history of the planet and its life forms. *"Planetary geologists"* extend their studies to the Moon, planets, and other solid bodies in the solar system.

giant impact theory: an explanation for the origin of the Moon from Earth debris which collected in space after a projectile the size of planet Mars smashed into a growing Earth.

gigapascal (GPa): conventional unit of pressure when discussing the deep Earth: 1 GPa = 10 kilobars = 10,000 times air pressure at sea level.

geomorphology: the study of the shape and form of the landscape, and how the nature of landforms relates to their origin, development, and change over time.

globular clusters: groups of thousands to millions of old stars held together in orbits about their own center of mass and arranged in almost symmetrical (spherical) clusters.

GRAIL Mission: U.S. orbital mission to the Moon, with twin spacecraft, launched in September 2011 as a component of the NASA Discovery Program. The GRAIL (Gravity Recovery and Interior Laboratory) Mission has two primary objectives: to determine the structure of the lunar interior, from crust to core; and to advance understanding of the thermal evolution of the Moon. These broad objectives are addressed by mapping the gravity field of the Moon to provide the structure of the crust and lithosphere, help understand the Moon's asymmetric thermal evolution, determine the subsurface structure of impact basins and the origin of mascons, ascertain the temporal evolution of crustal brecciation and magmatism, place constraints on the deep interior structure from tides, and place limits on the size of the possible inner core. As a secondary objective, GRAIL observations will be used to extend understanding of the internal structure and thermal evolution of other terrestrial bodies. *GRAIL Mission* webpage.

Hadley–Apennine: landing site of Apollo 15 on the Moon on July 31, 1971.

half-life: the amount of time required for half of the mass of a radioactive "isotope" to decay.

HED: the three linked stony "achondritic" meteorite groups known as the HEDs are howardites, eucrites, and diogenites. They come from asteroid "Vesta." (Data collected by NASA's Dawn Mission, in orbit around Vesta from 2011–2012, strengthened the association between Vesta and HED meteorites.)

heliocentric distance: the distance from the Sun.

helium: an element with atomic number 2; symbol: He. It is the second most common element in the Sun and outer planets, but rare on the rocky planets.

hematite: an iron-oxide mineral (Fe_2O_3) that may be the cause of the red color on Mars. Courser-grained gray hematite has the same chemical formula as the red variety, but a different crystalline structure. The deposits of gray hematite found in the Terra Meridiani region of Mars may suggest that water once circulated through the region's rock layers.

Hesperian epoch: the geologic history of Mars has been divided into three broad time periods, or epochs. From oldest to youngest, these are "Noachian," "Hesperian," and "Amazonian" (named after places on Mars). These epochs are defined by the number of impact craters on the ground surface; older surfaces have more craters. The Hesperian epoch extends from about 3.5–3.7 to about 2.9–3.3 billion years ago. The actual timing of the epochs is not known because we have different models of the rates of meteorite falls on Mars.

highland "soil": sediment on the surface of the lunar highlands; composed of broken rock and mineral fragments, and glass produced by impact.

highlands: oldest exposed areas on the surface of the Moon; extensively cratered, and chemically distinct from the maria.

Hubble constant: the expansion rate of our universe or rate at which our universe is stretching apart. As of October 2012, the measurement of the Hubble constant is 74.3 km per second per mega "parsec" (a megaparsec is roughly three million light-years) announced by astronomers using NASA's Spitzer Space Telescope. This value agrees with a 2011 supernovae study, and improves by a factor of three on a 2001 study. (See the *Spitzer news article* from the Jet Propulsion Laboratory.)

hydrogen: an element with atomic number 1; symbol: H. It is the most abundant element in the Solar System, making up 90% of the Sun. (See also, "deuterium," one of the naturally occurring "isotopes" of hydrogen.) Hydrogen, carbon, nitrogen, and oxygen are essential for life.

hydrosphere: the portion of Earth, or other planet, that is water, including liquid water, ice, and water vapor on the surface, underground, or in the "atmosphere."

IDP: see *Interplanetary dust particles.*

igneous: rocks or processes involving the formation and solidification of hot, molten magma.

ilmenite: opaque mineral found in basalt; nearly pure iron-titanium oxide ($FeTiO_3$).

Imbrian: the geologic history of Earth's Moon has been divided into five broad time periods, or systems. From oldest to youngest, these are "pre-Nectarian," "Nectarian," "Imbrian," "Eratosthenian," and "Copernican." These systems are defined by the number of impact craters on the ground surface (more craters = older surface) correlated to the absolute ages determined for the lunar rock samples returned to Earth by the U.S. "Apollo" missions and unpiloted Soviet "Luna" missions. The Imbrian system extends from 3.85 billion to 3.2 billion years ago.

impact: (see *Crater*) the forceful striking of one body, such as a meteorite, against another body such as a moon or planet.

inclination: Angle between the plane of the object's orbit and the "ecliptic" (defined by Earth's orbit).

inclusions: fragments of older material (e.g., minerals, rock) that have been enclosed within another rock.

incompatible elements: elements that are not incorporated into common rock-forming minerals during magma crystallization, hence they become enriched in the residual magma and in the rocks finally formed from it. Examples are potassium, rubidium, strontium, hafnium, thorium, uranium, and lanthanum, samarium, europium, and the rest of the rare earth elements (*REE*).

interplanetary dust particles (IDPs): particles typically ranging in size from 1–50 μ (microns), though some are larger and most are between 5 and 15 μm, moving in interplanetary space. They come from asteroids or comets, and reach Earth-crossing orbits; IDPs have been collected in the stratosphere using NASA ER2 aircraft. IDPs are grouped into two main types: chondritic porous (also known as anhydrous) and chondritic smooth (also known as hydrated).

interstellar medium: the material between the stars, consisting of gas, dust, and cosmic rays (high-energy charged particles, moving at nearly the speed of light).

intrusive: refers to igneous rocks formed underground.

ion microprobe: (see *Secondary Ion Mass Spectrometer* [*SIMS*]), a precision instrument used to quantitatively analyze the elements and isotopes of materials at microscopic scales. For more, see the PSRD article: *Ion Microprobe*.

iron: an element with atomic number 26; symbol: Fe. Iron is one of the most abundant elements in the rocky planets. It is the most abundant element in the metallic cores of the inner planets.

iron meteorite: iron *meteorites* are made, almost completely, of iron and nickel metal. They are chemically distinguished and grouped according to the abundances of the trace elements such as gallium and germanium, as well as nickel. Initially, irons were classified into four groups and were given Roman numerals I, II, III, and IV. Today 12 groups are recognized and designated further by letters A through F according to concentrations of "siderophile" ("iron-loving") trace elements. Iron meteorites that do not fit into the groups are called ungrouped. The two iron-nickel alloys in iron meteorites are called kamacite (low-nickel content, usually up to 7.5 wt% nickel) and taenite (high-nickel content, ~20–50 wt% nickel). These alloys are rare in terrestrial rocks. (See also *Widmanstätten Pattern*.)

iron–wustite buffer: in geology, a buffer is an assemblage of minerals that constrain "oxygen fugacity" as a function of temperature. The iron–wüstite buffer (IW) is the iron metal–iron oxide reference buffer commonly used in cosmochemistry.

isochron: in radiometric age dating by isotopic analyses: The line in an isochron diagram, whose slope is proportional to the amount of time passed (it defines the crystallization age) and whose intercept gives the initial isotopic ratio of the sample when it crystallized. In relative age dating by crater counting: The line on a log–log plot (of cumulative number of craters per km^2 vs. crater diameter in kilometers) indicating the size distribution of impact craters found on a surface of a specified age, if no other processes have obliterated or altered the surface.

isotopes: two or more varieties of the same chemical element. Isotopes have different masses because they have different numbers of neutrons (e.g., ^{18}O and ^{16}O), thus they have different physical and chemical properties. Unstable isotopes undergo "radioactive" decay.

isotopic dating: see *Radiometric Dating*.

isotopic fractionation: the separation of isotopes or enrichment of one isotope relative to another, which can be a result of kinetic or equilibrium processes.

Kaguya mission: JAXA mission to the Moon; see *SELENE* (Kaguya).

Kelvin: the fundamental unit of temperature. It is not calibrated in terms of the freezing and boiling points of water, but in terms of energy itself. The number 0 K is assigned to the lowest possible temperature, called absolute zero. (At absolute zero all molecular motion stops.) The sizes of the "degree" are the same as on the "Celsius" scale. On the Kelvin scale, the freezing point of water is 273 K and the boiling point is 373 K.

keV: A unit of energy equal to a thousand electron volts (eV). According to the Bureau International des Poids et Mesures: "The electron volt is the kinetic energy acquired by an electron in passing through a potential difference of one volt in vacuum. The electron volt is often combined with the SI prefixes."

KREEP: an acronym for a geochemical component in lunar rocks rich in potassium (K), rare-earth elements (REE), phosphorus (P), thorium, and other incompatible elements. These elements are not incorporated into common rock-forming minerals during magma crystallization; hence they become enriched in the residual magma and in the rocks finally formed from it.

landslide: general term to describe the process of mass movement (or avalanche) of material downslope by falling, sliding, or flowing, under the force of gravity. It also describes the landform (pile of debris) produced.

laser: an acronym for light amplification by the stimulated emission of radiation. Laser devices produce a narrow, continuous or pulsed beam of coherent light in a specific wavelength, which can be in the visible, infrared, or ultraviolet regions of the electromagnetic spectrum.

Late Heavy Bombardment: see *Lunar Cataclysm*.

latitude: the angular distance North or South from the Earth's equator measured in degrees on the meridian of a point; Equator being 0° and the poles 90° N and 90° S.

lava: fluid magma that flows onto the surface of a planet or moon; erupted from a volcano or fissure. Also, the rock formed by solidification of this material.

leaching: the loss or dissolving out of soluble substances from a rock by percolation of water or other fluid.

levee: zones in a lava flow where the lava between the zones is moving faster than the lava outside the zones.

light-year: distance that light can travel in 1 year. Light travels at a speed of 300,000 km/s, so this distance is equal to 9.46 × 1012 km.

liquidus: the highest temperature at which crystals and melt can co-exist in thermodynamic equilibrium. On a plot of temperature (y-axis) versus pressure (x-axis), everything is liquid (completely melted) above the liquidus curve and the crystallization order of minerals at a given pressure can be determined by following a straight line from the liquidus curve to the "solidus" curve.

lithology: the description of the physical characteristics of rocks including color, mineral compositions, grain sizes, grain shapes, texture, or fabric (relationship between grains).

lithophile: (rock-loving) element. Geochemists divide elements into four basic types (based on the Goldschmidt classification). Lithophile elements are those that are contained primarily in silicate minerals. The other types are "chalcophile," "siderophile," and "atmophile" elements.

lithosphere: layer in a planetary body consisting of the "crust" and upper "mantle."

longitude: the angular distance East or West, between the meridian of a particular place on Earth and that of Greenwich, England, expressed in degrees or time.

long-period: in astronomy, a comet with an orbital period greater than 200 years.

Luna: one of two successful series of Soviet unpiloted lunar missions. There were 24 missions to the Moon in the Luna series launched from 1959 to 1976. (The other series, Zond, had 5 lunar missions launched in 1965 to 1970.) The first image of the farside of the Moon was taken by the Luna 3 spacecraft in October 1959. Three Luna landers (Luna 16, 20, and 24) collected a total of approximately 300 g of lunar samples and returned them to Earth.

lunar: of or pertaining to the Moon.

lunar basin: an "impact crater" on the Moon larger than 300 km in diameter with two or more concentric rims and no central peak.

lunar cataclysm: term referring to the idea that a sharp increase in impact bombardment rate between about 3.92 and 3.85 billion years ago affected the inner solar system—most notably recorded in the numerous craters and large basins on the Moon. Also described as the late heavy bombardment.

lunar meteorites: meteorites that have been identified through their chemistry and minerals as being from the Moon. (See *Lunar Meteorites* compiled by Randy Korotev from Washington University, St. Louis, MO.)

Lunar Prospector: U.S. Discovery-class mission to the Moon launched on January 6, 1998. Its instruments were designed to provide global maps and data sets of the Moon's composition and magnetic and gravity fields from a low polar orbit. The mission lasted 18 months and ended in July 31, 1999, with a controlled crash landing into a crater at the Moon's south pole.

Lunar Reconnaissance Orbiter (LRO): NASA's LRO is a Moon-orbiting mission launched in the fall of 2008. Its low-polar orbit and instrument payload is acquiring new data in support of future human landing sites and resource exploration including polar volatiles (especially water ice). These data include high-resolution imaging and topography, lunar radiation environment, and global surface composition. The Lunar Crater Observation and Sensing Satellite (LCROSS) launched with LRO, then traveled independently for a controlled crash on October 9, 2009, into a permanently shadowed crater at the lunar south pole to support our search for water ice. An estimated 5.6% ± 2.9% by mass water ice was reported in the regolith at the LCROSS impact site, Cabeus crater.

mafic: term describing the chemical composition of an igneous rock composed of dark-colored, magnesium-rich and iron-rich minerals, generally with an average silica content between 45% and 55%. (See also *felsic* and *ultramafic*.)

magma: term applied to molten rock in the interior of a planet or moon. When it reaches the surface, magma is called lava.

magma ocean: term used to describe the thickness of magma thought to have covered the Moon 4.5 billion years ago. The "magma ocean concept" is the idea that the Moon melted substantially (probably completely) when it formed; this is an important theory in lunar science that is being applied to other planetary bodies, such as Mars. The timing, duration, cooling rates, convection, partial melting, lateral and vertical chemical variations, and overturn of the lunar magma ocean are topics of continuing research.

magnesium: an element with atomic number 12; symbol: Mg. Magnesium is contained in the minerals olivine and pyroxene, and is very abundant in the mantles of the inner planets.

magnesium number: The ratio of magnesium to iron, usually reported as the molar ratio Mg/(Mg+Fe), and abbreviated as Mg# or Mg′ or Mg*. This value is used commonly to discriminate between lunar rock types.

magnetic field: the region of "altered space" that will interact with the magnetic properties of a magnet. It is located mainly between the opposite poles of a magnet or in the energetic space about an electric charge in motion.

magnetite: a black, strongly magnetic iron-oxide mineral; Fe_3O_4.

magnetofossils: small grains of magnetite occurring in sedimentary rocks and thought to have been produced by bacteria. (Some bacteria, called magnetobacteria, make magnetite, presumably as an aid to determining directions.)

magnetometer: sensitive scientific instrument used to measure the intensity at various points of Earth's (or another planetary body's) magnetic field.

magnetosphere: the region around Earth dominated by Earth's magnetic field. In this region, the Earth's magnetic field lines guide the motions of charged particles (ions and electrons).

magnitude: astronomical brightness measured on a logarithmic scale, based on the ancient practice of noting that the brightest stars in the sky were of "first importance" or "first magnitude," the next brightest being "second magnitude," and so on. The human eye is a logarithmic detector, and in 1854 Pogson formalized this scale and defined a difference of 5 magnitudes to be exactly a factor of 100 in brightness. The scale is calibrated to the bright star Vega which is defined to have a magnitude of 0. For reference, the faintest naked-eye magnitude visible from a dark side is +6. Comet C/1996 B2 Hyakutake reached magnitude 0 in late March 1996.

mantle: a mostly solid layer of Earth lying beneath the crust and above the core; consisting mostly of iron, magnesium, silicon, and oxygen. On other planets and moons, the mantle is the layer between the crust and core. Mantles contain information about a planet's total composition, a key parameter in understanding planet formation and how the planets vary in composition with distance from the Sun. A second definition of mantle is a smooth blanket of surface material (generally fine-grained, sometimes layered) that was deposited by some process such as wind or water and that now obscures the underlying land.

mare basalt: rocks making up the dark, smooth, mare areas of the Moon.

mare "soil": sediment on the surface of the lunar maria; fragments of basalt rocks, broken mineral grains, and glass produced by impact.

maria (singular: mare): dark areas on the Moon covered by "basalt" lava flows.

Mariner 9: U.S. mission to Mars, launched in 1971, achieved global imaging of the surface, including the first detailed views of the Martian volcanoes, Valles Marineris, the polar caps, and the satellites Phobos and Deimos. The spacecraft gathered data on the atmospheric composition, density, pressure, and temperature and also the surface composition, temperature, gravity, and topography of Mars. *Mariner 9* website.

Mars Exploration Rovers: U.S. missions to Mars with twin rovers equipped with tools to study a diverse collection of rocks and soils that may hold clues to past water activity on Mars. MER-A, named "Spirit," launched on June 10, 2003, for a seven-month journey to a landing site in Gusev Crater and MER-B, named "Opportunity," launched on July 7, 2003, toward a landing site in Terra Meridiani.

Mars Express: The European Space Agency's first spacecraft to Mars, Mars Express launched in June 2003. The mission objectives are to search for subsurface water from orbit with a set of seven remote sensing instruments. The lander, Beagle 2, was released from the orbiter 6 days before Mars Express went into orbit around Mars, however, no signal was received from it despite repeated attempts at communication from a variety of spacecraft and Earth-based telescopes. Beagle 2 was declared lost on December 2003. Although the lander failed to fully deploy after it landed on the Martian surface, the orbiter has been successfully performing scientific measurements since early 2004 and is expected to continue through 2014.*

---

* For more details on Mars Express, we refer you to websites: http://mars.nasa.gov/programmissions/missions/present/express/ and https://en.wikipedia.org/wiki/Mars_Express.

Mars Global Surveyor: U.S. mission launched in 1996 to orbit Mars. Main instruments include cameras (*MOC*), a laser altimeter (*MOLA*), thermal emission spectrometer (*TES*), and magnetometer. During the primary mission (through January 2001), the spacecraft circled the planet once every 118 min at an average altitude of 378 km (235 miles). Four extended mission phases began February 1, 2001, and continued to November 2006 when the orbiter stopped communicating with Earth (due to solar panel problems). *Mars Global Surveyor* website.

Mars Mission, Phoenix: see *Phoenix Mars Mission.*

Mars Odyssey: U.S. orbital mission to Mars, part of NASA's Mars Exploration Program, launched in April 2001 and arrived at Mars in October 2001. The mission's goals are to map chemical elements and minerals, look for water, and analyze the low-orbit radiation environment using three primary instruments: Thermal Emission Imaging System (THEMIS), Gamma Ray Spectrometer (GRS), and Mars Radiation Environment Experiment (MARIE). During and after its science mission, the Odyssey orbiter will also support other missions in the Mars Exploration program. It is providing communications relay for U.S. and international landers, including the Mars Exploration Rovers launched in June and July 2003. *2001 Mars Odyssey* website.

Mars Orbiter Camera (MOC): camera system onboard the "Mars Global Surveyor" spacecraft. MOC consists of three cameras: A narrow angle system that provides grayscale high-resolution views of the planet's surface (typically, 1.5–12 m/pixel), and red and blue wide angle cameras. *Malin Space Science Systems.*

Mars Orbiter Laser Altimeter (MOLA): instrument onboard the "Mars Global Surveyor" spacecraft that has been obtaining "topographic" data since March 1999. MOLA transmits focused infrared laser pulses toward Mars at a rate of ten times per second and measures the round-trip reflection time to determine the range of the MGS spacecraft to the Martian surface. Scientists use the range measurements to construct precise topographic profiles and maps of Mars in unsurpassed detail.

Mars Pathfinder: the second in NASA's Discovery program of low-cost spacecraft with highly focused science goals, Mars Pathfinder landed successfully on Mars on July 4, 1997. Images, science results, and engineering data were released on the World Wide Web throughout, and beyond, the primary mission dates (July 4–August 3, 1997). The Jet Propulsion Laboratory (JPL), Pasadena, Californai, developed and managed the Mars Pathfinder mission for NASA's Office of Space Science, Washington, DC. JPL is a division of the California Institute of Technology (Caltech).

Mars Reconnaissance Orbiter (MRO): NASA's orbital mission to Mars launched in August 2005. MRO is conducting science investigations with a suite of instruments tasked for three purposes: global mapping, regional surveying, and high-resolution targeting of specific spots on the surface. Goals include studying the history of water on Mars, analyzing minerals, tracing how much dust and water are distributed in the atmosphere, and monitoring daily global weather. MRO also serves as a communications relay for landed missions.

Mars Science Laboratory (MSL): a NASA robotic lander (named Curiosity) designed to determine the habitability of Mars. MSL launched on November 26, 2011, and landed on Mars on August 5, 2012, as part of NASA's long-term Mars Exploration Program. MSL science goals are to determine whether Mars ever had an environment capable of supporting microbial life, to characterize the geology and climate of Mars, and to prepare for human exploration. MSL science instruments include three cameras, four spectrometers, two radiation detectors, and an environmental sensor. *MSL* homepage.

Martian meteorites: (See *SNC meteorites* and *Meteorite.*) Of the 24,000+ meteorites collected on Earth, only 30+ have been identified as meteorites from Mars. They are generally divided into three groups (all with relatively young ages slightly over 1 billion years old). The groups are the "shergottites," "nakhlites," and "chassignites." A Martian meteorite that does not fall into one of these groups is ALH 84001, a cumulate rock composed mostly of orthopyroxene, which is much older than all the others with an age of 4.5 billion years. A NASA-sponsored research team reported in 1996 that ALH 84001 may contain evidence of past primitive life on Mars—an idea that remains a topic of great debate and continued investigation. (See *Mars Meteorites* compiled by Ron Baalke from the Jet Propulsion Laboratory.)

mass spectrometer: instrument that produces and measures, usually by electrical means, a mass spectrum. It separates ions according to the ratio of their mass to charge, allowing scientists to determine the abundances of each isotope.

mass spectrum: the pattern of the relative abundances of ions of different atomic or molecular mass (mass-to-charge ratio) within a sample. It frequently refers to the measured relative abundances of isotopes of a given element.

megaregolith: term coined by William K. Hartmann in 1973 to describe the lunar-wide zone of fragmented material developed on the Moon by mixing of impact basin/crater ejecta (transported by ballistic trajectories) and local rock. Estimates of the thickness of this mixed zone are between 1.5 and 2.5 km.

mesosiderite: a class of stony-iron meteorites consisting of metal and fragments of igneous rocks. These meteorites formed as breccias, but most have been recrystallized during metamorphism.

metamorphic: rocks that have recrystallized in a solid state as a result of changes in temperature, pressure, and chemical environment.

metasomatism: the change in the composition of a rock as a result of the addition or removal of chemical elements, commonly due to the interaction of fluids with the rock. "Contact" metasomatism refers to chemical changes in a rock by direct contact (or near the contact) with magma.

meteor: the visible light in the sky produced when a meteoroid or small asteroid passes through Earth's atmosphere (can also refer to the glowing meteoroid itself).

meteorite: a natural solid object (metallic or stony/silicate) larger than 10 μm (microns) that has traveled through space, by natural means from the celestial body on which it is formed, and has landed on Earth or other planetary body or artificial body/satellite (larger than itself). Most meteorites come from "asteroids," but a small number found on Earth came from the Moon (see *Lunar meteorites*) or Mars (see *Martian meteorites*). Meteorite types include "iron," "stony iron," "chondrite," "carbonaceous chondrite," and "achondrite." A primitive meteorite is defined as a chondrite that experienced minimal heating and aqueous alteration on its parent asteroid. Primitive meteorites have the highest concentrations of "presolar grains." [See *"Meteors, Meteorites, and Impacts"* from The Eight Planets website.] Meteorites between 10 μm (microns) and 2 mm in size are called "micrometeorites." Meteorites collected on Earth are named after the location where they are found.

meteoritic bombardment: intensive and prolonged impacts of a surface by meteorites or other impactors.

meteoriticist: person specializing in the scientific studies of meteorites. They work at universities, government laboratories (such as at NASA Centers), or associated with important meteorite collections (such as natural history museums in many different countries, and the Vatican).

meteoritics: the scientific study of meteorites to better understand the origin, geologic history, and evolution of our solar system.

meteoroid: A 10 μm to 1-m-size natural solid object moving in interplanetary space. Meteoroids may be primary objects or derived by the fragmentation of larger celestial bodies, not limited to asteroids. Meteoroids between 10 μm (microns) and 2 mm in size are called "micrometeoroids."

methane: a compound composed of carbon and hydrogen: CH_4.

mineral: naturally occurring inorganic solid with a definite chemical composition and crystal structure.

mineralogy: the scientific study of minerals. This is the field within geology that focuses on the chemistry, structure, properties, and classification of minerals.

molecular cloud: large concentration of gas (roughly 75% hydrogen and 21%–24% helium with trace amounts of other molecules), dust, and mineral grains. Our solar system formed out of one approximately 4.55 billion years ago when the cloud contracted and our proto-Sun formed in the hot, dense center. The remainder of the cloud formed a swirling disk we refer to as the "solar nebula."

monomict: adjective used to describe a "breccia" formed from a single rock type. See also *polymict*.

moonquake: (see *Earthquake*) Sudden motion or trembling of the Moon caused by the abrupt release of slowly accumulated elastic energy in rocks.

mountain: a natural elevation of a planetary surface.

nakhlites: one of the three types of meteorites from Mars (the *SNC* meteorites). Nakhlites are cumulate igneous rocks characterized by a high abundance of pyroxene and lesser amounts of olivine. This type is named after the Nakhla meteorite, which fell in Egypt in 1911.

nanometer: unit of measure that is one-billionth of a meter. A nanometer-sized particle is smaller than a living cell and can only be seen with the most powerful microscopes.

NASA: U.S. federal agency; National Aeronautics and Space Administration. "NASA is an investment in America's future. As explorers, pioneers, and innovators, we boldly expand frontiers in air and space to inspire and serve America and to benefit the quality of life on Earth" from the NASA Strategic Plan.

native element: a chemical element that occurs naturally uncombined with any other elements. Native elements are divided commonly into three groups: metals, semimetals, and nonmetals. An example is native iron (kamacite), which is found primarily in meteorites.

nautical twilight: the time when the Sun is 6°–12° below the horizon. Sky begins to get noticeably bright.

Nd:YAG: an acronym for neodymium-doped yttrium aluminum garnet, a synthetic crystalline compound used as the active medium for certain solid-state lasers.

NEAR (Near-Earth Asteroid Rendezvous): U.S. Discovery-class mission launched on February 17, 1996. Primary scientific objectives were to return data on the bulk properties, composition, mineralogy, morphology, internal mass distribution, and magnetic field of 433 Eros as the spacecraft orbited the asteroid from February 14, 2000, to February 12, 2001. On February 12, 2001, the NEAR-Shoemaker spacecraft touched down on 433 Eros. Earlier in the mission, on June 27, 1997, NEAR-Shoemaker flew within 1200 km of the C-class asteroid 253 Mathilde, and it flew by the Earth on January 23, 1998. *NEAR* website.

nearside: the side of the Moon that always faces Earth.

nectarian: the geologic history of Earth's Moon has been divided into five broad time periods, or systems. From oldest to youngest, these are "pre-Nectarian," "Nectarian," "Imbrian," "Eratosthenian," and "Copernican." These systems are defined by the number of impact craters on the ground surface (more craters = older surface) correlated to the absolute ages determined for the lunar rock samples returned to Earth by the U.S. "Apollo" missions and unpiloted Soviet "Luna" missions. The Nectarian system extends from 3.9 billion to 3.85 billion years ago.

NIMS (near-infrared mapping spectrometer): an instrument carried to Jupiter on the "Galileo" planetary spacecraft, imaging in regions of the energy spectrum that humans cannot see. Refer to *NIMS* website at the Jet Propulsion Laboratory.

nitrogen: an element with atomic number 7; symbol: N. It is common in Earth's atmosphere and along with hydrogen, carbon, and oxygen is essential for life.

Noachian epoch: the geologic history of Mars has been divided into three broad time periods, or epochs. From oldest to youngest, these are "Noachian," "Hesperian," and "Amazonian" (named after places on Mars). These epochs are defined by the number of impact craters on the ground surface; older surfaces have more craters. The Noachian epoch extends from the birth of Mars to about 3.5–3.7 billion years ago. The actual timing of the epochs is not known because we have different models of the rates of meteorite falls on Mars.

noble gas: chemically inert gas that does not bond chemically to minerals (or has extremely low reaction rate), such as helium, neon, argon, krypton, xenon, and radon.

nongravitational forces: forces that alter the path of a comet, which are not caused by gravitational interaction with the Sun and planets. Typically, this refers to uneven outgasing from vents on the comet, which serve to act like rocket nozzles.

norite: igneous rock found in the lunar highlands composed of plagioclase and pyroxene.

nucleosynthesis: element formation by reactions inside stars.

nucleus: (see *comet*) kilometer-sized "dirty snowball" composed of dust ("refractory" material) and primarily water-ice which gives rise to all of the features observers associate with comets. As the nucleus approaches the Sun, the temperatures rise sufficiently to cause the "sublimation" of the ices. As the gas leaves the nucleus, it is able to drag some of the dust with it in the low gravity. Sunlight reflected off the dust is what we see as the yellowish *"coma"* and tail of the comet, and interaction of the solar radiation with the gases gives us the characteristic blue appearance of the plasma tail.

Ocean of Storms: landing site of Apollo 12 on the Moon on November 19, 1969; Oceanus Procellarum.

olivine: mineral found in basalt; ranges from Mg_2SiO_4 to Fe_2SiO_4. The magnesium-rich end-member of the olivine solid solution series is called forsterite and the iron-rich end-member is called fayalite.

Olympus Mons: largest shield volcano on Mars. Located in the Tharsis region at 18.5°N, 133.2°W, it is 624 km in diameter and rises 18 km over its surroundings. The volcano is rimmed by a 6-km-high scarp. The summit has nested circular depressions, calderas, formed by collapse after repeated withdrawal of magma and eruptions of lava. The last major summit volcanism on Olympus Mons ceased about 150–400 million years ago, based on crater counting.

Oort cloud: vast spherical swarm of comets orbiting our solar system from roughly 2,000 to 100,000 AU. The Oort cloud, while roughly spherical at the largest radius, is wedge-shaped where it merges with the outer planet region in the vicinity of the

Kuiper belt of comets. The cloud is divided into different regions of dynamical stability: the Kuiper belt (35–50 AU; affected by planetary perturbations), a dynamically inert region (50–2000 AU; not affected by gravity of planets or stars), the inner Oort cloud (2,000–15,000 AU; affected by galactic tidal forces), and the outer Oort cloud (15,000–100,000 AU; affected by stellar perturbations). The Oort cloud has never been observed, rather its existence is inferred from the careful analysis of the orbits of comets which come in from the cloud. Recent discoveries of objects in the Kuiper belt have confirmed its existence.

orange "soil": on the Moon, a mixture of very small dark orange and black glass balls which formed from quickly cooled lava droplets during a pyroclastic eruption.

ordinary chondrite: the most common class of meteorite to fall on Earth. They contain variable amounts of metal and chondrules in a matrix of mostly silicate minerals. See *chondrite.*

organic: pertaining to carbon-containing compounds. Organic compounds can be formed by both biological and nonbiological processes.

organism: any form of animal or plant life.

oxygen: an element with atomic number 8; symbol: O. It is the most common element in the crusts and mantles of the inner planets and rocky moons, making up all silicate minerals. Along with hydrogen, carbon, and nitrogen, oxygen is essential to life. There are three isotopes of oxygen, ^{16}O, ^{17}O, and ^{18}O. Graphs of the $^{17}O/^{16}O$ versus $^{18}O/^{16}O$ ratios (normalized to the Earth value *SMOW* and plotted as deviations from that value in parts per thousand) show that the data for Earth and Moon fall on the same line, called the terrestrial fractionation line. On these plots, $\Delta\,^{17}O$ (called "big delta O-17") is a measure of the vertical displacement of any data point from the terrestrial fractionation line.

oxygen fugacity: a measure of the availability of free or uncombined oxygen in an environment to react chemically; measured in tiny fractions of Earth's atmospheric pressure. For example, in most magmas, oxygen fugacity ranges from 10^{-10} to 10^{-18} atmospheres of pressure.

pahoehoe: basaltic lava with a smooth, billowy, or ropy surface.

pallasite: a stony-iron "meteorite" that is a mixture of isolated silicate crystals (usually olivine) surrounded by metal.

parallax: apparent motion of a nearby object as projected against more distant background objects due to the motion of the observer.

parent body: an object (e.g., asteroid) from which meteorites come.

parsec (pc): a unit of astronomical distance equal to 3.26 "light-years," or 206265 AU, which equals 3.08 × 1013 km.

partition coefficient: a ratio describing the concentration of a chemical element in a mineral crystal divided by its concentration in coexisting molten rock in which the mineral is forming. The term partition coefficient is also referred to as partition constant or partition ratio or distribution ratio.

Pascal: the standard unit of pressure is the pascal, abbreviated Pa, which is equivalent to 1 kg/m^2. The pressure at the surface of the Earth is 100,000 Pa. Pressures inside planets are very large numbers, usually expressed as Gpa (giga means billion).

Pathfinder: (see *Mars Pathfinder*)

peridotite: a course-grained igneous rock consisting largely of the mineral olivine, with pyroxenes and garnet or spinel. It is found in the "mantles" of terrestrial planets and "differentiated" asteroids.

perihelion: point in the elliptical orbit of a comet, or asteroid, or planet about the Sun in which it reaches its closest approach to the Sun. The opposite orbital position is called "aphelion."

petrology: the scientific study of rocks. This is the field within geology that focuses on the origin, occurrence, formation, structure, and composition of mineral assemblages, and classification of rocks.

Phoenix Mars Mission: NASA lander mission; it is the first Mars "Scout class" mission to study the planet's north pole ice cap, and the geologic history of water and habitability potential in the polar ice-rich soil. Phoenix launched in August 2007 and touched down on May 25, 2008. The lander's robotic arm will dig down into the subsurface to collect ice and soil samples for onboard analyses by a suite of instruments. *Phoenix Mars Mission* homepage.

photosynthesis: the process by which plants convert water and carbon dioxide into carbohydrates, using sunlight as the source of energy and the aid of chlorophyll.

phyllosilicates: also called sheet silicates, these minerals have layered structures that contain silicon and oxygen in some multiple of Si_2O_5.

pixel: a term used in digital image processing to describe a unique, single picture element (or, a single data point on an image) which has a unique color (or brightness) value.

plagioclase feldspar: common mineral; ranges from $NaAlSi_3O_8$ (albite) to $CaAl_2Si_2O_8$ (anorthite).

Planck's constant: a constant value: 6.626×10^{-27} erg-seconds.

planetesimals: bodies ranging in size from meters up to hundreds of kilometers in diameter that formed during the process that formed the planets by accretion. Most planetesimals accreted to form the planets.

plasma: the fourth state of matter beyond solid, liquid, and gas. Plasma may be the most common state of matter in the universe. Our Sun is plasma. In space science, plasma is hot, ionized gas—a mixture of electrons (negative electric charge) and protons (positive electric charge). Plasma conducts electrical currents and responds to electromagnetic fields.

plate tectonics: theory formulated in the late 1960s that states the Earth's crust and upper mantle (a layer called the lithosphere) is broken into moving pieces called plates. The formation of mountains and volcanoes, and the occurrence of earthquakes have been explained using this theory.

pluton: a body of igneous rock formed beneath the surface of a planet by the intrusion and crystallization of magma.

polycyclic aromatic hydrocarbons (PAH): one group of hydrocarbons (benzene) produced by decomposition or decay of organic matter. These are certainly aromatic: they stink!

polymict: adjective used to describe a "breccia" that has an assortment of fragments of many different rock compositions and textures. (See also *monomict*.)

porosity: the percentage of the total volume of rock or soil that consists of open spaces. (The space between particles is called pore space.)

pre-Nectarian: the geologic history of Earth's Moon has been divided into five broad time periods, or systems. From oldest to youngest, these are "pre-Nectarian," "Nectarian," "Imbrian," "Eratosthenian," and "Copernican." These systems are defined by the number of impact craters on the ground surface (more craters = older surface) correlated to the absolute ages determined for the lunar rock samples returned to Earth by the U.S. "Apollo" missions and unpiloted Soviet "Luna" missions. The pre-Nectarian system extends from 4.5 to 3.9 billion years ago.

presolar grain: preserved in primitive meteorites, presolar grains are stardust that formed in stellar outflows or ejecta in the "molecular cloud," out of which our solar system formed.

pressure ridges: long, narrow wavelike folds in the surface of lava flows; formed where lava may have buckled up against slower moving or stationary lava downstream.

primitive meteorite: a primitive meteorite is a "chondrite" that experienced minimal heating and aqueous alteration on its parent asteroid. Primitive meteorites have the highest concentrations of presolar grains.

protoplanetary disk: the general term used to describe the rotating disk of dense gas surrounding a newly formed star, out of which a planetary system evolves. We call the protoplanetary disk from which our solar system formed the "solar nebula."

protostar: a star in the process of formation which has not yet become hot enough in the core to initiate the process of nuclear fusion (107 K) to halt its gravitational collapse.

pulsar: variable radio star having stable, very short (around one second) periods of pulsations. Electrons moving rapidly in a pulsar's magnetic field produce narrow beams of radiation, which sweep around as the pulsar spins (analogous to sweeping search-light beams).

pyroclastic eruption: explosive eruption of lava producing and ejecting hot fragments of rock and lava.

pyroxene: a silicate mineral common in basalt and composed mostly of iron (Fe), magnesium (Mg), calcium (Ca), and silicon (Si). Composition varies as a mixture among $FeSiO_3$ (ferrosilite), MgSiO3 (enstatite), and $CaSiO_3$ (wollastonite).

pyroxenite: an igneous rock composed largely of pyroxene.

racemic: refers to a chemical compound made of equal amounts of L- and D-structures (which rotate the plane of polarization of light to the left or to the right, respectively), rendering it optically inactive under polarized light.

radioactivity: the spontaneous breaking apart, or decay, of unstable nuclei in "isotopes." The unstable radioactive isotope is called the parent, and the products of the decay of the parent are called the daughter isotopes.

radiocarbon, ^{14}C: the unstable, radioactive isotope of carbon having atomic number 6 and atomic mass 14. It is continuously produced in Earth's upper atmosphere by "cosmic ray" bombardment. In space, ^{14}C is produced when cosmic rays interact with oxygen in silicate minerals in rocks. The "half-life" of ^{14}C is 5730 years, which makes this radioactive isotope suitable for dating rocks and archaeological items as old as about 75,000 years.

radiometric dating: technique to determine the age of a rock by measuring the last time the rock being dated was either melted or disturbed sufficiently to rehomogenize its naturally occurring radioactive elements. This is accomplished because of the decay of long-lived radioactive isotopes. Some of the parent–daughter isotopes used as chronometers to determine the age of extraterrestrial rocks are aluminum-magnesium, uranium-lead, thorium-lead, rubidium-strontium, potassium-argon, samarium-neodymium, and lutetium-hafnium. (Adapted from the *U.S. Geological Survey*.)

R-process: the buildup of heavy "isotopes" inside massive stars by the rapid (hence the use of "*r*") absorption of neutrons by atomic nuclei. See also *s-process* and *supernova*.

rare earth elements (REE): fifteen closely related elements, the lanthanoid series, from atomic numbers 57 through 71 on the Periodic Table (some people define the series from atomic numbers 58 through 71). These elements usually occur together and are useful in geochemical modeling.

ray: streak of material blasted out and away from an impact crater.

recovery: in astronomy, the first time a periodic comet is observed on its inbound journey toward "perihelion." In meteoritics, the collection of a meteorite after its fall.

reflectance spectroscopy: the study of light as a function of wavelength that has been reflected or scattered from a solid, liquid, or gas. Also see *spectroscopy*.

refractory elements: any chemical element that vaporizes (i.e., boils) at high temperatures or condenses from a gas at high temperatures. Examples are calcium (Ca); aluminum (Al); uranium (U); titanium (Ti); the rare earth elements such as cerium (Ce), europium (Eu), and gadolinium (Gd); tungsten (W); zirconium (Zr); and thorium (Th). The opposite of refractory is "volatile."

refractory inclusion: inclusions, enriched in the "rare earths" and the other elements mentioned in the definition of refractory. These inclusions are often referred to as Ca-, Al-rich inclusions, or "CAIs." Most refractory inclusions contain the minerals spinel and melilite and/or hibonite.

regmaglypt: shallow depression, resembling a thumb print in clay, that is commonly seen on meteorites. Regmaglypts are formed by ablation from the surface by vortices of hot gas as a meteor falls through a planetary atmosphere.

regolith: (see *Sediment, Soil*) general term referring to the layers of fragmental and loose, incoherent, or unconsolidated rock, mineral, and glass fragments of any origin (residual or transported) that accumulate on the surface. On the Moon, this blanket of debris is produced by impacts.

remote sensing: the acquisition of information on some property of a target object or material without being in physical contact with it. In the broadest terms, this is done by using instruments or scanners to sense and record reflected or emitted energy from the target and then processing the data in order to better analyze and understand the property in question.

resolution: the degree to which fine details in an image are separated or resolved.

right ascension (RA): celestial coordinate which is equivalent to the longitude of an object on Earth. The starting point for longitude on Earth is at Greenwich, England, whereas in the celestial coordinate system it is at a point in the sky called the vernal equinox. The two great circles defined by the celestial equator and the ecliptic, which is defined as the plane of the solar system, cross each other at 2 points: at the vernal equinox (first day of spring) and the autumnal equinox (first day of autumn). The vernal equinox is the starting point for the RA coordinate system.

rille: long channel on the Moon crossing the surface of maria; probably formed either as an open channel in a lava flow, or as an underground tube carrying hot lava which collapsed as the lava flowed out.

Roche limit: the smallest distance at which a planetary object that has no internal strength can orbit another body without being torn apart by tidal forces (gravitational forces per unit mass exerted by the larger body). Earth's Roche limit is 18,470 km (11,470 miles).

rock: a naturally formed solid that is an aggregate of one or more minerals.

scale: the relationship of a distance on a map or model to the true distance in space; written as a ratio, such as 1:24,000.

Sea of Serenity: one of the maria on the Moon's nearside; Mare Serenitatis.

Sea of Tranquility: landing site of Apollo 11 on the Moon on July 20, 1969; Mare Tranquillitatis.

secondary crater: (see *Impact, Crater, Ejecta*) the relatively small craters formed when ejecta of a primary crater falls down and forcefully strikes the surface.

sediment: (see *Regolith*) solid rock or mineral fragments transported and deposited by wind, water, gravity, or ice; precipitated by chemical reactions; or secreted by organisms; accumulated as layers in loose, unconsolidated form.

sedimentary: rock formed when sediment is compacted and lithified.

SELENE (Kaguya): the Japan Aerospace Exploration Agency ("JAXA") launched the SELENE (Kaguya) mission to the Moon in September 2007. Its instruments were designed to provide scientific data from a polar orbit of lunar origin and evolution and to develop the technology for future lunar exploration. In addition to the main spacecraft two smaller relay satellites provided real-time radio transmissions from the far side of the Moon to Earth. The mission lasted 21 months and ended June 10, 2009, with a controlled crash onto the lunar nearside.

shergottites: one of the three types of meteorites from Mars (the "SNC" meteorites). The type is named after the Shergotty meteorite, which fell in India in 1865. Basaltic shergottites consist mainly of plagioclase and pyroxenes. They are fine-grained and probably formed as volcanic lava flows or shallow intrusions. Lherzolitic shergottites are related to the basaltic shergottites but are coarser grained, have larger amounts of olivine, and are considered to be cumulate rocks.

shock wave: a powerful compressional wave that lasts for an extremely brief time. Shock waves are created by explosive events, such as the impact of a meteorite on a planetary surface.

short-period: in astronomy, a comet with an orbital period less than 200 years.

siderophile: (iron-loving) element. Geochemists divide elements into four basic types (based on the Goldschmidt classification). Siderophile elements concentrate preferentially into metal (compared to silicate) or form alloys easily with iron. The other types are "chalcophile," "lithophile," and "atmophile" elements.

silicate: a variety of minerals that always contain silicon (Si) and oxygen (O).

silicon: an element with atomic number 14; symbol: Si. Silicon is the most abundant element besides oxygen in planets, and forms the basis for silicate minerals such as olivine, pyroxene, and plagioclase.

sill: a tabular, parallel-sided sheet of igneous rock, formed underground, that is parallel to the layered rocks around it.

SIMS (secondary ion mass spectrometry): a precision measurement technique used to quantitatively analyze the elements and "isotopes" of materials at microscopic scales (with a typical spatial resolution of 1 μm). A SIMS instrument is also called an ion microprobe. For more, see the PSRD article: *Ion Microprobe*.

SMOW (standard mean ocean water): a precise definition of the "isotopic" composition of water on Earth. It serves as a reference standard for comparing hydrogen and oxygen isotope ratios. For example, cosmochemists measure the $^{17}O/^{16}O$ and $^{18}O/^{16}O$ ratios in terms of deviations in parts per thousand from a standard (delta ^{18}O and delta ^{17}O). The usual standard is SMOW.

SNC meteorites: (see *Meteorite* and *Martian Meteorite*) general name for three groups of meteorites with relatively young ages (slightly over 1 billion years old) that came from Mars. The groups are the "shergottites," "nakhlites," and "chassignites." Another Martian meteorite that does not fall into one of these groups is ALH 84001, a cumulate rock composed mostly of orthopyroxene, which is much older than all the others with an age of 4.5 billion years. (See *Mars Meteorites* compiled by Ron Baalke from the Jet Propulsion Laboratory.)

SOHO mission: the Solar and Heliospheric Observatory (SOHO), is a project of international collaboration between ESA and NASA to study the Sun from its deep core to the outer corona and the solar wind. SOHO was launched in December 1995. *SOHO* website.

soil: (see *Regolith*) the upper layers of sediment on Earth that support plant growth.

sol: a local day on a planet, solar day. On Earth, for example, a sol lasts 24 h. On Mars a sol lasts 24 h and 39 min 35 s. During the Mars Pathfinder mission, scientists began counting sols with the successful landing of the spacecraft on July 4, 1997 (Sol 1), and images or data from the mission contain a reference to the sol on which they were taken. Space missions, in general, follow this protocol.

solar conjunction: a condition where the solar elongation is zero, or practically speaking, so small that the celestial object cannot be seen in the glare of the Sun.

solar elongation: the angle between the lines of sight to the Sun and to the celestial body in question.

solar flare: a region of exceptional brightness in an atmospheric layer of the Sun, often associated with sunspots of complex magnetic fields.

solar nebula: the immense rotating disk of gas and particles of floating cosmic dust from which our solar system "condensed" approximately 4.55 billion years ago. The generic term for such a rotating disk around a newly formed star is "protoplanetary disk."

solar power: energy derived from the Sun or sunlight for use as a source of electricity.

solar system: our Sun and all the objects (planets, moons, asteroids, and comets) that orbit the Sun.

solar wind: streams of "plasma"—charged particles (mainly ionized hydrogen and some helium but actually a mixture of all electrons and protons in the Sun)—ejected from the outer atmosphere (corona) of the Sun and moving outward all the time with velocities in the range 300–500 km/s. For amazing images and information about the Sun and solar wind see the *SOHO* website for the Solar and Heliospheric Observatory, a project of international cooperation between ESA and NASA.

solidus: the temperature at (and below) which material crystallizes (solidifies) completely. On a plot of temperature (y-axis) versus pressure (x-axis), everything is solid below the solidus curve and the crystallization order of minerals at a given pressure can be determined by following a straight line from the "liquidus" curve to the solidus curve.

solubility: the ability or tendency of one substance to dissolve into another at a given temperature and pressure.

source: (see *Eruption*) location where igneous matter (lava and gases) erupts onto the surface; vent, fissure, volcano, and so on.

spacecraft: vehicle capable of traveling in outer space.

space weathering: processes that change the physical structure, optical properties, or chemical or mineralogical properties of the surface of an airless planetary body from their original state. These processes include meteorite, micrometeorite, cosmic dust bombardment, solar wind ion implantation, and sputtering, in deep vacuum and temperatures approaching absolute zero (–273°C).

spectra: the intensity of electromagnetic radiation (such as ultraviolet, visible light, and infrared, etc.) across a range of wavelengths. A prism separates white light into the range of wavelengths we call a rainbow of colors. "Spectrum" is the singular form of the word.

spectroscopy: the study of spectra. Every atom can only emit or absorb certain energies or wavelengths. Scientists use instruments called spectrometers to study the location and spacing of reflectance, emission, and absorption (spectral) lines in specific wavelengths to detect any atoms or molecules that may be present in a target material (such as minerals on the surface of the Moon or a planet).

speed of light: a constant value: 299,792,458 m/s (186,212 miles per second).

spinel: $MgAl_2O_4$, magnesium aluminum oxide mineral, with Fe^{2+} able to substitute for Mg and with Cr or Fe^{3+} able to substitute for Al.

S-process: the buildup of heavy "isotopes" inside massive stars by the slow (hence the use of "s") absorption of neutrons by atomic nuclei. See also *r-process* and *supernova*.

sputtering: in the context of space weathering on the surfaces of airless planetary bodies, sputtering is the process in which atoms on the surface are knocked free by high-speed atomic particles in the solar wind; much higher-energy cosmic rays can also sputter surface materials.

Stardust Mission: the fourth in NASA's Discovery program of low-cost spacecraft with highly focused science goals, Stardust was comprised of a spacecraft and capsule that completed a 7-year, 3-billion-mile journey to a comet in 2006. The capsule returned to Earth, carrying an aerogel collector that captured particles from comet Wild 2 as well as interplanetary dust. The Stardust capsule is now on display at the Smithsonian's National Air and Space Museum.

stellar nucleosynthesis: the naturally occurring nuclear reactions in stars that make the elements heavier than hydrogen and helium. See also *supernova*.

stratigraphic column: a drawing showing the vertical sequence of rock. Geologists use this to interpret the relative sequence of past geological events.

stratigraphy: study of layered rock to understand the sequence of geological events. Normally, older layers or units are on the bottom unless the sequence has been overturned or disrupted.

subduction: the process describing when one "lithospheric" plate collides with and is overridden by, or descends under, an adjacent plate.

sublimation: phase transformation from solid to gas.

sulfur: an element with atomic number 16; symbol: S. Sulfur is common on the surface of Jupiter's moon Io.

supernova (Type II): also known as a core-collapse supernova, this is an enormously bright, energetic, catastrophic explosion that occurs at the end of the lifetime of a massive star whose core collapses. A star must have at least nine times the mass of the Sun to undergo a core-collapse supernova. The extreme heat generated by the explosion makes elements heavier than iron, this is called supernova nucleosynthesis. See also *r-process* and *s-process*, and *stellar nucleosynthesis*.

Taurus–Littrow: landing site of Apollo 17 on the Moon on December 11, 1972.

tectonics: general term referring to the large-scale change of rock in response to forces causing faulting and folding. The forces acting upon a rock mass are generally termed compressional (squeezing together), tensional (pulling apart), or shear (parallel sliding). Common landforms resulting from tectonic processes are mountain ranges, rift zones, faults, fractured rock, and folded rock masses.

tektite: tektites are small beads of silica glass that are usually black but can also be shades of brown or green. People used to think they came from the Moon, but now we know they are from the Earth. Tektites formed during impacts on Earth when plumes of vapor and melted rock pierced the atmosphere. The impact debris traveled briefly through space before reentering the atmosphere—spreading the tektites over huge areas called strewn fields. The largest tektites, which are found on land, are several centimeters across. The smallest, which are found in layers of deep-sea sediments, are only micrometers across.

terrain: area of the surface with a distinctive geological character.

thermocouple: a sensor for measuring temperature. It consists of two dissimilar metals, joined together at one end, which produce a small voltage proportional to the temperature.

thermodynamics: the study of heat and its transformation to mechanical and other forms of energy.

thin section: a thin slice or rock, usually only 30 μm thick. Thin sections are used to study rocks with a microscope.

thrust fault: low-angle "fault" in which rock above the fault plane moves up in relation to rock below.

tidal stresses: force per unit area acting on a planetary body resulting in periodic bulging (of the crust and, in the case on Earth, oceans) caused by the gravitational attraction of another object such as the Sun, a moon, or a planet. The alternate growth and decay of a tide in the crust of a planetary body can lead to frictional heating.

topography: the configuration (shapes, positions, arrangements) and relief (elevations, slopes) of the land surface.

Torino scale: the official scale (created by Professor Richard P. Binzel and formalized in Torino, Italy, in 1999) for categorizing the Earth impact hazard for asteroids and comets. Values range from 0 (no chance of collision with Earth) to 10 (collision with global climatic disasters). (*Torino Scale*)

transient crater: (see *Crater*) the initial crater excavated in the target rock during the first stage of an impact cratering event.

troctolite: igneous rock, found in the lunar highlands, composed of plagioclase and olivine.

troilite: iron-sulfide mineral with the chemical formula FeS.

ultramafic: term describing the chemical composition of an igneous rock composed almost entirely of magnesium-rich and iron-rich minerals (e.g., olivine, pyroxenes) and minimal (less than ~45 %) silica. (See also *felsic* and *mafic*.)

ultraviolet: invisible part of the light spectrum where wavelengths are shorter than the visible spectrum and longer than X-rays.

unconsolidated: materials loosely packed and not cemented together.

vent: an opening in the Earth's surface (or other planet or moon) through which lava, gases, and hot particles come out.

vesicle: bubble-shaped cavity in a volcanic rock formed by expanding gases.

Vesta: the brightest of the known asteroids with a diameter of ~530 km. Also called 4 Vesta because it was the fourth asteroid to be discovered (by Olbers in 1807). Vesta is the "differentiated" parent body of the "HED" meteorites. (Data collected by NASA's Dawn Mission, in orbit around Vesta from 2011–2012, strengthened the association between Vesta and HED meteorites.)

Viking: U.S. mission to Mars, composed of two spacecraft, launched in 1975. Viking 1 and Viking 2 both consisted of an orbiter and a lander. Primary mission objectives were to obtain high-resolution images of the Martian surface, characterize the structure and composition of the atmosphere and surface, and search for evidence of life. *Viking Mission* website from the National Space Science Data Center.

viscosity: an internal property of a fluid that offers resistance to flow. If a fluid has a high viscosity, then it strongly resists flow. An example of a low viscosity fluid is water.

volatile elements: chemical elements that vaporize (i.e., boil) at relatively low temperatures. Examples are water (H_2O), carbon dioxide (CO_2), carbon monoxide (CO), methane (CH_4), ammonia (NH_3), potassium (K), sodium (Na), and lead (Pb). The opposite of volatile is "refractory."

volcano: mountain formed from the eruption of "igneous" matter through a source vent.

Voyager: U.S. mission consisting of two spacecraft launched in 1977 to explore Jupiter, Saturn, Uranus, Neptune, and their moons, rings, and magnetic environments. It took 2 years for the crafts to reach Jupiter. The last image was taken in 1989 and now both crafts are headed for the outer solar system.

Wark–Lovering rims: thin mineral layers surrounding most coarse-grained CAIs. The layers are typically only a few micrometers thick and unaltered rims have the same sequence of high temperature minerals as in CAIs, including melilite, hibonite, perovskite, spinel, anorthite, pyroxene, and olivine. Multiple hypotheses exist for how these rims formed including condensation, flash heating, and reaction of the CAI with a Mg–Si-rich reservoir of gas or melt. The rims are named after their discoverers David Wark and John Lovering, University of Melbourne, Australia (1977).

weathering: the mechanical breakdown and chemical alteration of rocks and minerals at Earth's surface during exposure to air, moisture, and organic matter.

Widmanstätten pattern: a pattern found in "iron meteorites" due to a change in crystal structure of iron-nickel metal grains during cooling. This structural change produces a pattern of crystallographically oriented kamacite (low nickel content, usually up to 7.5 wt% nickel) plates in taenite (high-nickel content, ~20–50 wt% nickel).

Yarkovsky effect: an explanation for the shift in an asteroid's orbit around the Sun due to a thermal thrust caused by the body's uneven heating by sunlight. The illuminated side of an asteroid absorbs the sunlight and heats up. After a short delay produced by thermal inertia, the asteroid re-radiates the thermal energy and this causes a recoil spin that pushes the asteroid away from its current orbital path. In 2003, scientists used radar data from the Arecibo Observatory of Asteroid 6489 "Golevka" to obtain the first-ever measurements of the Yarkovsky effect. (Named after the Russian engineer who came up with the idea around the year 1900.)

Suggested Readings

Berge, G.L. and Gulkis, S. (1976), Earth-based radio observations of Jupiter: Millimeter to meter wavelengths, in *Jupiter* ed. T. Gehrels, University of Arizona Press, Arizona.

Bhattacharya, A.B. (2012), *Search for Extra Terrestrial Intelligence*, Lambert Academic Publishing, Germany.

Bhattacharya, A.B., Joardar, S., and Bhattacharya, R. (2008), *Astronomy and Astrophysics*, Infinity Science Press, MA.

Bhattacharya, A.B. and Raha, B. (2014), *Planet Jupiter and Its Galilean Satellites*, Lambert Academic Publishing, Germany.

Burns, J.A. et al. (2004), Jupiter's Ring-Moon System, in *Jupiter: The Planet, Satellites and Magnetosphere*, eds. F. Bagenal, T.E. Dowling, W.B. McKinnon, Cambridge University Press, Cambridge, UK.

Crowe, M. (1986), *The Extraterrestrial Life Debate, 1750–1900: The Idea of a Plurality of Worlds from Kant to Lowell*. Cambridge University Press, Cambridge, UK.

de Pater, I. (1990), Radio images of the planets. *Annual Review of Astronomy and Astrophysics*, 28: 347–399.

de Pater, I. (1999), The Solar System at Radio Wavelengths, in *the Encyclopedia of the Solar System*, eds. P. Weissman, L. McFadden, and T.V. Johnson. Academic Press, Cambridge, MA.

Dessler, A.J. (Ed.) (1983), Physics of the Jovian magnetosphere. in *Atmospheric Remote Sensing by Microwave Radiometry*, eds. M.A. Janssen, Cambridge University Press, Cambridge, UK, (1993).

Fischer, D. (1999), *Mission Jupiter: The Spectacular Journey of the Galileo Spacecraft*, New York: Copernicus.

Harrison, E. (1981), *Cosmology: The Science of the Universe*. Cambridge University Press, Cambridge, UK.

Lewis, J.S. and Prinn, R.G. (1984), *Planets and Their Atmospheres*. Academic Press, Orlando, FL.

Morrison, D. (1969), Venus: Absence of a phase effect at a 2-centimeter wavelength. *Science*, 163(3869): 815–7.

Lopes, R.M.C. and Spencer, J.R. (2007), *Io after Galileo: A New View of Jupiter's Volcanic Moon*, Springer, Berlin, Germany.

McConnell, S. (2003), *Galileo: Journey to Jupiter*, NASA/Jet Propulsion Laboratory. http://www.jpl.nasa.gov/missions/galileo/ (retrieved November 28, 2006).

McDonough, T.R. (1987), *The Search for Extraterrestrial Life: Listening for Life in the Cosmos*. Wiley, New York.

Muhleman D.O., Berge G.L., and Orton G.S. (1980), A model of the Venus atmosphere from radio, radar, and occultation observations. *Astrophysics Journal*, 234: 733–745.

Muhleman, D.O., Orton, G.S., and Berge, G.L. (1979), Thermophysics of the planet Mercury. *Space Science Reviews*, 38: 251–266.

Ridpath, I. (1987), *Messages from the Stars: Communication and Contact with Extraterrestrial Life*. Harper & Row, New York.

Sullivan W.T., III. (Ed.) (1984), in *The Early Years of Radio Astronomy: Reflections Fifty Years after Jansky's Discovery*, Cambridge University Press, Cambridge, UK.

Weast, R.C. (1983), *CRC Handbook of Chemistry and Physics*, 64th edn, F-141. CRC Press, Boca Raton, FL.

Index

Note: Page numbers followed by f and t refer to figures and tables, respectively.

T

U

V

100⁰⁰